Charles E Hobbs

Hobb's botanical hand-book of common local, English, botanical and pharmacopoeial names arranged in alphabetical order, of most of the crude vegetable drugs, etc., in common use

Charles E Hobbs

Hobb's botanical hand-book of common local, English, botanical and pharmacopoeial names arranged in alphabetical order, of most of the crude vegetable drugs, etc., in common use

ISBN/EAN: 9783742833617

Manufactured in Europe, USA, Canada, Australia, Japa

Cover: Foto ©berggeist007 / pixelio.de

Manufactured and distributed by brebook publishing software (www.brebook.com)

Charles E Hobbs

Hobb's botanical hand-book of common local, English, botanical and pharmacopoeial names arranged in alphabetical order, of most of the crude vegetable drugs, etc., in common use

C. E. HOBBS'

BOTANICAL HAND-BOOK

OF

COMMON LOCAL, ENGLISH, BOTANICAL AND PHARMACOPŒIAL NAMES
ARRANGED IN ALPHABETICAL ORDER,

OF MOST OF THE

CRUDE VEGETABLE DRUGS, ETC., IN COMMON USE:

THEIR PROPERTIES, PRODUCTIONS AND USES, IN AN ABBREVIATED FORM.

ESPECIALLY DESIGNED AS A REFERENCE BOOK
FOR DRUGGISTS AND APOTHECARIES.

IN THREE PARTS.

COMPILED AND PUBLISHED BY
CHARLES E. HOBBS.

BOSTON:
PRINTED BY CHAS. C. ROBERTS, 21 BRATTLE STREET.
1876.

Electrotyped by H. C. WHITCOMB & Co., Boston.

PREFACE.

It is the misfortune of common names, as applied to plants, that they are very apt to be used loosely. Sometimes the same name is given to widely different plants, and again a plant will have several common names. The greatest confusion prevails among the common names of our native plants. There are about a dozen known as Snake Root; three as Cheekerberry, and five as Yellow Root; hence we are not surprised to find four or five distinct plants in different localities called by the same common name. If those persons who hold in contempt the botanical names of plants, simply because they are derived from Latin and Greek, could be perplexed as every druggist and apothecary is, almost daily, by the indefiniteness of common names, they would gladly adopt the *definite* botanical one. Notwithstanding all this, common local names *are* used, and will continue to be used, by the far greater portion of the people, and it was with this fact in mind that the publication of this book was commenced. It often happens that there are two names, one as common as the other; in this case, each name will refer to the other, both giving the botanical name.

In the compilation of this work the author has consulted the best standard authorities and works on Materia Medica, as well as the medical publications of the day during the past twelve years, and the chief difficulty was what to leave out, rather than what to include, and not injure the book as a work of reference. An experience of eighteen years in the botanic drug business, materially assisted in the selection of matter for this work.

In the second part of the book will be found the botanical name, referring to the most common name that the drug or plant is known by, the third column giving the medical properties, or product of the same.

The third part gives the pharmacopœial names, as found in the United States, British, and German Pharmacopœias, referring to the common and botanical name.

To make the work practically useful the subject has been treated with plainness and simplicity as far as possible, and has at least indicated the vast field for improvement in the naming of plants in common use by giving one common English name instead of many (twelve for Eupatorium perfoliatum.)

As the work is, from its nature, a compilation, the only originality that can be claimed by the author is the selection and arrangement of his materials.

C. E. H.

WINTER HILL, SOMERVILLE, MASS.
    Feb. 1, 1876.

C. E. HOBBS'

BOTANICAL HAND-BOOK.

ENGLISH.

COMMON.	ENGLISH.	BOTANICAL.
Aaron's beard,	False fringe tree,	Rhus cotinus.
Abanga,	Palm fruit,	
Abele tree,	White Poplar tree,	Populus alba.
Abronia,		Abronia umbellata.
Abscess root,	American Greek valerian,	Polemonium reptans.
Absinthium,	Wormwood,	Artemisia absinthium.
Absus,		Acacia absus.
Acacia,	Gum Arabic,	Acacia vera.
Ach root,		Morinda tinctoria.
Achiote,	Annotto tree,	Bixa orellana.
Ackroot,	Indian name for walnut,	
Aconite,	Wolfsbane,	Aconitum Napellus.
Acorn,	Fruit of the oak,	Quercus,
Acrid lettuce,		Lactusa virosa.
Adam and Eve,	Putty root,	Aplectrum hyemale.
Adam's apple,	Lime tree fruit,	Citrus Limetta.
Adam's needle,		Yucca gloriosa.
Adder's leaf,	Adder's tongue,	Erythronium Americanum.
" mouth,	Stitch wort,	Stellaria media.
" meat,		Microstylis ophioglossoides.
" tongue,	Dogstooth violet,	Erythronium Americanum.
" " fern,		Ophioglossum vulgatum.
" violet,	Net leaf plantain,	Goodyera pubescens.
" wort,	Viper's bugloss,	Echium vulgare.
Adstringous bark,	of Brazil,	Acacia Jurema.
African lily,	Love flower,	Agapanthus.
" turmeric,		Canna speciosa.
Afternoon ladies,	Marvel of Peru,	Mirabilis longiflora.
Agar agar,	Ceylon Moss,	Gracilaria lichenoides.
Agaric, of the oak,		Boletus ignarius.
" purging,	Larch agaric,	" laricis.
" white,	" "	" "
Agathamosa,		Agathamosa pulchella.
Agave, American,	American aloe,	Agave, Americana.
" Virginian,	False aloe,	" Virgianica.
Ageratum,		Ageratum, Mexicana.
Agnus castus,	Chaste tree,	Vitex agnus castus.
Agrimony,		Agrimonia Eupatoria.
" Hemp,	Sweet smelling trefoil,	Eupatorium cannabinum.
" Water Hemp,	Swamp beggar's tick,	Bidens tripartita.

1*

COMMON.	ENGLISH.	BOTANICAL.
Ague bark,	Wafer ash,	Ptelea trifoliata.
" grass,	Unicorn root,	Aletris farinosa.
" root,	" "	" "
" weed,	Boneset,	Eupatorium perfoliatum.
" tree,	Sassafras,	Sassafras officinalis.
Ailantus,	Chinese sumac,	Ailantus glandulosa.
Air plant, magnolia,		Epidendrum conopseum.
Ajava seed,	Bishop's weed,	Ammi copticum.
Akasgia,	Ordeal poison of Africa.	
Albany beech drops,	.	Pterospora andromeda.
" hemp,	Canada nettle,	Urtica canadensis.
Alcanna,	Henne,	Lawsonia inermis.
Alconet,	American anchusa,	Batschia canescens.
Alcamphora,		Croton pedicipes.
Alcornoque of Spain,	Corktree.	
" bark,		Brysonima crassifolia.
Alder, American,	Tag alder,	Alnus serratula.
" Black,		Prinos verticillatus.
" Common,	Tag alder,	Alnus rubra.
" Dogwood, black,		Rhamnus frangula.
" European,		Alnus glutinosa.
" leaved dogwood,	Round leaved dogwood,	Cornus circinata.
" Red,	Tag alder,	Alnus rubra.
" smooth swamp,	Tag alder,	" "
" spotted,	Witch hazel,	Hamamelis Virginica.
" striped,	Black alder,	Prinos verticillatus.
" Tag,		Alnus rubra.
" white,	Sweet pepper bush,	Clethra alnifolia.
Ale cost,	Costmary,	Pyrethrum tanacetum.
Ale hoof,	Ground ivy,	Nepeta glechoma.
Alexanders,	Meadow parsnip,	Zizia aurea.
Algaroba bean,	Saint John's bread,	Ceratonia siliqua.
Alkanet,		Anchusa tinctoria.
Alkekengi berries,	Winter cherry,	Physalis alkekengi.
All bones,	yields Litmus,	Lichen roccella.
Alleghany fringe,	Climbing fumitory,	Adlumia cirrhosa.
All heal,	Valerian root,	Valeriana officinalis.
Alligator apple,	Water apple,	Anona palustris.
" pear,	Avocado pear,	Persea gratissima.
All seed,	Wild amaranth,	Amaranthus blitum.
Allspice,	Pimento,	Eugenia pimenta.
Almond, cutting,	Nephritic plant,	Parthenium integrifolium.
" Dwarf,	Flowering almond,	
" Flowering,		Amygdalus pumila.
Almonds, Bitter,		Amygdalus communis var amara.
" Guiana,	Brazil, Para nuts,	Bertholletia excelsa.
" Sweet,	Jordan almonds,	Amygdalus communis var dulcis.
Almug,	Sandal wood,	
Aloe, American,	Century plant,	Agave, Americana,
" False,		" Virginica.
" root,	Unicorn root,	Aletris farinosa.
" wood,		Aloexylon, agallochum.
Aloes,	the inspissated juice of aloe plants.	
" African,	Cape (of Good Hope) Aloes	Aloe Africana.
" Barbadoes,		" vulgaris.
" Bombay,	Socotrine aloes,	" Socotrina.
" Caballine,	Horse aloes,	" Guinensis.
" Cape,		" spicata.
" Curacoa,	Island of Curacoa aloes,	
" East Indian,	India aloes,	" vulgaris var Indica.

COMMON.	ENGLISH.	BOTANICAL.
Aloes Hepatic,	Barbadoes aloes,	Aloe vulgaris.
" Horse,	Fetid aloes,	" Guiniensis.
" India,		" vulgaris var indica.
" Mocha,		" Socotrina.
" Natal,		" ( ).
" Shining,	Cape aloes,	" Spicata.
" Socotrine.		" Socotrina.
" Turkey,	Socotrine aloes,	" "
" Zanzibar,	" "	" "
Aloes wood,	Aloe wood,	Aloexylon agallochum.
Althæa,	Marshmallow root,	Althæa officinalis.
" rose,	Hollyhock,	" rosea.
Alum root,	Cranesbill root,	Geranium maculatum.
" "	American Sanicle root,	Heuchera Americana.
Alyssum,.	Sweet alyssum,	Koniga maritima.
Amada ginger,		Curcuma amada.
Amadou,	German tinder,	Boletus fomentarious.
Amaranth,	Prince's feather,	Amaranthus hypochondriacus.
Amazon root,	Galangal root,	Alpina galanga.
Amber seed,	Musk seed,	Abelmoschus esculentus.
Amber tree,		Anthospermom.
Ambrette,	Musk seed,	Abelmoschus esculentus.
Ambrosia,	Hogweed,	Ambrosia elatior.
"	Roman wormwood,	" artemisifolia.
" tall,	Horse weed,	" trifida.
American Agave,	Century plant,	Agave Americana.
" Alcornoque,		Bowdictia virgilioides.
" Almond,	African almond,	Braejum stellatum.
" Aloe,	Century plant,	Agave Americana.
" Anchusa,	Alconet,	Batschia canescens.
" Angelica,		Archangelica atropurpurea.
" Aspen,	American poplar,	Populus tremuloides.
" Balm Gilead,		" candicans.
" Barberry,		Berberis vulgaris.
" Broomrape,		Orobanche Americana.
" Centaury,		Sabbatia angularis.
" China root,	False China root,	Pseudo smilax china.
" Coffee bean,	Kentucky coffee bean,	Gymnocladus Canadensis.
" Columbo,		Frasera Carolinensis.
" Dittany,	Mountain dittany,	Cunila mariana.
" Foxglove,		Digitalis purpurea.
" "	Bushy gerardia,	Gerardia pedicularia.
" Gamboge,		Hypericum baccatum.
" Gentian,		Gentiana Catesbei.
" Ginseng,		Panax quinquefolium.
" Greek Valerian	Abscess root,	Polemonium reptans.
" Hellebore,		Veratrum Viride.
" Hemlock,	Poison hemlock,	Cicuta maculata.
" Holly,		Ilex opaca.
" Indigo,	Wild Indigo,	Baptisia tinctoria.
" Ipecac,		Euphorbia Ipecacuanha.
" "	Indian physic,	Gillenia stipulacea.
" Ivy,		Ampelopsis quinquefolia.
" Kino root,	Cranesbill root,	Geranium maculatum.
" Larch,	Tamarack,	Larix Americana.
" Licorice,		Glycyrrhiza lepidota.
" Mandrake,	May apple,	Podophyllum peltatum.
" Mezereon,	Leatherwood,	Dirca palustris.
" Olibanum,	Jamaica cedar,	Juniperus Bermudiana.
" Olive,	Devilwood,	Olea Americana.

COMMON.	ENGLISH.	BOTANICAL.
American Pepper,	Capsicum,	Capsicum annuum.
" Plane tree,	Buttonwood tree,	Platanus occidentalis.
" Poplar,		Populus tremuloides.
" Saffron,	Dyers' saffron,	Carthamus tinctorius.
" Sanicle,	Alum root,	Heuchera Americana.
" "	Ground maple,	" acerifolia.
" Sarsaparilla,	Small spikenard,	Aralia nudicaulis.
" Senna,		Cassia Marilandica.
" Silver fir,	Balsam fir,	Abies balsamea.
" Sloe,		Prunus pygmæa.
" Spikenard,		Aralia racemosa.
" Thrift,	Marsh rosemary,	Statice Caroliniana.
" Tormentil,	Cranesbill root,	Geranium maculatum.
" Valerian,	Cypripedium of several species.	
" Eng., Valerian,		Valeriana officinalis.
" Water hemlock	American Hemlock,	Cicuta maculata.
" Woodbine,	" ivy,	Ampelopsis quinquefolia.
" Wormseed,	Oak of Jerusalem,	Chenopodinum anthelminticum.
" Yew,		Taxus canadensis.
Amethyst, blue,		Browallia elata.
Anacahuite wood,		Cordia Boisseri.
Anamita,	Mushrooms,	Agaricus muscarius.
Anchovy pear,		Grias cauliflora.
Anchusa,	Alkanet,	Anchusa tinctoria.
Anda seed,	of Brazil,	Anda Gomesei.
Anemone, Garden,		Anemone hortensis.
" meadow,	Pasque flower,	" pulsatilla.
" pulsatilla,	Meadow anemone,	" "
" Rue,		Thalictrum anemonoides.
" Wood,	Wind flower,	Anemone nemorosa.
" Yellow,		" vernalis.
Angelica, American,		Archangelica atropurpurea.
" Gard'n, Europe		" officinalis.
" Purple,	High angelica,	Angelica atropurpurea.
" tree,	Prickly elder,	Aralia spinosa.
" Wild,	British specie,	Angelica sylvestris.
Angelico,	American Lovage,	Ligusticum actæifolium.
Angle pod,	common name for	Gonolobus.
Angola weed,	Litmus weed,	Roccella fusiformis.
" pea,	Pigeon pea,	Cytisus cajan.
Angostura bark,		Galipea officinalis.
" False,	Nux vomica bark,	Strychnos nux vomica.
Augustura,	Angostura,	Galipea officinalis.
Angsava gum,	Red Dragon's blood,	
Anil,	The indigo plant,	Indigofera anil.
Animated Oats,		Avena sterilis.
Anime,	Gum resin anime,	Hymenæa courbaril.
Anise, common.	Anise seed,	Pimpinella anisum.
" Chinese,	Star anise seed,	Illicium anisatum.
" Florida,	Wild anise tree,	" Floridanum.
" root,	Sweet cicely,	Osmorrhiza longistylis.
" "		Hypogon anisatum.
" seed,	Common anise,	Pimpinella anisum.
" " Star.		Illicium anisatum.
Annatto,	a dye-stuff,	Bixa orellana.
Antimony, Vegetable,	Boneset,	Eupatorium perfoliatum.
Apalachian tea,	Leaves used as tea,	Prinos glaber.
Apple,	the fruit,	Pyrus malus.
" Alligator,	Water apple,	Anona palustris
" Adam's,	a species of Lime,	Citrus Limetta.

COMMON.	ENGLISH.	BOTANICAL.
Apple, Adam's,	Plantain fruit,	Musa paradisica.
" berry,		Billardiera longiflora.
" of Cain,		Arbutus unedo.
" Devil's,	Fruit of Europ'n Mandrake,	Mandragora officinalis.
" Indian,	Mayapple fruit,	Podophyllum peltatum.
" May,	American Mandrake,	" "
" Peru,	Thorn apple,	Datura stramonium.
" root,	Large, flowering spurge,	Euphorbia corollata.
" Thorn,		Datura stramonium.
" tree,		Pyrus malus.
Apricot,		Armeniaca vulgaris.
" So. American,	Mammea apple,	Mammea Americana.
Arabian Lavender,		Lavendula stoechas.
" Manna,		Tamarix Gallica.
" Senna,	Mecca senna,	Cassia lanceolata.
Arar tree,	yields sandarach,	
Arbor vitæ,	White cedar,	Thuja occidentalis.
Arbutus, Trailing,	Mayflower,	Epigæa repens.
Archel,		Lecanora parella.
Archil,	Orchil, Litmus,	Roccella tinctoria.
Archangel,	Angelica,	Angelica atropurpurea.
" green,	Bugle weed,	Lycopus, Virginicus and Europæus
Areca, nut,	Betel nut,	Areca catechu.
Argel,		Cynauchum Oleæfolium.
"		Solenostemma arghel.
Arica bark,	Cusco bark,	Cinchona pubescens Pelliticriana.
Aristolochia root,	Virginia snake root,	Aristolochia serpentaria.
Arnica,		Arnica montana.
Arnotta,	Annotto,	Bixa orellana.
Arrach,	Garden Orach,	Atriplex hortensis.
" Dog,		Chenopodium olidum.
" stinking,	Goosefoot,	" "
Arrow, Arum,		Peltandra Virginica.
" grass,		Triglochin maritima.
" head,		Sagittaria sagittifolia.
" Indian,	Wahoo,	Euonymus atropurpureus.
" poison,	Tieute, Upas tree,	Strychnos Tieute.
" plant,	yields arrow root,	Maranta arundinacea.
" wood,	Wahoo,	Euonymus atropurpureus.
" "		Viburnum dentatum.
" " Indian,	Florida dogwood,	Cornus Florida.
Arrow root, Bermuda,		Maranta arundinacea.
" " Brazilian,		Jatropha manihot.
" " Common,	Potato starch,	Solanum tuberosum.
" " E. Indian,		Curcuma angustifolia.
" " English,	Potato starch,	Solanum tuberosum.
" " Portland,		Arum maculatum.
" " Tahiti,	Tacca starch,	Tacca oceanica.
Arsmart,	Smartweed,	Polygonum hydropiperoides.
Artichoke,	Garden artichoke,	Cynara Scolymus.
" Jerusalem,	Earth apple, .	Helianthus tuberosus.
Artillery plant,		Pillea serpyllifolia.
Asafœtida,	Assafœtida,	Narthex Assafœtida.
Asarabacca,	European snake root,	Asarum Europæum.
" Broad leav'd,	Canada snake root,	" Canadense.
Ash bark,	Jaen Peruvian Bark,	Cinchona ovata vulgaris.
" berries, prickly, .		Xanthoxylum fraxineum.
" Bitter,	Quassia,	Simaruba excelsa.
" Black,		Fraxinus sambucifolia.
" Blue,		" quadrangulata.

COMMON.	ENGLISH.	BOTANICAL.
Ash Common,		Fraxinus polygamie.
" European,		" excelsior.
" Flowering,	Manna tree, European,	Ornus Europæa.
" maple,		Acer negundo.
" Mountain,	Round wood tree,	Sorbus Americana.
" Poison,	Fringe tree,	Chionanthus Virginica.
" Prickly,		Xanthoxylum fraxineum.
" " Southern,	Prickly elder,	Aralia spinosa.
" " "		Xanthoxylum Carolinianum.
" Red,		Fraxinus pubescens.
" Walnut,		Juglans paxinea.
" White,		Fraxinus Americana.
" Yellow,	Fustic tree,	Cladrastis tinctoria.
Ash weed,	Gout weed,	Ægopodium podograria.
" wort,		Cineraria heterophylla.
Ashy crown bark,	Peruvian bark,	Cinchona Cordifolia rotundifolia.
Asiatic poison bulb,		Crinum Asiaticum.
Aspaghul seed,	Spogel seed,	Plantago decumbens.
Asparagus,		Asparagus officinalis.
Aspen,	American poplar,	Populus tremuloides.
Asphodel,		Asphodelus ramosus.
"		Narthecium Americanum.
Asphaltum,	Bitumen,	
Aspic,	Spike lavender,	Lavendula spica.
Assacou,		Hura Brasiliensis.
Assafetida,	a gum resin,	Narthex assafœtida.
Assafœtida,	Assafetida,	" "
Asses' eyes,	the seeds of Cowhage,	Mucuna puriens.
Aster,	Red stalked aster,	Aster puniceus.
" Cape,		Agathœa amelloides.
" China,	China aster,	Callistephus chinensis.
" Silver,	Silver aster,	Chrysopsis argentea.
Asthma weed,	Lobelia herb,	Lobelia inflata.
Astringent gum,	Bengal kino,	Butea frondosa.
" root,	Cranesbill,	Geranium maculatum.
Atamasco lily,		Amaryllis atamasco.
Atlee galls,	galls procured from	Tamarix orientale.
Atwisha,	an East Indian poison,	Aconitum ferox.
Autumnal crocus,	Spanish saffron,	Crocus sativus.
" gentian,	Bastard gentian,	Gentiana amarella.
Ava kava,	a kind of pepper,	Macropiper methysticum.
Avens,	common name for	Geum.
" European,	Herb Bennet,	" urbanum.
" Purple,		" rivale.
" Water,	Purple avens,	" "
" White,	Chocolate root, Bennet,	" Virginianum.
" Yellow,	European avens,	" urbanum.
Avignon berries,	Yellow dye berries,	Rhamnus infectorius.
Avocado pear,		Persea gratissima.
Awl tree,	Indian mulberry,	Morinda citrifolia.
Awlwort,		Subularia aquatica.
Awn,	The beard of corn or grass.	
Ayapana,		Eupatorium ayapana.
Ayerayer,	The esculent fruit,	Lansium domesticum.
Azalia,	Swamp pink,	Azalea nitida.
Azedarach bark,	Pride of India,	Melia azedarach.
Bablach pods,	Acacia bambolah,	Mimosa çineraria.
Babool bark,	Babul tree,	Acacia horrida.
Bachelors' buttons,	Globe amaranth,	Gromphrena globosa.
" "	Ragged robin,	Lychnis flos cuculi.

COMMON.	ENGLISH.	BOTANICAL.
Bachelor's buttons,	Buttercups,	Ranunculus acris.
" "	Nux vomica,	Strychnos nux vomica.
Backache brake,	Female fern,	Aspidium Filix Fœmina.
" root,	Button snake root,	Liatris spicata.
Badgersbane,	Aconite,	Aconitum meloctonum.
Badiane,	Star anise,	Illicum anisatum.
Bael,	Bengal quince,	Ægle marmelos.
Bahama tea,	Sage tree,	Lantana camara.
Bairnwort,	Common daisy,	Bellis perennis.
Baldmoney,	Spicknel,	Meum athimanticum.
Balloon vine,	Heart seed,	Cardiospermum halicacabum.
Ballote,	Black horehound,	Ballota nigra.
Balmony,		Chelone glabra.
Balm, Field,	Catmint,	Nepeta cataria.
" of Gilead, Arabian	Balsam of Gilead,	Amyris Gileadensis.
" " " buds,	" poplar,	Populus candicans.
" " " fir,	" fir tree,	Abies balsamea.
" " " herb,	Sweet balm herb,	Dracocephalum Canariense.
" " " tree,	Balsam poplar,	Abies balsamea.
Balm, Indian,	Beth root,	Trillium pendulum.
" Lemon,		Melissa officinalis.
" mint,	Curled mint,	Mentha crispa.
" Red,	Oswego tea,	Monarda didyma.
" Stinking,	Pennyroyal,	Hedeoma pulgeioides.
" Sweet,		Dracocephalum Canariense.
Balsam, acouchi,		Icica heterophylla.
" apple,	Balsamina,	Momordica balsamina.
" " Wild,	Elaterium cucumber,	" elaterium.
" Bayce,		Balsamodendron pubescens.
" Calaba,	Tacamahac,	Calophyllum Calaba.
" Canada,	Oleo resin procured from	Abies balsamea.
" Capivi,	Balsam copaiba,	
" Carpathian,	Oleo resin from stone pine,	Pinus cembra.
" Copaiba,	Oleo resin procured from	Copaifera multijuga and species.
" Copaiva,	Balsam copaiba,	
" Copalm,	Sweet gum,	Liquidambar styraciflua.
" Fir,	Balsam Canada,	
" Garden,	Ladies' slippers,	Impatiens balsaminæ.
" Groundsel,		Senecio balsamitæ.
" Gurgun,	Gurgun Balsam copaiba,	Dipterocarpus turbinatus.
" Hungarian,		Pinus pumilo.
" of Gilead,		Amyris Gileadensis.
" " Mecca,	Balsam of Gilead,	" "
" Peru,	The balsamic exudation,	Myrospermum Peruiferum.
" " white,	Expressed, from the fruit,	" "
" poplar,		Populus balsamifera.
" Rhiga,	Carpathian balsam,	
" seed,	The seeds of	Myrospermum Peruiferum.
" spruce,	Fir balsam tree,	Abies balsamea.
" Sweet,	Life everlasting,	
" Tolu,	Balsamic exudation,	Myrospermum Toluiferum.
" tree,	Peruvian balsam tree,	
" Umri,		Humiria floribundum.
" vine,	Balsam apple,	Momordica balsamina.
" weed,	Jewel weed,	Impatiens pallida.
" "	Life everlasting,	Gnaphalium polycephalum.
" White,	Life everlasting,	" "
" wood,		Myroxylon.
Bamboo,		Bambusa arundinacea.
" brier,	Southern states sarsaparilla	Smilax sarsaparilla.

COMMON.	ENGLISH.	BOTANICAL.
Bamboo cane,	Bamboo,	Bambusa arundinacea.
Banana,		Musa sapientum.
Baneberry,	Herb Christopher,	Actæa spicata.
" Red,	Red cohosh,	" rubra.
" White,	White cohosh,	" alba.
Bang. Bhang.	The leaves and capsules,	Cannabis sativa var indica.
Banyan tree,	Indian fig tree,	Ficus Indica.
Baobab,	Monkey bread tree,	Adansonia digitata.
Barbadoes aloes,		Aloe vulgaris.
" cedar,		Cedrela odorata.
" cherry,		Malphigia glabra.
" flower fence,	Spanish carnation,	Cæsalpina pulcherrima.
" nuts,	Physic nuts,	Curcas purgans.
" pride,	Spanish carnation,	Cæsalpina pulcherrima.
" tar,	Liquid bitumen,	
Barberry,		Berberis vulgaris.
" Indian,		" aristata.
Bardana,	Burdock plant,	Arctium lappa.
Bardane,	" "	" "
Bark, in medicine,	refers to Cinchona,	
Bark, in tanning,	refers to Oak and Hemlock barks.	
Barley,		Hordeum distichon.
Barngrass,		Panicum crus-galli.
Barras,	Burgundy pitch,	Abies excelsa and var.
Barren strawberry,		Potentilla fragaria.
Barrenwort,		Epimedium alpinum.
Barwood,	Cam wood,	Baphia nitida.
Basil, common,		Ocymum basilicum.
" Sweet,	Basil common,	" "
" thyme,	Calamint,	Calamintha officinalis.
" Wild,	American dittany,	Cunila mariana.
Basket of gold,		Alyssum saxatile.
Basket willow,	Basket osier,	Salix viminalis.
Bass wood,		Tilia glabra.
" American,	Lime tree,	" Americana.
Bast tree,	Linden tree, American,	" "
Bastard alkanet,	Corn gromwell,	Lithospermum arvense.
" balm,		Melissa fuschii.
" bearsfoot,		Helleborus fœtidus.
" Brasilwood,		Comocladia dentata.
" cabbage tree,		Geoffroya.
" cedar,		Guazuma tomentosa.
" "	African mahogany,	
" china root,	American china root,	Smilax pseudo china.
" cinnamon,	Cassia bark,	Cinnamomum cassia.
" cress,		Lepidium campestre.
" dittany,	White fraxinella,	Dictamus alba.
" Fr. physic nut,	Wild cassada,	Jatropha gossypifolia.
" gentian,	Autumnal gentian,	Gentiana amarella.
" hellebore,	Helleborine,	Epipactis latifolia.
" hemp,	Hemp nettle, flowering,	Galeopsis tetrahit.
" horehound,	Black horehound,	Ballota nigra.
" Ipecac,		Asclepias curassavica.
" manchineel,		Cameraria latifolia.
" mustard,		Polanisia icosandra.
" Nicaragua wood	Peach wood,	Cæsalpina bijuga.
" parsley,	Hensfoot,	Caucalis daucoides.
" pellitory,	Sneezewort,	Achillea ptarmica.
" pimpernel,	Chaff weed,	Centunculus minimus.
" plantain,	Mudwort,	Limosella subulata.

COMMON.	ENGLISH.	BOTANICAL.
Bastard saffron,	Dyers' saffron,	Carthamus tinctorius.
" sarsaparilla,	German sarsaparilla,	Carex villosa and var.
" spignel,		Seseli montanum.
" St. Johnswort,		Hypericum coris.
" stone parsley,	Honewort,	Sison amomum.
" toad flax,	.	Thesium linophyllum.
" vetch,	Bitter wood vetch,	Orobus sylvaticus.
" wormwood,	Roman wormwood,	Ambrosia elatior.
Bawdmoney,		Meum athamanticum.
Bayberries, American,		Myrica cerifera.
" European,	the fruit of	Laurus nobilis.
Bayberry,	Wax myrtle,	Myrica cerifera.
" Jamaica,	Allspice,	Myrtus Pimenta.
" wax, 'tallow,'	American vegetable wax,	Myrica cerifera.
Bay bush buds,	Meadow fern buds,	Myrica gale.
Bay, laurel,	Bay tree,	Laurus nobilis.
" leaves,	" "	" "
" plum,	Guava,	Psidium pyriferum.
" Sweet,	Magnolia,	Magnolia glauca.
" tree,		Laurus nobilis.
" rum leaf,		Myrcia acris.
Bayberry tree,	Wild clove,	" "
Bdellium, African,		Heudelotia Africana.
" Indian,		Amyris commiphora.
Beach plum,		Prunus maritima.
Bead tree,	Pride of India,	Melia azedarach.
Beads,	The fruit (nut),	" "
Beaked hazel,		Corylus rostrata.
" violet,	Canker violet,	Viola rostrata.
Beam tree,		Pyrus aria.
Bean, Calabar,		Physostigma venosum.
" Castor,	Castor oil plant,	Ricinus communis.
" Common,	Common kidney bean,	Phaseolus vulgaris.
" Garden,	Dwarf, or field bean,	" nanus.
" St. Ignatius',	the seed,	Strychnos Ignatia.
" Tonka,	Tonquin bean,	Dipterix odorata.
" Vanilla,		Vanilla aromatica and var.
Bean trefoil,	Buckbean,	Menyanthes trifoliata.
" tree,	St. John's bread,	Ceratonia siliqua.
Bearbane,	Aconite,	Aconitum arctophonum.
Bearberry,	Mountain Cranberry,	Arctostaphylos Uva Ursi.
Bearbind,		Convolvulus arvensis.
Bearsbreech,		Acanthus spinosus.
Bears' bed,	Haircap moss,	Polytrichum juniperum.
" ear,		Primula auricula.
" foot,		Helleborus fœtidus.
" fright,	Hogwort,	Heptallon graveolens.
" grape,	The fruit of Uva Ursi,	Arctostaphylos Uva Ursi.
" grass,	Silk grass,	Yucca filimentosa.
" paw root,	Male fern root,	Aspidium Filix Mas.
" wort,		Meum athamanticum.
Beard grass,		Andropogon nutans.
" tongue,		Penstemon pubescens.
Beastsbane,	Aconitum,	Aconitum theriphonum.
Beaumont root,	Black root,	Leptandra Virginica.
Beauty of the night,	Marvel of Peru,	Mirabilis Jalapa.
Beaver poison,	Water hemlock,	Cicuta maculata.
" root,	Yellow pond lily,	Nuphar advena.
" tree,	Magnolia,	Magnolia glauca.
Bebeeru,	Green heart tree,	Nectandra Rodiœi,

COMMON.	ENGLISH.	BOTANICAL.
Beccabunga,	Brooklime,	Veronica beccabunga.
Bedeguar,	Dog rose galls,	Rosa canina.
Bedstraw,	Cleavers,	Galium aparine.
Bee balm,	Oswego tea,	Monarda didyma.
Beech, American,		Fagus ferruginea.
Beech drops,	Cancer root,	Orobanche Virginiana.
" " Albany,		Pterospora andromeda.
" White European,		Fagus sylvatica.
Beeflower,	Bee Orchis,	Ophrys apifera.
Beehive,	Snail plant,	Medicago scuttellata.
Beefsteak plant,	Strawberry geranium,	Saxifraga sarmentosa.
Beesnest plant,	Wild carrot,	Daucus carota.
Beesnettle,	Flowering nettle,	Galeopsis versicolor.
Beet, common,		Beta vulgaris.
Beggars' tick,		Bidens frondosa.
" lice,	Virginia mouse ear,	Cynoglossum Morrisoni.
Behen nuts,	Horse radish tree,	Moringa aptera.
Bela,	Bael, Bengal quince,	Ægle marmelos.
Belladonna,	Deadly nightshade,	Atropa Belladonna.
Bellflower,		Campanula rotundifolia.
Bell pepper,		Capsicum grossum.
Bellwort,		Uvularia perfoliata.
Bellyache root,	Angelica,	Angelica lucida.
Ben oil,		Guilandina Moringa.
Bendee,	Okra,	Hibiscus esculentus.
Bengal root,	Zedoary,	Zingiber casumunar.
" quince,		Ægle marmelos.
Benjamin bush,	Spicewood,	Benzoin odoriferum.
" tree,	Benzoin,	Styrax benzoin.
Benne,	Oily grain,	Sesamum orientale.
Bennet,	White avens,	Geum Virginianum.
Ben nuts,	Oily acorn,	Guilandina Moringa.
Bent grass,		Agrostis alba.
Benzoin,	Balsamic exudation of	Styrax benzoin.
Bergamot,	An oil from the fruit rind,	Citrus bergami.
" herb,		Mentha citrata.
Bermuda grass,	Dog grass,	
Besenna,	Bisenna,	Albizzia anthelmintica.
Beshan,	Balsam of Mecca,	Amyris Gileadensis.
Betel,	the leaves used,	Piper betel.
" nuts,	Areca nuts,	Areca catechu.
Beth root,	Trillium,	Trillium pendulum.
Bethelsdorf aloes,	Cape aloes,	
Betony,		Lycopus Virginicus.
" Wood,	Betony; common,	Betonica officinalis.
Bhang,	Bang,	Cannabis sativa var indica.
Bibernel,		Poterium sanguisorbia.
Bibiru bark,	Green heart tree,	Nectandra Rodiœi,
Bicolorata bark,	Pitaya Carthagena bark,	Cinchona Condaminea.
Bigarade,	Bitter or Seville orange,	Citrus bigaradia.
Bigbloom,	Magnolia,	Magnolia macrophylla.
Big leaved ivy,		Kalmia latifolia.
Bilberry,	Whortleberry,	Vaccinium vitis idœa.
Bindweed,	Man root,	Convolvulus panduratus.
"	Morning glory,	" purpurea.
"	Italian sarsaparilla,	Smilax aspera.
" Bracted,		Calystegia sepium.
Birch, Black,		Betula lenta.
" camphor,	a resinous substance in	" nigra.
" Cherry,	Black birch,	" lenta.

COMMON.	ENGLISH.	BOTANICAL.
Birch, Mahogany,	Black birch,	Betula lenta,
" Red,		" rubra.
" Spicy,	Black birch,	" lenta.
" Sweet,	" "	" "
" White,	Paper birch,	" alba var populifolia.
Birdcherry,	Cerasus padus,	Prunus avium.
Birdfoot violet,	Blue violet,	Viola pedata.
Bird lime,	a viscid substance of	Ilex opaca.
" "	a " " "	Viscum album.
" pepper,	Cayenne,	Capsicum baccatum.
" seed,	Canary, Hemp and Maw	Seeds.
" weed,	Knot grass,	Polygonum aviculare.
Birds' bill,		Trigonella ornithorhynchus.
" eye,	False hellebore,	Adonis vernalis.
" foot, common,		Ornithopus puppusillus.
" nest, plant,	Ice plant,	Monotropa uniflora.
" " root,	Wild carrot,	Daucus carota.
" tongue,	the seed of	Fraxinus excelsior.
Birth root,	Beth root,	Trillium pendulum.
" wort,	Virginia snake root,	Aristolochia serpentaria.
" " long,		" longa
" " round,		" rotunda.
" " thick,	Virginia snake root,	" serpentaria.
Bisenna,	Messenna,	Albizzia anthelmintica.
Bish,	Indian aconite,	Aconitum ferox.
Bishop's weed, American		Discopleura capillacea.
" " true,	Ajava seed,	Ammi copticum.
" leaves,	Water betony,	Scrophularia aquatica.
" "	. " figwort,	" "
" cap,	Mitre wort,	Mitella diphylla.
Bishopswort,	Black cummin,	Nigella sativa.
Bistort,	Officinal bistort,	Polygonum Bistorta.
Biting knot weed,	Water pepper,	" hydropiper.
" stone crop,	Small houseleek,	Sedum acre.
Bitter almond,		Amygdalus communis var amara.
" aloes,	Barbadoes aloes,	Aloe vulgaris.
" apple,	Colocynth,	Citrullus Colocynthis.
" ash,	Quassia,	Simaruba excelsa.
" "	Wahoo,	Euonymus atropurpureus.
" bark,	Florida bark,	Pinckneya pubens.
" "	Quassia bark,	Simaruba excelsa.
" blain,		Vandellia diffusa.
" bloom,	American centaury,	Sabbatia angularis.
" candytuft,		Iberis amara.
" cassava,	Tapioca plant,	Jatropha manihot.
" clover,	American centaury,	Sabbatia angularis.
" cress,		Cardamine amara.
" cucumber,	Colocynth,	Citrullus Colocynthis.
" cups,	Cups turned from quassia wood.	
" damson,		Simaruba officinalis.
" dogsbane,	Bitter root,	Apocynum androsæmifolium.
" gourd,	Colocynth,	Citrullus Colocynthis.
" grass,	Unicorn root,	Aletris farinosa.
" herb,	Balmony,	Chelone glabra.
" milkwort,		Polygala amara.
" nut,	Hickory,	Carya amara.
" oak,		Quercus cerris.
" orange,		Citrus vulgaris.
" polygala,	Ground flower,	Polygala rubella.
" quassia,	Quassia wood,	Simaruba excelsa.

COMMON.	ENGLISH	BOTANICAL.
Bitter root,		Apocynum androsæmifolium.
" " bark,	Bark of bitter root	" "
" "	Gentian root,	Gentiana lutea.
" stem,	Chiretta,	Agathotes Chirayta.
" stick,	"	" "
" sweet,	Woody nightshade,	Solanum Dulcamara.
" " false, bark,		Celastrus scandens.
" " climbing,	False bittersweet,	" "
" trefoil,	Buckbean,	Menyanthes trifoliata.
" weed,	Tall ambrosia,	Ambrosia trifida.
" wintergreen,	Pipsissewa,	Chimaphylla umbellata.
" wood,	Quassia,	Simaruba excelsa.
" "		Xylopia glabra.
" worm,	Buckbean,	Menyanthes trifoliata.
" wort,	Gentian,	Gentiana lutea.
Bitumen, liquid,	Petroleum.	
" solid,	Asphaltum,	Mineral tar.
Black alder,		Prinos verticillatus.
" " tree,	Persian berry,	Rhamnus frangula.
" ash,		Fraxinus sambucifolia.
" barley,	Barley wheat,	Hordeum vulgare, var.
" balsam,	Balsam Peru,	Myrospermum Peruiferum.
" bearberry,		Arctostaphylos alpina.
" berry,		Rubus villosus.
" " low,		" trivialis.
" " lily,		Ixia Chinensis.
" birch,		Betula lenta.
" birdweed,	Climbing buckwheat,	Polygonum convolvulus.
" bittervetch,		Orobus niger.
" borehound,	Black horehound,	Ballota nigra.
" boy resin,		Xanthorrhœa arborea.
" bryony,	Tamus,	Tamus communis.
" caco,	Yam root. Tara.	Colocasia esculenta.
" caraway,	Nutmeg flower,	Nigella sativa.
" cherry,	Wild black cherry	Prunos serotina.
" " poison,	Belladonna,	Atropa Belladonna.
" cohosh,		Cimicifuga racemosa.
" cummin,	Black caraway,	Nigella sativa.
" currant,	Corinthian grape,	
" "		Ribes nigrum.
" " wild,		" Floridum.
" cypress wood,	Virginia cypress,	Cupressus disticha.
" dogwood,	Black alder tree,	Rhamnus frangula.
" drink, Indian,		Ilex vomitoria.
" ebony tree,		Cocos fusiformis.
" flower,	Quafodil,	Melanthium Virginicum.
" gentian,		Libanotis vulgaris.
" ginger,	Unbleached ginger,	Zingiber officinalis.
" haw,	Sloe,	Viburnum prunifolium.
" hellebore,		Helleborus niger.
" henbane,		Hyosciamus niger.
" horehound,		Ballota nigra.
" huanuco bark,		Cinchona glandulifera.
" indian hemp,	·	Apocynum cannabinum.
" ipecac,		Psychotria emetica.
" jack oak,	Barren oak,	Quercus nigra.
" larch,	Tamarack,	Larix Americana.
" liquorice,	The hard extract,	Glycyrrhiza glabra.
" locust,		Robini pseudo acacia.
" masterwort,	Imperial masterwort,	Astrantia major.

COMMON.	ENGLISH.	BOTANICAL.
Black medick,	Dutch clover,	Medicago lupulina.
" moss,	Florida long moss,	Tillandsia usneoides.
" mustard,		Sinapis nigra.
" mulberry,		Morus nigra.
" mullein,		Verbascum nigrum.
" nightshade,	Garden nightshade,	Solanum nigrum.
" oak,	Quercitron,	Quercus tinctoria.
" oats,		Avena sativa nigra.
" pepper,		Piper nigrum. .
" poplar,		Populus nigra.
" poppy,		Papaver somniferum.
" pursley,	Black spurge,	Euphorbia maculata.
" ramthorn,		Rhamnus lycioides.
" root,		Leptandra Virginica.
" saltwort,	Sea milkwort,	Glaux maritima.
" sampson,	Red sunflower,	Rudbeckia purpurea.
" sanicle,		Sanicula Marilandica.
" snakeroot,	Black cohosh,	Cimicifuga racemosa.
" "	" sanicle,	Sanicula Marilandica.
" snakeweed,	Wild ginger,	Asarum Virginicum.
" spleenwort,		Asplenium adiantum nigrum.
" spruce,	produces spruce gum,	Abies nigra.
" spurge,		Euphorbia maculata.
" tamarinds,	Tamarinds,	Tamarindus Indica.
" tea,		Thea Bohea.
" thorn,	Nanny bush,	Viburnum lentago.
" "		Prunus spinosa.
" truffle,		Tuber griseum.
" turnip,		Leontice leontopetalum.
" walnut,		Juglans nigra.
" whortleberry,	Huckleberry,	Vaccinium myrtillus.
" willow,	Pussy willow,	Salix nigra.
Bladder campion,	Sea pink,	Cucubalus behen.
" fern,	Cup fern,	Polypodium fragile.
" fucus,	Seawrack,	Fucus versiculosus.
" green,	a color from Buckthorn bs.,	Rhamnus catharticus
" ketmia,	Flower of an hour,	Hibiscus trionum.
" nut,	Pistachio,	Pistacia vera.
" "		Staphylea trifolia.
" pod,		Physolobium.
" podded lobelia,	Lobelia herb,	Lobelia inflata.
" senna,		Colutea arborescens.
" wort,		Utricularia macrorhiza.
" wrack,	Seawrack,	Fucus versiculosus.
Blazing star,	Rattlesnakes' master,	Liatris squarrosa.
" "	Unicorn root,	Aletris farinosa.
Bleaberry,	Whortleberry,	Vaccinium myrtillus.
Bleeding heart,	Valerian, American,	Cypripedium pubescens and var
" "		Dicentra spectabilis.
Blessed herb,	Avens root, European,	Geum Urbanum.
" thistle,		Centaurea benedicta.
Blechnum,		Blechnum Virginicum.
Blight,	Mildew on leaves,	Rubigo alnea.
Blind nettle,	Stingless nettle,	Lamium album.
Blisterweed,	Crowfoot buttercup,	Ranunculus acris.
Blite,	Indian strawberry,	Blitum capitatum.
Blockwood,	Logwood,	Hæmatoxylon Campechianum.
Blood elder,		Sambucus Ebulus.
" geranium,	Cranesbill,	Geranium sanguineum.
" marigold,		Zinnia violacea.

2

COMMON.	ENGLISH.	BOTANICAL.
Blood root,		Sanguinaria Canadensis.
" staunch,	Canada fleabane,	Erigeron Canadense.
" tree,		Croton gossypifolium.
" twig,	Boxwood,	Cornus Florida.
" weed,	Bastard Ipecac,	Asclepias curassavica.
" wood,	Jamaica redwood,	Gordonia Hæmatoxylon.
" "		Eucalyptus citriodora.
" wort,	Hawkweed,	Hieracium venosum.
" " striped,	"	" "
Bloom grass,		Bromus pubescens.
Blooming spurge,	Large flowering spurge,	Euphorbia corollata.
Blue ash,		Fraxinus quadrangulata.
" balm,	Bee balm,	Monarda didyma.
" beech,	Hornbeam,	Carpinus Americana.
" bells,	Blue gentian,	Gentiana Catesbæi.
" "	American Greek valerian,	Polemonium reptans.
" berry,	Blue whortleberry,	Vaccinium frondosum.
" "	" cohosh,	Caulophyllum thalictroides.
" " cornel,	Rose willow,	Cornus sericea.
" bonnets,	Blue centaury,	Centaurea cyanus.
" bottles,	" "	" "
" cardinal,	" Lobelia,	Lobelia syphilitica.
" centaury,		Centaurea cyanus.
" chamomile,	Michaelmas daisy,	Aster Tradescantia.
" cohosh,		Caulophyllum thalictroides.
" curls,		Trichostema dichotoma.
" flag,		Iris versicolor.
" fringed gentian,		Gentiana crinita.
" gentian,		Gentiana Catesbæi.
" ginseng,	Blue cohosh,	Caulophyllum thalictroides.
" grass,		Poa pratensis.
" gum tree,	Fever tree,	Eucalyptus globulus.
" lobelia,		Lobelia syphilitica.
" mallows,	Common low mallows,	Malva rotundifolia.
" palmetto,		Chamærops hystrix.
" scarlet pimpernel,		Anagallis cœrulea.
" scullcap,		Scutellaria lateriflora.
" tangles,	Whortleberry,	Vaccinium frondosum.
" thistle,	Vipers' bugloss,	Echium vulgare.
" vervain,		Verbena spuria.
" violet,		Viola pedata.
" weed,	Vipers' bugloss,	Echium vulgare.
" whortleberry,	Blueberry,	Vaccinium frondosum.
Bluets,	"	" "
Blumea camphor,	Canton camphor,	Blumea balsamifera.
Blunt leaved dock,		Rumex obtusifolia.
Bocconia,	Celandine tree,	Bocconia cordata.
Bofareira,	Castor oil plant leaf,	Ricinus communis.
Bog bean,	Buckbean,	Menyanthes trifoliata.
" moss,		Sphagnum commune.
" myrtle,	Buckbean,	Menyanthes trifoliata.
" onion,	Dragon root,	Arum triphyllum.
" rush,		Killinga monocephala.
" " false,		Rhynchospora glomerata.
" " Water,		Schoenus mariscoides.
Bogota bark,	Carthagena bark,	Cinchona Condaminea.
Bohea,	Black tea,	Thea Bohea.
Bois de crabe,	Clove bark,	Canella giroffee.
Bonduc nuts,	Bonduc seed,	Guilandina bonduc.
Bondou,	Gum Senegal, ·	

COMMON.	ENGLISH.	BOTANICAL.
Boneset,	Thoroughwort,	Eupatorium perfoliatum.
" purple,	Queen of the meadow,	" purpureum.
Bonnet grass,	White top,	Agrostis alba.
" pepper,		Capsicum tetragonum.
Bonny dame,	Orache, garden,	Atriplex hortensis.
Bookoo,	Buchu,	Barosma crenata.
Boor tree,	Elder,	Sambucus nigra.
Borage,		Borago officinalis.
Boreal sour dock,	Mountain sorrel,	Oxyria reniformis.
Boston smilax,	Cape smilax,	Myrsiphyllum asparagoides.
Botany bay gum,		Xanthorrhœa arborea.
" " kino,	Brown gum tree,	Eucalyptus resinifera.
" " tea.		Smilax glycophilla.
Bottle tree, Australian,		Delabechea rupestris.
Bouncing bet,	Soapwort,	Saponaria officinalis.
Boundou,	Akasgia,	
Bountry,	Elder,	Sambucus nigra.
Bowman's root,		Euphorbia corrolata.
" "		Gillenia stipulaceæ.
" "		Leptandra Virginica.
Bowstring hemp,	African hemp,	Sanseviera Guinensis.
Bow wood,	Osage orange,	Maclura aurantiaca.
Box,		Buxus sempervirens.
" berry,	Checkerberry,	Gaultheria procumbens.
" Dwarf,		Buxus suffruticosa.
" elder,	Ash maple,	Acer negundo.
" Garden edging,		Buxus sempervirens.
" thorn,	Matrimony vine,	Lycium barbarum.
" wood,	Virginia dogwood,	Cornus Florida.
" " Jamaica,		Tecoma pentaphylla.
Boys' love,	Southernwood,	Artemisia abrotanum.
Braid root,		Justicia nasuta.
Brake,	Common brake,	Pteris aquilina.
" Buckhorn,	Royal flowering fern,	Osmunda regalis.
" Fern,	Polypody,	Polypodium vulgare.
" Rock,	Polypody,	" "
" root,	"	" "
" Sweet,	Male fern,	Aspidium Filix Mas.
" Winter,		Pteris atropurpurea.
Brandybottles,	Yellow pond lily flowers,	Nuphar advena.
Brazilian Arrow root,	Tapioca,	
" Beans,	Sassafras nuts,	Nectandra purchury.
Braziletto,		Cæsalpina Brasiliensis.
Brazil nuts,	Castana nuts,	Bertholletia excelsa.
" tea,	Paraguay tea,	Ilex Paraguensis.
" "		Lantana pseudothera.
" wood,		Cæsalpina echinata.
Bramble,	Blackberry vine,	Rubus fruticosus.
Bread fruit,		Artocarpus incisa.
" root,	Prairie turnip,	Psoralea esculenta.
Breast weed,	Lizards' tail,	Saururus cernuus.
Brier,		Smilax rotundifolia.
" herb,	Rock blackberry,	Rubus saxatilis.
" rose,	Flowering bramble,	" rosæfolius.
" Sweet,	Eglantine rose,	Rosa rubiginosa.
Brimstonewort,	Sulphurwort,	Peucedanum officinale.
Brinton root,	Black root,	Leptandra Virginica.
Briony,	Bryony,	Bryonia alba and dioica.
Bristol wall cress,		Arabis stricta.
Bristle stem sarsaparilla	Dwarf elder,	Aralia hispida.

COMMON.	ENGLISH.	BOTANICAL.
Bristly rose acacia,		Robinia hispida.
British tea,	Elm leaves,	Ulmus.
" tobacco,	Coltsfoot leaves,	Tussilago farfara.
Brittle gum,	Senegambia gum arabic,	Acacia albida.
" worts,	Diatoms.	
Broad leaved dock,		Rumex obtusifolius.
" " dogwood,	Green ozier,	Cornus circinata.
" " kalmia,	Broad leaved laurel,	Kalmia latifolia.
" " laurel,		" "
" wort,	Water betony,	Scrophularia aquatica.
Brompton stock,		Cheiranthus incanus.
" queens,		" "
Brome grass,	Broom grass,	Bromus pubescens and var.
Brook alder,	Black alder,	Prinos verticillatus.
" bean,	Buckbean,	Menyanthes trifoliata.
" lime,		Veronica beccabunga.
" liverwort,	Woodrow,	Marchantia polymorpha.
" weed,	Water pimpinel,	Samolus valerandi.
Broom corn,		Sorghum saccharatum.
" Dyers,		Genista tinctoria.
" flowers,	Broom herb,	Cytisus Scoparius.
" grass,		Bromus purgans and var.
" herb,		Cytisus Scoparius.
" Indigo,	Wild indigo,	Baptisia tinctoria.
" pine,	Yellow pitch pine,	Pinus palustris.
" rape, American,		Orobanche Americana.
" " One-flowr'd,		" uniflora.
" Yellow,	Wild indigo,	Baptisia tinctoria.
Brown gum tree,	Botany bay kino tree,	Eucalyptus resinifera.
" wort,	Yellow figwort,	Scrophularia vernalis.
Bruise root,	Horn poppy,	Chelidonium glaucum.
" wort,	Soap wort,	Saponaria officinalis.
Bryony, white,		Bryonia alba.
" Black,		Tamus communis.
" Red berried,		Bryonia dioica.
Bucco,		Agathosma pulchella.
Buchu,		Barosma crenata and species.
" long-leaved,		" serratifolia.
" short-leaved,		" crenata.
Buckbean,		Menyanthes trifoliata.
Buckeye,		Æsculus glabra.
" small,		" pavia.
Buckhorn brake,		Osmunda regalis.
Bucks'horn plantain,		Plantago coronopus.
Buckthorn,		Rhamnus catharticus.
" European,	European Black alder,	" frangula.
" Purging,	Buckthorn berries,	" catharticus.
Buckwheat,		Polygonum fagopyrum.
" Indian,		Fagopyrum Tartaricum.
Budwood,	Dogwood,	Cornus Florida.
Buffalo clover,		Trifolium Pennsylvanicum.
" berry,		Shepherdia argentea.
" currant,		Ribes aureum.
Bugbane,	Black cohosh,	Cimicifuga racemosa.
"	American Hellebore,	Veratrum viride.
Bugle,	Ground pine,	Ajuga chamæpitys.
" Bitter,		Lycopus Europæus.
" Common		Ajuga repens.
" Sweet,	Bugle weed,	Lycopus Virginicus.
" Water,	" "	" "

COMMON.	ENGLISH.	BOTANICAL.
Bugle weed,		Lycopus Virginicus.
" wort,	Bugle weed,	" "
Bugloss, common,	Borage,	Borago officinalis.
" small,		Anchusa Italica.
" Viper's,		Echium vulgare.
" Wild,	German madwort,	Asperugo procumbens.
Bugwort,	Black cohosh,	Cimicifuga racemosa.
Buku,	Buchu,	Barosma crenata.
Bulrush,	Typha,	Juncus effusus.
Bullace,		Prunus communis.
Bullet tree,		Sapota mulleri.
Bulldogs,	Snap dragon,	Antirrhinum majus.
Bullock's eye,	House leek,	Sempervivum tectorum.
" lungwort,	Mullein,	Verbascum Thapsus.
Bullsfoot,	Coltsfoot,	Tussilago farfara.
Bunch berry,	Dwarf Cornel,	Cornus Canadensis.
" flower,	Black flower,	Melanthium hybridum.
" pink,		Dianthus barbatus.
Bunweed,	Ragwort,	Senecio Jacobæa.
Burdock,		Arctium lappa.
" Prairie,		Silphium terebinthinaceum.
Burgamot, Wild,	Wild bergamot,	Monarda oblongata.
Burgundy pitch,		Abies, excelsa and picea.
Bur marigold,	Cuckold,	Bidens connata.
Burnet, Garden,		Sanguisorba officinalis.
" Great,		" media.
" Saxifrage,		" Canadensis.
" " Eurp'n,		Pimpinella Saxifraga.
Burning bush,	Wahoo,	Euonymus Americana.
Burrage,	Borage,	Borago officinalis.
Burr-flower,		Hydrophyllum Virginicum.
" grass,		Cenchrus echinatus.
" reed,		Sparganium ramosum.
" seed,	Burdock seed,	Arctium lappa.
" thistle,	Clotburr,	Xanthium strumarium.
" weed,	"	" "
" wort,	Buttercup,	Ranunculus acris.
Bush clover,		Lespedeza sessiliflora.
" honeysuckle,	Gravel weed,	Diervilla Canadensis.
" whortleberry,		Vaccinium dumosum.
Bushy gerardia,		Gerardia pedicularia.
Butea gum,		Butea frondosa.
Butchers' broom,	Knee holly,	Ruscus aculeatus.
Butter and eggs,	Double flowered variety of	Narcissus aurantius.
" bur,	Coltsfoot,	Tussilago officinalis.
" cup,	Acrid crowfoot,	Ranunculus bulbosus.
" " Garden,	Double flowered variety of	" aconitifolius.
" fly weed,	Pleurisy root,	Asclepias tuberosa.
" jags,	Lotus,	Lotus corniculatus.
" nut,	White walnut,	Juglans cinerea.
" nuts,	Souari,	Caryocar nuciferum.
" Canara,	Piney tallow, a solid oil,	Vateria Indica.
" tree,	Shea tree,	Bassia Parkii.
" weed,	Canada fleabane,	Erigeron Canadense.
" "		Senecio lobatus.
" wort,		Pinguicula elatior.
Button bush,		Cephalanthus occidentalis.
" snake root,	Water eryngo,	Eryngium aquaticum.
" " "		Liatris spicata.
" weed,		Diodia Virginica.

2*

COMMON.	ENGLISH.	BOTANICAL.
Buttonwood,	False sycamore,	Platanus occidentalis.
" " shrub,	Button bush,	Cephalanthus occidentalis.
Caballine aloes,	Horse aloes,	Aloe Guiniensis.
Cabbage,		Brassica oleracea.
" palm,		Areca oleracea.
" rose,	Pale rose,	Rosa centifolia.
" Skunk,	Skunk cabbage,	Symplocarpus fœtidus.
" tree,		Andira retusa.
" " bark, br'wn		" "
" " " yell'w		" inermis.
Cacao beans,	Chocolate nuts,	Theobroma cacao.
" butter,	the fixed oil from	" "
Cachew nut,		Anacardium occidentale.
Cactus,	Prickly pear,	Cactus opuntia,
" Night blooming,	Night blooming cereus,	" grandiflorus.
Cafta leaves,		Catha edulis.
Cahinca root,		Chiococca racemosa.
Cailleau,	Sage tree,	Lantana Camara.
Cajeput,	the leaves yield oil Cajeput,	Melaleuca Cajuputi.
Cajuput,	Cajeput,	" "
Cake saffron,	Pressed saffron,	Crocus sativa.
Calabar bean,	.	Phytostigma venenosum.
Calabash gourd,	Bottle gourd,	Lagenaria vulgaris.
Calamba root,	Colombo root,	Cocculus palmatus.
" wood,	False colombo,	Pareira medica.
Calamint,	Basil thyme,	Calamintha officinalis.
Calamus,	Sweet flag root,	Acorus Calamus.
Calandrina,		Calandrina discolor.
Calathian violet,	Marsh gentian,	Gentiana pneumonanthe.
Calendula,	Marigold,	Calendula officinalis.
California nutmeg,		Torreya California.
" wood,	Nicaragua wood,	Cæsalpina "
Calico bush,		Kalmia latifolia.
Calisaya bark,	Peruvian bark, yellow,	Cinchona calisaya.
" " spurious,		Cascarilla quepo.
Calla lily,		Richardia Africana.
Callahuala root,		Polypodium calaguala.
Calumbo,	Colombo root,	Cocculus palmatus.
Camelia,	Common camellia,	Camellia Japonica.
Camels' thorn,	yields Persian manna,	Alhagi camelorum.
" hay,		Andropogon Schœnanthus.
Camomile,	Chamomile,	Anthemis nobilis.
Camphor, Batavia,	a concrete oil,	Camphora officinarum.
" Borneo,		Dryobalanops Camphora.
" Canton,		Blumea balsamifera.
" Chinese,	Japan camphor,	Cinnamomum Camphora.
" Sumatra,		Dryobalanops Aromatica.
" tree,	Camphor laurel,	Laurus Camphora.
Cammock,	Rest harrow,	Ononis spinosa.
Campeachy wood,	Log wood,	Hæmatoxylon Campechianum.
Campion pink,	Bladder campion,	Cucubalus behen.
Cam wood,	a red dye wood,	Baphia nitida.
Canada balsam,	Oleo resin of	Abies balsamea.
" fleabane,	.	Erigeron Canadense.
" nettle,	Albany hemp,	Urtica Canadensis.
" pitch,	Hemlock gum,	Abies "
" root,	Pleurisy root,	Asclepias tuberosa.
" snake root,		Asarum Canadense.
" tea,	Checkerberry,	Gaultheria procumbens.

COMMON.	ENGLISH.	BOTANICAL.
Canada thistle,	Cursed thistle,	Cirsium arvense.
" turpentine,	Canada balsam,	Abies balsamea.
Canadian bur,	Hounds' tongue,	Cynoglossum officinale.
Canary archil,	Orchil,	Roccella tinctoria.
" bird flower,		Tropæolium peregrinum.
" grass,	Canary seed grass,	Phalaris Canariensis.
" " wild,	Ribbon grass,	" Americana.
" seed,		" Canariensis.
" weed,	Litmus weed,	Roccella tinctoria and var.
Cancer drops,	Beech drops,	Orobanche Virginiana.
" root,	" "	" "
" weed,	Wild sage,	Salvia lyrata.
" wintergreen,	Creeping wintergreen,	Gaultheria hispidula.
Candle berry,	Bayberry,	Myrica cerifera.
" " myrtle,	"	" "
" " tree,	Candlenuts,	Aleurites triloba.
" wood,		Gomphia Guianensis.
" "	Jamaica roswood,	Amyris balsamifera.
Candy carrot,		Athamanti Cretensis.
" tuft,	Bitter candytuft,	Iberis amara.
" " common,		" umbellata.
Cane,	Bamboo,	Bambusa arundinacea.
" brake,		Arundinaria.
" Rattan,	Rattan,	Calamus rotang.
Canella,		Canella alba.
Canker lettuce,	Round leaved pyrola,	Pyrola rotundifolia.
" root,	Marsh rosemary,	Statice Caroliniana.
" "	Goldthread root,	Coptis trifolia.
" "	Canker weed,	Prenanthes alba.
" violet,	Beaked violet,	Viola rostrata.
" weed,		Prenanthes alba.
Canna,	Indian cane,	Canna Indica.
" starch,	St. Kitt's arrow root,	" edulis and var.
Canoe birch,	Paper birch,	Betula papyracea.
" wood,	Tulip tree,	Liriodendron Tulipifera.
Cannon ball tree,		Courrupita Guianensis.
Caoutchouc,	India rubber,	Siphonia elastica and species.
Caper bush,		Capparis spinosa.
" spurge,	Mole plant,	Euphorbia lathyris.
Cape aloes,	Cape of Good Hope aloes,	Aloe Capensis.
" gooseberry,	Winter cherry,	Physalis alkekengi.
" jasmine,		Gardenia Florida.
" smilax,		Myrsiphyllum Asparagoides.
" tulip,		Homeria collina.
" weed,	Orchilla weed,	Roccella tinctoria.
Capsicum,	Red pepper,	Capsicum annuum.
Carabaya bark,	Peruvian bark,	Cinchona orata var Rufinervis.
Caraccas kino,	Jamaica Kino,	Coccoloba uvifera.
Caramania gum,	False tragacanth.	
Caranna,		Bursera acuminata.
. " gum,	Jamaica birch gum,	" gummifera.
Caraway seed,		Carum carui.
Carats,	the seed of	Erythrina Abyssinica.
Cardamom,	the fruit of	Elettaria Cardamomum.
" Bastard,		Alpina "
" Bengal,	Java Cardamom,	Amomum maximum.
" Ceylon,		Elettaria Cardamomum.
" Java,		Amomum maximum.
" Malabar,		Elettaria Cardamomum.

COMMON.	ENGLISH.	BOTANICAL.
Cardamom Nepal,	Java Cardamom,	Amomum maximum.
" Round,	Cluster Cardamoms,	" racemosum.
" Wild,		Elettaria Cardamomum, major.
Cardinal flower,	Red Lobelia,	Lobelia cardinalis.
" " Blue,	Blue "	" syphilitica.
" " Red,	Red "	" cardinalis.
" sage,	Mexican sage,	Salvia fulgens.
Cardleaf tree,	West Indian name for	Clusia of several species.
Cardoon,		Cynara cardunculus.
" Spanish,		Scolymus Hispanicus.
Cardus, Blessed,	Blessed thistle,	Centaurea benedicta.
" spotted,	" "	" "
Carib tea,	Goat weed,	Capraria biflora.
Caribæan bark,	St. Lucia bark,	Exostemma floribunda,
Carline thistle,		Carlina acaulis.
Carnation, pink,	Clove pink,	Dianthus Caryophyllus.
" Spanish,	Barbadoes Pride,	Cæsalpina pulcherrima.
Carnauba palm,	a Brazilian wax palm,	Corypha cerifera.
" wax,	produced by	" "
Carob tree,	St. John's bread,	Ceratonia siliqua.
Carolina allspice,		Calycanthus Floridus.
" bark,	Florida bark,	Pinckneya pubens.
" cedar,	Red cedar,	Juniperus Virginiana.
" jessamine,	Yellow jessamine	Gelseminum sempervirens.
" laurel cherry,	Winter laurel,	Prunus Caroliniana.
" pink,		Spigelia Marilandica.
" poplar,	American Balm gilead,	Populus balsamifera.
" potato,	Sweet potato,	Ipomea Batatus.
" vanilla,	Vanilla leaf,	Liatris odoratissama.
Carony bark,	Angostura bark,	Galipea officinalis.
Carpathian balsam,	Oleo resin of stone pine,	Pinus cembra.
Carpenters' herb,	Bugle weed,	Lycopus Virginicus.
" leaf,		Galax rotundifolia.
" square,	Scrofula plant,	Scrophularia Marilandica.
Carpetweed,		Mollugo verticillata.
Carrageen moss,	Irish moss,	Chondrus crispus.
Carragheen moss,	" "	" "
Carrion flower,	Ground Ivy,	Nepeta glechoma.
Carrot, Garden,	Cultivated Carrot,	Daucus carota.
" Weed,	Bastard wormwood,	Ambrosia elatior.
" Wild,		Daucus carota.
Carthagena bark,	Pale bark.	Cinchona condaminea.
" " fibrous,	Bogota bark,	" " var. lancifolia.
" " hard,	Velvet bark,	" cordifolia, var. vera.
Carthaginian apple,	Pomegranate fruit,	Punica Granatum.
Cascarilla,	a sub genera of Cinchona,	
" bark,		Croton Eleuteria.
" "	Wild rosemary, W. I.	" lineare.
Cashew nuts,		Anacardium occidentale.
" "	Marking nut,	Semecarpus anacardium.
Cassava plant,	Tapioca plant,	Jatropha manihot.
Cassia bark,	Bastard Cinnamon,	Cinnamomum Cassia.
" buds,	Cinnamon buds,	" Lourciril.
" stick tree,	Purging cassia,	Cassia Fistula.
Cassie,	the flowers yield a perfume,	Acacia Farnesiana.
Cassumuniar,	Zerumbet,	Zingiber cassumuniar.
Castor bean,	yields Castor oil,	Ricinus communis.
" oil plant,	Palma Christa,	" "
Catalpa tree,	the seed used,	Bignonia Catalpa.

COMMON.	ENGLISH.	BOTANICAL.
Catawba tree,	Catalpa tree,	Bignonia Catalpa.
Catarrh root,	Galangal,	Alpina galanga.
Cat brier,	common name for	Smilax.
Catch fly,	Wild pink,	Silene Virginica.
" "	Bitter root,	Apocynum androsæmifolium.
" weed,	Cleavers,	Galium aparine.
Catechu,	the inspissated juice of	Areca Catechu and Acacia C.
" Black, china,		Acacia Catechu.
" Pale,	Gambir,	Nauclea gambir.
" red, Bengal,		Acacia Catechu.
Caterpillar fern,	Harts' tongue fern,	Scolopendrium officinalis.
" plant,		Medicago circinata.
Catmint,		Nepeta cataria.
Catnip,	Catmint,	" "
Cat-o'nine-tail,	Cat tall flag,	Typha latifolia.
Cat thyme,		Teucrium marum.
Catgut,	Hoary pea,	Tephrosia Virginiana.
Cat tail flag,	Reed mace,	Typha latifolia.
Cats' eye,	Star scabious,	Scabiosa stellata.
Cats foot,	Canada snake root,	Asarum Canadense.
" "	Ground ivy,	Nepeta glechoma.
" milk,	Wartwort,	Euphorbia helioscopia.
" paw,	Ground ivy,	Nepeta glechoma.
" wort,	Catmint,	" cataria.
Cauliflower,		Brassica Florida.
Cayenne,	Capsicum,	Capsicum annuum.
" cinnamon,	Analogous to Ceylon cinnamon.	
" pepper,	African pepper,	Capsicum fastigiatum.
Cedar, apples,	Excrescences formed on	Juniperus Virginiana.
" Barbadoes,	Sweet scented cedar,	Cedrela odorata.
" Bastard,	Soymida,	Swietenia febrifuga.
" Bermuda,		Juniperus Bermudiana.
" Berry bearing,	yields oil of cade,	" oxycedrus.
" of Lebanon,	.	Cedrus Libani.
" Red,		Juniperus Virginiana.
" Spanish,	Incense wood,	Icica Guianensis.
" White,	White cedar,	Cupressus thyoides.
" " false,	Arbor vitæ,	Thuja occidentalis.
Cedrat,	the oil from the fruit rind,	Citrus medica.
Cedron seed,		Simaba cedron.
Celandine,	common name for	Chelidonium.
" Garden,		" majus.
" Great,		"
" lesser,		" minus.
" poppy,		Stylophorum diphyllum.
" tree,		Bocconia cordata.
" Wild,	Touch me not,	Impatiens pallida.
Celery,	Garden celery,	Apium graveolens.
Cembra nuts,	Stone pine nuts,	Pinus cembra.
Centaury, American,		Sabbatia angularis.
" Chili,		Erythræa Chilensis.
" European,		" centaurium.
" Greater,		Centaurea "
" Ground,		Polygala Nuttallii.
" lesser,	Common Europ'n centaury,	Erythræa centaurium.
" Red,	American centaury,	Sabbatia angularis.
Century plant,	" aloe,	Agave Americana.
Ceterach,	Milt waste,	Ceterach officinarium.
Cevadilla seed,		Veratrum Sabadilla.

COMMON.	ENGLISH.	BOTANICAL.
Ceylon cardamom,		Elettaria Cardamomum.
" cinnamon,		Cinnamomum Zelanicum.
" moss,		Gracilaria lichenoides.
Chafe weed,	Life everlasting,	Gnaphalium polycephalum.
Chaff weed,	Bastard pimpernel,	Centunculus minimus.
Chalcedonian lily,		Lilium Chalcedoncium.
Chamomile,		Anthemis nobilis.
" Garden,		" "
" German,	Chamomilla,	Matricaria chamomilla.
" Low,	Garden chamomile,	Anthemis nobilis.
" Roman,		" "
" Spanish,	Pellitory of Spain,	Anacyclus Pyrethrum.
" Wild,	May weed,	Maruta cotula.
Chanchi,	Ink plant,	Coriaria thymifolia.
Charas,	Churrus,	Cannabis sativa var Indica.
Charlock,		Sinapis arvensis.
Chaste tree,		Vitex agnus castus.
Chaulmugra seed,		Gynocardia odorata.
Chaw stick,		Gouania Domingensis.
Chay root,		Oldenlandia umbellata.
Checkerberry,	Spicy wintergreen,	Gaultheria procumbens.
"	Partridge berry vine,	Mitchella repens.
Cheese bowl,	Red poppy capsules,	Papaver rhœas.
" rennet herb,	Cleavers,	Galium aparine.
Cheeses,	Low mallows,	Malva rotundifolia.
Chequer flower,	Hermodactyle,	Colchicum variegatum.
Chequered daffodil,	Guinea hen flower,	Fritillaria meleagris.
Cherry, Bastard,		Cerasus pseudo cerasus.
" bay,	Cherry laurel,	" Lauro "
" birch,	Black birch,	Betula lenta.
" Bird,		Prunus padus.
" Black Wild,		Prunus serotina.
" Fowl,	Bird cherry,	" padus.
" gum,	obtained from	Cerasus and Prunus.
" Ground,	Yellow henbane,	Physalis viscosa.
" laurel,		Cerasus Lauro cerasus.
" pepper,		Capsicum cerasiforme.
" Wild,		Cerasus serotina.
Chervil,		Anthriscus cerefolium.
" Spanish,	Cicily root,	Myrrhus odorata.
Chestnut,	common name for	Castanea of several varieties.
" American,		" vesca var. Americana.
" European,		" "
" Horse,	Horse chestnut,	Æsculus Hippocastanum.
Chia seed,		Salvia Hispanica.
Chicken weed,	Litmus,	Roccella tinctoria.
Chick pea,		Cicer arietinum.
" vetch,	Sweet pea,	Lathyrus odoratus.
" weed,		Stellaria media.
" " Red,		Anagallis arvensis.
" wintergreen,		Trientalis Europæa.
Chicory,		Cichorium intybus.
Chiendent,	Dog grass,	Triticum repens.
Children's bane,		Cicuta maculata.
Chillies,		Capsicum annuum.
Chilly pepper,	Capsicum,	" "
China aster,		Aster Chinensis.
" "		Callistephus Chinensis.
" bark,		Buena hexandra.

COMMON.	ENGLISH.	BOTANICAL.
China grass,	Ramie grass plant,	Bœhmeria nivea.
" pink,	India pink,	Dianthus Chinensis.
" root,		Smilax China.
" " American,		Smilax pseudo China.
" "	Wild yam,	Dioscorea villosa.
Chinese box,		Euonymus Japonicus.
" nut galls,	Galls,	Distylium racemosum.
" rhubarb,		Rheum compactum.
" varnish,	exudation from	Rhus vernicifera.
" wax,	Japan wax,	" succedanum.
Chink,	Checkerberry,	Gaultheria procumbens.
Chinney weed,	Litmus weed,	Roccella tinctoria.
Chinquapin,		Castana pumila.
Chin wood,	Yew tree,	Taxus baccata.
Chirayta,	Chiretta,	Agathotes Chirayta.
Chiretta,		" "
Chocolate,	a paste prepared from	Theobroma Cacao.
" Indian,	Avens root,	Geum rivale.
" nuts,		Theobroma Cacao.
" root,	Avens root,	Geum rivale and var.
Choice dielytra,	Turkey corn,	Corydalis formosa.
Choke, berry,		Pyrus Arbutifolia.
" Black,		Cerasus hiemalis.
" cherry,		" Virginiana.
" "	June berry,	Aronia sanguinea.
" dog,	Angle pod,	Gonolobus obliquus.
Choy root,	Chay root,	Oldenlandia umbellata.
Christmas daisy,	Aster,	Callistephus chinensis.
" evergreen,	Festoon pine,	Lycopodium rupestre.
" rose,	Black hellebore,	Helleborus niger.
Christ's eye,		Inula oculis Christi.
" "	Clear eye,	Salvia verbenaca.
" thorn,		Paliurus aculeatus.
Chulchunchulli,		Ionidium microphyllum.
Churnstaff,	Wartwort,	Euphorbia helioscopia.
Churrus,	a resinous substance from	Cannabis sativa var. Indica.
Cicily, Sweet,		Osmorrhiza longistylis.
" " European,	Spanish chervil,	Myrrhis odorata.
Cicuta,	Water hemlock,	Cicuta maculata.
Cinchona,	common name for bark of	Cinchona of different species.
" Pale,	Pale bark,	" Condaminea and var.
" Red,	Red bark,	" succirubra.
" Yellow,	Calisaya bark,	" calisaya.
Cinnamon,	Ceylon cinnamon,	Cinnamomum Zeylanicum.
" bark, Chinese,		" aromaticum.
" Bastard,	Cassia bark,	" Cassia.
" Black,		Pimenta acris.
" buds,	the immature buds,	Cinnamomum Loureirii and var.
" Ceylon,	True cinnamon,	" Zeylanicum.
" Chinese,		" aromaticum.
" colored fern,		Osmunda Cinnamomea.
" leaves,		Cinnamomum tamala.
" Santa Fe,	Ishpingo,	Nectandra cinnamonoides.
" suet,	procured from fruit of	Cinnamomum Zeylanicum.
" True,	Ceylon cinnamon,	" "
" White,	Canella bark,	Canella alba.
" Wild,	" "	" "
" wood,	Sassafras wood,	Sassafras officinale.
Cinquefoil,	Five finger,	Potentilla Canadensis.

COMMON.	ENGLISH.	BOTANICAL.
Cinquefoil herb,	Five finger,	Potentilla Canadensis.
" root,	" "	" "
Citron,	the fruit. Cedrat.	Citrus medica.
" melon,	variety of watermelon,	Citrullus vulgaris.
Citronelle,	Balm,	Melissa officinalis.
" oil,		Andropogon Nardus.
Cives. Chives.	a variety of onion or leek,	Allium schœnoprasum.
Clabbergrass,	Cleavers,	Galium aparine.
Clammy locust,		Robinia viscosa.
" mustard,	Clammy weed, .	Polanisia viscosa.
" sage,	Clarry,	Salvia sclarea.
" weed,	Wormweed,	Polanisia viscosa.
Clapwort,	Broomrape,	Orobanche Americana.
Clarry. Clary.		Salvia sclarea.
Clear eye,	Wild clary,	" Verbenaca.
Clearing nut,		Strychnos potatorum.
Clearweed,	Stingless nettle,	Urtica pumila.
Cleavers, .		Galium aparine.
Cleaverwort,	Cleavers,	" "
Cleome,		Cleome pungens.
Cliffweed,	Maple leaf alum root,	Heucheria acerifolia.
Climbing bittersweet,	False bittersweet,	Celastrus scandens.
" buckwheat,	Blackbird weed,	Polygonum Convolvulus.
" fumitory,	Alleghany fringe,	Adlumia cirrhosa.
" hempweed,		Mikania scandens.
" orange root,	False bittersweet,	Celastrus scandens.
" staff tree,	" "	" "
" sumach,	Poison ivy vine,	Rhus Toxicodendron var. radicans.
Clingstones,	Fruits in which the stones and flesh adhere.	
Clivers,	Cleavers,	Galium aparine.
Clotbur,	Burr thistle,	Xanthium strumarium.
"	Burdock,	Arctium lappa.
Cloud berry,	High blackberry,	Rubus villosus.
" " root,	" " root,	" "
Clove bark,		Myrtus caryophyllus.
" "	Bois de crabe,	Canella giroffee.
" "		Cinnamomum culilawan.
Clove garlic,	Common garlic,	Allium sativum.
" nutmeg,	.	Agathophyllum aromaticum.
" pepper,	Pimento,	Eugenia Pimenta.
pink,	Carnation gilliflower,	Dianthus caryophyllus.
" stalks or stems,		Caryophyllus aromaticus.
" wild,		Eugenia acris.
Cloves,	the undeveloped flower,	Caryophyllus aromaticus.
Clover broom,	Wild indigo,	Baptisia tinctoria.
" Red,		Trifolium pratense.
" Sweet,		Melilotus leucanthe.
" Melilot,	Sweet clover,	" "
" White,	White "	Trifolium repens.
" " melilot,	Sweet "	Melilotus alba.
Clown heal,	Hedge nettle,	Stachys palustris.
Club moss,	yields lycopodium,	Lycopodium clavatum.
" rush,		Scirpus.
Clustered solomon seal,	Small Solomon seal,	Convallaria racemosa.
Cluster cherry,	Bird cherry,	Cerasus padus.
" pine.	yields Bordeaux turpentine,	Pinus pinaster.
Coakum,	Garget,	Phytolacca decandra,
Cobæa, Mexican,		Cobæa scandens.
Cob nut,	a kind of hazel,	Corylus avellana.

COMMON.	ENGLISH.	BOTANICAL.
Cob pink,		Dianthus hortensis.
Coca,		Erythoxylon coca.
Cocao,	Cacao,	Theobroma cacao.
" beans,	"	" "
Cocash root,	Red stalked aster,	Aster puniceus.
" weed,	Life root plant,	Senecio aureus.
Cocculus, Carolina,		Cocculus Carolinus.
" Indicus,	Oriental berries,	" Indicus.
Cochineal fig,	Prickly pear,	Opuntia cochinillifera.
" tree,		Quercus coccifera.
Cockle bur,	Agrimony,	. Agrimonia Eupatoria.
" Corn,	Lychnis githago,	Agrostemma githago.
" Indian,	Oriental berries,	Anamirta cocculus.
Cockmint,		Tanacetum balsamita.
Cockscomb,		Celosia cristata.
"		Erythrina crista galli,
" Red,	Prince's feather,	Amaranthus hypochondriacus.
Cockspur rye,	Ergot of rye,	Sclerotium clavus.
Cockup hat,	Queen's root,	Stillingia sylvatica.
Cocky baby,		Arum maculatum.
Cocoa nut,	Cocos,	Cocos nucifera.
" palm,	Cocoanut tree,	" "
" plum,		Chrysobalanus Icaco.
Cocowort,	Shepherds' purse,	Capsella bursa pastoris.
Cocum butter, "oil,"	Mangosteen,	Garcinia purpurea.
Codlins and cream,		Epilobium hirsutum.
Coffee,	the seed of	Caffea Arabica.
" corn,	Indian millet,	Sorghum vulgare.
" Green,	Raw coffee,	Caffea Arabica.
" pea,	Chickpea,	Cicer arietinum.
" tree, American,	Kentucky coffee bean,	Gymnocladus Canadensis.
Cohosh, Black,	Black cohosh,	Cimicifuga racemosa.
" Blue,	Blue cohosh,	Caulophyllum thalictroides.
" Red,	Red baneberry,	Actæa rubra.
" White,	White baneberry,	" alba.
Colchicum,		Colchicum autumnale.
Cold seeds,	the seed of Watermelon, Pumpkin, Squash, etc.	
Cold water root,	Cocash root,	Aster puniceus.
Cole seed,	Rape seed,	Brassica campestris.
" wort,	Common cabbage,	" oleracea.
Coliander,	Coriander seed,	Coriandrum sativum.
Colic root,	Wild yam,	Dioscorea villosa.
" "	Unicorn root,	Aletris farinosa.
" weed,		Corydalis cucullaria.
Collard,	Skunk cabbage,	Symplocarpus fœtidus.
Collinsia,		Collinsia verna.
Colocynth,		Citrullus Colocynthis.
" apple,	Colocynth fruit,	" "
Colomba,	Colombo root,	Cocculus palmatus.
" root,	" "	" "
Colombia bark,	Peruvian bark,	Cinchona condaminea pitayensis.
Colophony,	Rosin,	Pinus palustris and other species.
Coltsfoot,		Tussilago farfara.
" snake root,	Canada snake root,	Asarum Canadense.
Coltstail,	" fleabane,	Erigeron "
Colubrina,	Snake wood,	Strychnos colubrinum.
Columbia kino,	Jamaica kino,	Coccoloba uvifera.
Columbine,		Aquilegia vulgaris.
" Feathered,	Meadow rue,	Thalictrum dioicum.

COMMON.	ENGLISH.	BOTANICAL.
Columbine, Garden,		Aquilegia vulgaris.
" Wild,		" Canadensis.
Columbo,	Colombo root,	Cocculus palmatus.
" American,		Frasera Walteri.
Columbrina,	old name for Nux vomica,	Strychnos nux vomica.
Colza,	Rape seed,	Brassica campestris.
Comb flower,	Wild sunflower,	Helianthus annuum.
Comfrey,		Symphytum officinale.
" Saracens,	Ragwort,	Senecio Jacobœa.
Compass weed,		Silphium laciniatum.
Cone disk,	Thimble weed,	Rudbeckia laciniata.
" flower, purple,	Red sunflower,	" purpurea.
" " tall,	Thimble weed,	" laciniata.
Conessi bark,		Wrightia antidysenterica.
Conium,	Poison hemlock,	Conium maculatum.
Consumption brake,	Rattlesnake fern,	Botrychium fumarioides.
" weed,	Round leaved wintergreen,	Pyrola rotundifolia.
Contrayerva,	Jamaica contrayerva,	Dorstenia contrayerva and species.
Convulsion weed,	Ice plant, American,	Monotropa uniflora.
Coolweed,	Stingless nettle,	Urtica pumila.
Coolwort,		Mitella cordifolia.
Copaiva balsam,	Balsam copaiba,	Copaifera multijuga and species.
Copal, African,	a gum resin procured from	Goibourtia copallifera.
" Brazilian,	" " " "	Hymenæa courbaril.
" East Indian,	" " " "	Vateria Indica.
" Madagascar,	" " " "	Hymenæa verrucosa.
" Mexican,	" " " "	" courbaril.
Copalchi bark,		Croton pseudo china.
" "		Strychnos pseudo quinia.
Copalm,	Liquid ambar,	Liquidambar styraciflua.
Coquetta bark,	Bogota bark,	Cinchona condaminea lancifolia.
Coral berry,	Indian currant,	Symphoricarpus vulgaris.
" bloom,		Erythrina herbacea.
" moss,		Corallina officinalis.
" root,	Crawley root,	Corallorhiza odontorhiza.
" tree,		Erythrina crista galli.
Coriander,		Coriandrum sativum.
Cork,	the bark of	Quercus suber.
" tree,	Alcornoque,	" "
" wood,	"	" "
" "	Custard apple tree,	Anona palustris.
Corn,	Indian corn. Maize,	Zea Mays.
" chamomile,		Anthemis arvensis.
" cockle,		Lychnis githago.
" flag,	Round mandrake,	Gladiolus communis.
" gromwell,	False gromwell,	Onosmodium Virginianum.
" marigold,		Chrysanthemum.
" mustard,	Wild mustard,	Sinapis arvensis.
" poppy,	Red poppy,	Papaver Rhœas.
rose,	" "	" "
salad,	Lambs' lettuce,	Fedia radiata.
" " European,		Valeriana dentata.
" snake root,		Eryngium yuccefolium.
" " "	Water eryngo,	" aquaticum.
Cornel,	Dogwood tree,	Cornus Florida.
" Flowering,	Dogwood,	" "
" Male,	Cornelian cherry,	" Mascula.
" Round leaved,	Round leaved dogwood,	" Circinata.
Cornels,	Cornelian cherry,	" Mascula.

COMMON.	ENGLISH.	BOTANICAL.
Cornelian cherry,		Cornus Mascula.
Cornish lovage,		Ligusticum cornubiense.
Corpse plant,	Ice plant,	Monotropa uniflora.
Corsican moss,	Laurencia obtusa,	Fucus helminthocorton.
Cossoo,	Koosso,	Brayera anthelmintica.
Costmary,		Pyrethrum Tanacetum.
Cotton,		Gossypium herbaceum.
" Common Amer'n,	Short fibre cotton,	" nigrum.
Cotton grass,		Eriophorum polystachyon.
" rose,		Filago Germanica.
" Sea Island,	Long fibre cotton,	Gossypium album.
" thistle,	Musk thistle,	Onopordon acanthium.
" tree,		Populus lævigata.
" " gum,		Bombax pentandrum.
" " Indian,		" "
" weed,	Pearl flow'd life everlasting,	Gnaphalium margaritaceum.
" wood,	Necklace poplar,	Populus monilifera.
" "	Water poplar,	" angulata.
Couch grass,	Quack grass,	Agopyrum repens.
Cough root,	Beth root,	Trillium pendulum
" weed,	Life root plant,	Senecio aureus.
Coumarouma bean,	Tonka bean,	Dipterix odorata.
Coventry bells,	Canterbury bells,	Campanula media.
Countrymans' treacle,	Rue,	Ruta graveolens.
Cowbane,	Water hemlock,	Cicuta virosa.
Cowberry,	Red bilberry,	Vaccinium Vitis idæa.
Cowcabbage,	White pond lily,	Nymphæa odorata.
Cowhage,	hairs from the pod of	Mucuna pruriens.
" cherry,		Malphigia urens.
" East Indian,		Mucuna prurita.
" Wild,		Dolichos purpureus.
Cow herb,		Saponaria vaccaria.
" itch,	Cowhage,	Mucuna pruriens.
" lily,	Yellow pond lily,	Nuphar advena.
" parsley,		Heracleum sphondylium.
" parsnip,	Masterwort,	" lanatum.
" " Royal,	a variety of	" "
Cowrie pine,	yields cowrie resin,	Damarra australis.
Cowslip,	Virginia lungwort,	Mertensia Virginica.
" American,	Marsh marigold,	Caltha palustris.
" "		Dodecatheon media.
" primrose, Eng.		Primula veris.
Cowslips,	Marsh marigold,	Caltha palustris.
Cow tree,		Clusia galactodendron.
" weed,		Cicutaria vulgaris.
" wheat,	Poverty weed,	Melampyrum arvense.
Cows' tail,	Canada fleabane,	Erigeron Canadense.
Crab apple,		Pyrus caronaria.
" grass,	Finger grass,	Digitaria sanguinalis.
" "	Samphire,	Salicornia herbacea.
" tree,	Crab apple,	Pyrus caronaria.
Crabs' eyes,	the seed of	Abrus precatorius.
Cramp bark,	High cranberry,	Viburnum opulus.
Cranberry,		Oxycoccos macrocarpus.
" Common Am'n,		Vaccinium oxycoccos.
" High,	Cramp bark,	Viburnum opulus.
" Mountain,	Uva Ursi,	Arctostaphylos Uva Ursi.
Cranesberry,	Cranberry, American,	Vaccinium oxycoccos.
Cranesbill,		Geranium maculatum.

COMMON.	ENGLISH.	BOTANICAL.
Crawley,		Corallorhiza odontorhiza.
Cream fruit,		Roupellia grata.
" nuts,	Castana nuts,	Bertholletia excelsa.
" of tartar tree,		Adansonia Gregorii.
Creeper,	False grape,	Ampelopsis quinquefolia.
" Virginia,	American Ivy,	" "
Creeping blackberry,	Low blackberry,	Rubus trivialis.
" bugloss,		Lycopsis vesicularia.
" cinquefoil,	Five finger, European,	Potentilla reptans.
" cucumber,		Melothria pendula.
" leopardsbane,	Arnica,	Arnica scorpoides.
" wintergreen,	Cancer wintergreen,	Gaultheria hispidula.
Creyat,	analogous to chiretta,	Andrographis paniculata.
Cress,	Pepper cress,	Lepidium sativum.
" Garden,	" "	" "
" Meadow,		Arabis rhomboides.
" Penny,		Thlaspi arvense.
" Pepper,	Garden pepper cress,	Lepidium sativum.
" Water,		Nasturtium officinale.
Cretian balm,		Thymus Creticus.
Crinkle root,	Toothwort,	Dentaria diphylla.
Cross cleavers,		Galium circæzans.
" of Jerusalem,	Red campion,	Lychnis diurna.
" toes,	Lotus,	Lotus corniculatus.
" vine,	Trumpet flower,	Tecoma radicans.
" wort,	Boneset,	Eupatorium perfoliatum.
" "		Cruceata.
" "	Yellow balm,	Lysimachia quadrifolia.
Croton oil plant,		Croton Tiglium.
" seeds,	yield croton oil,	" "
Crow berry,		Empetrum nigrum.
" corn, root,	Unicorn root,	Aletris farinosa.
" foot,	Cranesbill,	Geranium maculatum.
" " common,		Ranunculus repens.
" " Acrid,	Buttercup,	" bulbosus.
" " buttercup,		" "
" " "		" acris.
" garlic,		Allium vineale.
" silk,	common name for	Confervæ in general.
Crown bark,		Cinchona officinalis.
" " Pale,	Loxa bark,	" condaminea var. vera.
" of the field,		Agrostemma coronaria.
" imperial,		Frittilaria imperialis.
Crude camphor,	Dutch camphor,	Camphora officinarum.
Crummock,	Skirret,	Sium sisarum.
Crystal wort,	Liverwort,	Hepatica triloba.
Cubeb pepper,	African black pepper,	Piper Afzelii.
Cubebs,	the unripe fruit of	" Cubeba and species.
" African,	African black pepper,	" Afzelii.
" Dutch E. Indian		" anisatum.
" Guinea,	African black pepper,	" Afzelii.
Cuckoo buds,	Buttercup,	Ranunculus bulbosus.
" bread,	Wood sorrel,	Oxalis acetosella.
" flower,		Cardamine pratensis.
" "	Ragged robin,	Lychnis flos-cuculi.
" pint,	Dragon root,	Arum maculatum.
Cuckold,		Bidens connata.
Cucumber,		Cucumis sativus.
" Bitter,	Colocynth,	Citrullus Colocynthis.

COMMON.	ENGLISH.	BOTANICAL.
Cucumber, Prickly,		Cucumis anguria.
" root, Indian,		Medeola Virginica.
" Squirting,	Wild cucumber,	Momordica Elaterium.
" tree,	Magnolia,	Magnolia acuminata.
" Wild,		Momordica Elaterium.
Cudbear,	a dye-stuff procured from	Lecanora tartarea.
Cudweed,	Life everlasting,	Gnaphalium Uliginosum.
"	Mouse ear,	" plantaginifolia.
" golden,		Tanacetum annuum.
Cuichunchulli root,	So. American Ipecac,	Ionidium microphyllum.
Culilawan,	Clove bark,	Cinnamomum culiliwan.
Culm,	the straw of grain,	
Culvers' physic,	Black root,	Leptandra Virginica.
" root,	" "	" "
Cumin,	Cummin,	Cuminum cyminum.
Cummin seed,		" "
" Black,		Nigella sativa.
" Sweet,	Cummin seed,	Cuminum cyminum.
" Wild,		Lagœcia cuminoides.
Cundurango,	Condor vine,	Asclepias mataperro.
Cup fern,	Bladder fern,	Polypodium fragile.
" lichen, Scarlet,		Lichen cocciferus.
" plant,	Indian cup plant,	Silphium perfoliatum.
" " Indian,	" " "	" "
Curacoa orange,	Bitter orange,	Citrus vulgaris (Bignradia).
" peel,	" "	" " "
Curana wood,		Icica altissima.
Curare,	Woorari,	Strychnos toxifera.
Curcuma,	Turmeric,	Curcuma longa.
" Ohio,	Goldenseal,	Hydrastis Canadensis.
Cureall,	Balm lemon,	Melissa officinalis.
"	Avens white,	Geum Virginianum.
Curled dock,	Yellow dock,	Rumex crispus.
" mint,	Balm mint,	Mentha crispa.
Currant,	common name for Corinthian grape.	
" Corinth,		Vitis Corinthiaca.
" Black,		Ribes nigrum.
" "	Commercial name for the dried fruit	Vitis Corinthiaca.
" leaf,		Mitella diphylla.
" Indian,		Symphoricarpus vulgaris.
" Red,		Ribes rubrum.
" White,	Garden var. of Red currant,	" album.
" Zante,	Corinthian grape,	Vitis Corinthiaca.
Curry leaf tree,		Bergera konigii.
Cursed thistle,	Cursed thistle,	Cirsium arvense.
Cusco bark,		Cinchona pubescens.
Cuscus,	Vettivert,	Andropogon muricatus.
Cusso,	Koosso,	Brayera anthelmintica.
Custard apple,	Papaw,	Uvaria triloba.
" "		Unona palustris.
Cutch,	Catechu,	Acacia Catechu.
Cutting almond,	Nephritic plant,	Parthenium integrifolium.
Cypress, bark,		Cupressus sempervirens.
" herb,		Santolina chamœcyparissus.
" nuts,		Cupressus sempervirens.
" powder,	Portland sago,	Arum maculatum.
" spurge,		Euphorbia cyparissias.
" standing,	Ipomopsis,	Gilia coronopifolia.
" vine,	Red jessamine,	Ipomœa quamoclit.

3

COMMON.	ENGLISH.	BOTANICAL.
Cypress, Virginia,		Cupressus disticha.
" wood, tree,		" sempervirens.
Cypruss wood,	Rhodium wood,	Convolvulus scoparius.
Daffodil,		Narcissus pseudo narcissus.
Dagger flower, plant,	Adam's needle,	Yucca gloriosa.
Dahlia,		Dahlia variabilis.
Dahoon holly,		Ilex vomitoria.
Daisy,		Bellis perennis.
" Common,	Daisy,	" "
" Blue,	Michaelmas daisy,	Aster Tradescantia.
" Ox eye,	White weed,	Leucanthemum vulgare.
" White,	" "	" "
Damar,		Agathis damarra.
Damarra turpentine,	Pinus damarra,	" "
Dames' violet,	Garden rocket,	Hesperis matronalis.
Dammer tree, white,	yields E. I. Copal,	Vateria Indica.
Damson,	a variety of plum,	Prunus domestica.
" Bitter,	Simaruba bark,	Simaruba officinalis.
" Mountain,	" "	" "
Dandelion,		Taraxacum Dens-leonis.
" Dwarf,		Krigia Virginica.
" Fall,		Leontodon autumnale.
Danes' blood,	Clustered bell flower,	Campanula glomerata.
Danewort,	English Dwarf elder,	Sambucus ebulus.
Dangleberry,	Squaw huckleberry,	Vaccinium stamineum.
Dark ash bark,		Cinchona nitida.
Darnel,	Ivrale,	Lolium perenne.
" .	Tare,	" tremulentum.
Date tree,		Phœnix dactylifera.
" palm,		" "
" plum,	Persimmon,	Diospyros Virginiana.
Dates,	fruit of Date palm,	Phœnix dactylifera.
David root,	Snowberry,	Chiococca racemosa.
Day flower,		Homerocallis flava.
" "		Commelina angustifolia.
" lily,	Funkia,	Homerocallis flava.
Deadly carrot,		Thapsia.
" nightshade,	Belladonna,	Atropa Belladonna.
" "	Garden nightshade,	Solanum nigrum.
Dead men's bells,	Foxglove,	Digitalis purpurea.
" nettle,	Henbit,	Lamium amplexicaule.
" "	Angelica,	Angelica atropurpurea.
" Sea apple,	Apple of Sodom,	Quercus Infectoria.
" tongue,	Hemlock drop wort,	Œnanthe crocata.
Deal pine,	White pine,	Pinus strobus.
Death head moss,		Usnea hirta.
" of man,	Poison hemlock,	Cicuta maculata.
Death's flower,	Periwinkle,	Vinca major.
Deer balls,	Puff balls,	Lycoperdon cervinum and var.
" berry,	Squaw vine,	Mitchella repens.
" "	Checkerberry,	Gaultheria procumbens.
" "	Dangleberry,	Vaccinium stamineum.
" food,	Frog leaf,	Brasenia hydropelta.
" grass,	Meadow beauty,	Rhexia Virginica.
" tongue,	Vanilla leaf,	Liatris odoratissima.
" "		Erythronium flavum.
" wort boneset,		Eupatorium urticifolium.
Dentellaria,		Plumbago Europœa.
Deutzia,		Deutzia gracilis and var.

COMMON.	ENGLISH.	BOTANICAL.
Devil's apple,	the fruit of	Mandragora officiualis.
" "		Datura Stramonium.
" apron,		Laminaria saccharina.
" bit,	False unicorn,	Helonias dioica.
" "	Button snake root,	Liatris spicata.
" " herb,	Wood scabious,	Scabiosa succissa.
" " root,	" "	" "
" bite,	American hellebore,	Veratrum viride.
" bones,	Wild yam root,	Dioscorea villosa.
" dung,	Assafetida,	Narthex Assafœtida.
" dye,		Indigofera.
" fig,	Prickly poppy,	Argemone Mexicana.
" gut,	Dodder,	Cuscuta Americana.
" in a bush,	Fennel flower,	Nigella Damascena.
" leaf,		Urtica urentissima.
" milk,	Wartwort,	Euphorbia helioscopia.
" shoestring,	Hoary pea,	Tephrosia Virginiana.
" tree,		Alstonia scholaris.
" turnip,	Red bryony,	Bryonia dioica.
" wood,	American olive,	Olea Americana.
Dew berry,	Low blackberry,	Rubus Canadensis.
" plant,	Sundew,	Drosera rotundifolia.
De Witt snake root,	Gall of the earth,	Prenanthes Fraseri.
Dextrine,	British gum,	
Dhak tree,	produces kino,	Butea frondosa.
Diamond plant,	Ice plant,	Mesembryanthemum crystallinum.
Dianthus,	Pink,	Dianthus armenia and var.
Diatoms,	Microscopic moving plants,	Diatomaceæ.
Didiscus, blue,		Didiscus cœrulea.
Diclytra,	Turkey corn,	Corydalis formosa.
"		Dicentra spectabilis.
Dill seed,		Anethum graveolens.
Dillisk,	Dulse,	Rhodymenia Palmata.
Dillydilweed,	May weed,	Anthemis cotula.
Dittander,	Pepperwort,	Lepidium latifolium.
Dittany, American,	Mountain dittany,	Cunila mariana.
" Bastard,	White fraxinella,	Dictamus fraxinella.
" European,	" "	" alba.
" Mountain,		Cunila mariana.
" of Crete,		Origanum Dictamus.
Divi divi,	Libi dibi,	Cæsalpina coriaria.
Divinum remedium,	Masterwort,	Imperatoria ostruthium.
Dock, Blunt leaved,		Rumex obtusifolius.
" Curled,	Yellow dock,	" crispus.
" Great water		" aquaticus.
" Narrow,	Yellow dock,	" crispus.
" Sour,	Mountain sorrel,	Oxyria reniformis.
" Spatter,	Yellow pond lily,	Nuphar advena.
" Sweet,	Bistort,	Polygonum Bistorta.
" Water,		Rumex Britannica.
" Yellow,		" crispus.
" " rooted,		" Britannica.
Dockmackie,	Maple guelder rose,	Viburnum acerifolium.
Dodder,	Devil's gut,	Cuscuta Americana.
" cake,	the refuse oil cake of	Camelina sativa.
" Greater,		Cuscuta Europæa.
Dogachamus,	Green osier,	Cornus circinata.
Dogberry tree,		" sanguinea.
Dog button,	Nux vomica.	Strychnos nux vomica.

COMMON.	ENGLISH.	BOTANICAL.
Dog fennel,	Mayweed,	Anthemis cotula.
" grass,		Triticum repens.
" mint,	Field thyme,	Clinopodium vulgare.
" poison,	Small hemlock,	Æthusa cynapium.
" rose,		Rosa canina.
" tree,	Dogwood,	Cornus Florida.
" weed,	Dog grass,	Triticum repens.
" wood,	Florida dogwood,	Cornus Florida.
" " alder leaved,	Green osier,	" circinata
" " broad "	" "	" "
" " round "	" "	" "
" " swamp,	Red osier,	" sericea.
" " Virginia,	Florida dogwood,	" Florida.
Dogs' bane,	Bitter root,	Apocynum androsæmifolium.
" finger,	Foxglove,	Digitalis purpurea.
" mercury,		Mercurialis perennis.
" mouth,	Snapdragon,	Antirrhinum majus.
" tooth violet,	Adders' tongue,	Erythronium Americanum.
Dollar leaf,	Round leaved pyrola,	Pyrola rotundifolia.
Dombeya turpentine,		Dombeya excelsa.
Donkeys' eyes,	seed of the cowhage plant,	Mucuna pruriens.
Doom bark,	Sassy bark,	Erythrophleum judiciale.
Door weed,	Hearts' ease,	Polygonum persicaria.
" "	Water pepper,	" hydropiper.
Dornboom,	South African gum tree,	Acacia horrida.
Double claw,	Martynia,	Martynia proboscidea.
" tansy,	Tansy, double-flowered,	Tanacetum crispum.
Doum palm,	Gingerbread tree,	Hyphæna thebaica.
Doves' foot,	Cranesbill,	Geranium sylvaticum.
Dragoness plant,	Wild lily of the valley,	Convallaria borealis.
Dragon head,		Dracocephalum Virginianum.
" root,		Arum triphyllum.
" tree,		Dracæna Draco.
" turnip,	Dragon root,	Arum triphyllum.
" wort,	Bistort,	Polygonum Bistorta.
Dragons' claw,	Crawley root,	Corallorhiza odontorhiza.
" blood,		Calamus Draco.
" " American		Pterocarpus Draco.
" " Canary,		Dracæna Draco.
" " Mexican,		Croton Draco.
" " W. Ind'n,		Pterocarpus Draco.
" eye,		Nephalium longanum.
" mouth,		Epidendrum macrochilum.
Drooping starwort,	False unicorn,	Helonias dioica.
Droopwort,		Spiræa filipendula.
Drop berry,	Solomon seal,	Convallaria Multiflora.
" seed,		Muhlenbergia diffusa.
Dropsy plant,	Balm lemon,	Melissa officinalis.
Dropwort common,		Spiræa filipendula.
" Hemlock,		Œnanthe crocata.
" water,		" phellandrium.
" Western,		Gillenia trifoliata.
Drumstick tree,	Cassia stick tree,	Cathartocarpus conspicua and var.
Drunkwort,	Virginia tobacco,	Nicotiana Tabacum.
Duckretter,	American hellebore,	Veratrum Viride.
Ducks' foot,	" Mandrake,	Podophyllum peltatum.
" meat,	a floating aquatic plant,	Lemua minor.
Duckweed,	Ducks' meat,	" "
Dulse,	a sea weed,	Rhodymenia palmata.

COMMON.	ENGLISH.	BOTANICAL.
Dulse, Red,		Iridœa edulis.
Dumb cane,		Caladium seguinum.
" nettle,	Dead nettle,	Lamium amplexicaule.
Dusty miller,		Cineraria maritima.
" "		Senecio cineraria.
Dutch camphor,	Japan tub camphor,	Camphora officinarum.
" clover,		Medicago Lupulina.
" medlar,	English medlar,	Mespilus Germanica.
" myrtle,	Meadow fern,	Myrica gale.
" pink,	Weld,	Reseda luteola.
" rush,	Scouring rush,	Equisetum hyemale.
Dutchmans' breeches,		Dicentra cucullaria.
" laudanum,		Passiflora rubra.
" pipe,		Aristolochia sipho.
" " plant,	Pipe vine,	Siphisia glabra.
Dwale,	Belladonna,	Atropa Belladonna.
Dwarf almond,	Flowering almond,	Amygdalus pumila.
" cassia,	Prairie senna,	Cassia chamæocrista.
" cedar, Red,		Juniperus prostata.
" cornel,	Bunchberry,	Cornus Canadensis.
" elder,		Aralia hispida.
" " European,	Danewort,	Sambucus ebulus.
" " berries,	"	" "
" gentian,		Gentiana acaulis.
" ground nut,	Ginseng,	Panax trifolia.
" mallow,	Low mallow,	Malva rotundifolia.
" milkwort,	Evergreen snakeroot,	Polygala paucifolia.
" nettle,		Urtica urens.
" Solomon seal,		Convallaria biflora.
" spleenwort,	Wall rue,	Asplenium Ruta muraria.
" stinger,	Dwarf nettle,	Urtica urens.
" umbel,	American Valerian,	Cypripedium.
" water lily,	Fringed bogbean,	Menyanthes nymphæoides.
" wild flax,	Purging flax,	Linum catharticum.
" yew,		Taxus Canadensis.
Dye berry,	Huckleberry,	Vaccinium myrtillus.
" tree, yellow,	Bebeeru tree,	Nectandra Rodiœi.
Dyers' alkanet,		Anchusa tinctoria.
" broom,		Genista tinctoria.
" cleavers,		Galium tinctorium.
" green weed,	Dyers' broom,	Genista tinctoria.
" lichen,		Roccella "
" madder,	Madder,	Rubia "
" oak,	produces nut galls,	Quercus infectoria.
" "		" tinctoria.
" saffron,		Carthamus tinctoria.
" weed,	Dyer's broom,	Genista tinctoria.
" "	Weld,	Reseda luteola.
" "	Woad,	Isatis tinctoria.
Dysentery root,	Virginia mouse car,	Cynoglossum Morrisoni.
" weed,	" " "	" "
Eagle vine,	Cundurango,	Asclepias mataperro.
" wood,	Aloes wood,	Aloexylon agallochum.
Ear drop,		Fuchsia Magellanica.
" wort,	Ceylonian plant,	Dysophylla auricularia.
Earth apple,	Jerusalem artichoke,	Helianthus tuberosus.
" club,	Broomrape,	Orobanche Americana.
" gall,	American hellebore,	Veratrum viride.
" nut,	Peanut,	Arachis hypogea.

COMMON.	ENGLISH.	BOTANICAL.
Earth nut,		Bunium bulbocastanum.
" smoke,	Fumitory,	Fumaria officinalis.
East India catarrh root,	Galangal root,	Alpina galanga.
East Indian arrow root,		Curcuma angustifolia.
" " balmony,	Chiretta,	Agathotes chirayta.
E. I. black hellebore,	Levant hellebore,	Helleborus orientalis.
" elemi,	Manilla elemi,	Canarium commune
" horse radish tree,	Ben nut tree,	Moringa pterygosperma.
" sarsaparilla,		Hemidesmus Indicus.
" squills,		Erythronium Indicum.
" tamarinds,		Tamarindus Indica.
Easter giant,	Bistort,.	Polygonum Bistorta.
Ebony,		Diospyros ebenum.
" Mountain,		Bauhinia variegata.
Eel grass,	Tape grass,	Vallisneria spiralis.
Eenhorn,	False unicorn,	Helonias dioica.·
Egg apple,	Egg plant,	Solanum melongena.
" plant,		" "
" squash,	.	Cucurbita ovifera.
Eglantine,	Sweet brier,	Rosa rubiginosa.
" gall,	Bedeguar,	" canina.
" rose,	Sweet brier,	" rubiginosa.
Egyptian bean,	Jamaica water lily,	Nelumbium specieosum.
" gum arabic tree,	Egyptian thorn,	Acacia Arabica.
" lotus,	" bean,	Nelumbium specieosum.
" privet,	Henne,	Lawsonia inermis.
" thorn,	Gum arabic tree,	Acacia Arabica.
Elaterium,	procured from fruit juice of	Momordica Elaterium.
" cucumber,	Wild squirting cucumber,	" "
Elder, Common,		Sambucus Canadensis.
" Dwarf,		Aralia hispida.
" " berries,	European dwarf elder,	Sambucus ebulus.
" " European,		" "
" European,		" nigra.
" flowers,		" Canadensis.
" German,		" nigra.
" Prickly,		.Aralia spinosa.
" Red berried,		Sambucus pubescens.
" rob,	the juice of elderberries,	" Canadensis.
Elecampane		Inula Helenium.
Elemi, African,	the gum resin of	Boswellia Frereana.
" American,	" . "	Icica Icicariba.
" Bengal,	" "	Amyris Elimifera.
" Brazilian,	" "	Icica Icicariba.
" East Indian,	" "	Canarium commune.
" Manilla,	" "	" "
" Mexican,	" "	Elaphrium Elemiferum.
Elephant apple,	Wood apple,	Feronia elephantium.
" foot,	Hottentot bread,	Testudinaria elephantipes.
" pepper,	Long pepper,	Piper longum.
Elephants' ear,	Common name for	Begonia.
Eleuthera bark,	Cascarilla,	Croton Eluteria.
Elk bark,	Magnolia,	Magnolia macrophylla.
" tree,	Sorrel tree,	Andromeda arborea.
" wood,	" "	. " "
" "	Magnolia,	Magnolia macrophylla.
Ellan wood,	Elder, European,	Sambucus nigra.
Ellhorn,	" "	" "
Elm, American,		Ulmus Americana.

COMMON.	ENGLISH.	BOTANICAL.
Elm, European,		Ulmus campestris.
" Red,	Slippery elm,	" fulva.
" Slippery,		" • "
" White,	American elm,	" Americana.
Emetic herb,	Lobelia herb,	Lobelia inflata.
" root,	Blooming spurge,	Euphorbia corallata.
" weed,	Lobelia herb,	Lobelia inflata.
Enchanters' herb,	Vervain, European,	Verbena officinalis.
" nightshade,		Circæa Canadensis and var.
Endive,	Chicory,	Cichorium endiva.
" Garden,	Endive,	"
Englishman's foot,	Plantain,	Plantago major.
English mercury,		Chenopodium Bonus Henricus.
" pepper,		Draba acaulis.
" walnut,		Juglans regia.
" water cress,	Hedge mustard,	Sisymbrium officinale.
Ergot,	the sclerotium of Claviceps purpurea replacing the grain of rye.	
" of corn,	Smut corn,	Ustilago maydis.
" of rye,	Ergot,	Sclerotium clavus.
Eryngo, Water,		Eryngium aquaticum.
" Wild,		" campestre.
Erythronium,	Adders' tongue,	Erythronium Americanum.
Eschscholtzia.	California poppy,	Eschscholtzia Californica.
Eternal flower,		Gnaphalium stœchas.
Ethiopian pepper,	Monkey pepper,	Unona Æthiopica.
Euphorbia,	American ipecac,	Euphorbia Ipecacuanha.
Euphorbium,	the concrete resinous juice of	" Canariensis.
European aspen,		Populus tremulus.
" avens,		Geum urbanum.
" goldenrod,		Solidago virgaura.
" holly,		Ilex aquifolium.
" water lily,		Nymphæ alba.
Evans root,	Avens root,	Geum rivale.
Evening beauty,	Marvel of Peru,	Mirabilis longiflora.
" primrose,		Œnothera biennis.
Evergreen,	Festoon pine,	Lycopodium rupestre and var.
" oak,		Quercus Ilex.
" privet,		Rhamnus alaternus.
" snake root,	Dwarf milkwort,	Polygala paucifolia.
Everlasting,	a common name for	Gnaphalium.
" rose-flowered,		Rodanthe manglesii.
" sweet-scent'd,		Gnaphalium polycephalum.
Eve's cup,	Side saddle plant,	Sarracenia flava.
Eye balm,	Goldenseal,	Hydrastis Canadensis.
Eyebright,		Euphrasia officinalis.
"		Euphorbia hypericifolia.
"	American centaury,	Sabbatia angularis.
"	Lobelia,	Lobelia inflata.
Eye root,	Goldenseal,	Hydrastis Canadensis.
Faham tea,		Angræcum fragrans.
Fair maids of France,	Garden buttercup,	Ranunculus Aconitifolius.
Fairy fingers,	Foxglove,	Digitalis purpurea.
" gloves,	"	" "
Fall poison,		Amianthium muscætoxicum.
False alder,		Prinos verticillatus.
" aloe,		Agave Virginica.
" angostura,		Strychnos nux vomica.
" "		Brusea antidysenterica.
" beech drops,		Monotropa hypopitys.

COMMON.	ENGLISH.	BOTANICAL.
False bittersweet,		Celastrus scandens.
" boneset,		Kuhnia Eupatoroides.
" box,	Dogwood,	Cornus Florida.
" calisaya,	Calisaya quepo.	Cinchona Amygdalifolia.
" china bark,		Biccia Australis.
" cinchona bark,	False china bark,	Biccia Australis.
" colombo,	Columbo wood,	Coscinium fenestratum.
" coltsfoot,	Canada snake root,	Asarum Canadense.
" crawley,	Albany beech drops,	Pterospora andromeda.
" dogwood,	Striped maple,	Acer Pennsylvanicum.
" flax,		Thlaspi campestre.
" foxglove,	Golden oak,	Gerardia Quercifolia.
" fringe tree,	Aaron's beard,	Rhus cotinus.
" grapes,	American Ivy,	Ampelopsis quinquefolia.
" gromwell,	Wild Job's tears,	Onosmodium Virginianum.
" hellebore,		Adonis vernalis.
" hemp,		Datisca hirta.
" indigo,		Baptisia Australis.
" jalap,	Four o'clocks,	Mirabilis Jalapa.
" johnswort,	Nit weed,	Sarothra Gentianoides.
" mallows,		Sida spinosa.
" manna,	Manna produced by	Alhagi maurorum.
" "	" " "	Eucalyptus mannifera.
" "	" " "	Larix cedras.
" "	" " "	" Europæa.
" "	" " "	Pinus Lambertina.
" "	" " "	Tamarix Gallica.
" "	" " "	Quercus Persica.
" mustard,	Clammy weed,	Polanisia graveolens and var.
" myrrh,	Bdellium,	Heudelotia Africana.
" peruvian bark,		Buena obtusifolia.
" pimpernel,		Pimpinella nigra.
" saffron,	American saffron,	Carthamus tinctorius.
" sandal, Red,		Adenanthera parvonia.
" sarsaparilla,	American sarsaparilla,	Aralia nudicaulis.
" star grass,	Unicorn root,	Aletris farinosa.
" sunflower,	Sneezewort,	Helenium autumnale.
" sycamore,	Buttonwood,	Platanus occidentalis.
" syringia,	Mock orange,	Philadelphus coronarius.
" unicorn,		Helonias dioica.
" valerian,	Life root,	Senecio aureus.
" Winter's bark,	Canella,	Canella alba.
" white cedar,	Arbor vitæ,	Thuja occidentalis.
Fan palm,		Borassus flabelliformis.
Farkleberry,		Vaccinium arboreum.
Feather few,	Feverfew,	Pyrethrum parthenium.
" foil,	Water violet,	Hottonia palustre.
" geranium,	Oak of Jerusalem,	Chenopodium botrys.
" grass,		Stipa avenacea.
" tree,	Aaron's beard,	Rhus cotinus.
Febrifuge bark,		Cedrela febrifuga.
" plant,	Feverfew,	Pyrethrum parthenium.
Felonwort,	Bittersweet herb,	Solanum Dulcamara.
Felwort,	Gentian,	Gentiana lutea.
Female agaric,	Oak agaric,	Boletus Agaricus.
" fern,	Common polypody,	Polypodium vulgare.
" "		Aspidium Felix Fœmina.
" "		Asplenium " "
" nervine,	Showy ladies' slipper,	Cypripedium spectabile.

COMMON.	ENGLISH.	BOTANICAL.
Female regulator,		Senecio gracilis.
Fenberry,	Cranberry,	Oxycoccos macrocarpus.
Fen grape,	"	" "
Fennel, Common,	Fennel seed,	Fœniculum vulgare.
" Large,		" officinalis.
" flower,		Nigella Damascena.
" Sweet,		Fœniculum dulce.
Fenugreek seed,	Fœnugreek seed,	Trigonella fœnum græcum.
Fernambuco wood,	Brazil wood,	Cæsalpina echinata.
Fern brake,	Buckhorn brake,	Osmunda regalis.
" bush,	Sweet fern,	Comptonia asplenifolia.
" Cinnamon colored,		Osmunda Cinnamomea.
" Flowercup,		Woodsia hyperborea.
" Flowering,	Buckhorn brake,	Osmunda regalis.
" gale,	Meadow fern,	Myrica gale.
" "	Sweet fern,	Comptonia asplenifolia.
" Grape,	Rattlesnake fern,	Botrychium Virginicum and var.
" Hartford,	Climbing fern,	Lygodium palmatum.
" Male,		Aspidium Filix Mas.
" " shield,	Male fern,	" " "
" Meadow,		Myrica gale.
" Rattlesnake,		Botrychium Virginicum and var.
" Rock,	Maiden hair,	Adiantum pedatum.
" root,	Polypody,	Polypodium vulgare.
" Royal flowering,	Buckhorn brake,	Osmunda regalis.
" Spleen,	Spleenwort fern,	Asplenium scolopendrium.
" "	Sweet fern,	Comptonia asplenifolia.
" Sweet,		" "
Feronia gum tree,	Elephant apple tree,	Feronia elephantum.
Fescue grass,		Festuca elatior.
Festoon pine,	Christmas evergreen,	Lycopodium rupestre.
Fetid hellebore,	Stinking black hellebore,	Helleborus fœtidus.
Fever bush,	Spice bush,	Laurus Benzoin.
" few,		Pyrethrum parthenium.
" " Bastard,		Parthenium hysterophorus.
" root,		Triosteum perfoliatum.
" "	Crawley root,	Corallorhiza odontorhiza.
" tree,	Florida bark,	Pinckneya pubens.
" "		Eucalyptus globulus.
" twig,	Bittersweet, false,	Celastrus scandens.
" twitch,	" "	" "
" weed,	Bushy Gerardia,	Gerardia pedicularia.
" wood,	Spice bush,	Laurus Benzoin.
" wort,	Boneset,	Eupatorium perfoliatum.
" "	Fever root,	Triosteum perfoliatum.
Field cypress,	Ground pine,	Ajuga chamæpitys.
" pink,		Caryophyllus arvensis.
" scabious,		Scabiosa arvensis.
" thyme,	Dogmint,	Clinopodium vulgare.
" weed,	Mayweed,	Anthemis cotula.
" wort,	"	" "
Fig,	the fruit of	Ficus carica.
" Indian,	Banyan tree,	" Indica.
" marigold,		Mesembryanthemum spectabile.
" tree,		Ficus carica.
Figwort,		Scrophularia Marilandica and var.
" herb,		" "
" knotty-rooted,		" nodosa.
" root,		" "

COMMON.	ENGLISH.	BOTANICAL.
Fig wort, Water,		Scrophularia aquatica.
Filbert,	Hazel nut,	Corylus of several species.
" American,		" Americana.
" European,		" Avellana.
Finger flower,	Foxglove,	Digitalis purpurea.
" grass,	Crab grass,	Digitaria sanguinalis.
" leaf,	Five finger,	Potentilla Canadensis.
Fir balsam tree,		Abies balsamea.
" common,	Silver fir tree,	" picea.
" club moss,		Lycopodium selago.
" rape,	False beech drops,	Monotropa hypopitys.
" "	Beech drops,	Orobanche Virginiana.
" tree,		Abies balsamea.
Fire pink,	Catch fly, wild pink,	Silene Virginica.
" weed,		Erechthites hieracifolius.
" "	Canada fleabane,	Erigeron Canadense.
" "	Female regulator,	Senecio gracilis.
Fish berries,	Cocculus Indicus,	Anamirta cocculus.
" mouth,	Balmony,	Chelone glabra.
" poison,		Lepidium piscidium.
Fitch,	Tare. Vetch,	Vicia sativa and var.
Fit root plant,	Ice plant,	Monotropa uniflora.
Fivefinger,	Cinque foil,	Potentilla Canadensis.
" European,		" reptans.
Fivefingers, root,	Ginseng root,	Panax quinquefolia.
Five leaves,	American Ivy,	Ampelopsis quinquefolia.
Flag, Blue,		Iris versicolor.
" lily,	Blue Flag,	" "
" Sweet,	Calamus,	Acorus Calamus.
Flannel flower,	Mullein,	Verbascum Thapsus.
Flax, Common,		Linum usitatissimum.
" drop,	Flax vine,	Cuscuta Europea.
" Mountain,	Senega root,	Polygala Senega.
" Purging,		Linum catharticum.
" seed,	Linseed,	" usitatissimum.
" vine,	Greater dodder,	Cuscuta Europea.
" weed,	Flax vine,	" "
" "		Linaria vulgaris.
" Wild, American,		Linum Virginianum.
Fleabane, Canada,		Erigeron Canadense.
" Marsh,	Plowman's wort,	Inula Marilandica.
" Philadelphia,		Erigeron Philadelphicum.
" Small,		Inula dysenterica.
" various leav'd,		Erigeron heterophyllum.
Flea seed,	Fleawort,	Plantago psyllum.
" wort,		" "
" "	Groundsel,	Senecio vulgaris.
Flesh colored asclepias,	White Indian hemp,	Asclepias incarnata.
Flimsy calisaya bark,		Cinchona Boliviana.
Flix weed,	Fine leaved hedge mustard,	Sisymbrium sophia.
Florentine orris,	Orris root,	Iris Florentina.
" "	Issue peas,	" "
Florida anise,		Illicium Floridanum.
" arrowroot,		Zamia integrifolium.
" balsam,		Amyris Floridiana.
" bark,		Pinckeya pubens.
" cornel,	Dogwood,	Cornus Florida.
" dogwood,	"	" "
Flower cup fern,		Woodsia hyperborea.

COMMON.	ENGLISH.	BOTANICAL.
Flower de luce,	Blue Flag,	Iris versicolor.
" gentle,		Amaranthus caudatus.
" of cassia,	Cinnamon buds,	Cinnamomum Loureirii.
" of an hour,	Bladder ketmia,	Hibiscus trionum.
" of Jove,	Ragged robin,	Lychnis flos Jovis.
Flowering almond,		Amygdalus pumila.
" aloe,	Mexican maguey,	Agave Americana.
" ash,	Manna tree,	Ornus Europæa.
" brake,	Buckhorn brake,	Osmunda regalis.
" bramble,	Brier rose,	Rubus rosæfolius.
" cornel,	Dogwood,	Cornus Florida.
" fern,	Buckhorn brake	Osmunda regalis.
" nettle,	Hemp nettle,	Galeopsis tetrahit.
" "	Bees' nettle,	" versicolor.
" willow,	Rose bay,	Epilobium angustifolium.
" wintergreen,	Dwarf milkwort,	Polygala paucifolia.
Fluellin,	Speedwell,	Veronica officinalis.
Flux root,	Pleurisy root,	Asclepias tuberosa.
" "	Blue gentian,	Gentiana Catesbæi.
Fly agaric,		Agaricus muscaricus.
" bane,	Plowman's spikenard,	Inula conyza.
" trap,	Sidesaddle plant,	Sarracenia purpurea.
" "	Bitter root,	Apocynum androsæmifolium.
" poison,		Amianthicum muscœtoxicum.
Fœnugreek seed,		Trigonella fœnum græcum.
Fog fruit,		Zapania nodiflora.
Foles foot,	Coltsfoot,	Tussilago farfara.
Folks' glove,	Foxglove,	Digitalis purpurea.
Food of the gods,	Assafetida,	Narthex Assafœtida.
Fools' parsley,		Æthusa divaricata.
" stones,		Orchis mascula.
" water cress,	Honewort,	Trinia vulgaris.
Forbidden fruit,	a variety of orange,	Citrus Paradisi.
Forefathers' cup,	Side saddle plant,	Sarracenia purpurea.
Forked fern,		Achrosticum aureum.
" spike,		Andropogon furcatus.
" stem,		Riccia fluitans.
Formosa Camphor,	Camphor oil,	Cinnamomum Camphora.
Forty knot,		Achyranthis repens.
Fountain tree,		Cedrus deodora.
Four o'clocks,	False jalap,	Mirabilis Jalapa and var.
Fox bane,		Aconitum Vulparia.
" berry,	Frost grape,	Vitis vulpina.
Foxglove,		Digitalis purpurea.
" False,	Golden Oak,	Gerardia quercifolia.
Foxgrape,	Herb Paris,	Paris quadrifolia.
Fox tail,	Club moss,	Lycopodium clavatum.
Frangipanni,		Plumieria rubra.
Frankincense,	Olibanum,	Boswellia serrata.
" American,		Pinus abies.
" African,		Daniellia thurifera.
" True, Arab'n,		Juniperus lycia.
French berries,		Rhamnus infectorius.
" birthroot,		Aristolochia pistolochia.
" rhubarb,		Rheum rhaponticum.
Friars' cap,	Aconite,	Aconitum Napellus.
" cowl,	"	" "
Fringed bog bean,	Dwarf water lily,	Menyanthes nymphæoides.
" polygala,	" milkwort,	Polygala paucifolia.
Fringe tree,		Chionanthus Virginica.

COMMON.	ENGLISH.	BOTANICAL.
Frog cheese,	Puff ball,	Lycoperdon proteus.
" grass,	Samphire,	Salicornia herbacea.
" leaf,	Water shield,	Brasenia hydropeltis.
" lily,	Yellow pond lily,	Nuphar advena.
" wort,	Crowfoot buttercup,	Ranunculus bulbosus.
Frost grape,		Vitis vulpina.
" plant,	Frost weed,	Helianthemum Canadense.
" weed,	Rock rose,	" "
" wort,	Frost weed,	" "
Fuchsia,	Ear drops,	Fuchsia Magellanica.
Fuellin,	Speedwell root,	Peucedanum oreoselinum.
Fullers' herb,	Soapwort,	Saponaria officinalis.
Fuminella,	Brazilian saffron,	
Fumitory,		Fumaria officinalis.
" Bulbous,		Corydalis bulbosa.
" Climbing,		Adlumia cirrhosa.
Furze,	Dyer's broom,	Genista tinctoria.
Fustic,	a dye-stuff,	Maclura "
" Old,		Morus "
" tree,	Yellow ash,	Cladrastis "
" wood,		Maclura "
" Young,	the wood of	Rhus cotinus.
Gaglee,	Wild turnip,	Arum maculatum.
Gag root,	Lobelia,	Lobelia inflata.
Galam butter,	Shea butter	Bassia latifolia.
" gum,		Acacia nebued, and Acacia vereck.
Galanga,	Galangal,	Alpina galanga.
Galangal,		" "
Galangale,	Galangal,	" "
Galangall,	"	" "
Galbanum,	a gum resin from	Ferula galbaniflua and var.
" Persian,	" " "	Opoidia galbanifera.
Gale, Sweet,	Meadow fern,	Myrica gale.
Galingale, English,	Sweet cyperus,	Cyperus officinalis and var.
Galipot,	Turpentine gum,	Abies excelsa and Abies picea.
Gall oak,	Dyers' oak,	Quercus Infectoria.
" of the earth,		Prenanthes Fraseri.
" " "	Albany beech drops,	Pterospora andromeda.
" weed,	Five flowered gentian,	Gentiana quinqueflora.
" wort,	Toad flax,	Linaria vulgaris.
Galls,	Nut galls,	
" Aleppo,	morbid excrescences on	Quercus infectoria.
" Atlee,	Tamarisk galls,	Tamarix orientalis.
" Chinese,		Distylium racemosum.
" "		Rhus semialata.
" Mecca,	Dead sea apples,	Quercus Infectoria.
" Tamarisk,		Tamarix orientalis.
" Turkey,	Aleppo galls,	Quercus Infectoria.
" White,	Myrobalans,	Emblica officinalis.
Gambeer,	Catechu, pale,	Nauclea gambir.
Gambier,	" "	" "
Gambir,	" "	" "
Gamboge,	gum resin derived from	Garcinia morella.
" American,	" " "	Hypericum baccatum.
" South Am'n,	" " "	Vismia Guaianensis.
" Ceylon,	" " "	Garcinia morella.
" Mexican,	" " "	Vismia Guaianensis.
" Mysore,	" " "	Garcinia pictoria.
" Siam,	" " "	" Cochinchinensis.
Gang flower,	Milkwort,	Polygala vulgaris.

COMMON.	ENGLISH.	BOTANICAL.
Gánja,	the woody stems of	Cannabis Indica.
Garantogen,	American Ginseng,	Panax quinquefolium.
Garb,	Weeping willow,	Salix Babylonica.
Garden angelica,		Archangelica officinalis.
" artichoke,		Cynaria Scolymus.
" azalea,		Azalea Indica.
" balsam,		Impatiens balsamina.
" carrot,		Daucus Carota.
" celandine,		Chelidonium majus.
" celery,		Apium graveolens var dulce.
" chamomile,		Anthemis nobilis.
" chicory,	Endive,	Cichorium endiva.
" columbine,		Aquilegia vulgaris.
" endive,		Cichorium endiva.
" lettuce,		Lactuca sativa.
" nightshade,	Black nightshade,	Solanum nigrum.
" patience,	Patience dock,	Rumex patientia.
" purslane,	Pursley,	Portulaca oleracea.
" rocket,	Dames' violet,	Hesperis matronalis.
" sage,		Salvia officinalis.
" sorrel,		Rumex acetosa.
" spurge,	Caper spurge,	Euphorbia lathyrus.
" sunflower,		Helianthus annuus.
" thyme,		Thymus vulgaris.
Gardeners' garters,	Canary grass,	Phalaris Canariensis.
Garland flower,	common name for	Hedychium.
Garget,		Phytolacca decandra.
Garlic,		Allium sativum.
" pear,		Cratæva gynandra.
" sage,		Teucrium scorodonia.
" shrub,		Bignonia alllacea.
Garnet berry,	Red currant,	Ribes rubrum.
Gaskins, .	Sweet cherries,	Cerasus avium.
Gaybine,	Kalandana,	Convolvulus nil.
Gay feather,	Button snake root,	Liatris spicata.
Gazania,		Gazania splendens.
Gem fruit,	Mitre wort,	Tiarella cordifolia.
Gentian,		Gentiana lutea.
" Black,		Libanotis vulgaris.
" Blue,	American gentian,	Gentiana Catesbæi.
" Five flowered,	Gall weed,	" quinqueflora.
" Marsh,		" pneumonanthe and var.
" Soapwort,		" saponaria.
" Southern,	Blue gentian,	" Catesbæi.
" White,		Laserpitium latifolia.
Georgia bark,	Florida bark,	Pinckneya pubens.
Geranium,	Rose geranium,	Pelargonium odoratissimum.
" Garden,		" "
" Spotted,	Cranesbill,	Geranium maculatum.
German chamomile,		Matricaria Chamomilla.
" contrayerva,	White swallow wort,	Asclepias vincetoxicum.
" ivy,		Senecio scandens.
" pellitory,		Anacyclus officinarium.
" "	Sneezewort,	Achillea Ptarmica.
" sarsaparilla,	Sea sedge,	Carex arenaria.
" tinder,	Spunk,	Boletus fomentarius.
" valerian,		Valeriana officinalis.
Germander,	Ground pine,	Teucrium chamædrys.
"	Wood sage,	" Canadense.
Giant solomon seal.		Convallaria multiflora.

COMMON.	ENGLISH.	BOTANICAL.
Giant whortleberry,		Vaccinium corymbosum.
Gilia,	Cypress gilia,	Gilia coronopifolia.
Gill-go-over-the-ground,	Ground ivy,	Nepeta glechoma.
Gilliflower,		Dianthus caryophyllus.
Gillrun,	Ground ivy,	Nepeta glechoma.
Gillyflower,	Stock,	Mathiola incana.
Ginger,		Zingiber officinalis.
" African,		Amomum Zingiber.
" Black,	African ginger,	" "
" bread tree,	Doum palm,	Hyphæua thebaica.
" grass,		Andropogon schœnanthus.
" Indian,	Canada snake root,	Asarum Canadense.
" Jamaica,		Amomum Zingiber.
" Race,	common name for crude root	" "
" White,	Bleached ginger,	" "
" Wild,	Canada snake root,	Asarum Canadense.
Gingo tree,	Ginkgo tree,	Salisburia adiantifolia.
Ginkgo tree,	Maidenhair tree,	" "
Ginseng, American,		Panax quinquefolium.
" Asiatic,		" schinseng.
" Blue,	Blue Cohosh,	Caulophyllum thalictroides.
" White,	White gentian,	Triosteum perfoliatum.
" Yellow,	Blue cohosh,	Caulophyllum thalictroides.
Ginsing,	Ginseng,	Panax quinquefolium.
Gipsy wort,	Bugle weed,	Lycopus Virginicus.
Girasol,	Heliotrope,	Heliotropum Peruvianum.
Girasole,	Jerusalem artichoke,	Helianthus tuberosus.
Gladiolus,	Sword lily,	Gladiolus communis.
Gladwine,	Orris root,	Iris fœtidissima.
Glasswort,	Samphire,	Salicornia herbacea.
" prickly,	Saltwort,	Salsola kali.
Glidewort,	Hemp nettle,	Galeopsis tetrahit.
Globe amaranth,		Gomphrena globosa.
" "	.	Ranunculus acris.
" berries,	Yew tree fruit,	Taxus baccatus.
" crowfoot,		Trollius Europæus.
" flower,	Button bush,	Cephalanthus occidentalis.
" "		Trollius laxus.
" hyacinth,	Grape hyacinth,	Muscari botryoides and var.
" thistle,		Echinops multiflorus.
Glory pea,		Clianthus Dampieri.
Glycine,	Wistaria,	Wistaria Sinensis.
Goa powder,	obtained from a species of	Cæsalpina.
Goatsbane,	Aconite,	Aconitum tragoctonum.
"		Prunus capricida.
Goat's beard,	Vegetable oyster,	Tragopogon porrofolius and var.
" bush,		Castela Nicolsoni.
" foot,	Ash weed,	Ægopodium podograria.
" leaf,	Honeysuckle,	Lonercera Peryclymenum.
" root,		Ononis natrix.
" rue,	Hoary pea,	Tephrosia Virginiana.
" "		Galega officinalis.
" thorn,	Tragacanth,	Astragalus tragacantha.
" weed,	Carib tea,	Capraria biflora.
God tree,		Eriodendron anfrectuosum.
Gold cup,	Crowfoot buttercup,	Ranunculus acris.
Golden apple,	Orange,	Citrus aurantium.
" "	Quince,	Cydonia vulgaris.
" bough,	Misletoe,	Viscum flavescens.
" chain,	Laburnum tree,	Cytisus laburnum.

COMMON.	ENGLISH.	BOTANICAL.
Golden club,	Tawkin,	Orontium aquaticum.
" coreopsis.	Nutalls weed,	Coreopsis tinctoria.
" crocus,	Spring crocus,	Crocus vernus.
" daisy,	Ox eye daisy,	Leucanthemum vulgare.
" motherwort,	Life everlasting,	Gnaphalium sylvaticum.
" oak,	False foxglove,	Gerardia quercifolia.
Golden rod, common,	name given to the several var. Solidago.	
" " European,		Solidago virgaurea.
" " Hard leav'd,		" rigida.
" " Rigid "		" "
" " sweet set'd,		" odora.
" seal,		Hydrastis canadensis.
" senecio,	Life root,	Senecio aureus.
" thistle,		Scolymus.
Goldicups,		Ranunculus acris.
Gold of pleasure,	Wild flax,	Camelina sativa.
Goldthread,		Coptis trifolia,
" Chinese,		" tecta.
Goldylocks,		Linosyris vulgaris.
Gollindrinera,		Euphorbia prostrata.
Gombo,	Okra,	Hibiscus esculentus
Good King Henry,	Goosefoot,	Chenopodium Bonus Henricus.
Goose berry,		Ribes grossularia.
" foot,		Chenopodium olidum.
" "	Wormseed,	" anthelminticum.
" grass,	Cleavers,	Galium aparine.
" tongue,	German pellitory,	Achillea Ptarmica.
Gooses hare,	Cleavers,	Galium aparine.
Gopher wood,		Lawsonia alba.
Gourd,		Cucurbita lagenaria.
" Common bottle,		Lagenaria vulgaris.
Gout weed,	Ashweed,	Ægopodium podograria.
" "	Wild angelica,	Angelica sylvestris.
" " Shrubby,	Carib tea,	Capraria biflora.
Gowans,	Daisy flowers,	Bellis perennis.
Grains, ambrette,	Musk seed,	Abelmoschus esculentus.
" paradise,		Amomum Grana Paradisi and var.
" tigli,	the seed of	Croton Tiglium.
Granadilla,		Passiflora edulis.
Grapes,		Vitis vinifera.
" Corinthian,	Black currants,	" Corinthiaca.
Grape hyacinth,		Muscari moschatum.
" vine,		Vitis vinifera.
Grass cloth plant,	Ramie plant,	Bœhmeria nivea.
" tree gum,	Gum Acroid,	Xanthorrhœa hastilis.
" Arrow,		Triglochin.
" Artificial,	name of various fodder plants, as clover, lucerne, sanfoin.	
" Awned hair,		Muhlenbergia capillaris.
" Ballach,		Orchis.
" Barley,		Hordeum.
" Barn,		Panicum Crus-galli.
" Barnyard,	Cocks foot grass,	Panicum Crus-galli.
" Bastard knot,		Corrigiola littoralis.
" " millet,		Paspalum.
" Bear,		Yucca filimentosa.
" Beard,		Andropagon nutans.
" Bengal,		Setaria Italica.
" Bent,		Agrostis of several species.
" " white,	Fiorin,	Agrostis alba.
" Bermuda,		Cynodon Dactylon.

COMMON.	ENGLISH.	BOTANICAL.
Grass Black,		Alopecurus agrestis.
" " oat,		Stipa avenacea.
" Blue,		Poa compressa.
" " eyed,		Sisyrinchium.
" " joint,		Calamagrostis Canadensis.
" Bottle,	Green foxtail grass,	Setaria viridis, and glauca.
" " brush,		Elymus hystrix
" Bristle tailed,		Chæturus.
" Bristly fox tail,		Setaria glauca.
" Brome,		Bromus pubescens and other var.
" Broom corn,		Sorghum saccharatum.
" Burdock,		Lappa racemosa.
" Burr,		Cenchrus.
" Canary,		Phalaris Canariensis.
" Capon's tail,		Festuca Myurus.
" Carnation,		Carex glauca, and other var.
" Cats' tail,	Timothy,	Phleum pratensis.
" Chess or cheat,		Bromus secalinus.
" China,		Bœhmeria nivea.
" Cleaver, .	old name for Clover,	Trifolium pratense.
" Cocks' comb,		Cynosurus echinatus.
" Cocks' foot,	Barnyard grass,	Panicum Crus-galli.
" " "		Dactylis glomerata.
" Comb fringe,	of New Holland,	Dactyloctenium rudulans.
" Cord,		Spartina stricta.
" Couch,	Dog grass,	Triticum repens.
" Cotton,		Eriophorum polystachyon.
" Cow,	Red clover,	Trifolium pratense.
" "	Knot grass,	Polygonum aviculare.
" Crab,		Eleusine Indica.
" "	Finger grass,	Panicum sanguinale.
" Crested hair,	.	Kœleria cristata.
" " Dogs' tail	Leghorn straw,	Cynosurus cristata.
" Cuckoo,		Luzula campestris.
" Darnel,	Rye grass,	Lolium perenne and tremulentum.
" Deer,	Meadow beauty,	Rhexia Virginica.
" Dew,	Cocks' foot grass,	Dactylis glomerata.
" Ditch,		Ruppia maritima.
" Dog,	Couch grass,	Triticum repens.
" Dog,		" caninum.
" " bent,		Agrostis canina.
" Dogs' tail,	Crab grass,	Eleusine Indica.
" " tail,		Cynosurus cristatus.
" " tooth,	Dog grass,	Triticum caninum,
" Dropseed,	Wind bent grass,	Muhlenbergia spica venti.
" Eel,		Vallisneria spiralis.
" Elephants,		Typha Elephantina.
" Egyptian,		Dactyloctenium Ægyptiacum.
" False redtop,		Poa serotina.
" Feather,		Stipa pennata.
" Fescue,		Festuca pratensis and elatior.
" Fescue, sheeps,	.	" ovina.
" Finger,		Panicum sanguinale.
" Fiorin,	White bent grass,	Agrostis alba, and vulgaris.
" Five leaved,	Cinque foil,	Potentilla reptans.
" Flea,		Carex pulicaris.
" Float or flote,	Manna grass,	Glyceria fluitans.
" Fodder,		Chilochloa Bœhmeri.
" Four leaved,	Herb Paris,	Paris quadrifolia.
" Fowl meadow,	False redtop,	Poa serotina.

COMMON.	ENGLISH.	BOTANICAL.
Grass, Fox tail,		Setaria glauca.
" French,	Sainfoin,	Onobrychis sativa.
" " sparrow,	Asparagus substitute,	Ornithogalum pyrenaicum.
" Frog,	Glasswort,	Salicornia herbacea.
" Gallow,	Indian hemp,	Cannibis sativa.
" Gama,	Sesame grass,	Tripsacum dactyloides.
" Ghohona,	a reputedly poison grass,	Paspalum scrobiculatum.
" Ginger,		Andropogon schœnanthus.
" Goats' beard,		Ægopogon pusillus.
" Goose,	Cleavers,	Galium aparine.
" "	Silverweed,	Potentilla anserina.
" "	Knot grass,	Polygonum aviculare.
" Great goose,		Asperugo procumbens.
" Green foxtail,	Green setaria,	Setaria viridis.
" Grip,	Cleavers,	Galium aparine.
" Guinea,	Panic grass,	Panicum jumentorum.
" Hair,	Bent grass,	Agrostis scabra.
" "		Aira.
" Hard,	Cocksfoot grass,	Dactylis glomerata.
" Hares' tail,	Oval spiked lagurus,	Lagurus ovata.
" Hassock,		Aira cœspitosa.
" Hedgehog,		Echinochloa echinata.
" " "		Cenchrus echinatus.
" Herd,		Agrostis dispar.
" Herds, of N. Eng.	Timothy grass,	Phleum pratense.
" " Pennsylvania	Bent grass,	Agrostis vulgaris.
" Holy,	Seneca grass,	Hierochloa borealis.
" Horn,		Ceratochloa pendula.
" " of plenty,		Cornucopia cucculata.
" Hungarian,	Hungarian millet,	Panicum Germanicum.
" Indian,	Wood grass,	Sorghum nutans.
" Italian millet,		Setaria Italica.
" June,		Poa pratensis.
" Kangaroo,		Anthristiria australis.
" Kentucky blue,	Spear grass,	Poa pratensis.
" Knot,		Polygonum aviculare.
" "	Dog grass,	Triticum repens.
" Knot,		Agrostis stolonifera.
" " German,	Gravel chickweed,	Scleranthus annuus.
" Lemon,		Andropogon citratum.
" Lob or lop,		Bromus mollis.
" Long,		Macrochloa tenacrissima.
" Love,	common name for	Eragrostis, of many varieties.
" Low spear,		Poa annua.
" Lyme,		Elymus geniculatus and Virginicus.
" Maiden hair,		Briza media.
" Manna,	Float grass,	Glyceria fluitans.
" Marl,		Trifolium pratense.
" Marram,		Elymus arenarius.
" Marsh bent,	Fiorin,	Agrostis alba.
" Marsh,		Spartina stricta and var.
" " hedgehog,		Carex flava.
" Mat,		Nardus stricta.
" Meadow,		Poa pratense.
" " foxtail,		Alopecurus pratensis.
" Melic,		Melica nutans and var.
" Millet,	Indian millet,	Sorghum vulgare.
" "		Milium effusum.
" "	Millet,	Panicum miliaceum.
" Millet,		Setaria Italica.

4

COMMON.	ENGLISH.	BOTANICAL.
Grass, Monkey,	Para grass,	Attalea funifera.
" Moor,	Blue sesleria,	Sesleria cærulea.
" Mountain,		Andropogon bicornis.
" Mouse ear scorpion	Forget-me-not,	Myosotis palustris.
" Mouse tail,		Festuca Myurus.
" " "		Alopecurus agrestis.
" Myrtle,	Sweet flag,	Acorus Calamus.
" Naked beard,		Gymnopogon racemosus.
" Nit,		Gastridium australe.
" Nut,		Cyperus hydra.
" Oat,		Arrenatherum avenaceum.
" "	Common Oats,	Avena sativa.
" of the Andes,	Oat grass,	" "
" of Parnassus,		Parnassia Caroliniana.
" Old witch,		Panicum capillare.
" One glumed,		Monachne.
" Orange,		Hypericum sarothra.
" Orchard,		Dactylis glomerata.
" Pampas,		Gynerium argenteum.
" Pale manna,		Glyceria pallida.
" Panic,		Panicum, of several varieties.
" Para,	name of the fibre of	Attalea funifera.
" Penny,		Rhinanthus Crista-galli.
" Perennial rye,		Lolium perenne.
" Pepper,	-	Pilularia globulifera.
" Pigeon's,		Verbena officinalis.
" Poverty,		Aristida dichotoma.
" Prickly,		Echinochloa.
" Pudding,		Mentha pulegium.
" Purple bent,		Calamagrostis brevipilis.
" Quack,	Dog grass,	Triticum repens.
" Quake,		Briza maxima.
" Quaking,		" " and media.
" Quick or quitch,		Triticum repens.
" Rattlesnake,		Glyceria Canadensis.
" Ray, of France,	Tall oat grass,	Avena elatior.
" Ray,	Rye grass,	Lolium perenne.
" Red top, False,	Fowl meadow grass,	Poa serotina.
" Red top,	Burden's fine top,	Agrostis vulgaris.
" Reed,		Arundo.
" " bent,		Calamagrostis.
" " Canary,		Phalaris arundinacea.
" " Indian,		Cinna arundinacea.
" Rib,		Plantago lanceolata.
" Ribbon,	Striped grass,	Phalaris colorata.
" "		Diagraphis arundinacea.
" "	Wild canary,	Phalaris Americana.
" Rice cut,		Leersia oryzoides.
" Rope,		Restio.
" Rot,		Pinguicula vulgaris.
" Rough stalked meadow,		Poa trivialis.
" Rough,	Orchard grass,	Dactylis glomerata.
" " cocksfoot,	" "	" "
" Rush,		Vilfa.
" " Salt,		Spartina juncea.
" Rye, Italian,		Lolium Italicum.
" Rye,	Darnel grass,	" perenne.
" "		Secale cereale.
" Sand,		Tricuspis purpurea.
" Salt marsh,		Spartina stricta and species.

COMMON.	ENGLISH.	BOTANICAL.
Grass Scorpion,		Myosotis arveusis.
" Scotch,		Panicum molle.
" Scurvy,		Cochlcaria officinalis.
" Scutch,	Bermuda grass,	Cynodon dactylon.
" Sea meadow,		Glyceria maritima.
" " hard,		Ophiurus.
" " lyme,		Elymus arenarius.
" " mat,		Ammophila arenaria.
" " sand,	Beach grass,	Calamagrostis arenaria.
" " spur,		Glyceria distans.
" Seneca,	Holy grass,	Hierochloa borealis.
" Sesame,	Gama grass,	Tripsacum dactyloides.
" Shave,	Scouring rush,	Equisetum hyemale.
" Sheeps' fescue,		Festuca ovina.
" Shelly,	Dog grass,	Triticum repens.
" Shere,	Sedge,	Carex of many var.
" Shore,		Littorella lacustris.
" Shrubby,		Thamnochortus.
" Sickle,		Polygonum arifolium.
" Silk,		Eriocoma cuspidata.
" "	Bear grass,	Yucca filimentosa.
" Slender,		Leptochloa.
" Small,		Microchloa.
" Soft,		Holcus.
" " meadow,	Velvet grass,	Holcus lanatus.
" Sour,		Panicum leucophæum.
" Sparrow,		Asparagus officinalis.
" Spear, Common,	Kentucky blue grass,	Poa pratensis.
" " Meadow,		Glyceria nervata.
" Spelt,	.	Triticum Spelta.
" Spike,		Brizopyrum spicatum.
" Spiked,	Marsh arrow grass,	Triglochin maritima.
" " quaking,		Brizopyrum boreale.
" Spring,		Anthoxanthum.
" Spurt,		Scirpus maritimus.
" Squirrel tail,		Hordeum jubatum.
" Stander,		Orchis mascula.
" Star,		Callitriche.
" "		Hypoxys erecta.
" "	Unicorn root,	Aletris farinosa.
" Striped,		Digraphis arundinacea and var.
" "	a var. of Reed canary grass	Phalaris colorata.
" Sweet,	Manna grass,	Glyceria fluitans.
" Swines',		Polygonum aviculare.
" Sword,	Sword lily,	Gladiolus communis.
" "		Arenaria segetalis.
" "		Melilotus segetalis.
" Tall fescue,		Festuca elatior.
" Three leaved,		Trifolium pratense and var.
" Tickle,	Hair grass,	Agrostis scabra.
" Timothy,	Herds' grass,	Phleum pratense.
" Toad,		Juncus bufonius.
" Toothache,		Ctenium Americanum.
" tree,		Xanthorrhœa hastilis.
" Turtle,		Zostera maritima.
" Tussock,		Dactylis cæspitosa.
" Twitch,	Dog grass,	Triticum repens.
" Two penny		Lysimachia nummularia.
" Vanilla,	Seneca grass,	Hierochloa borealis.
" Velvet,	Meadow soft grass,	Holcus lanatus.

COMMON.	ENGLISH.	BOTANICAL.
Grass Vernal,		Anthoxanthum odoratum.
" " sweet scented,		" "
" Vipers',		Scorzonera hispanica.
" Water scorpion,		Myosotis palustris.
" Water star,		Leptanthus palustris.
" Wheat,		Triticum of several varieties.
" Whip,		Scleria triglomerata.
" White,	English bent grass, .	Agrostis alba.
" "	Cut grass. False rice,	Leersia oryzoides and Virginica.
" Whitlow,		Draba verna.
" "		Saxifraga tridactylites.
" Wild Oat,		Danthonia.
" Willard's bromus,	Chess, or Cheat,	Bromus secalinus.
" Wind,		Apera spica venti.
" Wire,	Crab grass,	Eleusine Indica.
" "	Blue grass,	Poa compressa.
" " bent,	Wind grass,	Apera spica venti.
" Wood,		Sorghum (andropogon) nutans.
" " meadow,		Poa nemoralis.
" " reed,		Cinna arundinacea
" Wooly,		Lasiagrostis.
" " bearded,		Erianthus alopecuroides.
" Worm,		Spigelia Marilandica.
" "	.	Sedum album.
" Yard,	Crab grass,	Eleusine Indica.
" Yellow eyed,		Xyris Caroliniana.
" " oat,		Avena flavescens.
Gravel chickweed,	Knawel,	Scleranthus annuus.
" grass,	Cleavers,	Galium aparine.
" plant,	Mayflower,	Epigæa repens.
" root,	Queen of the meadow,	Eupatorium purpureum.
" weed,	Mayflower,	Epigæa repens.
" "	Bush honeysuckle,	Diervilla Canadensis.
" "	Wild Job's tears,	Onosmodium Virginianum.
Gray bark,		Cinchona nitida.
" beard tree,	Fringe tree,	Chionanthus Virginicus.
Grease wood,		Abione canescens.
Great leopardsbane,		Doronicum pardilianches.
" ragweed,	Tall ambrosia,	Ambrosia trifida.
" stinging nettle,	Stinging nettle,	Urtica dioica.
" valerian, Wild,		Valeriana officinalis.
" water dock,		Rumex aquaticus.
Greek nuts,	Almonds,	Amygdalus communis.
" valerian,		Polemonium cæruleum.
" " American,	Abscess root,	" reptans.
Green amaranth,	Pigweed,	Amarantus albus.
" archangel,	Bugle weed, Bitter,	Lycopus Europæus and var.
" brier,		Smilax rotundifolia.
" broom,	Dyers' broom,	Genista tinctoria.
" dragon,		Arum Dracontium.
" gages,		Prunus domestica.
" heart tree,	Bebeeru tree,	Nectandra Rodiœl.
" hellebore,	American Hellebore,	Veratrum viride.
" milkweed,		Accrates viridiflora.
" ozier. Osier,	Green Osier,	Cornus circinata.
" sauce,	Wood sorrel,	Oxalis acetosella.
" valerian,	American Greek valerian,	Polemonium reptans.
" weed,	Dyers' broom,	Genista tinctoria.
" withe,		Vanilla Claviculata.
" wood,	Dyers' broom,	Genista tinctoria.

COMMON.	ENGLISH.	BOTANICAL.
Grenadier,	Pomegranate,	Punica Granatum.
Grey bark,		Cinchona nitida.
Groats,	Oats deprived of their husks,	
Gromwell,	Stone seed,	Lithospermum arvense.
" False,	Wild Job's tears,	Onosmodium Virginianum.
Ground apple,	Chamomile flowers,	Anthemis nobilis.
" berry,	Checkerberry,	Gaultheria procumbens.
" bread,	Sow bread,	Cyclamen Europæum.
" centaury,		Polygala Nuttallii.
" cherry,	Yellow henbane,	Physalis viscosa.
" flower,	Bitter polygala,	Polygala rubella.
" heel,	Speedwell,	Veronica officinalis.
" hemlock,	Dwarf yew,	Taxus Canadensis.
" holly,	Pipsissewa,	Chimaphilla umbellata.
" ivy,		Nepeta glechoma.
" laurel,	Mayflower,	Epigæa repens.
" lemon,	American mandrake,	Podophyllum peltatum.
" lily,	Beth root,	Trillium pendulum.
" maple,	American sanicle,	Heuchera acerifolia.
" moss,	Hair cap moss,	Polytrichum juniperum.
" nut,	Pea nut,	Arachis hypogæa.
" oak, ·	Germander,	Teucrium chamædrys.
" pine,		Ajuga chamæpitys.
" "		Lycopodium complanatum.
" plum,		Astragalus caryocarpus,
" raspberry,	Goldenseal,	Hydrastis Canadensis.
" squirrel pea,	Twin leaf root,	Jeffersonia diphylla.
" thistle,		Carlina acaulis.
" vine,	Twinflower,	Linnæa borealis.
Groundie swallow,		Senecio vulgaris.
Groundsel,		" vulgaris.
" tree,	Pencil tree,	Baccharis halimifolia.
Grouseberry,	Checkerberry,	Gaultheria procumbens.
Guaco,		Mikania guaco.
Guaiac,	Lignum vitæ,	Guaiacum officinale.
Guaiacum,	Guaiac,	" " "
Guarana,		Paullinia sorbilis.
Guava,	Bay plum,	Psidium pyriferum.
Guayva,	Guava,	" "
Guelder, Maple,	Dockmackle,	Viburnum acerifolium.
" rose,	Snowball,	Viburnum roseum.
Guernsey lily,		Nerine sarniensis.
Guiana almond,		Caryocar tomentosum.
" bark,		Cascarilla macrocarpa.
Guinea corn,		Sorghum Vulgare var. cernuum.
" cubebs,	African black pepper,	Piper afzelii.
" · grains,	Grains paradise,	Amomum, Grana Paradisi and var.
" hen flower,		Fritillaria meleagris.
" palm,	Oil palm,	Elais Guiniensis.
" pepper,	Cayenne pepper,	Capsicum annuum.
" "	Zauzibar pepper,	" fastigiatum.
" sorrel,		Hibiscus sabdariffa.
Guirila,	Insect powder,	Pyrethrum roseum and var.
Guilancha bark,		Cocculus cordiferum.
Gulf weed.	Laver,	Sargassum bacciferum.
Gum acroid,	a gum resin procured from	Xanthorrhœa hastilis and var.
" ammoniacum.	·	Dorema ammoniacum.
" "		Ferula tingitand.
" arabic,		Acacia vera and other varieties.
" anime,	So. Am'n Locust tree gum.	Hymenæa courbaril.

4*

COMMON.	ENGLISH.	BOTANICAL.
Gum Artificial,	Dextrin,	
" assafœtida,		Narthex Assafœtida.
" Australian,	Wattle gum,	Acacia decurrens.
" Barbary,		" gummifera.
" Bassora,		" Bassora and var.
" benzoin,	Balsamic exudation of	Styrax benzoin.
" Black boy,	Gum acroid,	Xanthorrhœa hastilis.
" British,	Dextrin,	
" Butea,	Kino, Bengal,	Butea frondosa and var.
" Botany bay,		Xanthorrhœa arborea.
" Cape,	Gum Arabic,	Acacia karroo.
" caranna,		Bursera gummifera.
" "		Amyris caranna.
" Cedar,	resembles olibanum,	Widdringtonia juniperoides.
" Cherrytree,	obtained from	Cerasus and Prunus.
" cistus,		Cistus ladaniferus.
" copal,	see Copal,	
" Doctors,	West India hog gum,	Rhus metopium.
" "	Tragacanth,	Astragalus verus and var.
" dragon,	"	" Creticus.
" "	Dragon's blood, American,	Pterocarpus Draco.
" East Indian,	Gum Arabic,	Acacia Arabica and var.
" elastic,	Caoutchouc,	Siphonia elastica.
" elemi,	Gum resin of	Amyris Elemifera.
" "	see Elemi,	
" galbanum,		Galbanum officinalis.
" gamboge,	see Gamboge,	
" gedda,	Gum Arabic,	
" Grass tree,	Gum acroid,	Xanthorrhœa hastilis.
" Guaiac,		Guaiacum officinalis.
" guttI,	Gamboge,	
" hackmatac,	Gum tamarack,	Larix Americana.
" hemlock,		Abies Canadensis.
" hog,	Tragacanth,	Astragalus verus and var.
" " West India,		Rhus metopium.
" India,	Gum Arabic,	Acacia Arabica and var.
" Ivy,	exudes from old stems of	Hedera Helix.
" Juniper,	Sandarach,	Thuja articulata.
" kino,	see Kino,	
" lac,	see Lac,	Aleurites laccifera.
" ladanum,	a gum resin from	Cistus creticus and var.
" mesquite,		Algarobia glandulosa.
" myrrh,	see myrrh,	Balsamodendron Myrrha.
" Orenburgh,	Manna of Briancon,	Abies Larix.
" Peruvian,	Bassorin,	
" sandarach,		Thuja articulata.
" sassa,	False tragacanth,	Igna sassa.
" Senegal,		Acacia Senegal and var.
" Shellac,	see Lac,	Aleurites laccifera, etc.
" spruce,	Chewing gum,	Abies nigra.
" Tacamahac,		Amyris tomentosum.
" tamarack,		Larix Americana.
" tragacanth,		Astragalus verus.
" Turkey,	Gum Arabic,	
" Wattle,	Australian gum,	Acacia decurrens.
" yellow,	Gum acroid,	Xanthorrhœa hastilis and var.
" tree,	Opossum tree,	Liquidambar styraciflua.
" "		Eucalyptus robusta.
" "	Gum Arabic tree,	Acacia of several varieties.
" plant,	Comfrey,	Symphytum officinale.

COMMON.	ENGLISH.	BOTANICAL.
Gum wood,	the wood of	Eucalyptus.
Gunbright,	Scouring rush,	Equisetum hyemale.
Gunjah,	the dried flower branches of	Cannabis Indica.
Gutta percha,		Isonandra gutta.
Gutter tree,	Dogberry tree,	Cornus sanguinea.
Gypsie weed,	Bugle weed,	Lycopus Virginicus.
Hackberry,	Nettle tree,	Celtis occidentalis.
Hackmatac,	Tamarack,	Larix Americana.
Hackmetack,	"	" "
Hagberry,	Bird cherry,	Cerasus padus and var.
Hair bell,	Bellflower,	Campanula rotundifolia.
" cap moss,		Polytrichum Juniperum.
" strong,	Sow fennel,	Peucedanum officinale.
Halish,	Hashish,	Cannabis Indica.
Hard Carthagena bark,		Cinchona cordifolia vera.
Hardhack,		Spiræa tomentosa.
"	Stone root,	Collinsonia Canadensis.
Hardock,	Burdock,	Arctium lappa.
Harebell,	Bellflower,	Campanula rotundiflora.
Hareburr,	Burdock,	Arctium lappa.
Hares bane,	Aconite,	Aconitum lagoctonum.
" beard,	Mullein,	Verbascum Thapsus.
" ear,		Bupleurum rotundifolium.
" foot,	Clover,	Trifolium arvense.
Harts horn bush,	Buckhorn brake,	Osmunda regalis.
" horn plantain,	Bucks' horn plantain,	Plantago coronopus.
" tongue fern,		Scolopendrium officinarum.
Hartwort,		Laserpitium siler.
Harvest bells,	Calathian violet,	Gentiana pneumonanthe.
" lice,	Cuckold,	Bidens connata.
Hashish,	Bang,	Cannabis Indica.
Haskwort,		Campanula latifolia.
Hautbois,		Fragraria elatior.
Haw Black,	Sloe,	Viburnum prunifolium.
Hawk beard,		Crepis virens.
" bit,	Hawk weed,	Hieracium venosum.
" weed,	Striped blood wort,	" "
Haws,	the fruit of	Crategus Oxycantha.
Hawthorn,		" coccinea.
" English,		" oxycanthus.
Haymaids,	Ground ivy,	Nepeta glechoma.
Hay saffron,	name for loose saffron,	Crocus sativa.
Hazel,	Filbert,	Corylus Americana and avellana.
" crottles,	Lung moss,	Sticta pulmonaria.
" nut,	Filbert,	Corylus Americana.
" " Witch,		Hamamelis Virginica.
" wort,	Asarabacca,	Asarum Europæum.
Headache tree,		Premna integrifolia.
Head betony,		Pedicularis Canadensis.
Healall,		Prunella vulgaris.
"	Scrofula plant	Scrophularia Marilandica.
"	Stone root,	Collinsonia Canadensis.
Healing herb,	Comfrey,	Symphytum officinalis.
Heart liver leaf,		Hepatica acutiloba.
Heartsease,	Pansy,	Viola tricolor.
"		Polygonum persicaria.
Heart seed,	Balloon vine,	Cardiospermum halicacabum.
Heart snake root,	Canada snake root,	Asarum Canadense.
Heath,		Erica.
" corn,	Buckwheat,	Polygonum fagopyrum.

COMMON.	ENGLISH.	BOTANICAL.
Heath pea,		Orobus atropurpurea.
Heather,		Erica vulgaris.
"		Calluna vulgaris.
Hebrew manna,		Alhagi maurorum.
Hedgebell,	Bearbind,	Convolvulus arvensis.
Hedge berry,	Sweet cherry,	Cerasus avium.
" garlic,		Alliaria officinalis.
" hog cactus,		Echinocactus Texensis.
" hogs,		Medicago intertexta.
" hyssop,		Gratiola aurea.
" maids,	Ground ivy,	Nepeta glechoma.
" mustard,	English water cress,	Sisymbrium officinale.
" "		Erysimum Nasturtium.
" nettle,	Clown heal,	Stachys palustris.
Heliotrope, Garden,	Vanilla scented heliotrope,	Heliotropium Peruvianum and var.
Hellebore, American,		Veratrum viride.
" Black,		Helleborus niger.
" False,	American hellebore,	Veratrum viride.
" Fetid,	Skunk Cabbage,	Symplocarpus fœtidus.
" "		Helleborus fœtidus.
" Green,	American hellebore,	Veratrum viride.
" "		Helleborus viridis.
" Stinking,	Fetid hellebore,	" fœtidus.
" Swamp,	American hellebore,	Veratrum viride.
" White,		" album.
" " Am'n,		" viride.
" Winter,		Helleborus hyemalis.
Hellweed,	Flax vine,	Cuscuta Europæa.
Helmet flower,	Scull cap,	Scutellaria lateriflora.
" pod,	Twin leaf root,	Jeffersonia diphylla.
Hemlock,	common name for	Conium maculatum.
" American,	the Hemlock tree,	Abies Canadensis.
" bark,		" "
" dropwort,		Œnanthe crocrata.
" Fir,	Balsam fir,	Abies balsamea.
" gum,	a gum resin obtained from	" Canadensis.
" Ground,	Dwarf yew,	Taxus "
" parsley,		Selinum Canadense.
" Poison,		Conium maculatum.
" pitch,	Hemlock gum,	Abies Canadensis.
" Small leaved,	Fools' parsley,	Æthusa cynapium.
" spruce,		Abies Canadensis.
" storksbill,		Erodium cicutarium.
" Water,		Cicuta aquatica.
" " Am.		• " maculatum.
" " fine leaved,		Phellandrium aquaticum.
" Wild,	Poison hemlock,	Conium maculatum.
Hemp,	common name of	Cannabis sativa.
" African,	Bow string hemp,	Sanseviera Guineensis.
" agrimony,	Sweet smelling trefoil,	Eupatorium cannabinum.
" Bengal,	Hemp, Madras,	Crotalaria juncea.
" Black Indian,		Apocynum cannabinum.
" Common Am.		Cannabis sativa.
" Indian, east,		" " var Indica.
" Madras fibre,		Crotalaria juncea.
" nettle,	Flowering nettle,	Galeopsis tetrahit.
" seed,		Cannabis sativa.
" weed,		" "
" White Indian,		Asclepias incarnata.
Hen and chickens,	Daisy,	Bellis perennis.

COMMON.	ENGLISH.	BOTANICAL.
Henbane,		Hyosciamus niger.
" Black,		" "
" White,		" alba.
" yellow,	Ground cherry,	Physalis viscosa.
Henbit,	Dead nettle,	Lamium amplexicaule.
Henne,	yellow dye of the E. Indies,	Lawsonia inermis.
Hen's foot,	Bastard parsley,	Caucaulis daucoides.
Hep tree,	Hip tree,	Rosa canina.
Herb Barbara,	Winter cress,	Barbarea vulgaris.
" Bennet,	White avens,	Geum Virginianum.
" Blessed,		" urbanum.
" Christopher,	Baneberry,	Actæa spicata.
" "	Buckhorn brake,	Osmunda regalis.
Herb Eve,	Wart cress, ·	Coronopus Ruellii.
" Margaret,	White weed,	Leucanthemum vulgare.
" mastich,		Thymus mastichina.
" " Syrian,	Cat thyme,	Teucrium marum.
" of grace,	Rue,	Ruta graveolens.
" Paris,	Fox grape,	Paris quadrifolia.
" Peter,	Cowslip primrose,	Primula veris.
" Robert,	·	Geranium Robertanium.
" sophia,	Hedge mustard, fine-leaved,	Sisymbrium sophia.
" trinity, ·	Liverwort,	Hepatica triloba.
" two pence,	Moneywort,	Lysimachia nummularia.
Hercules' club,	Yellow prickly ash, ·	Xanthoxylon clava Herculis.
" "	Prickly elder,	Aralia spinosa.
Herds' grass,	Timothy grass,	Phleum pratense.
Hermodactyle,	Chequer flower,	Colchicum variegatum.
Herons' bill,	Storks' bill,	Erodium cicutarium.
Hickory,	Shellbark hickory,	Hicorya sulcata.
"	Shag bark walnut,	" alba.
High angelica,	Purple angelica,	Angelica atropurpurea.
Highbelia,	Blue cardinal,	Lobelia syphilitica.
High cranberry,	Cramp bark,	Viburnum opulus. ·
" " bark,	" "	" "
" healall,		Pedicularia gladiata.
" mallow,	common mallow,	Malva sylvestris.
Hig taper,	Mullein,	Verbascum Thapsus.
Hillberry,	Checkerberry,	Gaultheria procumbens.
Hilwort,	European pennyroyal,	Mentha pulegium.
Hindberry,	Garden raspberry,	Rubus Idæus.
Hindheel,	Tansy, common,	Tanacetum vulgare.
Hini,	Black root,	Leptandra Virginica.
Hip fruit,	Hips,	Rosa Canina.
" seed,	·	" "
" tree,	Dog rose,	" "
" wort,	Navelwort,	Cotyledon umbilicus.
Hippo,	Blooming spurge,	Euphorbia corollata.
" Indian,	Indian physic,	Gillenia trifoliata.
Hips,	the fruit of the dog rose,	Rosa canina.
Hive vine,	Squaw vine,	Mitchella repens. ·
Hoarhound,	Horehound,	Marrubium vulgare.
Hoary pea,	Goats' rue,	Tephrosia Virginiana.
Hobble bush,		Viburnum lantanoides.
Hog apple,	May apple,	Podophyllum peltatum.
" bean,	Henbane,	Hyosciamus niger.
" bed,	Ground pine,	Lycopodium complanatum.
" fennel,	Sulphur weed,	Peucedanum officinarium.
" nut,	Pig nut,	Carya porcina, and species.
" plum,		Spondias entra.

COMMON.	ENGLISH.	BOTANICAL.
Hog weed,		Ambrosia elatior.
" "		" artemisiæfolia.
" wort,		Heptallon graveolens.
Holewort,	Round birthwort,	Corydalis bulbosa.
Hollow tooth herb,		Galeopsis grandiflora.
" wort,		Corydalis bulbosa.
Holly, American,		Ilex opaca.
" European,		" aquifolium.
" bay,	Swamp laurel,	Gordonia lasianthus.
" rose,		Cistus salvifolius.
" "		Helianthemum roseum.
Hollyhock,		Althæa rosea.
" rose,	Resurrection plant,	Selaginella lepidophylla.
Holmes' weed,	Scrofula plant,	Scrophularia Marilandica.
Holy Ghost,	Angelica, European,	Archangelica officinalis.
" herb,	Vervain, "	Verbena officinalis.
" thistle,	Blessed thistle,	Centaurea benedicta.
Honesty,		Lunaria biennis.
Honewort,	Fools' water cress,	Trinia vulgaris.
"	Bastard stone parsley,	Sison amomum.
Honey berry,		Melicocca paniculata.
" bloom,	Bitter root,	Apocynum androsæmifolium.
" bread,	Saint John's bread,	Ceratonia siliqua.
" locust,		Gleditschia triacanthos.
" lotus,	White melilot,	Melilotus alba.
Honeysuckle,		Lonicera caprifolium.
"		Caprifolium perfoliatum.
" Bush,		Diervilla Canadensis.
Hooded milfoil,	Bladderwort,	Utricularia macrorhiza.
" willow herb,	Scullcap, European,	Scutellaria galericulata.
Hoodwort,	"	" lateriflora.
Hook heal,	Healall,	Prunella vulgaris.
" weed,	"	" "
Hoop ash,	Hag berry,	Cerasus padus.
" petticoat,	Medusa's trumpet,	Narcissus bulbocodium.
" tree,	Pride of China,	Melia azedarach.
Hop,		Humulus Lupulus.
" hornbeam,	Iron wood,	Ostrya Virginica.
" marjoram,	Dittany of Crete,	Origanum Dictamnus.
" medick,	Dutch clover,	Medicago lupulina.
" tree,	Wafer ash,	Ptelea trifoliata.
" trefoil,	Dutch clover,	Medicago lupulina.
" vine,		Humulus Lupulus.
" Wild,	Red berried bryony,	Bryonia dioica.
Hops,	the strobiles of	Humulus Lupulus.
Horehound,	White horehound,	Marrubium vulgare.
" Black,		Ballota nigra.
" Stinking,	Black horehound,	" "
" Water,	Bugle weed, bitter,	Lycopus Europæus.
" White,	Horehound,	Marrubium vulgare.
" Wild		Eupatorium teucrifolium.
Horestrang,	Sow fennel,	Peucedanum officinale.
Hornbeam,	Ostrya Virginica.	Carpinus Americana. .
" Swamp,	Tupelo,	Nyssa multiflora.
Hornwort,		Ceratophyllum dermersum.
Horn seed,	Ergot of rye,	Sclerotium clavus.
Horn of plenty,		Fedia cornucopia.
Horned poppy,	Bruise root,	Chelidonium glaucum.
Horse aloes,		Aloe Guiniensis.
" balm,	Stone root,	Collinsonia Canadensis.

COMMON.	ENGLISH.	BOTANICAL.
Horse bane,	Fine leaved water hemlock,	Œnanthe Phellandrium.
" bean,		Vicia faba.
" blob,	Marsh marigold,	Caltha palustris.
" cane,	Tall ambrosia,	Ambrosia trifida.
" cassia,	West India Purging cassia,	Cassia Brasiliana.
" chestnut,		Æsculus Hippocastanum.
" chire,	Germander,	Teucrium chamædrys.
" fleaweed,	Wild Indigo,	Baptisia tinctoria.
" flower,		Melampyrum sylvaticum.
" fly weed,	Wild indigo,	Baptisia tinctoria.
" gentian,	Fever root,	Triosteum perfoliatum.
" ginseng,	" "	" "
" heal,	Elecampane,	Inula Helenium.
" hoof,	Coltsfoot,	Tussilago farfara.
" knob,		Centaurea nigra.
Horse mint, American,		Monarda punctata.
" " European,		Mentha sylvestris.
" " round leav'd		" rotundifolia.
" " sweet,		Cunila mariana.
" mushroom,		Agaricus arvensis.
" nettle,		Solanum Carolinense.
" pipe,	Scouring rush,	Equisetum hyemale.
" radish,		Cochlearia armoracia.
" " tree,	Oily acorn,	Guilandina Moringa.
" " " nuts,	" "	" "
" " Water,		Cochlearia aquatica.
" sugar,	Sweet leaf,	Hopea tinctoria.
" tail,	Canada fleabane,	Erigeron Canadense.
" "	Scouring rush,	Equisetum hyemale.
" thyme,		Calamintha clinopodium.
" weed,	Tall ambrosia,	Ambrosia trifida.
" "	Stone root,	Collinsonia Canadensis.
" "	Hardhack,	Spiræa tomentosa.
" "	Canada fleabane,	Erigeron Canadense.
Hotela,		Hotcia Japonica.
Hottentot bread,	Elephants' foot,	Testudinaria elephantipes.
" "		Dioscorea.
Hounds' berry tree,	Dog berry tree,	Cornus sanguinea.
" tongue,		Cynoglossum officinale.
" "	Vanilla leaf,	Liatris odoratissima.
House leek,		Sempervivum tectorum.
" " common,		" "
" " small,		Sedum acre.
" " tree,		Æonium arboreum.
Hove,	Ground ivy,	Nepeta glechoma.
Huamilles bark,		Cinchona condaminea var vera.
Huanuco bark,	Pale Peruvian bark,	" nitida and micrantha.
Huckleberry,		Vaccinium myrtillus.
"		" resinosum.
Humble plant,		Mimosa pudica.
Hundred leaf rose,		Rosa centifolia.
" eyes,	Periwinkle,	Vinca major and Vinca minor.
Hunger flower,	Whitlow grass,	Draba incana.
" weed,		Ranunculus arvensis.
Huntsmans' cup,	Side saddle plant,	Sarracenia purpurea.
Husks, of the ancients,	Saint Johns' bread,	Ceratonia siliqua.
Hurr burr,	Burdock,	Arctium lappa.
Hyacinth bean, Purple,		Dolichos lablab.
" Garden,		Hyacinthus orientale.
Hyena poison,	Wolveboon,	Hyænanche capensis.

COMMON.	ENGLISH.	BOTANICAL.
Hya hya,	Cow tree,	Clusia galactodendron.
Hydrangea,		Hydrangea arborescens and species
" Garden,		" hortensia.
Hyeble,	Dwarf elder,	Aralia hispida.
Hyperuic wood,	Lima wood,	Cæsalpina echinata.
Hyssop,		Hyssopus officinalis.
" Hedge,		Gratiola aurea.
" Wild,	Vervain,	Verbena hastata.
Icajou,	Ordeal poison of Africa,	
Iceland moss,		Cetraria Islandica.
Ice plant, American,	Fit root plant,	Monotropa uniflora.
" " European,		Mesembryanthemum crystalinum.
" vine,	Pareira brava, False,	Cissampelos Pareira.
Immortelles,		Gnaphalium orientale.
Immortal flower,		Helichrysum bracteatum.
Imperial masterwort,		Astrantia major.
" "		Imperatoria ostruthium.
Imphee,	African sugar cane,	Holchus saccharatus.
Incense, Indian,	Olibanum,	Boswellia serrata.
" male,	"	" "
" tree,	Spanish cedar,	Icica Guianensis.
" wood,	" "	" "
India berries,	Cocculus Indicus,	Anamirta Cocculus.
" pink,	Chinese pink,	Dianthus Chinensis.
" rubber,	Gum elastic,	Siphonia elastica.
" "	Caoutchouc,	Ficus elastica.
" "		Hovea Guianeusis.
" "		Artocarpus integrifolia.
Indian aconite,	Nepal aconite,	Aconitum ferox.
" apple,	May apple,	Podophyllum peltatum.
" arrow,	Wahoo,	Euonymus atropurpureus.
" " root,		Maranta arundinacea.
" " wood,	Wahoo,	Euonymus atropurpureus.
" bael,	Bengal quince,	Ægle marmelos.
" balm,	Beth root,	Trillium pendulum.
" bark,	Sweet bay,	Magnolia glauca.
" barley-caustic,	Cevadilla seed,	Veratrum Sabadilla.
" bay,	Bay tree,	Laurus nobilis.
" berry,	Cocculus Indicus,	Anamirta Cocculus.
" black drink,	South sea tea,	Ilex vomitoria.
" buckwheat,		Fagopyrum Tartaricum.
" camphor,	Sumatra camphor,	Dryobalanops camphora and species
" chocolate,	Avens root,	Geum rivale.
" copal,		Vateria Indica.
" corn,	Maize,	Zea Mays.
" cress,	Nasturtion,	Tropæolum majus.
" cucumber,		Medeola Virginica.
" cup plant,		Silphium perfoliatum.
" currant,	Coral berry,	Symphoricarpus vulgaris.
" dates,	Tamarinds,	Tamarindus Indica.
" datura,	White flowered datura,	Datura alba.
" dream,	Rock brake,	Pteris atropurpurea.
" elm,	Slippery elm,	Ulmus fulva.
" fig,	Banyan tree,	Ficus Indica.
" geranium,	Ginger grass,	Andropogon Schœnanthus.
" ginger,	Canada snake root,	Asarum Canadensis.
" grass,	Wood grass,	Sorghum nutans.
" hazel nut,	Bonduc nuts,	Guilandina bonducella.
" hemp,	Cannabis Indica,	Cannabis sativa, var. Indica.
" " American,		" "

COMMON.	ENGLISH.	BOTANICAL.
Indian hemp, Black,	Black indian hemp,	Apocynum cannabinum.
" " Foreign,	Cannabis Indica,	Cannabis sativa, var. Indica.
" " White,		Asclepias incarnata.
" hippo,	Indian physic,	Gillenia trifoliata.
" hydrocotyle,	" pennyroyal,	Hydrocotyle Asiatica.
" ipecac,	Ceylon ipecac, -	Asclepias asthmatica.
" kale,		Arum esculentum.
" lettuce,	American Columbo,	Frasera Walteri.
" "	Round leaved pyrola,	Pyrola rotundifolia.
" liquorice,		Abrus precatorius.
" mallow,		Abutilon Avicennæ.
" mallows,	False mallows,	Sida spinosa.
" millet,	Coffee corn,	Sorghum vulgare.
" mulberry,	Awl tree,	Morinda citrifolia.
" nard,	Spikenard of the ancients,	Nardostachys Jatamansi.
" olibanum,	Frankincense,	Boswellia serrata.
" paint, red,	Blood root,	Sanguinaria Canadensis.
" " yellow,	Goldenseal root,	Hydrastis Canadensis.
" pennywort,		Hydrocotyle Asiatica.
" physic,		Gillenia trifoliata.
" "	American ipecac,	" stipulacea.
" "	Black indian hemp,	Apocynum cannabinum.
" " small flow'd,		Gillenia stipulacea.
" pink,	Carolina pink,	Spigelia Marilandica.
" pipe,	Ice plant, American,	Monotropa uniflora.
" plant,	Goldenseal,	Hydrastis Canadensis.
" plantain,		Cacalia tuberosa.
" poke,	American hellebore,	Veratrum Viride.
" posey,	White balsam,	Gnaphalium polycephalum.
" potato,		Apios tuberosa.
" red root,	Spirit weed,	Lacnanthes tinctoria.
" reed,		Cinna arundinacea.
" rice,		Zizania aquatica.
" root,	Spikenard, American,	Aralia racemosa.
" salt,	Cane sugar,	Saccharum officinarium.
" "	the powd'r on sumac berries	Rhus glabrum.
" sanicle,	White sanicle,	Eupatorium ageratoides.
" sarsaparilla,		Hemidesmus Indicus.
" shamrock,	Beth root,	Trillium pendulum and species.
" shoe,	Yellow ladies' slipper,	Cypripedium pubescens.
" shot,	Canna seed,	Canna Indica.
" soap plant,	Soap berry,	Sapindus marginatus.
" spikenard,		Nardus Jatamansi.
" strawberry,	Blite,	Blitum capitatum.
" "		Fragraria Indica.
" tobacco,	Lobelia.	Lobelia inflata.
" turnip,	Dragon root,	Arum triphyllum.
Indigo,	a blue dyestuff, the product of	{ Indigofera anil, argentea and tinctoria.
"	a blue dyestuff produced by	Galega tinctoria.
"	" " "	" Polygonum tinctorium.
"	" " "	" Wrightia tinctoria.
" berry,		Randia latifolia.
" broom,	Wild Indigo plant,	Baptisia tinctoria.
" plant,	produces Indigo,	Indigofera anil and species.
" "	.	Polygonum tinctorium.
" weed,	Wild Indigo plant,	Baptisia tinctoria.
" Wild, false,	" " "	" "
" Yellow,	" " "	" "
Ink berry,		Prinos glaber.

COMMON.	ENGLISH.	BOTANICAL.
Ink nut,		Terminalia cherbula.
" plant, New Granada,	Chanchi,	Coriaria thymifolia.
" root,	Marsh rosemary,	Statice Caroliniana.
Insect powder,	the powdered blossoms of	Persian feverfew.
" " Caucasian,		Pyrethrum roseum and species.
" " Dalmatian,		" cinerariæfolium.
" " German,		" roseum and species.
" " Persian,		" carneum and species.
Ipecac,		Cephælis Ipecacuanha.
" American,		Euphorbia "
" "		Gillenia Stipulacea and trifoliata.
" Bastard,		Asclepias curassavica.
" Black,	Peruvian ipecac,	Psychotria emetica.
" Ceylon,		Asclepias asthmatica.
" Malabar,		Randia dumetorum.
" Peruvian,	Black ipecac,	Psychotria emetica.
" Spurge,	American ipecac,	Euphorbia Ipecacuanha.
" Striated,	Black ipecac,	Psychotria emetica.
" Undulated,	White ipecac,	Richardsonia scabra.
" White,		" "
" Wild,	Fever root,	Triosteum perfoliatum.
" "		Cephælis Ipecacuanha.
" " American,	American ipecac,	Euphorbia "
Ipecacuanha,		Cephælis "
Iris,	Orris,	Iris Florentina.
Irish broom,	Broom herb,	Cytisus Scoparius.
" moss,		Chondrus crispus.
Iron bark tree,	Botany bay kino,	Eucalyptus resinifera.
" tree,		Siderodendron.
" weed,		Vernonia fasciculata.
" "		" angustifolia.
" wood,	Hop hornbeam,	Ostrya Virginica.
" "	Blue beech,	Carpinus Americana.
" wort,		Galeopsis ladanum.
" " German,		Sideritis hirsute.
Ishpingo,	Santa Fe cinnamon,	Nectandra cinnamomoides.
Isle of France cinnamon,		Laurus cupularis.
Issue peas,	prepared from	Iris Florentina.
Italian juice,	Extract of liquorice,	Glycyrrhiza glabra.
Itch weed,	American hellebore,	Veratrum viride.
Ivory palm,	Vegetable ivory,	Phytelephas macrocarpa.
" plum,	Checkerberry,	Gaultheria procumbens.
" Vegetable,		Phytelephas macrocarpa.
Ivy,	common name for	Hedera Helix.
" American,		Ampelopsis quinquefolia.
" Big leaved,		Kalmia latifolia.
" English,		Hedera Helix.
" German,		Senecio scandens.
" Ground,		Nepeta glechoma.
" gum,	exudes from the old stems of	Hedera Helix.
" leaved toad flax,		Linaria cymbalaria.
" Poison,		Rhus Toxicodendron, var radicans.
Iwarancuse,		Anatherum muricatum.
Jaborandi,		Piper reticulatum.
Jacinth,	Hyacinth,	Hyacinthus orientalis.
Jackalls' kost,		Hydnora Africana.
Jack by the hedge,		Sisymbrium alliaria.
Jacket bark,	Peruvian bark,	Cinchona.
Jack fruit tree,	yields India rubber,	Artocarpus integrifolia.
" in a box,		Hernandia sonora.

COMMON.	ENGLISH.	BOTANICAL.
Jack in the pulpit,	Dragon root,	Arum triphyllum.
" of the buttery,	Small house leek,	Sedum acre.
" tree,	Jack fruit tree,	Artocarpus integrifolia.
" wood,	Jack fruit tree,	" "
Jacobs' ladder,		Smilax peduncularis.
" "	American Greek valerian,	Polemonium reptans.
Jaen bark,		Cinchona villosa.
Jaffua moss,	Ceylon moss,	Gracilaria lichenoides.
Jaggery,	Crude palm sugar,	Caryota urens.
Jagong,	Indian corn,	Zea Mays.
Jalap,		Exogonium Purga and Ipomœa
" Cancer,	Garget,	Phytolacca decandra. [Jalapa.
" False,	Four o'clocks,	Mirabilis Jalapa.
" Male,		Ipomœa Orizabensis.
" plant,	False jalap,	Mirabilis Jalapa.
" Tampico,		Ipomœa simulans.
" Vera Cruz,	Jalap,	Ipomœa Purga.
" Wild,		Convolvulus panduratus.
Jamaica bark,		Cinchona triflora. (Exostemma.)
" bayberry,	Bay rum leaf tree,	Myrcia acris.
" birch,	yields caranna gum,	Bursera gummifera.
" cedar,		Juniperus Bermudiana.
" contrayerva,		Dorstenia contrayerva.
" dogwood,		Piscidia erythrina.
" ginger,		Amomum Zingiber.
" kino,	product of sea side grape,	Coccoloba uvifera.
" mignonette,	Henna plant,	Lawsonia inermis.
" pepper,	Allspice,	Eugenia Pimenta.
" redwood,	Blood wood,	Gordonia Hæmatoxylon.
" rosewood,		Amyris balsamifera.
" spikenard,		Ballota suaveolens.
" water lily,	Egyptian bean,	Nelumbium speciosum.
" yellow thistle,	Prickly poppy,	Argemone Mexicana.
Jamestown weed,	Thorn apple,	Datura Stramonium.
Japan camphor,		Cinnamomum Camphora.
" "		Camphora officinarum.
" isinglass,	procured from seaweed,	Gelidum corneum.
" lacquer,		Rhus vernicifera.
" "		Stagmaria verniciflua.
" lily,		Lilium lancifolium.
" quince,		Cydonia Japonica.
" wax,	produced by berries of	Rhus succedaneum.
Japanese pepper,		Xanthoxylon alatum.
Jasmine,		Jasminum officinale.
" Cape,		Gardenia Florida.
" Chili,		Mandevilla suaveolens.
" flowers, white,		Jasminum officinale.
" Red,	Cypress vine,	Ipomœa quamoclit.
" " West India,	Frangipanni,	Plumicria rubra.
Jatamansi,	Musk root,	Nardostachys Jatamansi.
Jaundice berry,	Barberry,	Berberis vulgaris.
" root,	Goldenseal root,	Hydrastis Canadensis.
Java almond tree,	yields Gum elemi,	Canarium commune.
" pepper,	Cubebs,	Piper Cubeba.
Jersey live long,		Gnaphalium luteo-album.
" tea,	Red root bark tree,	Ceanothus Americana.
Jerusalem artichoke,		Helianthus tuberosus.
" cherry,		Solanum pseudo-capsicum.
" cowslip,	Lungwort,	Pulmonaria officinalis.
" cross,		Lychnis Chalcedonica.

COMMON.	ENGLISH.	BOTANICAL.
Jerusalem Oak,		Chenopodium anthelminticum.
" " leaves,		" "
" " seed,	American wormseed,	" "
" sage,		Phlomis tuberosa.
" "	Lungwort,	Pulmonaria officinalis.
" star,	Vegetable oyster,	Tragopogon porrifolia.
" tea,	Paraguay tea,	Ilex Paraguaiensis.
" "		Chenopodium Ambrosioides.
Jessamine,		
" Cape,		Bouvardia triphylla.
" Chili,	Savannah flower,	Echites suaveolens.
" Shrubby,		Jasminum fruticans.
" Yellow,		Gelsemium sempervirens.
Jesuits' bark,	old name for Peruvian bark of several species.	
" nuts,		Trapa nutans.
" powder,	Peruvian bark,	
" tea,		Ilex Paraguaiensis.
Jew bush,		Pedilanthus tithymaloides.
Jewel weed,	Touch me not,	Impatiens pallida.
" " speckled,		" fulva.
Jews' ear,	a fungus on Elder and Elm,	Peziza auricula.
" harp plant,	Beth root,	Trillium pendulum.
" mallow,		Corchorus olitorius.
" manna,	Indian manna,	Alhagi maurorum.
" pitch,	Bitumen,	
Jimpson seed,	Thorn apple seed,	· Datura Stramonium.
" weed,	" "	" "
Jimson weed,	" "	" "
Joan silver pin,	Red corn poppy,	Papaver Rhœas.
Job's tears,		Coix lachryma.
" " Wild,	False gromwell,	Onosmodium Virginianum.
Joe pye,	Queen of the meadow,	Eupatorium purpureum.
" "	.	" verticillatum.
Johnswort,		Hypericum perforatum.
" False,	Nit weed,	Sarothra Gentianoides.
Johnny jumper,	Pansy,	Viola tricolor.
Joint weed,		Polygonum articulatum.
Jonquil,		Narcissus jonquilla.
" flower,	Louisiana squill,	Crinum Americanum.
Joseph's coat,		Amaranthus tricolor.
Jove's fruit,	Persimmon,	Diospyrus Virginiana.
Judas' tree,	Red bud,	Cercis Canadensis.
" " European,		Arbor Judæ.
Jujube,	the fruit pulp,	Rhamnus zizyphus.
July flower,		Cheiranthus annuus.
Jumble beads,	Love pea,	Abrus precatorius.
June berry,	Choke cherry,	Aronia ovalis and var.
" "		" botryapium.
Juniper,	common name for	Juniperus communis.
" bark,	.	" "
" berries,	the fruit of juniper,	" "
" bush,		" "
" gum,	Sandarach,	Thuja articulata.
" leaves,		Juniperus Sabina.
" wood,		" oxycedrus.
Juno's tears,	European vervain,	Verbena officinalis.
Jupiter's beard,	Common house leek,	Sempervivum tectorum.
" eye,	" " "	" "
" nuts,	Walnuts, European,	Juglans regia.
Juribali,	African mahogany,	Swietenia Senegalensis.

COMMON.	ENGLISH.	BOTANICAL.
Jurebeba,		Solanum paniculatum.
Justices' weed,		Eupatorium Hyssopifolium.
" "		" leucolepsis.
Jute,	the fibre of	Corchorus capsularis.
Kaladana seed,	Gaybine,	Convolvulus nil.
Kalumb,	Colombo,	Cocculus palmatus.
Kamala,	Kameela,	Rottlera tinctoria.
Kameela,		" "
Kangaroo apple,		Solanum laciniatum.
Karat seed,	Carats,	Erythrina Abyssinica.
Kassamah,	Galangal,	Alpina galanga.
Kat,	Cafta leaves,	Catha edulis.
Kedlock,	White mustard,	Sinapis alba.
Kelp,	Sea wrack burned in the open air yields kelp.	
Kemps,	Rib grass,	Plantago media.
Kentucky coffee tree,	American coffee bean,	Gymnocladus Canadensis.
" mahogany,	" " "	" "
" yellow wood,	Yellow ash,	Virgilia lutea. (Cladrastis.)
Kermes oak,	Cochineal oak,	Quercus coccifera.
Kernelwort,	Figwort root,	Scrophularia nodosa.
Khus khus,	Vittivert,	Andropogon muricatus.
Kidney bean, Carolina,	Woody Wistaria,	Wistaria frutescens.
" liver leaf,		Hepatica Americana.
" root,	Queen of the meadow root,	Eupatorium purpureum.
" vetch,	Ladies' fingers,	Anthyllis vulneraria.
" wort,	Navel wort,	Umbilicus pendulina.
Kill lamb,	Sheep laurel,	Kalmia angustifolia.
Kings' clover,	Melilot,	Melilotus officinalis.
" cup,	Crowfoot buttercup,	Ranunculus bulbosus.
" cure,	Wintergreen,	Chimaphilla umbellata.
" "	Spotted wintergreen,	" maculata.
" fern,	Buckhorn brake,	Osmunda regalis.
" spear,		Asphodelus luteus.
" tree,		Strychnos atherstonia.
Kinnikinnick,	Mountain cranberry,	Arctostaphylos Uva Ursi.
" bark,	the bark of	Cornus sericea.
" Indian,	a mixture of tobacco, leaves of sumach and twigs of willow.	
Kino,	the inspissated juice of	Pterocarpus marsupium and species
" African,		" erinaceus.
" American root,	Cranesbill root,	Geranium maculatum.
" Bengal,	Red astringent gum,	Butea frondosa.
" Botany Bay,	Iron bark tree, gum,	Eucalyptus resinifera.
" East Indian,	astringent juice of	Pterocarpus marsupium.
" Jamaica,		Coccoloba uvifera.
Kipper nut,		Bunium bulbocastanum.
Knap weed,		Centaurea calcitrapa.
Knawel,	Gravel chickweed,	Scleranthus annuus.
Knee holly,	Butchers' broom,	Ruscus aculeatus.
Knights' spur,	Larkspur,	Delphinium consolida.
Knitback,	Comfrey,	Symphytum officinale.
Knob grass,	Stone root,	Collinsonia Canadensis.
" root,	"	" "
Knot berry,		Rubus Chamæmorus.
" grass,		Polygonum aviculare.
" "	Dog grass,	Triticum repens.
" " German,	Gravel chickweed,	Scleranthus annuus.
" " Hastate,	Sickle grass,	Polygonum arifolium.
" root,	Stone root,	Collinsonia Canadensis.
" weed,	Heartsease,	Polygonum Persicaria.
" " Spotted,	" "	" "

5

COMMON.	ENGLISH.	BOTANICAL.
Knotty brake,	Male fern,	Aspidium Filix Mas.
" rooted figwort,		Scrophularia nodosa.
Kokum butter,	Oil cocum,	Garcinia purpurea.
Koosso,	Flowers and unripe fruit of	Brayera anthelmintica.
Kousso,	Koosso,	" "
Kus kus,	Vettivert,	Andropogon muricatus.
Kussander,	Wild potato,	Convolvulus panduratus.
Labaria plant,		Dracontium polyphyllum.
Labdanum,	a resinous substance of	Cistus ladaniferus.
Labrador tea,		Ledum latifolium.
Laburnum tree,		Cytisus laburnum.
Lac. Shellac,	a gum resin, procured from	Croton lacciferum,
"	" " " "	Ficus Indica.
"	" " " "	" religiosa.
"	" " " "	Aleurites laccifera.
"	" " " "	Erythrina monosperma.
"	" " " "	Rhamnus jujuba.
" dye,	a color obtained from stick lac.	
" Garnet,	Shellac, garnet colored.	
" Gum,	"	
" Lump,	Seed lac melted and run into cakes.	
" Orange,	Shellac, orange colored.	
" Ruby,	" ruby colored.	
" Seed,	Lac separated from the twigs.	
" Shell,	Seed lac melted and run into thin scales.	
" Stick,	consists of twigs encrusted with lac, as deposited by insects.	
" White,	Bleached seed lac.	
Lace bark,		Lagetta lintearia.
Lacmus,	Litmus,	Roccella tinctoria.
Lactucarium,	Lettuce opium,	Lactuca sativa.
" German,	" "	" virosa.
Ladanum,	a gum resin,	Cistus Creticus and other var.
Ladies' bedstraw,		Galium verum.
" bower,	Travellers' joy,	Clematis vitalba.
" fingers,	Kidney vetch,	Anthyllis vulneraria.
" glove,	Foxglove,	Digitalis purpurea.
" mantle,		Alchemilla vulgaris and other var.
" slipper,	American Valerian,	Cypripedium of several species.
" " showy,		" spectabile.
" slippers,	Garden balsam,	Impatiens balsamina.
" smock,	Cuckoo flower,	Cardamine pratensis.
" thumb,	Heartsease,	Polygonum Persicaria.
" tresses,		Spiranthes autumnalis.
Ladlewood,		Cassine colpoon.
Lads' love,	Southernwood,	Artemisia abrotanum.
Lake weed,	Water pepper,	Polygonum hydropiper.
Lambkill,	Sheep laurel,	Kalmia angustifolia.
Lambs' lettuce,	Corn salad,	Fedia radiata.
" "	Rib grass,	Plantago media.
" quarter,	Beth root,	Trillium pendulum.
" "		Chenopodium album.
" toes,	Kidney vetch,	Anthyllis vulneraria.
" tongue,	Adders' tongue,	Erythronium Americanum.
Lance wood,		Dignetia quitarvensis.
Land larch,	Wild garlic,	
Larch, American,	Tamarack,	Larix Americana.
" European,		" Europæa.
" agaric,	White agaric,	Boletus laricis.
Large flowering spurge,	Blooming spurge,	Euphorbia corollata.
" periwinkle,		Vinca major.

COMMON.	ENGLISH.	BOTANICAL.
Large spotted spurge,		Euphorbia hypericifolia.
" valerian,	Spikenard of Crete,	Valeriana phu.
Lark heel,	Larkspur,	Delphinium consolida.
Larks' claw,	"	" "
Larkspur,		" "
" Rocket,		" ajacis.
Laurel, ·		Magnolia.
" Alexandrian,	Poonwood tree,	Caulophyllum inophyllum.
" American,	Sheep laurel,	Kalmia of several species.
" Bay,		Laurus nobilis.
" berries,	Bay laurel berries,	" "
" Broad leaved,		Kalmia latifolia.
" leaves,	Bay laurel leaves,	Laurus nobilis.
" Cherry,		Prunus Lauro-cerasus.
" Mountain,	American laurel,	Kalmia latifolia.
" Narrow leaved,		" angustifolia.
" Sheep,		" "
" Swamp,		" glauca.
" Sweet,	Florida star anise,	Illicium Floridanum.
Lavender, Arabian,		Lavendula Stœchas.
" Common,		" vera.
" cotton,	Cypress herb,	Santolina chamœcyparissus.
" Garden,		Lavendula vera.
" Spike,		" spica.
" thrift,	Marsh rosemary,	Statice Caroliniana.
Laver,	Gulf weed,	Sargassum bacciferum.
Lavose,	Lovage,	Ligusticum levisticum.
Lead plant,		Amorpha canescens.
" wort,		Plumbago Europæa.
Leaf cup,		Polymnia nvadalia.
Leather bush,	Leatherwood,	Dirca palustris.
" flower,		Clematis viorna.
" leaf,		Andromeda calyculata.
" wood,	Moosewood,	Dirca palustris.
Leaver wood,	Leatherwood,	" "
Leek,		Allium porrum.
" House,	see House leek,	Sempervivum tectorum.
Lemon,	the fruit of	Citrus Limonnm.
" balm,		Melissa officinalis.
" grass,		Andropogon citratum.
" thyme,		Thymus citriodorus.
" tree,		Citrus Limonum.
" walnut,	Butternut,	Juglans cinerea.
" verbena,		Verbena triphylla.
Lentil,	the seed of	Ervum lens.
Lentisk,	Mastich tree,	Pistacia Lentiscus.
Leopardsbane,	Arnica,	Arnica montana.
Leopard wood,		Brosimum Aubletii.
Lettuce, Common,		Lactuca sativa.
" Garden,		" "
" liverwort,	Round leaved pyrola,	Pyrola rotundifolia.
" opium,	Lactucarium,	Lactuca sativa and Lactusa virosa.
" strong scented,		" virosa.
" Wild,		" elongata.
" "	Round leaved pyrola,	Pyrola rotundifolia.
" White,	Lions' foot,	Nabalus albus.
Levant hellebore,		Helleborus orientalis.
" nut,	Cocculus Iudičus,	Anamirta cocculus.
" wormseed,		Artemisia contra. (Siebieri.)
Lever wood,	Ironwood,	Ostrya Virginica.

COMMON.	ENGLISH.	BOTANICAL.
Levose,	Lovage,	Ligusticum levisticum.
Libi dibi,	Divi divi,	Cæsalpina coriaria.
Licorice bush,	Love pea,	Abrus precatorius.
" root,	Liquorice,	Glycyrrhiza glabra and species.
Life everlasting,		Gnaphalium.
" " Pearl flow'd,		" margaritaceum.
" of man,	American Spikenard,	Aralia racemosa.
" root plant,		Senecio aureus.
Light Jalap,	Male Jalap,	Convolvulus (Ipomœa) Orizabensis
Lignum vitæ,		Guaiacum officinale.
Lilac,		Lilaca vulgaris.
" African,	Pride of India,	Melia azedarach.
" White,		Syringia vulgaris var. alba.
Lily, Common white,	Meadow lily,	Lilium candidum.
" Meadow,		" "
" of the valley,		Convallaria majalis.
" St. James',		Amaryllis formosissima.
" Tiger,		Lilium tigrinum.
" White,		" candidum.
" " pond,		Nymphæa odorata.
" " of the valley,		Lilium candidum.
" Yellow pond,		Nuphar advena.
" tree,	Chinese magnolia,	Magnolia yulans.
Lima bark,		Cinchona nitida.
" wood,	Pernambuco wood,	Cæsalpina echinata.
Lime, the fruit,		Citrus Limetta.
" grass,		Elymus villosus.
" tree,	Basswood,	Tilia glabra and Tilia Americana.
" "	Ogeechee tree,	Nyssa capitata.
" " European,		Tilia Europœa.
Limewort catchfly,		Silene Armeria.
Linden,	common name for	Tilia of several species.
" American,	Basswood,	Tilia Americana.
" European,	Lime tree,	" Europœa.
" flowers,		" "
Linseed,	Flaxseed,	Linum usitatissimum.
Ling,	Heather,	Calluna vulgaris.
Lions' ear,	Motherwort,	Leonurus cardiaca.
" foot,		Nabalus albus.
" "	Gall of the earth,	Prenanthes Fraseri.
" snap,	Snap dragon,	Antirrhinum majus.
" tail,	Motherwort,	Leonurus cardiaca.
" tooth,	Dandelion herb,	Taraxacum Dens Leonis.
Lip fern,		
Liquid amber,	Oriental sweet gum,	Liquidambar orientale.
" copal,	Piney varnish,	Vateria Indica.
" storax,	a balsam prepared from	Liquidambar orientale.
" styrax,	Storax,	" "
Liquorice,		Glycyrrhiza glabra.
" American,		" lepidota.
" Brazilian,		Periandra dulcis.
" buds,	Love pea,	Abrus precatorius.
" root,		Glycyrrhiza glabra and species.
" " Calabria,	Wild Sicily root,	" echinata.
" Russian,		" glabra var. glandulifera
" Spanish,		" typica.
" Wild,	American sarsaparilla,	Aralia nudicaulis.
Lithy tree,		Viburnum lantana.
Litmus,	Orchilla weed,	Roccella tinctoria.
Little clotbur,		Xanthium strumarium.

COMMON.	ENGLISH.	BOTANICAL.
Little buckeye,		Æsculus pavia.
" good,	Churn staff,	Euphorbia helioscopia.
" pollom,	Dwarf milkwort,	Polygala paucifolia.
" snowball,	Buttonwood shrub,	Cephalanthus occidentalis.
" water lily,	Water shield,	Brassenia hydropeltis.
Live forever,	Orpine,	Sedum telphinum.
" "	Life everlasting,	Guaphalum polycephalum and var.
" long,	Orpine,	Sedum telphiuum.
Live oak,		Quercus virens.
Liver leaf,	Liverwort,	Hepatica Americana.
" " Heart,		" acutiloba.
" " Kidney,		" Americana.
" lily,	Blue flag,	Iris versicolor.
" moss,	Liverwort,	Hepatica Americana.
" mushroom,	on oak trees,	Boletus hepatica.
" weed,		Hepatica Americana.
" wort,		Marchantia polymorpha.
" "		Hepatica Americana, and acutiloba,
" " Noble,		" "
Lizards' tail,	Breast weed,·	Saururus cernuus.
" tongue,		Sauroglossum.
Lobe berry,	Sea side grape,	Coccoloba uvifera.
Lobelia,		Lobelia inflata.
" Acrid,		" urens.
" Blue,	Blue cardinal flower,	" syhilitica.
" Brown,	name given to the pul. seed,	" inflata.
" Garden,		" cardinalis.
" Green,	the pulverized herb,	" inflata.
" Red,	Red cardinal flower,	" cardinalis.
" Spiked,		" spicata.
Loblolly,	Old field pine,	Pinus tædæa.
Locust bean,	Saint John's bread,	Ceratonia siliqua.
" berry,		Malpighia coriacea.
" bloom,	Tilia flowers,	Tilia Europæa.
" plant,	American senna,	Cassia Marilandica.
" tree,	Black locust tree,	Robina pseudo-acacia.
" " So. Am'n,	produces anime,	Hymenæa courbaril.
Lofty quassia,	Quassia,	Picræna excelsa.
Loggerheads,		Centaurea nigra.
Logwood, Honduras,		Hæmatoxylon Campechianum.
" Campeachy,		" "
" Jamaica,		" "
" St. Domingo,		" "
London pride,		Saxifraga umbrosa.
Long birthwort,		Aristolochia longa.
Long moss,		Tillandsia usneoides.
" pepper,		Piper longum.
Loose strife,	Purple willow herb,	Lythrum salicaria.
" " Yellow,	Yellow " "	Lysimachia vulgaris.
Lopez root,		Morus Indica.
" "		Toddalia aculeata.
Lop seed,		Phryma leptostachya.
Lords and ladies,	Dragon root,	Arum triphyllum.
Lordwood,		Liquidambar orientale.
Lotos,		Zizyphus Lotus.
Lotus,		Nelumbium luteum.
" tree,	Persimmon,	Diospyros Virginiana.
Louse berry tree,		Euonymus Europæus.
" wort,	Bushy gerardia,	Gerardia pedicularia.
" " foxglove,		Pedicularis Canadensis.

COMMON.	ENGLISH.	BOTANICAL.
Louisiana squill,		Crinum Americanum.
Lovage, American,		Ligusticum actæifolium.
" European,		" levisticum.
Love apple,	Tomato,	Solanum lycopersicum.
" flower,	African lily,	Agapanthus.
" grove,		Nemophila insignia.
" in a mist,	Fennel flower,	Nigella Damascena.
" in a puff,	Balloon vine,	Cardiospermum halicacabum.
" lies bleeding,		Amarantus melancholicus.
" pea,		Abrus precatorius.
" vine,	Dodder,	Cuscuta Americana.
Low balm,	Oswego tea,	Monarda didyma.
" blackberry,		Rubus trivialis.
" blueberry,		Vaccinium Pennsylvanicum.
" centaury,		Hypericum parviflorum.
" chamomile,		Anthemis nobilis.
" mallow,		Malva rotundifolia.
Loxa bark,		Cinchona condaminea var. vera.
Lucerne,		Medicago sativa. -
Lung flower,	Marsh gentian,	Gentiana pneumonanthe.
" moss,		Lichen pulmonarius.
Lungs of the oak,	Maple lungwort,	Sticta pulmonaria.
Lungwort,		Pulmonaria officinalis.
" American,		Variolaria faginea.
" Bullocks,	Mullein,	Verbascum Thapsus.
" Cows,	"	" "
" Maple,		Sticta pulmonaria.
" Virginia,	Virginia cowslip,	Pulmonaria Virginica and var.
Lupine,		Lupinus perennis and var.
Lupuline,	a yellow powder on the strobiles of Humulus Lupulus.	
Lustwort,	Sundew,	Drosera rotundifolia and species.
Luteolin,		Reseda luteola.
Lycoperdon nuts,		Elaphomyces variegatus.
Lycopodium,	the sporules of	Lycopodium clavatum and species.
Lyre tree,	Tulip tree,	Liriodendron Tulipifera.
Macanet grains,		Cerasus mahaleb.
Mace,	the aril of the nutmeg,	Myristica fragrans.
Mad apple,	Thorn apple,	Datura Stramonium.
" "	Dead sea apple,	Quercus infectoria.
Madar bark,	.	Asclepias gigantea.
Madeira nut,		Juglans regia.
" vine,		Boussingaultia baselloides. ،
" wood,	Mahogany,	Swietenia mahogani.
Madder,	the root of,	Rubia tinctoria.
" Bengal,		" cordifolia.
" Chili,		" relbun.
" Dyers',	Madder,	" tinctoria.
" Indian,	Chayroot,	Oldenlandia umbellata.
" Wild,		Galium mollugo.
Mad-dog-weed,	Water plantain,	Alisma plantago.
" " "	Scullcap,	Scutellaria lateriflora.
Madnep,	Cow parsnep,	Heracleum lanatum.
Madweed,	Scullcap,	Scutellaria lateriflora.
Madwort,	Wild flax,	Camelina sativa.
"	Sweet alyssum,	Alyssum maritimum.
" German,		Asperugo procumbens.
Magnolia,	name applied to the cucumber, and umbrella trees.	
" Big leaved,		Magnolia macrophylla.
" Chinese,		" yulan.
Maguey,	Flowering aloe,	Agave Americana.

COMMON.	ENGLISH.	BOTANICAL.
Mahernia,		Mahernia verticillata.
Mahogany,		Swietenia mahogani.
" African,	Juribali,	" Senegalensis.
" birch,	Black birch,	Betula lenta.
" Mountain,	" "	" "
" Spanish,	Mahogany,	Swietenia mahogani.
" tree,	"	" "
Mahonia,		Berberis aquifolium.
Maiden hair,		Adiantum capillus veneris.
" " American,		" pedatum.
" " tree,	Ginkgo tree,	Salisburia adiantifolia.
Maids' hair,	Yellow cleavers,	Galium verum.
Maize,	Indian corn,	Zea Mays.
Malabar leaf,		Cinnamomum malabathrum.
" nuts,		Justicia adhatoda.
Malacca bean,	Marking nut,	Semecarpus Anacardium.
Malambo bark,	Matias bark,	Croton Malambo.
Malay apple,		Jambosa Mallaccensis.
" "	Rose apple,	Eugenia Jambos.
Male agaric,	Larch agaric,.	Boletus laricis.
" fern,		Aspidium Filix Mas.
" "	Buckhorn brake,	Osmunda regalis.
" lavender,	Spike lavender,	Lavendula spica.
" jalap,	Orizaba jalap,	Convolvulus Orizabensis.
" nervine,		Cypripedium pubescens.
" nutmeg,	Wild nutmeg,	Myristica moschata.
" shield fern,		Aspidium Filix Mas.
Maleguetta pepper,	Grains paradise,	Amomum Grana Paradisi and spec.
Mallaguetta "	" "	" " " " "
Mallow, Common,		Malva sylvestris.
" High,	Common mallows,	" "
" Indian,		Abutilon Indicum.
" Low,		Malva rotundifolia.
" Marsh,		Althæa officinalis.
" wort,		Callirrhoe digitata.
" Yellow,		Abutilon condalum.
Malt,	the baked, germinated seeds	of Hordeum vulgare and other spec.
Maltese cross,		Lychnis Chalcedonia.
" sponge,	Scarlet mushroom,	Cyromorum coccineum.
Malva,	Common mallows,	Malva sylvestris and other species.
Mammea apple,	So. American apricot,	Mammea Americana.
Manchineel,		Hippomane mancinella.
Mandarin orange,		Citrus Bigaradia and var.
Mandioca,	Cassava,	Jatropha manihot.
Mandragora root,	European mandrake,	Atropa mandragora.
Mandrake, American,	Mayapple,	Podophyllum peltatum.
" European,		Atropa mandragora.
Mangle,		Avicennia tomentosa.
" wurzel,		Beta vulgaris macrorhiza.
Mango,		Mangifera Indica.
" ginger,		Curcuma Amada.
Mangosteen,	Cocum butter " oil,"	Garcinia purpurea.
Mangrove,		Rhizophora Mangle.
Man-in-the-ground,	Wild potato,	Convolvulus panduratus.
Manna,	the concrete juice in flakes	of Fraxinus ornus and other species
" Arabian,	the product of	Tamarix Gallica.
" Australian,	" " "	Eucalyptus mannifera.
" croup,	the prepared seed of	Glyceria fluitans.
" False,	see False manna,	
" India, Hebrew,		Alhagi maurorum.

COMMON.	ENGLISH.	BOTANICAL.
Manna of Briancon,	produced by the Larch during combustion.	
" Persian,	Indian manna,	Alhagi maurorum.
" seed,	Manna grass seed,	Glyceria fluitans.
" tree,	Flowering ash,	Fraxinus ornus and other species.
Man-of-the-earth,	Wild potato,	Convolvulus panduratus.
Man root,	" "	" "
Many root,		Ruella tuberosa.
Manzanita,		Aretostaphylos glauca.
Maple, Birds' eye,		Acer saccharinum.
" Black,		" nigrum.
" flower,		Bellis perennis.
" Ground,	American sanicle,	Heuchera accrifolia.
" guelder rose,	Dockmackie,	Viburnum accrifolium.
" leaf alum root,	American sanicle,	Heuchera accrifolia.
" lungwort,		Sticta pulmonaria.
" Red,		Acer rubrum.
" Rock,	Sugar maple,	Acer saccharinum.
" Striped,	Moosewood,	" striatum.
" sugar tree,		" saccharinum.
" Swamp,	Red maple,	" rubrum.
March,	old name for Parsley,	Apium Petroselinum.
Marcory,	Queens' root,	Stillingia sylvatica.
Maracaibo bark,	Bogota bark,	Cinchona Condamiuca.
Marestail,		Hippuris vulgaris.
"	Scouring rush,	Equisetum hyemale.
Margosa tree,		Melia Indica.
Marigold,	common name for	Calendula of several species.
" African,		Tagetes erecta.
" French,		" patula.
" Garden,		Calendula officinalis.
" Marsh,	American cowslip,	Caltha palustris.
Marjoram,	Sweet marjoram,	Origanum marjorana.
" Common,	Wild "	" vulgare.
" Hop,	Dittany of Crete,	" Dictamnus.
" Knotted,	Sweet marjoram,	" marjorana.
" Sweet,		" "
" Spanish,		" Creticum.
" Wild,		" vulgare.
" Winter, Sweet,		" heracleoticum.
Marking nut,	Cashew nut,	Semecarpus anacardium.
Marmalade tree,		Lucuma mammosum.
Marsh beetle,	Reed mace,	Typha latifolia.
" cleavers,		Galium palustre.
" clover,	Buckbean,	Menyanthes trifoliata.
" dandelion,		Leontodon palustris.
" ilvefinger,		Comarum palustris.
" fleabane,	Plowmans' wort,	Conyza Marilandica.
" gentian,		Gentiana pneumonanthe.
" "		" ochroleuca.
" hibiscus,		Hibiscus palustris.
" mallow,		Althæa officinalis.
" marigold,	American cowslip,	Caltha palustris.
" parsley,		Selinum palustre.
" pestle,	Reed mace,	Typha latifolia.
" root,	Marsh rosemary,	Statice Caroliniana.
" rosemary,		" "
" " Europ'n,		" Limonium.
" smallage,	Marsh parsley,	Selinum palustre.
" tea,		Ledum palustre.
" trefoil,	Buckbean,	Menyanthes trifoliata.

COMMON.	ENGLISH.	BOTANICAL.
Marsh turnip,	Dragon root,	Arum triphyllum.
" violet,		Viola palustris.
" water cress,		Nasturtium palustre.
" wort,	variety of cranberry,	Oxycoccos palustre.
Martagon lily,	Turks' cap lily,	Lilium martagon.
Martynia,		Martynia proboscidea.
Marvel of Peru,	False jalap,	Mirabilis Jalapa and other species.
Mary bud,	Marigold,	Calendula officinalis.
" gold,	see Marigold,	
" thistle,		Carduus marianus.
Massoy bark,		Cinnamomum Kiamis.
Masterwort,	Cow parsnip,	Heraclcum lanatum.
"	Purple angelica,	Angelica atropurpurea.
"		Imperatoria ostruthium.
" Black,		Astrantia major.
" Imperial,	Black masterwort,	" ' "
Mastich,	the gum resin of	Pistacia Lentiscus.
Mata,		Eupatorium incarnatum.
Mate,	Paraguay tea,	Ilex Paraguaiensis.
Matias bark,	·	Croton Malambo.
Matico,		Artanthe elongata.
Matrimony vine,		Lycium barbarum.
Maudlin,	Maudlin tansy,	Achillea Ageratum.
" wort,	Ox eye daisy,	Leucanthemum vulgare.
Maurandia,		Maurandia Barclayana.
Maw-seed,	Poppy seed, blue,	Papaver somniferum.
May-apple,	American mandrake,	Podophyllum peltatum.
May duke,	a kind of cherry,	
" flower,	Trailing arbutus,	Epigæa repens.
" lily,	Lily of the valley,	Convallaria majalis.
" pops,		Passiflora incarnata.
" queen moss,	Hair-cap moss,	Polytrichum Juniperum.
" weed,		Anthemis Cotula.
" wort,	Mayweed,	" "
" wreath,		Spiræa hypericifolia.
Mazzards,	Sweet cherries,	Cerasus avium.
Meadow anemone,	Pulsatilla,	Anemone pulsatilla.
" beauty,	Deer grass,	Rhexia Virginica.
" bouts,	Marsh marigold,	Caltha palustris.
" bloom,	Buttercup,	Ranunculus bulbosus.
" cabbage,	Skunk cabbage,	Symplocarpus fœtidus.
" cress,		Arabis rhamboides.
" fern,		Myrica gale.
" " burrs,		" "
" garlic,		Allium Canadense.
" lily,	Common white lily,	Lilium candidum.
" parsnip,	Alexanders,	Zizia aurea.
" pink,		'Azalea nitida.
" pride,	American Columbo,	Frasera Walteri and species.
" queen,	Spiræa,	Spiræa ulmaria.
" Queen of the,	Gravel root,	Eupatorium purpureum.
" root,	Marsh rosemary,	Statice Caroliniana.
" rue,		Thalictrum dioicum.
" saffron,	Colchicum,	Colchicum autumnale.
" sage,	Wild sage,	Salvia lyrata.
" scabish,	Cocash,	Aster puniceus.
" silkweed,		Asclepias fibrosa.
" sorrel,	Common sorrel,	Rumex acetosa.
" star,	Stichwort,	Stellaria palustris.
" sweet,		Spiræa ulmaria.

COMMON.	ENGLISH.	BOTANICAL.
Meadow sweet,	Indian physic,	Gillenia trifoliata.
" "	Hardhack,	Spiræa tomentosa.
" wort,	Meadow sweet,	" ulmaria.
Mealbark tree.		Cycas caffræa.
Mealberry	Bearberry,	Arctostaphylos Uva Ursi.
Mealy bush,		Andromeda pulverulenta.
" starwort,	Unicorn root,	Aletris farinosa.
" tree,	Arrow wood,	Viburnum dentatum.
Mechameck,	Wild potato,	Convolvulus panduratus.
Mechoacan,	" "	" "
Median lemon,	Citron,	Citrus medicago.
Medick,	Hop medick,	Medicago lupulina.
Medlar,		Aronia ovalis.
" English,	Dutch medlar,	Mespilus Germanica.
Medusa's trumpet,		Narcissus bulbocodium.
Melegueta pepper,	Grains paradise,	Amomum Grana Paradisi and var.
Melilot,	.	Melilotus officinalis.
" clover,		" "
" trefoil,	Dutch clover,	Medicago lupulina.
" Sweet,	Sweet "	Melilotus leucantha.
Melon,		Cucumis Melo.
" cactus,	,	Melocactus communis.
" Cantaloupe,	a variety of	Cucumis Melo.
" Musk,		" "
" Water,		" citrullus.
Mendo,	the wild sweet potato,	
Mercury herb,		Mercurialis annua.
" " English,		Chenopodium Bonus Henricus.
" weed,		Acalypha Virginianica.
Mesenna bark,		Albizzia anthelmintica.
Mexican poppy,		Argemone Mexicana.
" sage,	Cardinal sage,	Salvia fulgens.
" tea,		Chenopodium ambrosioides.
Mexico seed,	Castor bean,	Ricinus communis.
Mezereon,		Daphne Mezereum.
" American,	Leatherwood,	Dirca palustris.
" Sweet scented,		Daphne odora.
Mezquit grass,		Bouteloua hirsuta and species.
Michaelmas daisy,		Aster Tradescantia.
Mignonette,		Reseda odorata and var.
" Wild,	Dyers' weed,	" luteola.
Milfoil,	Yarrow,	Achillea Millefolium.
Milk ipecac,	Blooming spurge,	Euphorbia corollata.
" parsley,		Peucedanum sylvestre.
" pursley,	Blooming spurge,	Euphorbia corollata.
" sweet,	Cleavers,	Galium aparine.
" tree,	Cow tree,	Clusia galactodendron.
" "	'	Tanghinia lactaria.
" thistle,	Common thistle,	Cnicus lanceolatus.
" vetch,		Astragalus glaux.
" weed, Common,	Silkweed,	Asclepias Cornuti.
" "	Blooming spurge,	Euphorbia corollata.
" Wandering,	Bitter root,	Apocynum androsæmifolium.
" wort, bitter,	" polygala,	Polygala amara.
" " common,		" incarnata.
" willow herb,	Purple willow herb,	Lythrum salicaria.
Millet,		
" Common,	Millet grass,	Panicum miliaceum.
" Hungarian,	Hungarian grass,	" Germanicum.
" Indian,	Common millet,	" miliaceum.

COMMON.	ENGLISH.	BOTANICAL.
Millet, seed,	a variety of grain,	Sorghum miliaceum.
" Turkish,	Guinea corn,	Sorghum vulgare var. cernuum.
" White,	White melilot,	Melilotus alba.
" Yellow,		" vulgaris.
Mill mountain,	Purging flax,	Linum catharticum.
Milt waste,	Ceterach,	Ceterach officinarium.
Mimosa bark extract,		Acacia melanoxylon.
Mint,	Spearmint,	Mentha viridis.
Mishmee bitter,	Chinese goldthread plant,	Coptis tecta.
Mist flower,		Conoclinium cœlestinum.
Mistletoe,		Viscum flavescens and other spec.
" of the oak,		Loranthus Europæus.
Mistress of the night,	Tuberose,	Polianthes tuberosa.
Mitre wort,	Coolwort,	Tiarella (Mitella) cordifolia.
" "	Gem fruit,	" " "
Moccasin flower,	American valerian,	Cypripedium, of several species.
" plant,	" "	" " "
" root,	" "	" " "
Mock orange,	Syringa,	Philadelphus coronarius.
" "	Carolina cherry laurel,	Prunus Caroliniana.
Mohawk weed, .	Bellwort, •	Uvularia perfoliata.
Moldavian balm,		Dracocephalum Moldavica.
Molucca balm,		Molucella lævis.
" grains,	Grains Tigli,	Croton Tiglium.
Monarda,	Horsemint,	Monarda punctata.
Monesia,	an extract derived from	Chrysophyllum glycyphlæum.
Money flower,	Honesty,	Lunaria biennis.
" wort,	Herb two pence,	Lysimachia nummularia.
Monkey bread tree,	Baobab tree, ,	Adansonia digitata.
" face tree,	yields kameela,	Rottlera tinctoria.
" flower,		Mimulus ringens.
" "	American valerian,	Cypripedium of several species.
" pepper,	Ethiopian pepper,	Unona Æthiopica.
Monkshood,	Aconite,	Aconitum Napellus and other spec.
" American,	"	" uncinatum.
Monks' rhubarb,		Rumex alpinus.
Moon flower,	Buckbean,	Menyanthes trifoliata.
" fruit pine,		Lycopodium lucidulum.
" seed,	Yellow parilla,	Menispermum Canadense.
" " root,	" "	" "
" trefoil,		Medicago arborea.
" wort,	Rattlesnake fern,	Botrychium lunaroides and species
" "		Rumex luparia.
Moorberry,	variety of cranberry,	Oxycoccos palustris.
Moosewood,	Leatherwood,	Dirca palustris.
"	Striped maple,	Acer striatum.
Morning glory,		Ipomœa (Convolvulus) purpureus.
" " Common,		Convolvulus purpureus.
" " Dwarf,		" spithameus.
Mortification root,	Marsh mallow root,	Althæa officinalis.
Moschatel,	Musk crowfoot,	Adoxa moschatellina.
Moss berry,	Cranberry,	Oxycoccos macrocarpus.
" bush,		Andromeda hypnoides.
" campion,		Silene acaulis.
" crop,		Eriophorum Virginicum.
" locust,	Rose acacia,	Robinia hispida.
" pink,		Phlox subulata.
" rose,		Rosa muscosa.
Mother cinnamon,		Cinnamomum albiflorum.
" clove,	the fruit of the clove tree,	Caryophyllus aromaticus.

COMMON.	ENGLISH.	BOTANICAL.
Mother of rye,	Ergot of rye,	Sclerotium clavus.
" of thousands,	Ivy leaved toadflax,	Linaria cymbalaria.
" of thyme,	Wild thyme,	Thymus serpyllus.
" of wheat,	Cow wheat,	Melampyrum arvense.
Motherwort,		Leonurus cardiaca.
Moth mullein,		Verbascum blattaria.
Mould,		Ascophora mucedo.
Mountain arnica,	Common arnica,	Arnica montana.
" ash, Am'n,		Sorbus Americana.
" " Europ'n,		Sorbus (Pyrus) acuparia.
" balm,	Rose balm,	Monarda coccinea.
" box,	Bearberry,	Arctostaphylos Uva Ursi.
" brake,		Pteris atropurpurea.
" bugle,		Ajuga pyrimidalis.
" calamint,		Melissa grandiflora.
" cranberry,	Bearberry,	Arctostaphylos Uva Ursi.
" damson,	Simaruba,	Simaruba (Quassia) officinalis.
" dittany,		Cunila mariana.
" elder,		Sambucus racemosa.
" flax,	Senega root,	Polygala Senega.
" fringe,	Climbing fumitory,	Adiumia cirrhosa.
" globe flower,	Button bush,	Cephalanthus occidentalis.
" holly,		Ilex aquifolium.
" laurel,	American laurel,	Kalmia of several species.
" "	Wild rose bay,	Rhododendron maximum.
" liquorice,		Trifolium alpinum.
" mahogany,	Black birch,	Betula lenta.
" mint,		Monarda didyma.
" "		Pycnanthemum montanum.
" parsley,	Speedwell root,	Athamanta oreoselinum.
" pink,	Mayflower,	Epigæa repens.
" rhubarb,	Monks' rhubarb,	Rumex alpinus.
" sorrel,	Boreal sour dock,	Oxyria reniformis.
" "	Wood sorrel,	Oxalis acetosella.
" sweet,	Red root,	Ceanothus Americana.
" sumach,		Rhus copallinum.
" tea,	Checkerberry,	Gaultheria procumbens.
" thyme,		Acinos vulgare.
" "		Calamintha alpina.
" tobacco,	Arnica,	Arnica montana.
Mourning bride,	Sweet scabish,	Scabiosa atropurpurea.
" widow,	" "	" "
Mouse bane,	Aconite,	Aconitum.
" blood wort,		Hieracium pilosella.
" ear,		Gnaphalium plantaginifolia.
" "	Mouse blood wort,	Hieracium pilosella.
" " chickweed,		Cerastium vulgatum.
" " cress,		Arabis thaliana.
" tail grass,		Festuca myurus.
" thorn,		Centaurea myacantha.
Mouth root,	Goldthread root,	Coptis trifolia.
" smart,	Brooklime,	Veronica beccabunga.
Moving plant,		Desmodium gyrans.
" plants,	Brittle worts,	Diatomaceæ.
Moxa,		Artemisia moxa and other species.
" Chinese,		" Chinensis.
" English,	prepared pith of sunflower,	Helianthus annuus.
Mudar bark,		Asclepias gigantea.
Mud plantain		Heteranthera reniformis.
Mudwort,		Limosella subulata.

COMMON.	ENGLISH.	BOTANICAL.
Mugweed, Golden,		Galium cruciatum.
Mugwort,		Artemisia vulgaris.
" Indian,		" hirsuta.
" West Indian,	Bastard feverfew,	Parthenium hysterophorus.
Mulberry, Black,		Morus nigra.
" Indian,		Morinda citrifolia.
" Paper,		Broussonetia papyrifolia.
" Red,		Morus rubra.
" White,		" alba.
Mullein,		Verbascum Thapsus.
" Moth,		" blattaria.
" pink,		Lychnis coronaria.
Muruxi bark,	Alcornoque bark,	Byrsonima crassifolia.
Muscovado sugar,	Raw sugar,	
Musenna,		Rottlera Schimperi.
Mushroom,		
" Eatable,		Amanita muscaria and other spec.
" Poison,		" aurantiaca.
Musk crowfoot,	Moschatel,	Adoxa moschatellina.
" leaves,		Abelmoschus esculenta.
" mallow,	Musk seed plant,	Hibiscus abelmoschus.
" melon,		Cucumis melo.
" plant,		Mimulus moschata.
" root,	Sumbul plant,	Euryangium (Ferula) Sumbul.
" "	Indian nard,	Nardostachys Jatamansi.
" seed,		Abelmoschus moschatus.
" thistle,	Cotton thistle,	Onopordon acauthium.
" "	Sow thistle,	Sonchus oleraceus.
" tree,		Eurybia argophylla.
" Vegetable,	Musk plant,	Mimulus moschata.
" wood,		Moschoxylum Swartzii.
Musquash root,	Water hemlock,	Cicuta maculata.
Mustard, Bastard,		Polanisia icosandra.
" Clowns',	Bitter candytuft,	Iberis amara.
" False clammy,	Clammy weed,	Polanisia viscosa.
" Hedge,	English water cress,	Sisymbrium officinale.
" "		Erysimum Nasturtium.
" " Fine leav'd,		
" Mithridate,	Pennycress,	Thlaspi arvense.
" Red,	Black mustard seed,	Sinapis nigra.
" Scripture,	Mustard tree,	Salvadora Persica.
" seed, Black,		Sinapis nigra.
" " White,		" alba.
" " Yellow,	White mustard,	" "
" Tower,		Turritis glabra.
" Treacle,		Erysimum cheiranthoides.
" tree,		Salvadora Persica.
" Wild,		Sinapis arvensis.
Myosotis,	Forget-me-not,	Myosotis arvensis.
Myriad leaf,	Water milfoil,	Myriophyllum verticillatum.
Myrobalans,	White galls,	Emblica officinale.
Myrrh,	a gum resin of	Balsamodendron Myrrha.
" seed,		Myrospermum pubescens.
Myrtle,		Myrtus communis.
" Bog,	Buckbean,	Menyanthes trifoliata.
" Candleberry,	Bayberry,	Myrica cerifera.
" Dutch,	Meadow fern,	" gale.
" "	a broad leaved variety of	Myrtus communis.
" flag,	Sweet flag,	Acorus Calamus.
" Jews,	a three leaved variety of	Myrtus communis.

COMMON.	ENGLISH.	BOTANICAL.
Myrtle leaf,	Cape smilax,	Myrsiphyllum Asparagoides.
" Sweet,	Sweet flag,	Acorus Calamus.
" Wax,	Bayberry,	Myrica cerifera.
" West Indian,		Eugenia Luma.
Nailwort,	Whitlow grass,	Draba incana and species.
Naked ladies,	Colchicum,	Colchicum autumnale.
Nannyberry,		Viburnum lentago.
Nannybush,		" "
Narcissus,		Narcissus of several species.
" False,		Allium senescens.
" Japan,		Nerine sarniensis.
" Poets',		Narcissus poeticus.
Nard,	East Indian Spikenard,	Nardostachys Jatamansi.
" Common,	Mat grass,	Nardus stricta.
" Indian,		Nardostachys Jatamansi.
Narrow dock,	Yellow dock,	Rumex crispus.
" leaved laurel,		Kalmia angustifolia.
" " Va. thyme,		Pycnanthemum Virginicum.
Napoleon plant,		Nelumbium codophyllum.
Naseberry,	Sapodil,	Achras Sapota.
Nasturtion,	see Nasturtium,	
Nasturtium,	Common water cress,	Nasturtium officinale.
"	Indian cress,	Tropæolum majus.
Navelwort,		Cotyledon umbilicus.
"	Kidneywort,	Umbilicus pendulinus.
Neb-neb,	the pods,	Acacia vera.
Necklace poplar,	Cottonwood tree,	Populus monilifera.
" seed,		Elæocarpus granitrus.
" tree,	West Indian bead tree,	Ormosia dasycarpa.
" weed,	Baneberry,	Actæa alba.
Neckweed,	Brooklime,	Veronica beccabunga.
"	Indian hemp,	Cannabis sativa.
Nectarine,	a smooth skinned variety of	Amygdalus Persica (glabra.)
Neem bark,	Nim bark,	Melia Azadirachta.(Indica.)
Negro corn,	W. I. name for Guinea corn,	Sorghum vulgare, var.
" head,	Vegetable ivory,	Phytelephas macrocarpa.
" vine,		Gonolobus hirsutus.
Nepal aconite,	Indian aconite,	Aconitum ferox.
Nephritic plant,	Cutting almond,	Parthenium integrifolium.
Neroli,	the oil of orange blossoms,	Citrus aurantium.
Nerve root,	American valerian,	Cypripedium flavum.
Nervine,	" "	" "
Nest root,	Ice plant,	Monotropa uniflora.
Net leaf plantain,		Goodyera pubescens.
Nettle leaf vervain,	White vervain,	Verbena urticifolia.
Nettle,		Urtica dioica and other species.
" Bees,		Galeopsis versicolor.
" Common,		Urtica dioica.
" Dwarf,		" urens.
" Great stinging,	Common nettle,	" dioica.
" Stingless,	Cool weed,	" pumila.
" tree,	Sugarberry tree,	Celtis occidentalis.
" White,	Blind nettle,	Lamium album.
Networt,	Net leaf plantain,	Goodyera pubescens.
New Jersey tea,	Red root,	Ceanothus Americana.
" Zealand flax,		Phormium tenox.
" " spinach,		Tetragonia expansa.
Nicaragua wood,	Peach wood,	Cæsalpina christa.
" "	an inferior Brazil wood,	" echinata.
Nic wood,	Nicaragua wood,	" christa.

COMMON.	ENGLISH.	BOTANICAL.
Nickar tree,	American coffee bean,	Gymnocladus Canadensis.
Nicker tree, Yellow,	Bonduc,	Guilandina bonduc.
Niepa bark,		Samadera Indica.
Nigella, Bastard,	Corn cockle,	Agrostemma (Lychnis) githago.
" seed,	Black cummin,	Nigella sativa.
Niger seed,		Guizotia oleifera.
Night blooming cereus,		Cactus (Cereus) grandiflorus.
" flower,	Night jasmine,	Nyctanthes arbortristis.
" henbane,		Ilyosciamus scopolia.
" jasmine,		Nyctanthes arbortristis.
Nightshade,	Solanum,	
" American,	Garget,	Phytolacca decandra.
" Black,	Garden nightshade,	Solanum nigrum.
" Common,	" "	" "
" Deadly,	Belladonna,	Atropa Belladonna.
" "	Garden nightshade,	Solanum nigrum.
" Enchanters',		Circæa luteana.
" Fetid,	Henbane,	Ilyosciamus niger.
" Garden,	Black nightshade,	Solanum nigrum.
" Three leav'd,	Trillium,	Trillium of several species.
" vine,	Bittersweet,	Solanum Dulcamara.
" Woody,	"	" "
Nim bark,	Margosa tree bark,	Melia Indica.
Nine bark,		Spiræa opulifera.
Ninety knot,	Knot grass,	Polygonum aviculare.
Ninsin root,	Ginseng root,	Panax quinquefolium.
Nipplewort,		Lapsana communis.
Nitweed,	False Johnswort,	Sarothra Gentianoides.
Noah's ark,	American valerian,	Cypripedium pubescens.
Noble epine,	Hawthorn, English,	Cratægus oxyacantha.
" fumitory,		Corydalis nobilis.
" liverwort,		Hepatica· Americana.
" pine,	Wintergreen,	Chimaphilla umbellata
" yarrow,		Achillea nobilis.
Nodding lily,		Lilium Canadensis.
None such,	Dutch clover,	Medicago lupulina.
" so pretty,	Pearl flow'd life everlasting,	Gnaphalium margaritaceum
Nolitangere,	Touch me not,	Impatiens Noli-tangere.
Noon day flower,	Goats' beard,	Tragopogon pratensis and species.
" flower,	" "	" "
" tide,	" "	" "
Norway spruce fir,	Norway pine,	Abies excelsa.
" pine,		Pinus resinosa.
Nosebleed,	Yarrow,	Achillea millefolium.
Noseburn tree,		Daphnopsis tenuifolia.
Notchweed,	Goose foot,	Chenopodium vulvaria.
Nuns' whipping rope,		Amarantus melancholicus.
Nut, Acajou,	Cashew nut,	Anacardium occidentale.
" Ar,	Hawk nut,	Bunium flexuosum.
" Areca,	Betel nut,	Areca Catechu.
" Bambarra,	the seed of	Voandzeria subterranea.
" Barbadoes,	" "	Curcas purgans.
" Beazor,	Bonduc nut,	Guilandina bonduc.
" Bedda,	Myrobalans,	Terminalia belliriea.
" Beech,	the fruit of	Fagus ferruginea.
" Ben,	the winged seed of	Moringa pterygosperma.
" Behen,		" aptera.
" Betel,	the seed of	Areca Catechu.
" Bitter,	Hickory nut,	Carya amara.
" Bladder,		Staphylea trifolia.

COMMON.	ENGLISH.	BOTANICAL.
Nut, Bladder, African,		Royena.
" Bomba,	Palm oil tree fruit,	Elais Guiniensis.
" Bonduc,	the seed of	Guilandina bonduc.
" Brazil,	Castana nut,	Bertholletia excelsa.
" Bread,	the fruit of	Brosimum alicastrum.
" " Monkey,	Baobab tree,	Adansonia digitata.
" Buffalo,	the fruit of	Pyrolaria oleifera.
" Butter,	White walnut,	Juglans cinerea.
" "	the seed of	Caryocar nuciferum.
" Cacao,	Chocolate nut,	Theobroma Cacao.
" Candle,	the seed of	Aleurites triloba.
" Cashew,	" " "	Anacardium occidentale.
" Castana,	" " "	Bertholletia excelsa.
" Chest,		Castana vesca (Americana.)
" Chocolate,	Cacao,	Theobroma Cacao.
" Chop,	Calabar bean,	Phytostigma venenosum.
" Clearing,		Strychnos potatorum.
" Cob,	Hazelnut,	Corylus Avellana barcelonensis.
" Cocoa,	the fruit of	Cocos nucifera.
" Cola,	the seed of	Cola acuminata.
" Coquilla,	the fruit of	Attalea funifera.
" Corozo,	Vegetable ivory nut,	Phytelephas macrocarpa.
" Cypress,		Cupressus sempervirens.
" Drinkers',	Clearing nut,	Strychnos potatorum.
" Earth, (Peanut)	the tuber of	Arachis hypogæa.
" "		Bunium bulbocastanum.
" Eboe,	the seed of	Dipterix oleifera.
" Elk,	Buffalo nut,	Pyrolaria oleifera.
" Euhæan,	Chestnut,	Castana vesca.
" French,	Walnut,	Juglans regia.
" " physic,		Curcas multifidus.
" Goora,	Cola nut,	Cola acuminata.
" Ground,	Pea nut,	Arachis hypogæa.
" " Dwarf,		Panax trifolium.
" Hara,	the drupes of	Terminalia citrina.
" Haugh,	Earth nut,	Arachis hypogæa.
" Hawk,		Bunium flexuosum.
" Hazel,	Witch hazel nut,	Hamamelis Virginica.
" "	Filbert,	Corylus Avellana and species.
" Hickory,	Shellbark,	Hicorya sulcata and other species.
" "	Shagbark,	Carya alba and other species.
" Hog,		" porcina.
" " Jamaica,		Omphalea diandra.
" " pea,		Amphicarpœa monoica.
" Hurr,	Ink nut,	Terminalia cherbula.
" Illinois,	Pecan nut,	Carya olivæformis.
" Ink,		Terminalia cherbula.
" Ivory,	Vegetable ivory,	Phytelephas macrocarpa.
" Jesuits',	Water nuts,	Trapa nutans.
" Jupiter's,	Walnuts,	Juglans regia.
" Keena,	the fruit of	Calophyllum Calaba.
" Kipper,		Bunium flexuosum.
" Kisky Thomas,	White walnut,	Juglans cinerea.
" Kola,	Cola nut,	Cola acuminita.
" Kundoo,	the fruit of	Carapa touloucouma.
" Levant,	Cocculus Indicus,	Anamirta cocculus.
" Lumbang,	Candle nut,	Aleurites triloba.
" Malabar,		Justicia adhatoda.
" Malacca,	Cashew nuts,	Anacardium occidentale.
" Manilla,	Pea nut,	Arachis hypogæa.

COMMON.	ENGLISH.	BOTANICAL.
Nut, Marany,	Marking nuts,	Semecarpus anacardium.
" Marking,		" "
" Mocker,	White heart hickory,	Carya tomentosa.
" Monkey bread	Baobab,	Adausonia digitata.
" Mote,		Carapa touloucouma.
" Oil,	Butternut,	Juglans cinerea.
" "	West Indian name for castor oil bean.	
" Olive,	the fruit of	Elæocarpus Hinau.
" Para,	Brazil nut, (Castana)	Bertholletia excelsa.
" Pecan,		Carya olivæformis.
" Pea,		Arachis hypogæa.
" Physic,	Barbadoes nut,	Curcas purgans.
" Pig,	Hognut,	Hicorya porcina.
" "	Hawknut,	Bunium flexuosum.
" Pine,	the seed of Stone pine,	Pinus pinea (cembra.)
" Pistachio,	the edible seed of	Pistacia vera.
" Poison,	the poisonous seed of	Strychnos Nux Vomica.
" Purging,	Barbadoes physic nuts,	Curcas purgans.
" Quandang,		Fusanus acuminatus.
" Rattle,		Nelumbium luteum.
" Ravensara,	Clove nutmeg,	Agathophyllum aromaticum.
" Rush,		Cyperus esculentus.
" Sapucaia,		Lecythis Zabucajo and other spec.
" Sardian,	Chestnut,	Castana vesca.
" Sassafras,	Pichurim bean,	Nectandra puchury.
" Shagbark,		Carya alba and other species.
" Shellbark,		Hicorya sulcata and other species.
" Singhara,	the fruit of various species of Trapa.	
" Snake,	the seed of	Ophiocaryon paradoxum.
" Soap,		Mimosa abstergens.
" Souari,	Guiana almonds,	Caryocar tomentosum.
" "	Butternuts,	" butryosum.
" Spanish,		Moræa sisyrinchium.
" Taqua,	Ivory nuts,	Phytelephas macrocarpa.
" Vegetable ivory,	" "	" "
" Vomit,	Nux Vomica,	Strychnos Nux Vomica.
" Water,	the various species of	Trapa.
" Wood,	Hazelnut,	Corylus Avellana.
" Yer,	Peanut,	Arachis hypogæa.
" Zibel,	Stone pine nuts,	Pinus pinea (cembra.)
Nutall's weed,	Golden coreopsis,	Coreopsis tinctoria.
Nutgrass,	Coco grass,	Cyperus rotundus.
Nutmeg,	the kernel of the fruit of	Myristica fragrans.
" American,		Monodora myristica.
" Brazilian,		Cryptocarya moschata
" Calabash,	American nutmeg,	Monodora myristica.
" Californian,		Torreya Californica.
" Clove,		Agathophyllum aromaticum.
" Jamaica	American nutmeg,	Monodora myristica.
" Long,		Myristica fatua.
" Madagascar,	Clove nutmeg,	Agathophyllum aromaticum.
" Male,		Myristica tomentosa.
" Penang,		" fragrans.
" Peruvian,	the seed of	Laurelia sempervirens.
" Plume,		Atherosperma moschata.
" Santa Fe,		Myristica Otoba.
" Stinking,	Californian nutmeg,	Torreya Californica.
" Wild,	Male nutmeg,	Myristica tomentosa.
" Wood,	the Palmyra palm,	Borassus flabelliformis.
Nutmeg flower,	Small fennel flower,	Nigella sativa.

6

COMMON.	ENGLISH.	BOTANICAL.
Nux Metella,	Nux Vomica,	Strychnos Nux Vomica.
Nux Vomica,		" " "
Nya paua,	Ayapana,	Eupatorium Ayapana.
Oak,	common name for the different species of Quercus.	
" African,		Oldfieldia Africana.
" agaric,	Female agaric,	Boletus agaricus.
" American, Turkey,		Quercus obtusiloba.
" apple,	Nutgall. Oak leaf gall,	
" Barren,	Black jack oak,	Quercus nigra.
" bark, Black,		" "
" " Red,		" Rubra.
" " White,		" alba.
" Bear,	Black scrub oak,	" ilicifolia.
" Bitter,		" cerris.
" Black jack,	Barren oak,	" nigra.
" Botany Bay,		Casurina torulsa.
" Burr,		Quercus macrocarpa.
" Champion,		Ambrina ambrosioides.
" Chinquapin,	Dwarf chestnut oak,	Quercus prinoides.
" Chestnut,		" Prinus and other var.
" Common European,		" Robur.
" Cork,		" Suber.
" currant,	a kind of gall on oak leaves,	" pedunculata.
" Cypress,		" ' var.
" Durmast,		" sessiflora pubescens.
" Dyers',		" tinctoria.
" English,	European oak,	" Robur.
" Evergreen,	Holly oak,	" Ilex.
" Female,		" pedunculata.
" Green,	Oak wood impregnated with the spawn of Peziza æruginosa.	
" Gall,		Quercus infectoria.
" gall,	Nutgall,	" '
" He,		Casuarina stricta.
" Holly,		Quercus Ilex.
" Indian,	Teak tree,	Tectonia grandis.
" Iron,		Quercus cerris and obtusiloba.
" Italian,		" Æsculus.
" Jerusalem,		Chenopodium Botrys.
" Kermes,		Quercus coccifera.
" Laurel,	Shingle oak,	" imbricaria.
" Live,		" virens.
" Lungs of the,		Sticta pulmonaria.
" Male,		Quercus sessiflora.
" nutgall,		" infectoria.
" of Jerusalem,	American wormseed,	Chenopodium anthelminticum.
" Pin,		Quercus palustris.
" Poison,		Rhus Toxicodendron.
" Post,	American Turkey oak,	Quercus obtusiloba.
" Red,		" rubra.
" Scarlet,		" coccinea.
" Scrub,		" Catesbæi and ilicifolia.
" Scrubby,		Lophira Africana.
" Sea,	Bladder wrack,	Fucus vesiculosus.
" She,		Casuarina quadrivalvis
" Shingle,	Laurel oak,	Quercus imbricaria.
" Silk bark,		Grevillea robusta.
" Spanish,		Quercus falcata.
" Stone,		Lithocarpus javensis.
" Swamp,		Quercus Prinus.
" " White,		" bicolor.

COMMON.	ENGLISH	BOTANICAL.
Oak, Water,		Quercus aquatica.
" White,		" alba.
" Willow,		" Phellos.
" Wainscoat,	Bitter oak,	" cerris.
" Yellow,	Quereitron,	" coccinea var. tinctoria.
Oats,	the seeds of	Avena sativa
Ochra,	Musk leaf,	Abelmoschus esculentus.
Ogeechee tree,		Nyssa coccinea.
Oil nut,	Butternut,	Juglans cinerea.
" plant,	Oily grain,	Sesamum orientale.
" "	Castor oil plant,	Ricinus communis.
Oil, Almond,	the expressed oil of	Amygdalus communis.
" Andiroba,	Carap oil,	Carapa Guianensis.
" Arachis,	Peanut oil,	Arachis hypogœa.
" Aspic,	Oil of spike lavender,	Lavendula spica.
" Baneoul,	Lumbang oil,	Aleurites triloba.
" Bay,	the fixed oil from the berries,	Laurus nobilis.
" Beech nut,	" " " " " nuts of	Fagus ferruginea and species.
" " " Europ'n,	" " " " " "	" sylvatica.
" Ben,	" " " " " seed,	Moringa pterygosperma.
" Benne,	the oil of the seed,	Sesamum Indicum and orientale.
" Camphor,	a limpid oil obtained from	Dryobalanops aromatica.
" Carap,	the solid fixed oil of	Carapa Guianensis.
" Cashew apple,	" vesicatory oil,	Anacardium occidentale.
" " nut,	" edible fixed oil,	" "
" Castana,	" expressed fixed oil of	Brazil nuts.
" Castor,	" " " "	Ricinus communis.
" Cebadilla,	" fixed, fatty oil,	Veratrum sabadilla.
" Cheerojce,	" " oil,	Buchanania latifolia.
" Cacao,	Cacao butter,	Theobroma Cacao.
" Cocoanut,	the expressed, fixed oil of	Cocos nucifera.
" Cocum,	" solid, fixed oil of	Garcinia purpurea.
" Cohune,	" fixed oil of the kernels,	Attalea cohune.
" Colza,	" " " " seed,	Brassica campestris.
" Corooka,	an Indian medicinal oil distilled from Argemone Mexicana.	
" Coondi,	Mote oil,	Carapa Guianensis.
" Cotton seed,	the expressed seed oil of	Gossypium.
" Crab,	see Carap oil,	
" Croton,	the fixed oil of croton seed,	Croton Tiglium and other species.
" Cucumber seed,	see Sassa oil,	
" Cumaru,	see Tonquin oil,	
" Dippel's,	a volatile oil by destructive distillation of bones.	
" Domba,	Poon seed oil,	
" Flaxseed,	see Linseed oil,	
" Earth nut,	see Peanut oil,	
" Epic,	the fixed, seed oil of	Bassia latifolia.
" Exile,	" " oil from the kernels of Thevetia nerifolia.	
" Fusel,	" essential oil distilled from wine and spirits.	
" Florence,	see Olive oil,	
" Ginger grass,	see Grass oil,	
" Gingelly,	the oil expressed from the seed of Sesamum Indicum.	
" Grass,	" volatile oil from Andropogon schœnanthus and other spec.	
" Ground nut,	see Peanut oil,	
" Hempseed,	the fixed, drying oil by expression, Cannabis sativa.	
" Hutsyellow,	" " oil of Niger seed,	Guizotia oleifera.
" Jatropha,	" oil of Barbadoes nuts,	Curcus purgans and other species.
" Kanairi,	" fixed oil of the fruit,	Canarium commune.
" Keena,	" oil obtained from	Calophyllum inophyllum.
" Kekune,	Spanish walnut oil,	Aleurites triloba.
" Keora,	an Eastern perfumery oil from Pandanus odoratissimus.	

COMMON.	ENGLISH.	BOTANICAL.
Oil, Kikuel,	Mustard tree, oil of seed,	Salvadora persica.
" Kokum,	see Cocum oil.	
" Kossumba,	a fixed oil from	Carthamus tinctoria.
" Kruine,	a crude, elastic gummy substance from Borneo.	
" Kundah,	a fixed oil from	Carapa Guianensis.
" Limbolee,	a fixed oil of Curry leaf seed, Bergera Konigii.	
" Linseed,	the fixed, drying oil of Flax seed, Linum usitatissimum.	
" Lumbang,	Spanish walnut oil,	Aleurites triloba. •
" Macuja,	a kind of Palm oil,	Acrocomia sclerocarpa.
" Madia,	a fixed oil of the seed,	Madia sativa.
" Mahowa seed,	see Epic oil.	
" Margosa,	the solid, fixed oil of the seed, Melia azidirachta.	
" Marking nut,	an acrid, vesicatory oil of the pericarp, Semecarpus anacardium	
" Marmotte,	a fixed oil from the kernel of Prunus Brigantiaca.	
" Mustard seed,	the fixed oil of	Sinapis nigra and other species.
" Nahor,	an oil obtained from the seed of Mesua ferrea.	
" Namur,	see Grass oil.	
" Nemaur,	see " "	
" Napala,	see Jatropha oil.	
" Napoota,	an East African oil from	Agati grandiflora.
" Narpaulah,	a fixed oil from Croton seed allied to Tiglium.	
" Neem,	see Margosa oil.	
" Nut,	Commercial name for peanut oil.	
" "	" " " walnut oil.	
" Olive,	the fixed oil expressed from the pericarp of Olea Europæa.	
" Oondee,	see Poon seed oil.	
" Ouabe,	the lubricating oil from seeds of Omphalea diandra.	
" Palm,	the fixed, oil of Palm fruit, Elais Guiniensis.	
" Palma Christa,	Castor oil,	Ricinus communis.
" Pand,	the volatile, perfumery oil from Michelia Champaca.	
" Pandang,	" " " " " Pandanus odoratissimus.	
" Patawa,	an oil procured from	Œnacarpus Batava.
" Peanut,	the expressed, fixed oil of	Arachis hypogæa.
" Phoolwa,	Vegetable butter from	Bassia butyracea.
" Physic nut,	see Jatropha oil.	
" Piney,	Piney tallow, from the pænoe tree, Vateria Indica.	
" Pinhoen,	the purgative oil of Barbadoes nuts, Curcas purgans.	
" Pinnacottoy,	see Poon seed oil.	
" Piquay,	the concrete oil from the fruit pulp, Caryocar Brasiliense.	
" Poonga,	Poon gum oil.	
" Poon gum,	the fixed oil of the seed of Sapindus emarginatus.	
" Poon seed,	" " Calophyllum inophyllum.	
" Pootingee,	" " the fruit, " spuria.	
" Poppy seed,	" " poppy seed, Papaver somiferum.	
" Portia nut,	" " Thespesia populnea.	
" Provence,	an esteemed kind of olive oil.	
" Pumpkin seed,	the expressed oil of the seed, Cucurbita Pepo.	
" Ramtil,	the fixed oil of niger seed, Guizotia oleifera.	
" Rape seed,	" " the seed, Brassica napus and other species.	
" Rosin,	obtained from pine resin.	
" Rusa,	see oil of Lemon grass.	
" Salad,	see Olive oil.	
" Sapucaia.	the expressed oil of Brazil nuts, Lecythis Zabucajo.	
" seed,	an indefinite name applied to seeds yielding oil.	
" Seneca,	Petroleum oil, in its crude state.	
" Seringa,	obtained from the India rubber tree, Siphonia elastica.	
" Serpolet,	distilled from	Thymus serpyllum.
" Sesamum,	obtained from blacktil seed, a variety of Sesamum orientale.	
" Shanghae.	a fixed oil procured from	Brassica Chinensis.
" Siri,	the same as oil of Lemon grass.	

COMMON.	ENGLISH.	BOTANICAL.
Oil, Soap nut,	Poon gum oil.	
" Spurry,	an oil obtained from	Spergula sativa.
" Sasa,	Cucumber seed oil,	Cucumis sativus.
" Sweet,	name given to Olive oil,	
" Teel,	the oil expressed from the seed of Sesamum Indicum.	
" Tallicoonah,	Kundah oil.	
" Tonquin,	a perfumery oil expressed from the seed, Dipterix odorata.	
" Tumika,	the concrete oil of	Diospyros embryopteris.
" Uggur,	a fragrant oil distilled from the wood of Aquillaria agollocha.	
" Valisaloo,	see Ramtil oil.	
" Virgin,	a name given to the finest Olive oil.	
" Walnut,	the fixed oil of	Juglans regia and other species.
" " Belgaum,	Candleberry nut oil,	Aleurites triloba.
" " Spanish,	" " "	" "
" Wood,	the balsam like production of Dipterocarpus turbinatus.	
" Yamadou,	the fixed oil by expression of the seed, Myristica sebifera.	
Oil of allspice,	the aromatic oil by distillation of Eugenia Pimenta.	
" Almond, bitter,	the volatile oil procured by distillation of Bitter almond kernels	
" " sweet,	the fixed oil obtained from the kernel of Amygdalus communis.	
" aloes,	the oil distilled from common aloes.	
" amber,	distilled from amber resin.	
" anda seed,	the oil expressed from the seeds of Anda Brasiliensis.	
" angelica,	from the root of	Archangelica officinalis.
" anise seed,	" fruit (seed) of	Pimpinella anisum.
" " Star,	" capsules of	Illicium anisatum.
" arnica,	" roots and flowers of Arnica montana.	
" arbor vitæ,	' fresh tops of	Thuja occidentalis.
" assafœtida,	" gum resin of	Narthex assafœtida.
" asarabacca,	" roots of	Asarum Europæum.
" balm,	distilled from the fresh herb Melissa officinalis.	
" balsam fir,	" " balsam (oleo-resin) Abies balsamea.	
" " Peru,	" " " Myrospermum Peruiferum.	
" bay,	" " leaves of Myrcia acris.	
" bayberries,	" " berries of Bay laurel.	
" ben,	from the seed of	Moringa pterygosperma.
" benne,	Teel oil,	Sesamum Indicum and orientale.
" bergamot,	by expression or distillation of the fruit rind, Citrus Bergami.	
" bitter almonds,	distilled from the marc of Bitter almond kernels.	
" birch bark,	" " " bark of Betula alba.	
" " " black,	" " " " " " lenta.	
" black pepper,	the oleo-resin of	Piper nigrum.
" buchu,	distilled from the leaves of Barosma crenata and other species	
" cacao,	Cacao butter obtained from Theobroma Cacao.	
" cade,	a kind of tar distilled, *per desensum*, from Juniperus Oxycedrus.	
" Canada snake root	by distillation, from the roots of Asarum Canadense.	
" cajeput,	the volatile oil of the leaves, Melaleuca minor and other spec.	
" calamus,	" " " sweet flag root, Acorus Calamus.	
" camphor,	a limpid oil obtained from	Dryobalanops aromatica.
" "	the volatile oil of	Camphora officinarum.
" Canada fleabane,	" " "	Erigeron Canadense.
" capsicum,	the oil, "by ether," from Capsicum annuum and other species.	
" caraway,	the volatile oil from the fruit (seed) of Carum carui.	
" cardamon,	" " " " seed of Elettaria Cardamomum.	
" cascarilla,	" " " " bark of Croton Eleuteria.	
" cassia,	the heavy, volatile oil distilled from Cassia bark.	
" cassie,	a fragrant, volatile from the flowers of Acacia Farnesiana.	
" cayenne pepper,	see oil of Capsicum.	
" cedar,	the volatile oil obtained from Juniperus Virginiana.	
" cedrat,	" " " from the fruit rind of Citrus medica.	
" celery,	from the fruit (seed) of	Apium graveolens.

COMMON.	ENGLISH.	BOTANICAL.
Oil of chamomile,	a volatile oil obtained from the flowers of Anthemis nobilis.	
" checkerberry,	the heavy, volatile oil from Gaultheria procumbens.	
" cherry,	distilled from the bark of Wild cherry, Cerasus serotina.	
" " laurel,	" " " leaves of Cerasus Lauro-cerasus.	
" cinnamon,	the heavy, volatile oil from Cinnamomum Zeylonicum.	
" " leaf,	the fragrant, volatile oil from the cinnamon leaf.	
" citron,	" " " " " leaves and rind of Citrus medica.	
" citronelle,	" " " " " Andropogon Nardus.	
" cloves,	" heavy, " " " cloves, Caryophyllus aromatica.	
" copaiba,	the volatile oil from the oleo-resin of Copaifera multijuga, etc.	
" copaiva,	see oil of Copaiba.	
" cocculus Indicus,	the oil of the seeds by evaporating a strong tincture.	
" coriander,	the product by distillation of the fruit of Coriandum sativum.	
" cucumber seed,	the expressed oil from the seed of Cucumis sativus.	
" cubeb,	the volatile medicinal oil distilled from Cubebs.	
" cummin,	" " " " Cummin (fruit) seed.	
" dill seed,	the volatile oil distilled from the fruit of Anethum graveolens.	
" ergot,	an oil procured by evaporation of the Etherial tincture.	
" erigeron,	see oil of Fleabane.	
" eucalyptus,	procured from the leaves of Eucalyptus piperita.	
" "	a perfumery oil from " citriodora.	
" "	the volatile oil of the Peppermint tree.	
" euphorbia,	a fixed oil obtained from the seed of Euphorbia lathyris.	
" fennel,	the volatile, medicinal oil from Fœniculum dulce, and vulgare.	
" fern,	Oil of Male Fern, Aspidium Filix Mas.	
" fleabane,	the volatile oil by distillation of Erigeron Canadense.	
" fir,	" " " " " of Abies pectinata.	
" fireweed,	the oil distilled from the plant, Erechthites hieracifolius.	
" Gaultheria,	the volatile oil obtained from the plant, Gaultheria procumbens.	
" garlic,	the stimulant, volatile oil obtained from Allium sativum.	
" gentian,	the volatile oil obtained from Gentiana lutea.	
" geranium,	the volatile perfumery oil from Pelargonium odoratissimum.	
" "	the commercial name for Grass oil.	
" ginger,	the oil, "by ether," of ginger root, Zingiber officinale.	
" " grass,	see Grass oil, Andropogon schœnanthus.	
" golden rod,	the volatile oil obtained from Solidago odora.	
" grain,	see Fusel oil.	
" guaiacum,	the volatile oil from Guaiacum officinale.	
" hedeoma,	the volatile oil of American pennyroyal.	
" hemlock,	" " " distilled from Abies (Pinus) Canadensis.	
" henbane,	an oil by distillation from Hyosciamus niger.	
" hop,	an acrid oil by expression of Humulus lupulus.	
" horseradish,	by distillation from the roots of Cochlearia armoracia.	
" horsemint,	the essential oil of Monarda punctata.	
" hyssop,	the oil, "by ether," of the herb, Hyssopus officinalis.	
" iva,	an oil obtained from Achillea moschata.	
" jasmine,	the perfumery oil, "by enfleurage," of jasmine flowers.	
" juniper berries,	the medicinal oil distilled from Juniper berries.	
" " wood,	see oil of Cade.	
" khuskhus,	the volatile, perfumery oil from Andropogon muricatus.	
" laurel,	the volatile oil from the berries of Laurus nobilis.	
" lavender,	the fragrant, volatile oil from the flowers of Lavendula vera.	
" " Spike,	see oil of Spike.	
" lemon,	the volatile oil obtained from the fruit rind, Citrus Limonum.	
" " grass,	" " " " " Andropogon citratus.	
" " thyme,	" " " " " Thymus citriodorus.	
" lily,	the fragrant, perfumery oil, "by enfleurage," of Lilium candidum	
" lobelia,	the oil, "by ether," from the seed of Lobelia inflata.	
" mace, expressed,	a fixed oil by expression, from mace.	
" "	the volatile oil from mace, Myristica fragrans.	

COMMON.	ENGLISH.	BOTANICAL.
Oil of male fern,	the oil, "by ether," from the root of Aspidium Filix Mas.	
" marjoram,	the oil, by distillation, of Origanum Marjorana.	
" "	see oil of Origanum.	
" massoy,	the volatile oil obtained from Cinnamomum Kiamis.	
" melissa,	see oil of Balm.	
" " Indian,	see oil of Lemon grass.	
" matico,	distilled from the leaves of Artanthe elongata.	
" meadow sweet,	a product of salacine.	
" mezereon,	the acrid, volatile oil from the roots of Daphne Mezereum.	
" mustard,	an oil expressed from the seed of Sinapis alba, and nigra.	
" "	the volatile oil distilled from the marc of mustard seed.	
" myrbane,	an artificial flavor of Bitter almonds, Nitro-benzole.	
" myrrh,	the volatile oil distilled from Balsamodendron Myrrha.	
" myrtle,	" " " " " Myrtle blossoms.	
" narcissus,	a perfumery oil, by enfleurage, of Narcissus flowers.	
" neroli, "bigarade,"	a " " from the blossoms of Citrus Bigaradia.	
" " " petale,"	the fragrant volatile oil from the flowers of Citrus Aurantium.	
" " "petit grain,"	by distillation of the leaves and unripe fruit of Citrus.	
" nutmeg,	the volatile oil from nutmegs.	
" " expressed,	the fixed oil obtained from Myristica fragrans.	
" onion,	an acrid, medicinal volatile oil of Allium cepa.	
" origanum,	the volatile oil from Origanum vulgare and thymus vulgare.	
" orange flowers,	see oil of Neroli.	
" " peel,	the volatile oil from the fruit rind, Citrus.	
" orangette,	the oil of Sweet marjoram.	
" parsley,	the volatile oil from the roots of Petroselinum sativum.	
" partridge berry,	see oil of Checkerberry.	
" patchouly,	the fragrant oil from the leaves of Pogostemnon patchouly.	
" petit grain,	Oil of Orange leaves, Citrus Aurantium and vulgare.	
" pennyroyal,	the oil distilled from the fresh herb, Hedeoma pulegioides.	
" " Europ'n,	" " " " " Mentha pulegium.	
" peppermint,	an oil distilled from fresh flowering herb, Mentha piperita.	
" pichurim bean,	see oil of Sassafras nuts.	
" pimento,	see oil of Allspice.	
" potatoes,	see Fusel oil.	
" prickly ash,	see oil of Xanthoxylum. [Genista Canariensis.	
" rhodium,	the volatile oil distilled from rosewood, Convolvulus scoparius,	
" rosemary,	" " " " " flow'g tops of Rosmarinus officinalis	
" roses,	" " " obtained from the petals of Rosa centifolia and	
" rosewort,	by distillation from the root of Rhodiola rosea. [other species.	
" rue,	the volatile oil from the fresh herb, Ruta graveolens.	
" safflower,	a fixed oil from Carthamus tinctoria.	
" sage,	the volatile oil distilled from Salvia officinalis.	
" sandal,	" " " " " Santalum myrtifolium and other	
" sassafras,	" " " " " Sassafras officinale. [species.	
" " nuts,	obtained by expression from the seeds, Nectandra puchury.	
" savine,	by distillation from the fresh tops, Juniperus sabina.	
" senna,	by " " leaves, Cassia of several species.	
" solidago,	see oil of Goldenrod.	
" spearmint,	by distillation from the fresh herb, Mentha viridis.	
" spike,	by " " " " Lavendula spica.	
" spikenard,	Commercial name for Grass oil.	
" spruce,	the oil, by distillation, of the leaves of Abies (Pinus) nigra.	
" star anise,	the volatile oil, by distillation, of Star anise fruit.	
" stillingia,	prepared from the alcoholic tincture of Stillingia sylvatica.	
" sunflower seed,	by expression from the seed of Helianthus annuus.	
" sweet almonds,	by " " " kernel of Amygdalus communis.	
" " bay,	the volatile oil of laurel, Laurus nobilis.	
" " marjoram,	see oil of Marjoram.	
" tansy,	by distillation from Tansy herb, Tanacetum crispum and vulgare	

COMMON.	ENGLISH.	BOTANICAL.
Oil of tar,	obtained by distilling tar.	
" theobroma,	Cacao butter, expressed from chocolate nuts.	
" thyme,	the oil, by distillation, of the plant, Thymus vulgaris.	
" tobacco,	" " " " " leaves, Nicotiana Tobacum.	
" tonquin,	by expression, from the seed of Dipterix odorata.	
" turpentine,	distilled from the oleo-resin of Pinus palustris and other spec. .	
" valerian,	by distillation, of the roots of Valeriana officinalis.	
" verbena,	by " from the plant, Aloysia citriodora.	
" "	Commercial name for Oil Lemon grass.	
" vertiver,	see oil of Khuskhus.	
" wintergreen,	see oil of Checkerberry.	
" wormseed,	the volatile oil from the seed, Chenopodium anthelminticum.	
" wormwood,	" ` " " " " herb, Artemisia Absinthium.	
" yarrow,	" " " " " " Achillea Millefolium.	
" xanthoxylum,	an oil, " by ether," from the berries of Xanthoxylum fraxineum	
" "	" " " " bark of " "	
Oily acorn,	Ben nuts,	Guilandina moringa.
Okra,	Musk leaf,	Abelmoschus esculentus.
Olcott root,	Bloody dock,	Rumex sanguineus.
Old field balsam,		Gnaphalium polycephalum
" man,	Southernwood,	Artemisia abrotanum.
" " cactus,		Cereus senilis.
" man's beard,	Fringe tree,	Chionanthus Virginica.
" " "	Travellers' joy,	Clematis vitalba.
" maids' pink,	Soapwort,	Saponaria officinalis.
' " Robert's root,	Squaw weed root,	Senecio obovatus.
Oleander,	Rose laurel,	Nerium oleander.
Olibanum,	Frankincense, Ancient,	Juniperus lycia.
" African,.	"	Plosslea floribunda.
" Arabic,	"	Juniperus oxycedrus and lycia.
" Indian,		Boswellia serrata.
Olive,	the fruit of the olive tree,	Olea Europæa, sativa.
" American,	Devil's wood,	" Americana.
" Barbadoes,		Bontia dapnoides.
" bark,		Olea Europæa.
" " tree,	W. I. wild olive,	Bucida buceras.
" gum,	Lecca gum,	Olea Europæa.
" Italian,		" longifolia.
" leaves,		" Europæa.
" Fragrant,	used to flavor tea leaves,	" fragrans.
" Oil,	see oil Olive,	" Europæa.
" resin,	a gum resin from the bark of olive trees.	
" Spanish,		Olea latifolia.
" Spurge,	Mezereon,	Daphne Mezereum.
" Wild,	Oleaster.	
" "	False fringe tree,	Rhus cotinus.
One berry,	Herb Paris,	Paris quadrifolia.
" " leaves,	Squaw vine,	Mitchella repens.
" blade,	Dwarf Solomon's seal,	Convallaria bifolia.
" seeded cucumber,		Sicyos angulatus.
Onion,		Allium cepa.
" Bog,	Dragon root,	Arum triphyllum.
" Sea,	Squills,	Scilla maritima. [ferum.
Opium,	the concrete juice from unripe Poppy capsules, Papaver somni-	
" Lettuce,	Lactucarium,	Lactuca sativa and Lactuca virosa.
" poppy,	the opium producing poppy, Papaver somniferum.	
Opopanax,	the concrete juice of	Pastinaca opopanax.
Opossum tree,	Sweet gum tree,	Liquidambar styraciflua.
Orache, Garden,		Atriplex hortense.
" Sea,	Sea Purslane,	" halamoides.

COMMON.	ENGLISH.	BOTANICAL.
Orange,	the fruit of	Citrus Aurantium and vulgaris.
" apple,	" " "	" " " "
" Bergamot,		" bergamia.
" berries,	Issue peas. Small unripe oranges.	.
" Bitter,		Citrus vulgaris.
" "		" Bigaradia.
" buds,	Orange flowers.	
" Curacoa,	Bitter orange.	Citrus Aurantium.
" flowers,	the blossoms of	" " and vulgaris.
" grass,		Hypericum sarothra.
" leaves,	the leaves of the tree,	Citrus Aurantium and vulgaris.
" Mandarin,		" Bigaradia myrtifolia (Sinen-
" peas,	Issue peas. Small unripe oranges.	[sis.
" peel, Bitter,	the rind of the fruit,	Citrus vulgaris.
" " Sweet,	" " " "	" Aurantium.
" root,	Goldenseal root,	Hydrastis Canadensis.
" Seville,	Bitter orange,	Citrus vulgaris (Bigaradia.)
" swallow wort,	Pleurisy root,	Asclepias tuberosa.
" Sweet,		Citrus Aurantium.
" vine,		Cucurbita Aurantia.
Orangettes,	Curacoa oranges,	
Orcauette,	Alkanet root,	Anchusa tinctoria.
Orchil,	Archil,	Roccella "
Orchilla weed,	Canary weed,	" "
Orchis, Bee,	Bee flower,	Ophrys apifera.
" Male,	yields salep,	Orchis mascula.
" Man,		Acerns anthropophora.
Ordeal bark, African,	Sassy bark.	
" bean of Calabar,	Calabar bean,	Phytostigma Venenosum.
" poison, African,	Akasgia.	
" root,		Strychnos.
" tree,	Sassy bark tree,	Erythrophlæum Guineense.
Orenburg gum,	Manna of Brianeon. '	
Oriental berries,	Cocculus Indicus,	Anamirta cocculus.
" sassafras,		Sassafras parthenoxylon.
Origanum,	Wild marjoram,	Origanum vulgare.
Orpine,	Live forever,	Sedum telephium.
Orris root,		Iris Florentina.
" " European,		" tuberosa.
" " German,		" Germanica.
Osage apple,	Osage orange,	Maclura aurantica.
" Orange,		" "
Osier, Basket,		Salix viminalis.
" bark, Green,	Round leaved dogwood,	Cornus circinata.
" " Red,	Swamp dogwood,	" sericea.
Oswego tea,	.	Monarda didyma.
Otaheite apple,		Spondias dulcis.
Ourari,	Urari poison,	Strychnos toxifera.
Ova-ova,	Ice plant, American,	Monotropa uniflora.
Owler,	Alder,	Alnus glutinosa.
Oxadoddy,	Black root,	Leptandra Virginica.
Oxalis,	Wood sorrel,	Oxalis acetosella.
Oxbalm,	Stone root,	Collinsonia Canadensis.
Oxeye,	Sneezewort,	Helenium autumnale.
"		Anacyclus radiatus.
" daisy,	White weed,	Leucanthemum vulgare.
Oxheal,	Bears' foot,	Helleborus fœtidus.
Oxlip,		Primula elatior.
Oxtongue,	Bugloss,	Anchusa officinalis.
Oyster root,	Goats' beard,	Tragopogon porrofolius and spec.

COMMON.	ENGLISH.	BOTANICAL.
Ozier, Green,	Round leaved dogwood,	Cornus circinata.
" Red,	Swamp dogwood,	" sericea.
Paddock pipes,		Equisetum limosum.
" stools,	see Agaric,	Boletus of several species.
Paddy,	Unhusked rice,	Oryza sativa.
Pæonia,		Pæonia officinalis,
Pæonoe tree,	Piney tallow tree,	Vateria Indica.
Painted cup,		Euchroma coccinea.
" gaillardia,		Gaillardia picta.
Pale bark,	Peruvian bark,	Cinchona condaminea.
" " Crown,	Loxa bark,	" " var. vera.
" catechu,	Gambir,	Nauclea Gambir.
" gentian,		Gentiana lutea.
" laurel,	Swamp laurel,	Kalmia glauca.
" rose,	Cabbage rose,	Rosa centifolia.
" touch me not,		Impatiens fulva.
Palma Christa,	the castor oil plant,	Ricinus communis.
Palmetto,		Sabal (Chamærops) Palmetto.
" Cabbage,		" Adansonii.
" Dwarf,	Palmetto of N. and S. Car.,	" Palmetto.
" Saw,	" " Southern coast,	" serrutata.
Palm fruit,	yields Palm oil,	Elais Guiniensis.
" oil,	the fixed oil from Palm fruit,	" "
" tree,		Chamærops Palmetto.
" " Fan,		Borassus flabelliformis.
Palmyra palm,	Nutmeg wood,	" "
" wood,	Cocoanut tree,	Cocos nucifera.
Palsy wort,	American cowslip,	Caltha palustris.
Pampas grass,		Gynerium argenteum.
Panama hat palm,	Jipijapa,	Carludorica Palmata.
Pan copal,	African copal,	Guibourtia copallifera.
Panna,		Aspidium athamanticum.
Pansy,	Heartsease,	Viola tricolor.
Papaw,	Custard apple tree,	Uvaria triloba.
"		Carica Papaya.
" American,		Asimina triloba.
Paper birch,	White birch,	Betula alba var. populifolia.
" mulberry,		Broussonetia papyrifera.
" sponk,		Racodium papyraceum.
Pappoose root,	Blue cohosh root,	Caulophyllum thalictroides.
Para cress,		Spilanthes oleracea.
Paradise seed,	Grains paradise,	Amomum Grana Paradisi and var.
Paraguay tea,		Ilex Paraguaiensis.
Para nuts,		Bertholletia excelsa.
Paratoda bark,		Canella auxiliaris.
Pareira Brava,		Chondodendron tomentosum.
" " Brown,	White Pareira,	Abuta rufescens.
" " False,		Cissampelos Pareira.
Parilla, Yellow,		Menispermum Canadensis.
Paris wort,	Beth root,	Trillium of several species.
Park leaves,		Androsæmum officinale.
Parrots' bill,		Clianthus puniceus.
" corn,	Saffron seed,	Carthamus tinctoria.
Parsley,		Apium Petroselinum.
" Common,		Petroselinum sativum.
" Fools',		Æthusa divaricata.
" Poison,	Poison hemlock,	Conium maculatum.
" piert,		Alchemilla arvense.
" Sea,	Lovage,	Ligusticum levisticum.
" Spotted,	Poison hemlock,	Conium maculatum.

COMMON.	ENGLISH.	BOTANICAL.
Parsnip, or Parsnep,	Garden parsnip,	Pastinaca sativa.
" Cow,	Masterwort,	Heracleum lanatum.
" " Royal,	a variety of	" "
" Meadow,		Zizia aurea.
Partridge berry,	Checkerberry,	Gaultheria procumbens.
" "	Squaw vine,	Mitchella repens.
" " vine,	" "	" "
" pea,	Prairie senna,	Cassia chamæcrista.
" wood,		Andira inermis.
Pasque flower,	Pulsatilla,	Anemone pulsatilla.
Passion flower,		Passiflora cærulea.
Passions,	Patience,	Rumex patientia.
Pastel leaves,	Woad,	Isatis tinctoria.
Patchouly,		Pogostemon Patchouly.
Patience,		Rumex patientia.
" dock,	Bistort,	Polygonum Bistorta.
" Garden,		Rumex patientia.
Patagonia mint,		Mentha rotundifolia.
Paul's betony,	Bugle weed,	Lycopus Virginicus.
Pauson,	Blood root,	Sanguinaria Canadensis.
Pea,	the seed of	Pisum sativum.
" Beach,		Lathyris maritimus.
" Coffee,	Chick pea,	Cicer arietinum.
" Common,	Field pea,	Pisum sativum.
" Guiana,		Ormosia coccinea.
Peacocks' tail fern,	Maidenhair fern,	Adiantum melanocaulon.
Peach,	the fruit,	Amygdalus Persica.
Peach meats,	the kernels of the fruit,	" "
" pits,	Peach meats,	" "
" tree,		" "
" wood,		Cæsalpina bijuga and other species
" wort,	Heartsease,	Polygonum persicaria.
Peanut,		Arachis hypogæa.
Pear,		Pyrus communis.
" Balsam,		Momordica charantia.
" leaf wintergreen,	Round leaved pyrola.	Pyrola rotundifolia.
Pearl barley,	the prepared seed of	Hordeum distichon.
" flowered life everlasting,		Gnaphalium margaritaceum.
" moss,	Irish moss,	Chondrus crispus.
" plant,		Lithospermum officinale.
" wort,		Sagina procumbens.
Pearmain,	a variety of apple,	Pyrus malus, var.
Pease,	the seeds of the varieties of	Pisum sativum.
Pea tree,		Sesbania.
Pecan,	Soft shelled hickory,	Corya olivæformis.
Pegwood,	Spindle tree,	Euonymus atropurpureus.
Pelargonium,	Rose geranium,	Pelargonium odoratissimum.
Pelican flower,	Virginia snake root,	Aristolochia serpentaria.
Pellitory,		Anacyclus Pyrethrum.
" American,		Parietaria Pennsylvanica.
" bark,	Prickly ash bark,	Xanthoxylum fraxineum.
" European,	Sneezewort,	Achillea Ptarmica.
" German,	"	" "
" root of Spain,	Pyrethrum,	Anacyclus Pyrethrum.
" Wall,		Parietaria officinalis.
Pencil flower,		Stylosanthes elatior.
" tree,	Groundsel tree,	Baccharis halimifolia.
" wood,	Red cedar,	Juniperus Virginiana.
Penny cress,		Thlaspi arvense.
" leaves,	Navelwort,	Cotyledon umbilicus.

COMMON.	ENGLISH.	BOTANICAL.
Pennyroyal,		Hedeoma pulegioides.
" European,		Mentha pulegium.
" tree,		Satureja viminea.
Pennywort,	Navelwort,	Cotyledon umbilicus.
" Indian,		Hydrocotyle Asiatica.
" Thick leaved,		" vulgaris.
Pentstemon, Bearded,		Chelone barbata.
Peony, Common,		Pæonia officinalis.
" Fragrant,		" albiflora.
" Tree,		" montana.
Pepper,	common name for Black pepper.	
" African,	Zanzibar pepper,	Capsicum fastigiatum.
" African black,	.	Piper afzelii.
" American,	Garden pepper,	Capsicum annuum.
" Anise,		Xanthoxylon mantchuricum.
" Bell,		Capsicum grossum.
" Betel,		Chavica Betel.
" Bird,		Capsicum baccatum.
" Bitter,		Xanthoxylon Daniellii.
" Black,		Piper nigrum.
" " African,		" afzelii.
" Bonnet,		Capsicum tetragonium.
" bush,	White bush,	Andromeda paniculata.
" Cayenne,		Capsicum annuum.
" corn,	name given to whole black pepper.	
" Chinese,		Xanthoxylon piperitum.
" cress,		Lepidium Virginicum.
" Cubeb,	African black pepper,	Cubeba.
" dulse,		Laurencia pinnatifida.
" elder,	Matico,	Artanthe elongata.
" Ethiopian	Monkey pepper,	Habzelia Æthiopica.
" Goats',	Bird pepper,	Capsicum frutescens.
" grass,		Lepidium sativa and other species.
" Guinea,	Cayenne pepper,	Capsicum annuum.
" Jamaica,		Eugenia Pimenta.
" Japanese,		Xanthoxylon piperitum (alatum.)
" Java,		Cubeba officinalis.
" Long,	the fruit spikes of	Piper (Chavica) longum.
" " India,		Chavica Pepuloides.
" " Java,		" officinarum.
" Malaguetta,	Grains Paradise,	Amomum Grana paradisi.
" Monkey,		Habzelia (Unona) Æthiopica.
" Mountain,	the seeds of	Capparis sinaica.
" Negro,	Monkey pepper,	Habzelia Æthiopica.
" Poor man's,		Lepidium latifolium.
" rod,		Croton humilus.
" root,	Tooth root,	Dentaria diphylla.
" Star,		Xanthoxylon Daniellii.
" tree,	Winter's bark tree,	Drymis aromatica.
" Wall,	Small house leek,	Sedum acre.
" Water,		Polygonum hydropiper.
" White,	the fruit of Piper nigrum deprived of their skins.	
" Wild,	the fruit of	Vitex trifolia.
" wort,	Pepper grass,	Lepidium sativa and other species.
Pepperidge bush,	Barberry,	Berberis vulgaris.
Peppermint,		Mentha piperita.
" Australian,		" Australis.
" Tasmanian,		Eucalyptus amygdalina (Piperita.)
" tree,		" " "
Perennial flax,		Linum perenne.

COMMON.	ENGLISH.	BOTANICAL.
Periwinkle,		Vinca minor, and major.
" Madagascar,		" rosea.
Pernambuco wood,	Lima wood,	Cæsalpina echinata.
Persian berries,		Rhamnus infectorius.
" feverfew,		Pyrethrum roseum and other spec.
" lilac,		Syringia persica.
" lily,	Persian fritillary,	Fritillaria persica.
Persimmon,	Date plum,	Diospyros Virginiana.
" bark,		" "
Persio,	see Cudbear,	
Peruvian bark,	Cinchona bark,	Cinchona of many species.
" " false,		Buena obtusifolia, etc.
" " Carthagena,	Pale bark,	Cinchona Condaminea.
" " Loxa,		" " var. vera.
" " Pale,		" "
" " Red,		" succirubra.
" " Yellow,	Calisaya bark,	" calisaya.
" cinnamon,		Laurus Quixos.
" ipecac,	Black ipecac,	Psychotria Ipecacuanha (emetica.)
" mastich,		Schinus molle.
Pestilence weed,		Tussilago petasites.
Peter's wort,		Ascyrum hypericoides.
Petit grain, Oil,	see oil of Petit grain,	
Petty morrel,	Spikenard,	Aralia racemosa.
" whin,	Rest harrow,	Ononis spinosa.
Petunia,		Petunia violacea and other species.
Pewter wort,	Scouring rush,	Equisetum hyemale.
Phacelia,		Phacelia viscida.
Pheasants' eye,		Adonis autumnalis.
" "	Plumed pink,	Dianthus plumaris.
Phlox,		Phlox Drummondi and other spec.
Physic, Culver's,	Black root,	Leptandra Virginica.
" Indian,	see Indian physic,	
" nuts,	Barbadoes nuts,	Curcas purgans and other species.
" " French,		Curcas multifidus.
" root,	Black root,	Leptandra Virginica.
Picac,	Blooming spurge,	Euphorbia corollata.
Pichurim beans,	Sassafras nuts,	Nectandra puchury.
Pickaway anise,	Wafer ash,	Ptelea trifoliata.
Pickerel weed,		Unisema deltifolia.
Pickpocket,	Shepherd's purse,	Capsella bursa-pastoris.
Pickpurse,	" "	" " "
Picris,	Hawkweed picris,	Picris hieracioides.
Pie plant,	Rhubarb, American,	Rheum Rhaponticum.
Pigeon berry,	Garget,	Phytolacca decandra.
" grass,	Vervain, European,	Verbena officinalis.
" pea,	Angola pea,	Cytisus cajan.
" root,		Claytonia lanceolata.
" tree,	Prickly elder,	Aralia spinosa.
" weed,	Pigeon grass,	Verbena officinalis.
" wood,	Zebra wood,	Omphalobium Lambertii.
Pigmy weed,		Tillæa ascendens.
Pig nut,		Hicorya porcina.
" weed,	Green amaranth,	Amarantus albus.
" "		Chenopodium.
Pile lotus,		Lotus hirsutus.
Pilewort,	Princes' feather,	Amarantus hypochondriacus.
"	Buttercup,	Ranunculus bulbosus.
"	Fireweed,	Erechthites hieracifolius.
Pillwort,		Pilularia globulifera.

COMMON.	ENGLISH.	BOTANICAL.
Pilot weed,	Compass plant (weed,)	Silphium laciniatum.
Pimento,	Allspice,	Eugenia (myrtus) Pimenta.
Pimpernel,		Pimpinella saxifraga.
" Blue,	Scullcap,	Scutellaria lateriflora.
" Italian,	Garden burnet,	Sanguisorba officinalis.
" Red,	Red chickweed,	Anagallis arvensis.
" Scarlet,	" "	" "
" Water,	Brook lime,	Veronica beccabunga.
Pimpinel,		Pimpinella saxifraga.
Pimpinella,	Great pimpinella,	" magna.
Pinckney bark,	Florida bark,	Pinckneya pubens.
Pine,		
" apple,		Ananassa sativa.
" bark, White,		Pinus strobus.
" Black,		" austriaca.
" Broom,	Yellow pitch pine,	" palustris.
" Cluster,		" pinaster and maritima.
" Cowrie,		Dammara Australis.
" Ground,		Ajuga chamæpitys.
" Loblolly,	Old field pine,	Pinus tadæa.
" Long leaved,	Yellow pitch pine,	" palustris.
" Norway,	Red pine,	" resina.
" nuts,	the seed of stone pine,	" cembra and pinea.
" Pitch,		" palustris (rigida.)
" pitch, ·	see Thus.	
" sap,	False beech drops,	Monotropa hypopitys.
" Scotch,		Pinus sylvestris.
" Stone,		" cembra and pinea.
" tops,	Hemlock leaves,	" (abies) Canadensis.
" tulip,	Pipsissewa,	Chimaphila umbellata.
" weed,		Hypericum sarothra.
" White,		Pinus strobus.
" wool,	the leaf fibre of	" sylvestris.
" Yellow pitch,		" palustris.
Piney tallow,	a solid, oily substance of	Vateria Iudica.
" thistle,		Atractylis gummifera.
" tree,		Calophyllum angustifolium.
" varnish,	Liquid copal,	Vateria Indica.
Pingo pingo,	a fibrous plant,	Ephedra Americana.
Pink catchfly,	Catch fly,	Silene Virginica.
Pink,		Dianthus.
" Carolina,		Spigelia Marilandica.
" China,		Dianthus Chinensis.
" Indian,	Carolina pink,	Spigelia Marilandica.
" Mountain,	Trailing arbutus,	Epigæa repens.
" root,		Spigelia Marilandica.
" " West Indian,		" anthelmia.
" snake root,		Ophiorhiza mitreola.
" Swamp,		Azalea nitida.
" weed,	Knot weed,	Polygonum persicaria.
" Winter,	Trailing arbutus,	Epigæa repens.
Pin weed,	Hemlock storksbill,	Erodium cicutarium.
Pinxter flower,		Azalea nudiflora.
Pipe plant,	Ice plant, American,	Monotropa uniflora.
" stem,		Andromeda nitida.
" tree,	Syringia,	Syringa vulgaris.
" vine,		Siphisia glabra.
Pipi pods,		Cæsalpina pipi.
Pipperidge,	Barberry,	. Berberis vulgaris.
Pipsissewa,		Chimaphila umbellata.

COMMON.	ENGLISH.	BOTANICAL.
Pipsissewa, Spotted,		Chimaphila maculata.
Pistachio,	Witch hazel,	Hamamelis Virginica.
" nuts,		Pistachia vera.
Pistacia galls,	Mastich tree galls,	" lentiscus.
Pitaya bark,		Cinchona (Pitaya) Condaminea.
Pitch, Black,	the residuum after distilling tar.	
" Burgundy,	the resin procured from	Abies excelsa and picea.
" Canada,	Hemlock gum,	" (Pinus) Canadensis.
" pine,		Pinus palustris (rigida.)
" seed plant,		Pittosporum tobira.
Pitcher plant,		Nepenthes.
" "	Side saddle plant,	Sarracenia purpurea.
" " California,		Darlingtonia Californica.
Plaintain,	the fruit resembles Bananas,	Musa paradisiaca.
Plane tree,	Button wood,	Platanus occidentalis.
" " Oriental,		" orientalis.
Plantain,		
" Common,		Plantago major.
" Dwarf,		" Virginica.
" Heart leaved,	Water plantain,	" cordata.
" Hartshorn,		" coronopus.
" Net leaved,		Goodyera pubescens.
" Robert's,		Erigeron bellidifolium.
" Snake,		Plantago lanceolata.
" tree,	see Plaintain,	Musa paradisiaca.
" Water,		Plantago cordata.
" White,		Goodyera repens.
Pleurisy root,		Asclepias tuberosa.
Plum,	Prune,	Prunus domesticus.
" Beach,		" maritima.
" French,	Prune,	" domesticus.
Plowman's spikenard,		Inula conyza.
" wort,	Marsh fleabane,	Conyza Marilandica.
Pocan bush,	Garget,	Phytolacca decandra.
Pock wood,	Guaiac wood,	Guaiacum officinalis.
Poinsettia,		Euphorbia pulcherrima.
Pointed cleavers,		Galium asprellum.
Poison ash,	Poison sumach,	Rhus venenata.
" "		Amyris toxifera.
" "	False fringe tree,	Rhus cotinus.
" berry,	Red cohosh,	Actæa rubra.
" bulb,		Crinum Asiaticum.
" dogwood,	Poison sumach,	Rhus vernix.
" elder,	" "	" "
" flag,	Blue flag,	Iris versicolor.
" hemlock,		Conium maculatum.
" ivy,		Rhus Toxicodendron var. radicans.
" nut,	Nux vomica,	Strychnos Nux Vomica.
" oak,		Rhus Toxicodendron.
" parsley,	Conium, spotted,	Cicuta (conium) maculata.
" root,	Poison hemlock,	Conium maculatum.
" sumach,		Rhus vernix.
" tobacco,	Henbane,	Hyoscyamus niger.
" vine,	Poison ivy,	Rhus Toxicodendron var. radicans.
" "		" radicans.
" wood,	Poison sumach,	" vernix.
Poke root,	Garget root,	Phytolacca decandra.
" Indian,	American hellebore,	Veratrum viride.
" Virginia,	Garget root,	Phytolacca decandra.
" weed,	" "	" "

COMMON.	ENGLISH.	BOTANICAL.
Polar plant,	Compass weed,	Silphium laciniatum.
Polecat weed,	Skunk cabbage,	Symplocarpus fœtidus.
Polyanthus,	Cowslip primrose,	Primula veris.
Polypod, Rock,	Polypody,	Polypodium vulgare.
Polypody,		" "
Pond dog wood,	Button bush,	Cephalanthus occidentalis.
" lily, White,	White pond lily,	Nymphæa odorata.
" " Yellow,	Yellow pond lily,	Nuphar advena.
" spice,		Tetranthera geniculata.
Pomegranate, .	the fruit of	Punica Granatum.
" bark,	the rind of the fruit,	" "
" root, bark,		" "
" peel,	the rind of the fruit,	" "
Pompion,	Pumpkin,	Cucurbita Pepo.
Pool root,	White sanicle root,	Eupatorium ageratoides.
" "	" snake root,	" · aromaticum.
" wort,	" " "	" "
Poon wood tree,	Alexandrian laurel,	Calopyllum inophyllum.
Poor man's weatherglass,	Red chickweed, .	Anagallis arvensis.
" " pharmacetty,	Shepherds' purse,	Capsella bursa-pastoris.
" " pepper,	Pepperwort,	Lepidium sativum.
" " rhubarb,	Meadow rue,	Thalictrum dioicum.
" robin,	Cleavers,	Galium aparine.
Poplar,	Abele tree,	Populus alba.
" American,	.	" tremuloides.
" Balsam,		. " balsamifera.
" Black,		" nigra.
" buds,		" "
" · European,	European aspen,	" tremulus.
" Tacamahac,	Balsam poplar,	" balsamifera.
" White,	Aspen poplar,	" alba.
" "	Tulip tree,	Liriodendron Tulipifera.
" Yellow,	" "	" "
Portland sago,	a starch obtained from	Arum maculatum.
Portulaca,	Garden portulaca,	Portulaca grandiflora.
Portugal,	the essential oil of sweet orange peel.	
Poppy,		Papaver somniferum.
" Black,		" " var. nigrum.
" Blue,	the seed, called Maw seed,	" " "
" California,		Eschscholtzia Californica.
" capsules,	the seed capsules,	Papaver somniferum.
" heads,	Poppy capsules,	" "
" Horned,		Glaucium luteum.
" Mexican,	Prickly poppy,	Argemone Mexicana.
" Opium,	the opium producing poppy,	Papaver somniferum.
" · Red,	Corn poppy,	" rhœas.
Potato,		Solanum tuberosum.
" pea,	Indian potato,	Apios tuberosa.
" Sweet,		Convolvulus (Ipomæa) batatus.
" Wild,	Man root,	" panduratus.
Poverty weed,	Cow wheat,	Melampyrum arvense.
" "	Life everlasting,	Gnaphalium polycephalum.
Prairie burdock,		Silphium terebinthinaceum.
" dock,	Cutting almond,	Parthenium integrifolium.
" "	Rosin weed,	Silphium gummiferum.
" grub,	Ptelea bark,	Ptelea trifoliata.
" hyssop,	Narrow leaved Va. thyme,	Pycnanthemum Virginicum.
" indigo,		Baptisia alba.
" pine,	Button snake root,	Liatris spicata.
" senna,		Cassia chamæcrista.

COMMON.	ENGLISH.	BOTANICAL.
Prairie turnip,	Bread root,	Psoralea esculenta.
" weed,	Rosin weed,	Silphium gummiferum.
Prickly ash,		Xanthoxylum fraxineum.
" " Northern,		" "
" " Southern,		" Carolinianum.
" " Yellow,		" clava Herculis.
" elder,		Aralia spinosa.
" lettuce,	Acrid lettuce,	Lactuca virosa.
" pear cactus,		Cactus opuntia.
" poppy,	Mexican poppy,	Argemone Mexicana.
" yellow wood,	Southern prickly ash,	Xanthoxylum Carolinianum.
" " " W. I.	Yellow " "	" clava Herculis.
Prick-madam,	White stone crop,	Sedum album.
Prickwood,		Euonymus Americana.
Pride of China,	Bead tree,	Melia Azedarach.
" " India,	" "	" "
" " the meadow,	Meadow sweet,	Spiræa ulmaria.
" tree,	Bead tree,	Melia Azedarach.
" weed,	Canada ficabane,	Erigeron Canadense.
Priests' crown,	Dandelion,	Taraxacum Dens-leonis.
" pintle,	Dragon root,	Arum triphyllum.
Prim,	Privet,	Ligustrum vulgare.
Primwort,	" .	" "
Primrose,	.	Primula officinalis.
" Chinese,		" sinensis.
" Common,		" vulgaris.
" Cowslip,	English cowslip primrose,	" veris.
" Evening,	Scabish,	Œnothera biennis.
" Peerless,		Narcissus biflora.
Princes' feather,	Amaranth,	Amarantus caudatus.
" "	Pilewort,	" hypochondriacus.
" pine,	Pipsissewa,	Chimaphila umbellata.
Privet,		Ligustrum vulgare.
Privy,	Privet,	" "
Procession flower,	Milkwort,	Polygala incarnata.
Provins rose,	French rose,	Rosa Gallica.
Prunells,	the stoned fruit of	Prunum Brignolense.
Prunes,	the dried fruit of	" domesticus.
Ptelea,	Wafer ash,	Ptelea trifoliata.
Public house plant,	Asarabacca,	Asarum Europæum.
Puccoon, Red,	Blood root,	Sanguinaria Canadensis.
" Yellow,	Goldenseal root,	Hydrastis "
Pucha pat,	Patchouly,	Pogostemon Patchouly.
Puchurim,	Sassafras nut,	Nectandra puchury.
Pucker needle,	Shepherds' needle,	Scandix pecten.
Pudding grass,	European pennyroyal, -	Mentha pulegium.
" pipe tree,	Purging cassia,	Cassia Fistula.
Puff-ball,		Lycoperdon proteus.
"	Dandelion,	Taraxacum Dens-leonis.
"		Bovista nigrescens.
Puke weed,	Lobelia,	Lobelia inflata.
Pull pipes,		Equisetum.
Pulsatilla,	Meadow anemone,	Anemone Pulsatilla.
Pulse,	the common name for Peas, beans, etc.	
Pumpkin,	.	Cucurbita Pepo.
" seed,	Common pumpkin seed,	" "
Punic apple,	Pomegranate fruit,	Punica Granatum.
Punk,	Touch wood,	Agaric.
Purging agaric,	Larch agaric,	Boletus laricis.
" berries,	Buckthorn berries,	Rhamnus catharticus.

7

COMMON.	ENGLISH.	BOTANICAL.
Purging buckthorn,	Buckthorn berries,	Rhamnus catharticus.
" cassia,		Cassia Fistula.
" " W. I.,	Horse cassia,	" Brasiliana.
" flax,		Linum catharticum.
" . nuts,	Barbadoes nuts,	Curcas purgans.
" . root,	Blooming spurge,	Euphorbia corollata.
Purple angelica,		Angelica atropurpurea.
" archangel,	Bugle weed,	Lycopus Virginicus.
" avens,		Geum rivale.
" boneset,	Queen of the meadow,	Eupatorium purpureum.
" .cone flower,	Red sunflower,	Rudbeckia purpurea.
" everlasting,		Xeranthemum annuum.
" foxglove,		Digitalis purpurea.
" fringe,	Smoke tree,	Rhus cotinus.
" leptandra,	Black root,	Leptandra Virginica.
" perilla,		Perilla Nankinensis.
" stramonium,		Datura tatula.
" willow herb,	Loose strife,	Lythrum salicaria.
Purslain, Milk,	Blooming spurge,	Euphorbia corollata.
" White,	" "	" "
Purslane,		Portulaca oleracea.
" Garden,	Wax pinks,	" grandiflora.
" speedwell,		Veronica peregrina.
Pursley,	Purslane,	Portulaca oleracea.
" Black,	Large spotted spurge,	Euphorbia maculata.
" Milk,	" " "	" . "
Purvain,	Vervain,	Verbena hastata.
Pussy willow,	Black willow,	Salix nigra.
Putcha pat,	Patchouly,	Pogostemon Patchouly.
Putty root,		Aplectrum hyemale.
Pycnanthemum,		Pycnanthemum pilosum.
Pyramid flower,	American columbo,	Frasera Carolinensis.
" plant, ·	" "	" "
Pyrethrum,	Pellitory of Spain,	Anacyclus Pyrethrum.
Pyrola,	Pipsissewa,	Pyrola (Chimaphila) umbellata.
" Round leaved,		" rotundifolia.
Quack grass, ·	Couch grass,	Agopyrum repens.
Quafodil,	Black flower,	Melanthium Virginicum.
Quaker button,	Nux vomica,	Strychnos Nux Vomica.
Quaking asp,	Aspen poplar,	Populus tremuloides.
" aspen,	" "	" "
" grass,		Briza maxima.
Quamash, ·		Camassia esculenta.
Quassia,	Bitter wood,	Simaruba excelsa.
Queen of the meadow,	Meadow sweet,	Spiræa ulmaria.
" " "	Purple boneset,	Eupatorium purpureum.
" " " rt.,	" "	" . "
" " . prairie,		Spiræa lobata.
Queens' delight,	Stillingia,	Stillingia sylvatica.
" root, ·	"	" "
Quercitron,	Black oak bark,	Quercus tinctoria.
Quickens,	Dog grass,	Triticum repens.
Quick in the hand,	Celandine, Wild,	Impatiens pallida.
Quickset,	Hawthorn,	Cratægus oxyacantha.
Quilmai,		Echites Chilensis.
Quince,		Cydonia vulgaris.
" Bengal,	Bael, ·	Ægle marmelos.
" cores,	the cores of the fruit,	Cydonia vulgaris.
" seed,		"
Quinquinia bark,	False Peruvian bark,	Buena obtusifolia and hexandra.

COMMON.	ENGLISH.	BOTANICAL.
Quinsy berry,	Black currant,	Ribes nigrum.
" wort,		Asperula cynanchica.
Quillai bark,	Soap tree bark,	Quillaya saponaria.
Quillaya bark,	" " "	" "
Quill wort,		Isoetes lacustris.
Quitch,	Dog grass,	Triticum repens.
Quitel,	Black root,	Leptandra Virginica.
Quiver leaf,	American poplar,	Populus tremuloides.
" tree,		Aloe dichtoma.
Rabbit berry,	Buffalo berry,	Shepherdia argentea.
Rabbits' foot,	Field clover,	Trifolium arvense.
" mouth,	Snap dragon,	Antirrhinum majus.
" root,	American sarsaparilla,	Aralia nudicaulis.
Raccoon berry,	" mandrake,	Podophyllum peltatum.
Race ginger,	name applied to crude ginger root.	
Radish,	Garden radish, ,	Raphanus sativus.
" Common garden,		" "
" Horse,	Horse radish,	Cochlearia armoracia.
" Water,		Nasturtium amphibium.
" Wild,	Charlock,	Sinapis arvensis.
Ragged cup,	Indian cup plant,	Silphium perfoliatum.
" lady,	Fennel flower,	Nigella Damascena.
" robin,		Lychnis flos-cuculi.
" sailor,	Princes' feather,	Polygonum orientale.
Ragweed,		Ambrosia elatior.
" Common,	Am'n Roman wormwood,	" artemisiæfolia.
" Giant,	Tall ambrosia,	" trifida.
" Great,	" "	" "
Ragwort,	Life root plant,	Senecio aureus.
"		" Jacobæa.
Rainbow weed,	Sage willow,	Lythrum salicaria.
Raisins,	Dried grapes,	Vitis vinifera.
" Corinthian,	Black currants,	" Corinthiaca.
Raisin tree,	Red currant bush,	Ribes rubrum.
Ramgoat,		Fagara microphylla.
Ramtil,		Guizotia oleifera.
Rampion,		Phyteuma obiculare.
Ramshead,	American valerian,	Cypripedium arietinum.
Ramsons,		Allium ursinum.
Ranstead weed,	Toad flax,	Antirrhinum Linaria vulgaris.
Rape seed,	Cole seed,	Brassica campestris.
Rapper dandies,	the fruit of bearberry,	Arctostaphylos Uva Ursi.
Raspberry,		Rubus strigosus.
" Garden,		" idæus.
" Ground,	Goldenseal root,	Hydrastis Canadensis.
" Red, .		Rubus strigosus.
" " leaves,		" "
" Rose flower'g,		" odoratus.
Ratsbane,	Nux vomica,	Strychnos Nux Vomica.
Rattan,		Calamus rotang and other species.
Rattle-box,		Crotalaria sagittalis.
Rattle-bush,	Wild indigo,	Baptisia tinctoria.
Rattle-nut,	Yellow water lily, .	Nelumbium luteum.
Rattle-root,	Black cohosh,	Cimicifuga racemosa.
Rattle-weed,	" "	" "
Rattlesnake fern,		Botrychium fumarioides and other
" flag,	Corn snake root,	˒ Eryngium yuccefolium. [species.
" grass,	Quaking grass,	Briza maxima.
" herb,	Baneberry,	Actæa spicata.
" leaf,	Net leaf plantain,	Goodyera pubescens.

COMMON.	ENGLISH.	BOTANICAL.
Rattlesnakes' master,	Water eryngo,	Eryngium aquaticum.
" "	Button snake root,	Liatris spicata.
" "	Blazing star,	" squarrosa.
" "	False aloe,	Agave Virginica.
" plantain,	White plantain,	Antennaria plantaginoum.
" "		Goodyera repens.
" root,	Lions' foot,	Nabalus albus.
" "	Beth root,	Trillium pendulum.
" "	Black cohosh,	Cimicifuga racemosa.
" violet,	Adders' tongue,	Erythronium Americanum.
" "		Viola ovata.
" weed,	Hawkweed,	Hieracium venosum.
" "	Scrofula weed,	Goodyera pubescens.
" "	Water eryngo,	Eryngium aquaticum.
Ravencheeney,	see Gamboge.	
Red alder,	Tag alder,	Alnus rubra.
" astringent gum,	Bengal kino,	Butea frondosa.
" balm,	Oswego tea,	Monarda didyma.
" baneberry,	Red cohosh,	Actœa rubra.
" bark,	" Peruvian bark,	Cinchona succiruba.
" " Cusco,	" " · "	" scorbiculata.
" " Lima,	" " "	" ovata.
" bay,	Magnolia,	Magnolia glauca.
" bean,	Love pea,	Abrus precatorius.
" berry,	Giuseng, ₎	Panax quinquefolium.
" " snake root,	Red cohosh,	Actœa rubra.
" " tea,	Checkerberry,	Gaultheria procumbens.
" berried elder,		Sambucus pubescens.
" bilberry,	Cowberry,	Vaccinium Vitis-idœa.
" bryony,		Bryonia dioica.
" buckeye,		Æsculus pavia.
" bud,	Judas tree,	Cercis Canadensis.
" cabbage,		Brassica oleracea rubra.
" campion,	White soap root,	Lychnis diurna.
" cardinal flower,	Red lobelia,	Lobelia cardinalis.
" Carthagena bark,	Bogota bark,	Cinchona condaminea.
" cedar,		Juniperus Virginiana.
" centaury,	American centaury,	Sabbatia angularis.
" cherry,		Prunus cerasus.
" chickweed,		Anagallis arvensis.
" clover,		Trifolium pratense.
" cockscomb,	Princes' feather,	Amarantus hypochondriacus.
" cohosh,	Red baneberry,	Actœa rubra.
" columbine,		Aquilegia Canadensis.
" currant,	Common Garden Currant,	Ribes rubrum.
" elm,	Slippery elm,	Ulmus fulva.
" head,	Bastard ipecac,	Asclepias curassavica.
" hot poker-plant,		Tritoma uvaria.
" ink plant,	Garget (berries,)	Phytolacca decandra.
" Jasmine,	Cypress vine,	Ipomœa quamoclit.
" " W. I.,	Frangipanni,	Plumieria rubra.
" knees,	Water pepper,	Polygonum hydropiper.
" laurel,	Magnolia,	Magnolia.
" legs,	Bistort,	Polygonum Bistorta.
" lobelia,	Red cardinal flower,	Lobelia cardinalis.
" maple,		Acer rubrum.
" morrocco,	Pheasants' eye,	Adonis autumnalis.
" mulberry,		Morus rubra.
" oak,		Quercus rubra.
" osier,	Swamp dogwood,	Cornus sericea.

COMMON.	ENGLISH.	BOTANICAL.
Red paint root,	Blood root,	Sanguinaria Canadensis.
" pepper,	Capsicum,	Capsicum annuum and other spec.
" pimpernel,	Red chickweed,	Anagallis arvensis.
" pollom,	Checkerberry,	Gaultheria procumbens.
" poppy,	Corn poppy,	Papaver rhœas.
" puccoon,	Blood root,	Sanguinaria Canadensis.
" raspberry,		Rubus strigosus.
" rattle,	Louse wort,	Pedicularis Canadensis.
" rod,	Swamp dogwood,	Cornus sericea.
" rood,	" . "	" "
" root,	Jersey tea,	Ceanothus Americana.
" "	Blood root,	Sanguinaria Canadensis.
" " Indian,	Spirit weed,	Lacnanthes tinctoria.
" " bark,	Jersey tea,	Ceanothus Americana.
" rose,		Rosa Gallica.
" sandal-wood,	Red saunders,	Pterocarpus Santalinus.
" santal,	" "	" "
" saunders,		" "
" seed,	False unicorn,	Helonias dioica.
" shanks,	Herb Robert,	Geranium Robertanium.
" "	Heartsease,	Polygonum persicaria.
" squills,		Scilla maritima.
" stalked aster,	Cocash,	Aster puniceus.
" sunflower,		Rudbeckia purpurea.
" top grass,		Agrostis vulgaris.
" weed,	Garget,	Phytolacca decandra.
" willow,	Swamp dogwood,	Cornus sericea.
" wood,	Red saunders,	Pterocarpus Santalinus.
Red-wood, Bahama,	Redwood tree,	Soymoida febrifuga.
" " -		Ceanothus colubrinus.
" California,		Sequoia gigantea.
" Jamaica,	Blood wood,	Gordonia Hæmatoxylon.
" of Turkey,	Cornelian cherry,	Cornus mascula.
" tree,		Soymoida febrifuga.
" " Indian,		Swietenia "
Redoul,	Tanners' sumach,	Coriaria myrtifolia.
"		Rhus Coriaria.
Reed, Common,		Phragmites communis.
Reedmace,	Cat tail flag,	Typha latifolia.
Reindeer moss,		Lichen rangiferinus.
Reine-claude,	Green gage plums,	Prunus domesticus.
Relbun,	a yellow dye-stuff,	Calceolaria arachnoides.
Resin of thapsia,		Thapsia garganica.
Rest harrow,		Ononis spinosa.
Resurrection plant,		Selaginella lepidophylla.
" "	Rose of Jericho,	Anastatica hierochuntica.
Rhapontic root,	French rhubarb,	Rheum rhaponticum.
Rhatany root,		Krameria triandria.
" Para,	Brazilian rhatany,	" argentea.
" West Indian,	Savanilla root,	" ixina.
Rheumatic weed,	Sampson's snake root,	Aster æstivus.
Rheumatism root,	Twin leaf root,	Jeffersonia diphylla.
" weed,	Spotted pipsissewa,	Chimaphila maculata.
Rhodes wood,	Jamaica rosewood,	Amyris balsamifera.
Rhodium wood,		Convolvulus (Rhodorrhiza) scopa-
Rhubarb,		Rheum. [parious.
" American,		" Rhaponticum (hybridum.)
" Bucharian,	Russian Rhubarb,	" Russicum.
" Chinese,		" Palmatum.
" English stick,		" Rhaponticum.

COMMON.	ENGLISH.	BOTANICAL.
Rhubarb, Crimea,	.	Rheum Rhaponticum.
" European,		" "
" French,		. " " (undulatum).
" Garden,		" "
" India,		" Sinense Indicum.
" Monks',		Rumex alpinus.
" Russian,		Rheum Russicum.
" Tartarian,		" leucorrhizum.
" Turkey,	.	" Palmatum.
Ribbon grass,	Wild canary grass,	Phalaris Americana.
Rib grass,		Plantago lancifolia.
" "	Lambs' lettuce,	" media.
" wort,	Snake plantain,	" lanceolata.
Rice,		Oryza sativa.
" Indian,	Wild rice,	Zizania aquatica.
" paper,		Aralia Japonica (papyrifera).
Richleaf,	Stone root,	Collinsonia Canadensis.
Richweed,	Black cohosh,	Cimicifuga racemosa.
"	Stone root,	Collinsonia Canadensis.
"	Tall ambrosia,	Ambrosia trifida.
"	Stingless nettle,	Urtica pumila.
Riga balsam,	Carpathian balsam,	Pinus cembra.
Rigid goldenrod,		Solidago rigida.
Ring worm bush,		Cassia alata.
Ripple grass,	Ribwort,	Plantago lanceolata.
Robert's plantain,	.	Erigeron bellidifolium.
Robin-run-away,	Ground ivy,	Nepeta glechoma.
Robin-run-in-the-hedge,	" "	" " .
Robins' rye,	Haircap moss,	Polytrichum Juniperum.
Rock blackberry,	Brier herb,	Rubus saxatilis.
" brake,	Polypody,	Polypodium vulgare.
" " .		Pteris atropurpurea.
" cress,	Mouse ear cress,	Arabis thaliana.
" fern,	Maidenhair, American,	Adiantum pedatum.
" maple,	Sugar maple,	Acer saccharinum.
" moss,		Gyrophora.
" polypod,	Polypody,	Polypodium vulgare.
" rose,	Frost weed,	Cistus Canadensis.
" "		Helianthemum Canadensis.
" tripe,		Umbilicaria.
" weed,	Herb Robert,	Geranium Robertanium.
Rocket,		Hesperis.
" candytuft,		Iberis coronaria.
" larkspur,		Delphinium ajacis.
" Sea,		Bunias Americana.
" Yellow, herb,	Winter cress,	Barbarea vulgaris.
Rod, Aaron's,	House leek,	Sempervivum tectorum.
" Red,	Rose willow,	Cornus sericea.
Rogation flower,	Milkwort,	Polygala vulgaris (incarnata).
Rohun-bark,	Soymida Red wood,	Swietenia febrifuga.
Roman chamomile,		Anthemis nobilis.
" fern,		Blechnum borealis.
" wormwood,		Artemisia pontica.
" " Am'n,		Ambrosia artemisiæfolia (elatior).
- " . "		" elatior.
Rope bark,	Leatherwood,	Dirca palustris.
Ropewind,	Bind weed,	Convolvulus purpurea.
Rose,		Rosa.
" acacia,	Moss locust,	Robinia hispida.
" apple,		Eugenia Jambos.

COMMON.	ENGLISH.	BOTANICAL.
Rose African,	Corn poppy,	Papaver Rhœas.
" Ayrshire,		Rosa arvensis.
" balm,		Monarda coccinea.
" bay,	Willow herb,	Epilobium angustifolium.
" "		Rhododendron ponticum.
" Bengal,		Rosa Indica, var.
" betty,	Robert's plantain,	Erigeron bellidifolium.
" Bourbon,		Rosa Indica, var.
" Brier,	Dog rose,	" Canina.
" Cabbage,	Pale "	" centifolia.
" campion,		Agrostemma coronaria.
" Cherokee,		Rosa sinica.
" Chinese,	Shoe black plant,	Hibiscus rosa sinensis.
" Christmas,	Black hellebore,	Helleborus niger.
" Cinnamon,		Rosa cinnamomea.
" colored silkweed,	White Indian hemp,	Asclepias incarnata.
" Corn,	Corn poppy,	Papaver rhœas.
" Cotton,		Filago Germanica.
" Damask,		Rosa Damascena.
" Dog,		" canina.
" elder,		Virburnum opulus.
" Evergreen,		Rosa sempervirens.
" French,	Red rose,	" Gallica.
" geranium,		Pelargonium odoratissimum.
" Guelder,	Snowball,	Viburnum roseum.
" Holly,		Helianthemum.
" Hundred leaved,	Pale rose,	Rosa centifolia.
" laurel,		Kalmia latifolia.
" "	Oleander,	Nerium oleander.
" Malabar,		. Hibiscus rosa Malabarica.
" Mallow,		" moschatus.
" Monthly,		Rosa Indica.
" Moss,		" centifolia, var.
" Musk,		" moschata.
" of China,	Shoe black plant,	Hibiscus rosa sinensis.
" of Jericho,		Anastatica lepidophylla.
" of Sharon,		Hybiscus Syriacus.
" Pale,		Rosa centifolia.
" pink,	American centaury,	Sabbatia angularis.
" Provins,	Red rose,	Rosa Gallica.
" Prairie,		" setigera.
" Red,		" Gallica.
" Sage,		Turnera ulmifolia.
" Scotch,		Rosa spinosissima.
" South sea,	Oleander,	Nerium oleander.
" Sun,		Helianthemum.
" Sweet brier,		Rosa rubiginosa.
" Swamp,		" Carolina.
" White,		" alba.
" Wild,		" blanda, and alba.
" Yellow,		" sulphurea.
Rose willow,	Swamp dogwood,	Cornus sericea.
" " Red,	" "	" "
Rosewood,	Rhodium wood,	Convolvulus scoparius.
" African,		Pterocarpus erinaceus.
" Canary,		Rhodorrhiza (Convolvulus) scopa-
" Jamaica,		Amyris balsamifera. [ria.
Rosewort,		Rhodiola rosea.
Rosey bush,	Hardhack,	Spiræa tomentosa.
Rosemalloes,	Liquid storax,	Liquidambar orientale.

COMMON.	ENGLISH.	BOTANICAL.
Rosemary,		Rosmarinus officinalis.
" Garden,	-	" "
" Marsh,		Statice Caroliniana.
" Wild,	Marsh tea,	Ledum palustre.
Rosin weed,		Silphium gummiferum.
Rottlera,	Kameela,	Rottlera tinctoria.
Roucou,	Annatto,	Bixa orellana.
Rouge plant,		Rivina tinctoria.
Rough boneset,	Wild horehound,	Eupatorium teucrifolium.
" gentian,	Blue gentian,	Gentiana Catesbœi.
" parsnip,	Opopanax,	Pastinaca opopanax.
" root,	Button snake root,	Liatris spicata.
" sunflower,		Helianthus divaricatus.
" woodbine,		Lonicera hirsuta.
Round birthwort,		Aristolochia rotunda.
" heart plant,		Thapsia trifoliata.
" leaved cornel,	Round leaved dogwood,	Cornus circinata.
" " dogwood,	" " "	" "
" " plantain,	Common plantain,	Plantago major.
" " pyrola,		Pyrola rotundifolia.
" " wintergreen,		" "
" mandrake,	Sword lily,	Gladiolus communis.
" ramsom,	" "	" "
" tree,	Mountain ash, ·	Sorbus (Pyrus) Americana.
" wood,	" "	" " "
Rowan tree,	" "	Pyrus Americana.
Roxbury wax work,	False bittersweet,	Celastrus scandens.
Royal cow parsnip,	a variety of	Heracleum lanatum.
" flowering fern,	Buckhorn brake,	Osmunda regalis.
" yellow bark,	. ·	Cinchona calisaya vera.
Ruby wood,	Red sandal wood,	Pterocarpus santolinus.
Rue,		Ruta graveolens.
" Garden,		" "
" Goats',		Tephrosia Virginiana.
" Meadow,		Thalictrum dioicum.
Rum cherry,	Wild black cherry,	Cerasus Virginiana.
Rupture wort,		Herniaria glabra.
Rush, Common,	Bulrush,	Juncus effusus.
" Dutch,	Scouring rush,	Equisetum hyemale.
" Polishing,	" "	" "
" Scouring,		" "
Rusat,	Barberry of India,	Berberis lycium.
Russia seeds,	Manna croup,	Glyceria (Festuca) fluitans.
Ruswut,	Barberry of India,	Berberis lycium.
Rutabaga	Swedish turnip,	Brassica campestris rutabaga.
Rye,		Secale cereale.
" Robins',	Haircap moss,	Polytrichum Juniperum.
" Smut of,	see Ergot,	Sclerotium clavus.
" Spurred,	" "	" "
Sabadilla seed,	Cevadilla,	Veratrum sabadilla.
Sacred bean,	Rattlenut,	Nelumbium luteum.
Saddle flower,		Darlingtonia Californica.
" plant,	Side saddle plant,	Sarracenia purpurea.
Sad tree,	Night flower,	Nyctanthes arboratristis.
Safflower,	Dyers' saffron,	Carthamus tinctorius.
Saffron,		Crocus sativus.
" African,		Lyperia crocea.
" American,	Dyers' saffron,	Carthamus tinctorius.
" Brazilian,	Fuminella,	
" Bastard,	Dyers' saffron,	Carthamus tinctorius.

COMMON.	ENGLISH.	BOTANICAL.
Saffron, Ceylon,		Crocus orientalis.
" Dyers',		Carthamus tinctorius.
" Meadow,	Colchicum,	Colchicum autumnale.
" Sicilian,		Crocus odorus.
" Spanish,		" sativus.
" thistle,	American saffron,	Carthamus tinctorius.
" wood,		Elæodendron croceum.
Sagackhomi,	Bearberry,	Arctostaphylos Uva Ursi.
Sagapenum,	a gum resin,	
Sage,		Salvia officinalis.
" Apple,	a gall found on Crete sage,	
" Indian,	Boneset,	Eupatorium perfoliatum.
" leaf mullein,	Jerusalem sage,	Phlomis tuberosa.
" Lyre leaved,	Meadow sage,	Salvia lyrata.
" Meadow,		" "
" Nettle,		" urticifolia.
" tree,	Bahama tea,	Lantana camara.
" Vervain,		Salvia Claytonia.
" Wild,		" lyrata.
" willow,	Purple willow herb,	Lythrum salicaria.
" Wood,	Germander,	Teucrium Canadensis
Sago,	prepared farina from the pith of Sagus Rumphii.	
" palm,		" "
" Pearl,	Sago,	" "
Sailor plant,	Strawberry geranium,	Saxifraga sarmentosa.
Sainfoin,	French grass,	Onobrychis sativa.
Saint Andrew's cross,		Ascyrum crux andræa.
" Ann's bark,		Cinchona scorbiculata genuina.
" Catherine prunes,		Prunus domesticus.
" Christopher herb,	Buckhorn male fern,	Osmunda regalis.
" Ignatius' bean,		Strychnos Ignatia.
" James' weed,	Shepherds' purse,	Capsella bursa-pastoris.
" " wort,	Ragwort,	Senecio Jacobæa.
" John's bread,		Ceratonia siliqua.
" " dogsbane,		Apocynum hypericifolium
" " wort,		Hypericum perforatum.
" Kitt's arrow root,	Canna starch,	Canna edulis and var.
" Lucia bark,	Caribæan bark,	Exostemma floribunda.
" " wood,	Perfumed cherry wood,	. Cerasus mahaleb.
" Martin's wort,		Sauvagesia erecta.
" Martha's wood,	Peach wood,	Cæsalpina bijuga and other species●
' Peter's wort,	Cowslip primrose,	Primula veris.
" Thomas' tree,		Bauhinia tomentosa.
Salad,	Garden lettuce,	Lactuca sativa.
Salep,	the prepared bulbs,	Orchis mascula and other species.
Saloop,	Sassafras,	Sassafras officinalis.
Salsafy,	Vegetable oyster,	Tragopogon porrifolius.
Salt rheum weed,	Balmony,	Chelone glabra.
" rock moss,	Irish moss,	Chondrus crispus.
" wort,		Salicornia annua.
" " prickly,	yields barilla,	Salsola kali.
Sampfen wood,	Brazil wood,	Cæsalpina sappan.
Samphire,	Glasswort,	Salicornia herbacea.
Sampson root,		Gentiana ochroleuca.
" snake root,		" "
" " "		Aster æstivus.
Sandbox tree,		Hura crepitans.
Sand-myrtle,	Sleek leaf,	Leiophyllum buxifolium.
Sand-nettle,		Janipha stimulosa.
Sand plum,	Beech plum,	Prunus maritima.

COMMON.	ENGLISH.	BOTANICAL.
Sand weed,	Spurry,	Spergula arvensis.
" wort,		Arenaria marina.
Sandalwood,	the fragrant wood,	Santalum freycinetianum.
" Red,		Pterocarpus santalinus.
" White,		Santalum album.
" Yellow,		" freycinetianum.
Sandarach,	a gum resin from	Thuja articulata.
Sanders wood,	Red sandal wood,	Pterocarpus santalinus.
Sangree root,·	Virginia snake root,	Aristolochia Serpentaria.
Sangrel,	" " "	" "
Sanicle, American,	Alum root,	Heuchera Americana.
" Black,		Sanicula Marilandica.
" European,		" Europœa.
" root,		" "
" White,	·	Eupatorium ageratoides.
" Wood,	Sanicle, European,	Sanicula Europœa.
Santa Fe tea,		Alstonia theœformis.
Sanvitalia,		Sanvitalia procumbens.
Sapan wood,		Cæsalpina sappan.
Sap green,	the green color of Buckthorn berries.	
Sapodil,		Achras sapota.
Sapodilla plum,		" "
Sappan,		Cæsalpina sappan.
Saracen's corn,	Buckwheat,	Polygonum fagopyrum.
" woundwort,		Senecio saracenicus.
Sarcocolla,	a vegetable product of	Penæa sarcocolla and other species
Sarsaparilla,	Officinal Sarsaparilla,	Smilax officinalis " " "
" American,	Small spikenard,	Aralia nudicaulis.
" Australian,		Hardenbergia monophylla.
" Brazilian,		Smilax papyracea.
" Bristle stem,	Dwarf elder,	Aralia hispida.
" Caraccas,		Smilax papyracea.
" Country, E. I.		Hemidesmus Indicus.
" False,	Small spikenard,	Aralia nudicaulis.
" German,	Sea sedge,	Carex arenaria.
" Guatemala,		Smilax papyracea.
" Guayaquil,		" officinalis.
" Honduras,		" "
" Indian,		Hemidesmus Indicus.
" Italian,		Smilax aspera.
" Jamaica,		" officinalis.
" Lisbon,	Brazilian sarsaparilla,	" papyracea.
" Mexican,		" medica.
· " Rio Negro,	Brazilian sarsaparilla,	" papyracea.
" Spanish,		" cumanensis.
" Texas,	Yellow parilla,	Menispermum Canadense.
" Vera Cruz,	Mexican sarsaparilla,	Smilax medica.
" Virginia,		" Sarsaparilla.
" Wild,	American sarsaparilla,	Aralia nudicaulis.
" Yellow,	Yellow parilla,	Menispermum Canadense.
Sasa,	the oil of Cucumber seed,	
Sassafras,		Sassafras officinalis.
" Australian,		Antherosperma moschata.
" bark,		Sassafras officinalis.
" nuts,	Brazilian beans,	Nectandra puchury.
" Orinoco,	·	" cymbarum.
" pith,	the pith of the stems,	Sassafras officinalis.
" root,		" "
" Swamp,	Magnolia,	Magnolia glauca.
" White,		Laurus alba.

COMMON.	ENGLISH.	BOTANICAL.
Sassafras wood,		Sassafras officinalis.
Sassy bark,	Doom bark,	Erythrophleum judiciale (Guinien-
Satin flower,		Lunaria rediviva. [sis).
" "	Chickweed,	Stellaria media.
" wood,		Swietenia chloroxylon.
Satyr,	Orchis,	Orchis bracteata.
Sauce alone,	Pepperwort,	Lepidium sativa and other species.
Saunders,	Sandalwood,	Santalum.
" Red,	Red sandalwood,	Pterocarpus santalinus.
" White,	White sandalwood,	Santalinum album.
Savanna bark,	Alcornoque,	Quercus suber.
Savannah flower,	Chili jessamine,	Echites suaveolens.
Savanilla,	West India rhatany,	Krameria ixina.
Savin,		Juniperus Sabina.
Savory, Summer,		Satureja hortensis.
" Winter,		" montana.
Savoy,	Savoy cabbage,	Brassica oleracea major.
Savoyan,	Cleavers,	Galium aparine.
Sawgrass,		Schœnus effusus.
Saw wort,	Button snake root,	Liatris spicata.
" "		Serratula tinctoria.
Saxifrage, Chinese,	Beef steak plant,	Saxifraga sarmentosa.
" Burnet,		Pimpinella Saxifraga.
Saxifrax,	Sassafras,	Sassafras officinalis.
Scabious,	Canada fleabane,	Erigeron Canadense.
"	Devil's bit herb,	Scabiosa succisa.
". Sweet,	Philadelphia fleabane,	Erigeron Philadelphicum.
Scabish,	Evening primrose,	Œnothera biennis.
Scabwort,	Elecampane,	Inula Helenium.
Scaldweed,	Dodder,	Cuscuta Americana.
Scaly blazing star,	Blazing star root,	Liatris squarrosa.
" dragons' claw,	Crawley root,	Corallorhiza odontorhiza.
Scammony,	the resinous exudation of	Convolvulus Scammonia.
" Aleppo,	Scammony,	" "
" of France,		Cynanchum Monspellacum.
" Montpelier,	Scammony of France,	" "
" root,	Wild potato,	Convolvulus panduratus.
" Smyrnia,		Periploca græca.
" Spurious,	Scammony of France,	Cynanchum Monspeliacum.
" Virgin,	"	Convolvulus Scammonia.
" Wild,	Wild potato,	" panduratus.
Scarlet berry,	Bittersweet,	Solanum Dulcamara.
" lightning,	Cross of Jerusalem,	Lychnis diurna.
" mushroom,	Maltese sponge,	Cyromorum coccineum.
" pimpernel,	Red chickweed,	Anagallis arvensis.
" rose balm,	Oswego tea,	Monarda didyma.
" runner,	Spanish bean,	Phaseolus multiflorus.
" sage,	Mexican sage,	Salvia fulgens (splendens).
Scent wood,		Alyxia buxifolia.
Scoke,	Garget.	Phytolacca decandra.
Scorpion grass,	Forget-me-not,	Myosotis arvensis.
" plant,		Genista scorpius.
" senna,		Coronilla emerus.
" · weed,		Myosotis palustris.
Scotch bonnets,		Capsicum tetragonum.
" scurvy grass,	Sea bindweed,	Convolvulus soldanella.
Scratch weed,	Cleavers,	Galium aparine.
Screw pod, Mimosa,		Strambo carpa-pubescens.
" tree,	Twisted stick, or horn,	Helicteres iscra.
Scrofula plant,		Scrophularia Marilandica.

COMMON.	ENGLISH.	BOTANICAL.
Scrofula plant,	Frostwort, -	Helianthemum Canadense.
" root,	Adders' tongue,	Erythronium Americanum.
" weed,	Net leaf plantain,	Goodyera pubescens.
Scrub oak,		Quercus Ilicifolia.
Scrubby grass,	Scurvy grass,	Cochlearia officinalis.
Scurvy "		" "
" "		Barbarea præcn.
Scullcap,		Scutellaria lateriflora.
" Blue,		Scutellaria lateriflora.
" European,		" galericulata.
" Side flowering,	Blue scullcap,	" lateriflora.
" White,	variety of Blue scullcap,	" "
Sea ash,	Southern prickly ash,	Xanthoxylum Carolinianum.
" beans,	the cowhage plant seed,	Mucuna pruriens.
" "		Entada scandens.
" bindweed,		Convolvulus soldanella.
" burdock,	Burr thistle,	Xanthium strumarium.
" celandine,	Horned poppy,	Chelidonium glaucium.
" colander,		Agarum Turneri.
" cole,	Sea rocket,	Bunias Americana.
" cypress,	Tamarisk,	Tamarix anglica.
" gilliflower,	Sea thrift,	Statice Caroliniana.
" grape,	Salt wort,	Salsala kali.
" holly,		Eryngium ovalifolium.
" kale,		Crambe maritima.
" lavender,	Marsh rosemary,	Statice Caroliniana.
" milkwort,	Black saltwort,	Glaux maritima.
" onion,	Squill,	Scilla "
" parsley,	Lovage,	Ligusticum Levisticum.
" pink,	Bladder campion,	Cucubalus behen.
" "	European thrift,	Statice armenia.
" rocket,		Bunias Americana.
" " purple,		Cakile maritima.
" sedge,	German Sarsaparilla,	Carex arenaria.
" side balsam,	Cascarilla,	Croton balsamifera.
" " beech,	Caribbee bark,	Exostemma Caribæum and other
" " grapes,	produces Jamaica kino,	Coccoloba uvifera. [species.
" thrift,	Marsh rosemary,	Statice Caroliniana
" wormwood,		Artemisia maritima.
" wrack,		Fucus versiculosus.
Sealroot,	Solomon seal,	Convallaria multiflora.
Sealwort,	" "	" "
Sebesten,		Cordia myxa.
Sebipira bark,		Sebipira major.
Sedge,		Carex.
" Sea,	Sweet flag,	Acorus Calamus.
" Sweet,	" "	" "
Seed box,		Ludwigia alternifolia.
Seeded plum,	Persimmon,	Diospyros Virginiana.
Seed lac,	see Lac,	
Self heal,		Prunella vulgaris, var. Pennsylva-
Seneca grass,	Vanilla grass,	Hierochloa borealis. [nica.
" root,	Senega snake root,	Polygala Senega.
" snake root,	" " "	" "
Senega root,	" " "	" "
" snake root,		" "
Senegal root,		Cocculus Bakis.
Seneka root,	Senega root,	Polygala Senega.
" snake root,	" "	" "
Senna,	the leaves of	Cassia acutifolia and other species.

COMMON.	ENGLISH.	BOTANICAL.
Senna, Aleppo,		Cassia obovata.
" Alexandrian,		" acutifolia.
" American,		" Marilandica.
" Bombay,	Wild India senna,	" angustifolia.
" Egyptian,	Alexandrian senna,	" acutifolia.
" herb,		Colutea vesicaria.
" husks,		Cassia acutifolia.
" Indian,		" elongata and angustifolia.
" Italian,	Aleppo senna,	" obovata.
" Mecca,	India senna,	" elongata.
" Mocha,	" "	" "
" Prairie,		" chamæcrista.
" Tinnevelly,	var. cultivated India senna,	" angustifolia.
" Tripoli,		" Æthiopica.
" West Indian,		" emarginata.
" Wild,	American senna,	" Marilandica.
" " European,		Globularia alypum.
Sensitive brier,		Schrankia uncinata.
" fern,		Onoclea sensibilis.
" pea,	Prairie senna,	Cassia chamæcrista.
" plant,		Mimosa pudica and other species.
" "		Cassia nictitans.
" senna,	Prairie senna,	" chamæcrista.
Septfoil,	Tormentilla,	Potentilla Tormentilla.
Serpentaria,	Virginia snake root,	Aristolochia Serpentaria.
Serpent cucumber,	a variety of	Cucumis Melo.
" melon,	" "	" "
Service tree,	Mountain ash, American,	Sorbus Americana.
Sesame leaves,	Benne plant,	Sesamum Indicum and species.
Settiswort,	Bearsfoot,	Helleborus fœtidus.
Setwall,	Valerian,	Valeriana officinalis.
Seven-bark,	Hydraugea,	Hydraugea arborescens.
Shad-bush,	Juneberry,	Aronia botryapium.
Shaddock,	a variety of orange,	Citrus decumana.
Shad-tree,	Juneberry,	Aronia ovalis and other species.
Shag-bark tree,	Hickory tree,	Hicorya (carya) alba.
" walnut,	" nut,	" " "
Shallot,		Allium ascalonicum.
Shamrock,	Wood sorrel,	Oxalis acetosella.
" .	Clover,	Trifolium repens.
" Water,	Buckbean,	Menyanthes trifoliata.
Shave grass,	Scouring rush,	Equisetum hyemale.
Shea-tree,	Butter tree,	Bassia Parkia.
Sheep berry,	Nannybush,	Viburnum lentago.
" " bark,	"	" "
" burr,	Burr thistle,	Xanthium strumarium.
" laurel,		Kalmia angustifolia.
" poison,	Sheep laurel,	" "
" sorrel,		Rumex acetosella.
" weed,	Soapwort,	Saponaria officinalis.
Shellac,	see Lac,	
Shell-bark hickory,		Hicorya (carya) sulcata.
Shell-flower,	Balmony,	Chelone glabra.
Shepherds' club,	Mullein,	Verbascum Thapsus.
" knot,	Tormentilla,	Potentilla Tormentilla.
" needle,		Scandix pecten.
" purse,		Capsella bursa-pastoris.
" rod,		Dipsacus pilosus.
" staff,		" "
" weatherglass,	Red chickweed,	Anagallis arvensis.

COMMON.	ENGLISH.	BOTANICAL.
Shield-fern, Male,	Male fern,	Aspidium Filix Mas.
Shield-lichen,		Parmelia caperata.
Shield root,	Male fern,	Aspidium Filix Mas.
Shin leaf,	Round leaved pyrola,	Pyrola rotundifolia.
" wood,	Dwarf yew,	Taxus Canadensis.
Shittah-wood,	Egyptian gum tree,	Acacia Seyal.
Shittim-wood,	" " "	" "
Shoe-black plant,		Hibiscus Rosa sinensis.
Shoes-and-stockings,	Trefoil, Birds' foot,	Lotus corniculatus.
Shooting star,	American cowslip,	Dodecatheon media.
Shore weed,		Littorella lacustris.
Shot-bush,	Prickly elder,	Aralia spinosa.
Shot plant,	Indian shot,	Canna Indica.
Shovel weed,	Pickerel weed,	Unisema deltifolia.
Showy ladies' slipper,		Cypripedium spectabilis.
Shrubby althæa,	Rose of Sharon,	Hybiscus Syriacus.
" fern,	Sweet fern,	Comptonia asplenifolia.
" goat weed,	Carib tea,	Capraria biflora.
" jessamine,		Jasminum fruiticans.
" trefoil,	Wafer ash,	Ptelea trifoliata.
Shrub oak,	Scrub oak,	Quercus Catesbæi and Ilicifolia.
" yellow root,		Xanthorrhiza apiifolia.
Sicily root,	Sweet Cicily root,	Osmorrhiza longistylis.
Sickle grass,		Polygonum arifolium.
" pod,		Arabis Canadensis.
" senna,		Cassia Tora.
" weed,	Water pepper,	Polygonum hydropiper.
" wort,	Self heal,	Prunella vulgaris.
Side flowering scullcap,	Blue scullcap,	Scutellaria lateriflora.
Side-saddle flower,	Side-saddle plant,	Sarracenia purpurea.
" " Calif'nia,	Pitcher plant,	Darlingtonia Californica.
" plant,		Sarracenia purpurea.
Silk cotton tree,		Bombax ceiba.
" flower,		Albizzia (Acacia) Julibrissin.
" grass,		Yucca filamentosa.
Silkweed,	Common milkweed,	Asclepias Cornuti (Syriaca).
"	Pleurisy root,	" tuberosa.
" Common,	Milkweed,	" Cornuti.
" Rose colored,	White Indian hemp,	" incarnata.
" Swamp,	" " "	" "
Silky cornel,	Swamp dogwood,	Cornus sericea.
" swallowort,	Milkweed,	Asclepias Cornuti (Syriaca).
Silver aster,		Chrysopsis argentea.
" bark,	Huanuco bark,	Cinchona nitida and micrantha.
" bell tree,	Snowdrop tree,	Halesia tetraptera.
" crown bark,	Loxa bark,	Cinchona condaminea, var. vera.
" fir,		Abies picea.
" leaf,	Stillingia,	Stillingia sylvatica.
" "	Hardhack,	Spiræa tomentosa.
" "	Pearl flow'd life everlast'g,	Gnaphalium margaritaceum.
" "	Magnolia, Big leaved,	Magnolia macrophylla.
" pine,	Silver fir,	Abies picea.
" weed,		Potentilla anserina.
" "	Hardhack,	Spiræa tomentosa.
Simaruba bark,	Mountain damson,	Simaruba officinalis.
Simplers' joy,	Vervain,	Verbena hastata.
Single seeded cucumber,		Sicyos angulata.
Sintoc bark,		Cinnamomum Sintoc.
Skevish,	Philadelphia fleabane,	Erigeron Philadelphicum.
Skewer wood,	Spindle tree,	Euonymus.

COMMON.	ENGLISH.	BOTANICAL.
Skirret,		Sium sisarum.
Skoke,	Garget,	Phytolacca decandra.
Skunk cabbage,		Symplocarpus (Ictodes) fœtidus.
" weed,	Skunk cabbage,	" " "
Slavewood,	Mountain damson,	Simaruba officinalis.
Sleekleaf,	Sand myrtle,	Leiophyllum buxifolium.
Sleepwort,	Garden lettuce, ·	Lactuca sativa.
Sleepy nightshade,		Solanum somniferum.
Slipper root,	American valerian,	Cypripedium pubescens and species
Slippers,	Celandine, Wild,	Impatiens pallida.
Slippery elm,		Ulmus fulva.
Sloe,	Black haw,	Viburnum prunifolium.
" American,		Prunus pygmæa.
" tree blossoms,	Wild plum tree,	" spinosa.
Slovenwood,	Southernwood,	Artemisia abrotanum.
Smallage,	Wild celery,	Apium graveolens.
" Marsh,	Marsh parsley,	Selinum palustre.
Small bibernel,		Poterium sanguisorbia.
" burnet saxifrage,	Saxifrage,	Pimpinella saxifraga.
" cleavers,		Galium trifidum.
" cranesbill,		Geranium pusillum.
" fennel flower,	Nutmeg flower,	Nigella sativa.
" fleabane,		Inula dysenterica.
" hemlock,		Arethusa cynapium.
" house leek,	Biting stone crop,	Sedum acre.
" maize,	Turkish millet,	Sorghum vulgare, var. cernuum.
" periwinkle,		Vinca minor.
" pimpernel,	Saxifrage,	Pimpinella saxifraga.
Small-pox plant,	Side-saddle plant,	Sarracenia purpurea.
Small saxifrage,	Small burnet saxifrage,	Pimpinella saxifraga.
" seaside balsam,		Croton balsamiferum.
" Solomon seal,		Convallaria racemosa.
" spikenard,	American sarsaparilla,	Aralia nudicaulis.
Smart weed,		Polygonum punctatum.
Smellage,	Lovage,	Ligusticum levisticum.
Smilax, Boston,	Cape smilax,	Myrsiphyllum asparagoides.
Smoke plant,	False fringe tree,	Rhus cotinus.
" tree,	" " "	" "
" wood,	Travellers' joy,	Clematis vitalba.
Smooth alder,	Red alder,	Alnus rubra (serratula).
" sumach,		Rhus glabra.
Smut of corn,	Ergot of corn,	Ustilago Maydis.
" " rye,	" " rye,	Sclerotium clavus.
Snaffles,	Lousewort,	Pedicularis Canadensis.
Snagrel,	Virginia snakeroot,	Aristolochia Serpentaria.
Snail flower,		Phaseolus caracalla.
" plant,		Medicago scutellata.
Snake bite,	Beth root,	Trillium pendulum
" head,	Balmony,	Chelone glabra.
" leaf,	Adders' tongue,	Erythronium Americanum.
" lily,	Blue flag,	Iris versicolor.
" milk,	Blooming spurge,	· Euphorbia corollata.
" plantain,	Hawkweed,	Hieracium venosum. ·
" · "	Ribwort,	Plantago lanceolata.
Snakeroot,	common name for	Asarum Canadense.
" · Black,	Black cohosh,	Cimicifuga (Actæa) racemosa.
" "		Sanicula Marilandica.
" Button,	Water eryngo,	Eryngium aquaticum.
" "		Liatris spicata.
" Canada,		Asarum Canadense.

COMMON.	ENGLISH.	BOTANICAL.
Snakeroot, Coltsfoot,	Canada snakeroot,	Asarum Canadense.
" Corn,		Eryngium yuccefolium.
" "	Water eryngo,	" aquaticum.
" European,	Asarabacca,	Asarum Europæum.
" gentian,	De Witt snakeroot,	Prenanthes Fraseri.
" Heart,	Canada snakeroot,	Asarum Canadense. ·
" Rattle,	Black cohosh,	Cimicifuga (Actæa) racemosa.
" "	Beth root,	Trillium pendulum.
" "	Lions' foot,	Nabalus albus.
" Red river,	Texas snakeroot,	Aristolochia reticulata.
" Sampson,	Marsh gentian,	Gentiana ochroleuca.
" "	Rheumatic weed,	Aster æstivus.
" Seneca,	Senega snakeroot,	Polygala Senega
" Senega,	" root,	" "
" Seneka,	". "	" "
" Southern,		Asarum Canadense.
" Texas,		Aristolochia reticulata.
" Vermont,	Canada snakeroot,	Asarum Canadense.
" Virginia,		Aristolochia Serpentaria.
" White,		Eupatorium aromaticum.
Snakes' tongue,	Adders' tongue fern,	Ophioglossum vulgatum.
Snakeweed,	Virginia snakeroot,	Aristolochia Serpentaria.
"	Bistort,	Polygonum Bistorta.
" Poison,	Poison hemlock,	Cicuta (conium) maculata.
Snakewood,		Strychnos colubrinum.
Snap dragon,		Antirrhinum majus.
" "	Toad flax,	" linaria vulgaris.
" tree,		Justicia hyssopifolia.
Snapweed,	Touch-me-not,	Impatiens pallida.
"	Fever bush,	Laurus Benzoin.
Snapwood,	" "	" "
Snapping hazel,	Witch hazel,	Hamamelis Virginica.
Sneezeweed,	False sunflower,	Helenium autumnale.
Sneezewort,	" "	" "
"	German pellitory,	Achillea Ptarmica.
Snowball,		Viburnum roseum.
" Wild,	Red root,	Ceanothus Americana.
Snowberry,	Cahinca,	Chiococca racemosa.
"		Symphoricarpus racemosus.
Snowdrop,		Galanthus nivalis.
" Yellow,	Adders' tongue,	Erythronium flavum.
" tree,		Halesia tetraptera.
Snowflake,	White violet,.	Leucoium vernum and species.
Snowflower,		Chionanthus Virginica.
Snuff bean,	Tonka bean,	Dipterix odorata.
Soap bark,		Quillaya saponaria.
" berry,		Sapindus "
" nut,		Mimosa obstergans.
" pods,	the pods of several species of Cæsalpina.	
" root,	Soapwort,	Saponaria officinalis.
" "		Sapindus marginatus.
" tree bark,	Soap bark,	Quillaya saponaria.
Soapwort,		Saponaria officinalis.
" California,		Phalangium pomeridianum.
" gentian,		Gentiana saponaria.
" White,		Lychnis diurna and dioica.
Soapwood,		Clethra tinifolia.
Soda plant,		Salsola soda.
Sodom, Apple of,	Dead sea apple,	Quercus infectoria.
Soldiers' herb,	Matico,	Artanthe elongata.

COMMON.	ENGLISH.	BOTANICAL.
Solomon seal,		Convallaria polygonatum.
" " Giant,		" multiflora.
" " Small,		" racemosa.
Sophora,	Japan sophora,	Sophora Japonica.
Sorb,	Mountain ash,	Pyrus domesticus.
Sorgo. Sorghum,	Chinese sugarcane,	Sorghum nigrum.
Sorrel, Common,	Garden sorrel,	Rumex acetosa.
" Field,	Sheep sorrel,	" acetosella.
" French,		" scutatus.
" Garden,		" acetosa.
" Ladies',		Oxalis corniculata.
" Mountain,	Boreal sour dock,	Oxyria reniformis.
" Peruvian,		Oxalis crassicaulis.
" Round leaf,	Boreal sour dock,	Oxyria reniformis.
" Sheep,		Rumex acetosella.
" tree,		Andromeda arborea and mariana.
" Tree,	Moonwort,	Rumex lunaria.
" Wood,		Oxalis acetosella.
Souari nut,		Caryocar nuciferum.
Souchet,	root of Sweet cyperus	Cyperus longus (officinalis).
Sour berry,	Cranberry,	Vaccinium oxycoccos.
" bush,		Callicarpa Americana.
" cherry,		Cerasus acidus.
" dock,	Yellow dock,	Rumex crispus.
" " Boreal,	Mountain sorrel,	Oxyria reniformis.
" gourd,		Adansonia Gregorii.
" gum tree,	Tupelo,	Nyssa multiflora.
Souring,	Crab apple,	Pyrus caronaria.
Sour leaf,	Sorrel tree,	Andromeda arborea and species.
" sop,		Anona muricatus.
" tree,	Sorrel tree,	Andromeda arborea.
" trefoil,	Wood sorrel,	Oxalis acetosella.
" wood,	Sorrel tree,	Andromeda arborea.
South American kino,	Jamaica kino,	Coccoloba uvifera.
" sea tea,	Indian Black drink,	Ilex vomitaria.
Southern gentian,	Blue gentian,	Gentiana Catesbæi.
" prickly ash,		Xanthoxylum Carolinianum.
" " "	Prickly elder,	Aralia spinosa.
" " elder,	" "	" "
Southernwood,		Artemisia abrotanum.
" root,		Carlina acaulis.
" Tartarian,	Levant wormseed,	Artemisia contra.
Sowbank,		Chenopodium.
Sowberry,	Cranberry,	Vaccinium oxycoccos.
Sowbread,		Cyclamen Europæum.
Sowfennel,		Peucedanum officinale.
Sowthistle,	Milk thistle,	Sonchus oleraceus.
Soymoida bark,	Bastard cedar,	Swietenia febrifuga.
Spadic,	Coca,	Erythoxylon coca.
Spanish bean,	Scarlet runner,	Phaseolus multiflorus.
" broom,		Spartium junceum.
" cane,		Arundo donax.
" carnations,	Barbadoes pride,	Cæsalpina pulcherrima.
" cedar,	Incense wood,	Icica Guianensis.
" chamomile,	Pellitory of Spain,	Anacyclus Pyrethrum.
" chestnut,		Castana vulgaris.
" dagger,		Yucca aloifolia.
" juice,	Licorice extract,	Glycyrrhiza glabra.
" moss,	Long Florida moss,	Tillandsia usneoides.
" needles,		Bidens bipinnata.

COMMON.	ENGLISH.	BOTANICAL.
Spanish pepper,	Cayenne pepper,	Capsicum annuum and other species
" saffron,		Crocus sativus.
" sarsaparilla,		Smilax officinalis.
" soapwort,		Gypsophila struthium.
" tinder,		Echinops strigosus.
" toothpicks,	European wild carrot,	Daucus sylvestris.
Sparrow grass,	Asparagus,	Asparagus officinalis.
Spatter dock,	Yellow pond lily,	Nuphar advena.
Spatling poppy,	Bladder campion,	Cucubalus behen.
Spearmint,		Mentha viridis.
Spearwort,		Ranunculus flammula.
Speckled alder,		Alnus incana.
" jewels,		Impatiens fulva.
Speedwell,		Veronica officinalis.
" root,		Athamanta (Peucedanum) oreseli-
"	Brooklime,	Veronica beccabunga. [num.
" Tall,	Black root,	Leptandra Virginica.
Sphenogyne,		Sphenogyne speciosa.
Spelt,	Spelt wheat,	Triticum spelta.
Spica celtica.		Valeriana celtica.
" Indica.	Indian spikenard,	" (Nardus) Jatamansi.
Spice berry,	Checkerberry,	Gaultheria procumbens.
" birch,	Black birch,	Betula lenta.
" bush,	Feverbush,	Laurus Benzoin.
" root,		Geum odoratissimum.
" wood,	Feverbush,	Laurus Benzoin.
" "	"	Benzoin odoriferum.
Spicknel,	Bearswort,	Meum athamanticum.
Spicy wintergreen,	Checkerberry,	Gaultheria procumbens.
Spider flower,	Spiderwort,	Tradescantia Virginica.
Spiderwort,		" "
Spignel,	Bearswort,	Meum athamanticum.
Spignet,	Spikenard, American,	Aralia racemosa.
Spike lavender,		Lavendula spica.
Spiked alder,	White bush, (alder),	Clethra alnifolia.
" aloe,	Flowering aloe,	Agave Americana.
Spikenard,		Aralia racemosa.
"	Nardus Indica,	Valeriana (Nardus) Jatamansi.
" American,		Aralia racemosa.
" East Indian,		Nardus Jatamansi.
" False,		Smilacina racemosa.
" Indian,		Nardus Jatamansi.
" of Crete,	Large valerian,	Valeriana phu.
" of the ancients,		" (Nardostachys) Jatamansi
" Plowmans',		Inula conyza.
" small,	American sarsaparilla,	Aralia nudicaulis.
" tree,	Prickly elder,	" spinosa.
Spinach,		Spinacia oleracea.
" Mountain,	Garden orach,	Atriplex hortensis.
Spindle tree,		Euonymus Americanus.
Spiraea,	Meadow sweet,	Spiraea ulmaria.
Spirit weed,	Indian redroot,	Lachnanthes tinctoria.
Spleen amaranth,	Princes' feather,	Amarantus hypochondriacus.
Spleenwort,		Asplenium scolopendrium.
" Black,		" Adiantum nigrum.
" bush,	Sweet fern,	Comptonia asplenifolia.
" Common,		Asplenium trichomanes.
" fern,	Sweet fern,	Comptonia asplenifolia.
" "	Spleenwort,	Asplenium scolopendrium.
Split rock,	Mapleleaf alum root,	Heuchera acerifolia.

COMMON.	ENGLISH.	BOTANICAL.
Spogel seed,	Ispaghul seed,	Plantago decumbens.
Sponge tree,	Cassie flower tree,	Acacia Farnesiana.
Spoonhunt,	Laurel, Broad leaved,	Kalmia latifolia.
Spoonwood,	" " "	" " "
Spoonwort,	Scurvy grass,	Cochlearia officinalis.
Spotted alder,	Witch hazel,	Hamamelis Virginica.
" boneset,		Eupatorium maculatum.
" cardus,	Blessed thistle,	Centaurea benedicta.
" comfrey,	Lungwort,	Pulmonaria officiualis.
" cowbane,	Water hemlock,	Cicuta maculata.
" cranesbill,		Geranium maculatum.
" geranium,	Spotted cranesbill,	" "
" hemlock,	Water hemlock,	Cicuta (conium) maculatum.
" knotweed,	Heartease,	Polygonum persicaria.
" lungwort,		Pulmonaria officinalis.
" parsley,	Poison hemlock,	Conium maculatum.
" pipsissewa,		Chimaphila maculata.
" plantain,	Net leaf plantain,	Goodyera pubescens.
" pursley,	Black spurge,	Euphorbia maculata.
" spurge,	" "	" "
" wintergreen,	Spotted pipsissewa.	Chimaphila maculata.
Spring beauty,		Claytonia Virginica.
" bloom,	Swamp pink,	Azalea nitida.
" cress,		Cardamine rhomboides.
" crocus,	Garden crocus,	Crocus vernus.
" orange,		Styrax Americana.
Spruce,		Abies communis.
" Black,		" nigra.
" Double,	Black spruce,	" "
" fir,	Norway pine,	" excelsa.
" gum tree,	Black spruce,	" nigra.
" Hemlock,	Hemlock tree,	" (Pinus) Canadensis.
" Norway,		" excelsa.
" Red,		" pectinata.
" tops,		" (Pinus) communis.
" White,		Abies alba.
Spunk,	Species of agaric,	Boletus fomentarius.
"		" ignarius.
Spurge,	American ipecac,	Euphorbia Ipecacuanha.
" Black,		" maculata.
" Blooming,		" corollata.
" caper,		" lathyris.
" flax,	Mezereon,	Daphne Mezereum.
" ipecac,	American ipecac,	Euphorbia Ipecacuanha.
" Large flowering,	Blooming spurge,	" corollata,
" " spotted,		" hypericifolia.
" laurel,		Daphne laureola.
" olive,	Mezereon,	" Mezereum.
" Spotted,	Black spurge,	Euphorbia maculata.
Spurred rye,	Ergot,	Sclerotium clavus.
Spurrey,	Sandweed,	Spergula arvensis.
Spurry,	Spurrey,	Spergula arvensis.
Square, Carpenters',	Scrofula plant,	Scrophularia Marilandica.
" stalk,	" "	" "
Squash, Common C. N.,		Cucurbita verrucosa.
" Marrow,	Vegetable marrow,	" " medullosa.
" Winter,		" maxima.
Squaw berry,	Dangle berry,	Vaccinium stamineum.
" bush,	High cranberry,	Viburnum opulus.
" drops,		Orobanche uniflora.

COMMON.	ENGLISH.	BOTANICAL.
Squaw huckleberry,	Dangle berry,	Vaccinium stamineum.
" mint,	Pennyroyal,	Hedeoma pulegioides.
" root,	Blue cohosh,	Caulophyllum thalictroides.
" "	Black cohosh,	Cimicifuga racemosa.
" vine,		Mitchella repens.
" weed,		Senecio aureus var. obovatus.
" "	Red stalked aster,	Aster puniceus.
Squilily,		Pancratium maritimum.
Squill,	the bulb of	Scilla maritima.
Squirrel corn,		Dicentra (Corydalis) Canadensis.
" ear,	White plantain,	Goodyera repens.
" pea, Ground,	Twinleaf root,	Jeffersonia diphylla.
Squirting cucumber,	Wild cucumber,	Momordica Elaterium.
Staff tree, Climbing,	False bittersweet,	Celastrus scandens.
Staff-vine,	" "	" "
Stagger-bush,	Sorrel tree,	Andromeda mariana.
Stagger-grass,	Atamasco lily,	Amaryllis atamasco.
Stagger-weed,	Turkey corn,	Corydalis formosa.
"	Larkspur,	Delphinium consolida.
Staggerwort,	Ragwort,	Senecio Jacobæa.
Staghorn,	Velvet sumach,	Rhus typhinum.
"	Club moss,	Lycopodium clavatum.
Stammerwort,	Roman wormwood,	Ambrosia artemisiæfolia.
Standerwort,	Male orchis,	Orchis mascula.
Starchwort,	Dragon root,	Arum maculatum and triphyllum.
Star anise,		Illicium anisatum.
" apple,		Chrysophyllum Jamaicense.
" bloom,	Pink root,	Spigelia Marilandica.
" flower,	Cocash root,	Aster puniceus.
" grass,	Unicorn root,	Aletris farinosa.
" "		Hypoxis erecta.
" of Bethlehem,		Ornithogalum umbellatum.
" of the earth,	European avens,	Geum urbanum.
" of Jerusalem,	Goats' beard,	Tragopogon pratensis.
Star-root,	Unicorn root,	Aletris farinosa.
"		Hypoxa erecta.
" Blazing,		Liatris squarrosa.
Star thistle,	Knapweed,	Centaurea calcitrapa and species.
" scabious,		Scabiosa stellata.
Starwort,		Aster of many species.
" Drooping,	False unicorn,	Helonias dioica.
" Water,		Callitriche verna.
Staunchwort,	Kidney vetch,	Anthyllis vulneraria.
Staverwort,	Ragwort,	Senecio Jacobæa.
Stavesacre,		Delphinium Staphisagria.
Stave wood,	Quassia,	Simaruba amara.
Stay plough,	Rest harrow,	Ononis spinosa.
Stemless ladies' slipper,		Cypripedium acaule.
Steeple bush,	Hardhack,	Spiræa tomentosa.
Stephenotis,		Stephanotis floribunda.
Stepmother,	Pansy,	Viola tricolor.
Stichwort,	Stitchwort,	Stellaria media.
Stickadore,	Arabian lavender,	Lavendula Stœchas.
Stickseed,	Beggars' tick,	Bidens frondosa.
Stickwort,	Agrimony,	Agrimonia Eupatoria.
Stillingia,	Queens' root,	Stillingia sylvatica.
Stinging loasa,		Loasa lateritia.
" nettle,		Urtica dioica (urens).
Stingless, "		" pumila.
Stinkhorn,		Phallus impudicus.

COMMON.	ENGLISH.	BOTANICAL.
Stinking ash,	Wafer ash,	Ptelea trifoliata.
" balm,	Pennyroyal,	Hedeoma pulegioides.
" black hellebore,	Fetid hellebore,	Helleborus fœtidus.
" chamomile,	Mayweed,	Anthemis Cotula.
" goosefoot,		Chenopodium fœtidum.
" horehound,		Ballota nigra.
" nightshade,	Henbane,	Hyosciamus niger.
" orache,	Goosefoot,	Chenopodium olidum.
" poke,	Skunk cabbage,	Symplocarpus (Ictodes) fœtidus.
" prairie bush,	Wafer ash,	Ptelea trifoliata.
" weed,	Oak of Jerusalem,	Chenopodium anthelminticum.
" Willie,	Ragwort,	Senecio Jacobæa.
Stinkwort,	Thorn apple,	Datura Stramonium.
Stitchwort,	Chickweed,	Stellaria media.
"	Meadow star,	" palustris.
Stock,	Gilliflower,	Mathiola incana.
" Tenweeks,		" annua.
Stonebrake,	Rockbrake,	Polypodium vulgare.
Stonebreak,		Saxifraga.
Stonecrop, biting,	Small houseleek,	Sedum acre.
Stone clover,	Rabbits' foot,	Trifolium arvense.
" crottles,		Lichen (Parmelia) caperata.
" fern,	Wall rue,	Asplenium Ruta muraria.
" fruit,	a drupe like the cherry, peach, plum, etc.	
" linden,		Phillyrea media.
" mint,	Mountain dittany,	Cunila mariana.
" pine,		Pinus cembra.
Stoneroot,		Collinsonia Canadensis.
Stoneseed,	Gromwell,	Lithospermum arvense.
Stonewort,		Chara.
Storax,	Styrax,	Styrax officinalis.
" bark,	Sweet gum bark,	Liquidambar orientale.
Storks' bill,		Erodium cicutarium.
" " Rose scented,		Pelargonium roseum.
Stramonium,	Thorn apple,	Datura Stramonium.
" Purple,		" Tatula.
Strangle tare,	Dodder, Greater,	Cuscuta Europæa.
Strasburgh turpentine,	a variety of turpentine obtained from European silver fir tree.	
Strawberry, Common,		Fragraria vesca.
" bush,	Spindle tree,	Euonymus Americana.
" geranium,		Saxifraga sarmentosa.
" Mountain,		Fragraria Canadensis.
" Pineapple,		" grandiflora.
" shrub,	Spindle tree,	Euonymus Americana.
" spinach,	Indian strawberry,	Blitum capitatum.
" tree,	Strawberry tree of Europe,	Arbutus unedo.
" Wild,		Fragraria Virginiana.
Straw colored gentian,	Marsh gentian,	Gentiana ochroleuca.
Striated ipecac,	Peruvian ipecac,	Psychotria Ipecacuanha (emetica).
Striped abutilon,		Abutilon striatum.
" alder,	Witch hazel,	Hamamelis Virginica.
" bloodwort,	Hawkweed,	Hieracium venosum.
" dogwood,	Striped maple,	Acer Pennsylvanicum.
" maple,		" "
Strong scented lettuce,	yields lettuce opium,	Lactuca virosa.
Stubwort,	Wood sorrel,	Oxalis acetosella.
Styptic weed,		Cassia occidentalis.
Styrax,		Styrax officinalis.
" Liquid,	a balsam prepared from	Liquidambar orientale and other
Suber,	Cork wood,	Quercus suber. [species.

8*

COMMON.	ENGLISH.	BOTANICAL.
Succory,	Chicory,	Cichorum intybus.
" gum,	the gummy juice of	Chondrilla juncea.
" dock cress,	Nipplewort,	Lapsana communis.
" Wild,	Chicory,	Cichorum intybus.
Sugar,	the saccharine constituent from the sap of certain plants.	
" Beet,	the product of different varieties of Beta.	
" berry tree,	Nettle tree,	Celtis occidentalis.
" Cane,	Common sugar,	Saccharum officinarum.
" cane,		" "
" " Chinese,		Sorghum saccharatum.
" Date,	the product of the date,	Phœnix sylvestris.
" Grape,	" " " grape,	Vitis Vinifera and other species.
" Liquorice,	the uncrystallizable extract,	Glycyrrhiza glabra and other spe-
" Maple,	the product of the sugar maple,	Acer saccharinum. [cies.
" palm,	yields jaggery,	Phœnix sylvestris.
" tree,		Myoporum platycarpum.
Sulphurwort,	Sowfennel,	Peucedanum officinale.
Sultan, Sweet,		Centaurea moschata.
Sumach, (shumach),	the dried and cut leaves and shoots of Rhus coriaria.	
" American,		Rhus typhina.
" Chinese,		Ailanthus glandulosa.
" Climbing,	Poison ivy,	Rhus Toxicodendron, var. radicans
" Dwarf,	Mountain sumach,	" copallinum.
" Jamaica,		" Metopium.
" Pennsylvania,		" glabra.
" Poison,		" vernix.
" Stagshorn,	American sumach,	" typhina.
" Swamp,	Poison sumach,	" vernix.
" Tanners',		" coriaria.
" Upland,		" glabra.
" Venice,	Young fustic,	" cotinus.
" Velvet,	American sumach,	" typhina.
" Virginian,	" "	" "
Sumbul,	Musk root,	Euryangium Sumbul.
"	Indian nard,	Nardostachys Jatamansi.
Summer grass,	Seneca grass,	Hierochloa arborealis.
" savory,		Satureja hortensis.
Sundew,		Drosera rotundifolia.
" long,		" longifolia.
Sundrops,	Scabish,	Œnothera glauca.
Sunflower, Common,		Helianthus annuus.
" False,	Sneezewort,	Helenium autumnale
" Garden,		Helianthus annuus.
" Red,		Rudbeckia purpurea.
" Rough,		Helianthus divaricatus.
" Swamp,		Helenium autumnale.
" Tickseed,		Coreopsis tripteris.
" Wild,		Helianthus giganteus.
Sunrose,	Rock rose,	Helianthemum Canadense.
Supplejack,		Berchemia volubilis.
Susumber berries,		Solanum bacciferum.
Sutterberry bark,	Prickly ash,	Xanthoxylum fraxineum.
Swallow wort,	Milkweed,	Asclepias Cornuti (Syriaca).
" " Orange,	Pleurisy root,	" tuberosa.
" " White,		" vincetoxicum.
Swamp alder,	Red alder,	Alnus rubra.
" apple,		Azalea nudiflora.
" beggars' tick,		Bidens tripartita.
" cabbage,		Symplocarpus (Ictodes) fœtidus.
" cheeses,	Swamp apple,	Azalea nudiflora.

COMMON.	ENGLISH.	BOTANICAL.
Swamp dogwood,		Cornus sericea.
" "	Wafer ash,	Ptelea trifoliata.
" hellebore,	American hellebore,	Veratrum viride.
" hornbeam,	Tupelo,	Nyssa multiflora.
" honeysuckle,	Swamp pink,	Azalea nitida and other species.
" laurel,		Kalmia glauca.
" "	Holly bay,	Gordonia lasianthus.
" milkweed,	White Indian hemp,	Asclepias incarnata.
" oak,	" chestnut oak,	Quercus prinus.
" pear,		Aronia Canadense.
" pink,		Azalea nitida.
" redberry,	Common cranberry,	Vaccinium oxycoccos.
" robin,	Wild water arum,	Calla palustris.
" sassafras,	Magnolia,	Magnolia glauca.
" silkweed,	White Indian hemp,	Asclepias incarnata.
" sneezewort,		Helenium autumnale.
" sowfennel,		Peucedanum palustre.
" spleenwort,		Asplenium angustifolium.
" sumach,	Poison sumach,	Rhus vernix.
" sunflower,	Sneezewort,	Helenium autumnale.
" valerian,		Valeriana dioica.
" willow herb,		Lythrum verticillatum.
" wood,	Button bush,	Cephalanthus occidentalis.
" "	Leatherwood,	Dirca palustris.
Swan daisy,		Brachycome iberidifolia.
" weed,	Cocash,	Aster puniceus.
Sweat root,	Abscess root,	Polemonium reptans.
" weed,	Marsh mallow	Althæa officinalis.
Sweating plant,	Boneset,	Eupatorium perfoliatum.
Swedish turnip,	Rutabaga,	Brassica campestris rutabaga.
Sweet almond,		Amygdalus communis.
" alyssum,		Alyssum maritimum.
" balm,		Dracocephalum canariense.
" balsam,	Life everlasting,	Gnaphalium polycephalum.
" basil,	Common basil,	Ocynum basilicum.
" bay,	Laurel,	Laurus nobilis.
" "	Magnolia,	Magnolia glauca.
" berry,	White cranberry,	Oxycocca hispidula.
" birch,	Black birch,	Betula lenta.
" brake,	Male fern,	Aspidium Filix Mas.
" brier,	Sweetbrier rose,	Rosa rubiginosa.
" broom,	Butchers' broom,	Ruscus aculeatus.
" bugle,		Lycopus Virginicus.
" bush,	Sweet fern,	Comptonia asplenifolia.
" cane,	" flag,	Acorus Calamus.
" cassava,	Tapioca,	Janipha (Jatropha) manihot.
" cherry,	Gaskins,	Cerasus avium.
" chervil,	Sweet cicily,	Osmorrhiza longistylis.
" cicily,		" "
" " European,		Myrrhis odorata.
" clover,		Melilotus leucanthe.
" cyperus,	English galingale,	Cyperus longus (officinalis).
" elder,	Common elder,	Sambucus Canadensis.
" elm,	Slippery elm,	Ulmus fulva.
" fennel,		Fœniculum dulce.
" fern,		Comptonia asplenifolia.
" ferry,	Sweet fern,	" "
" flag,		Acorus Calamus.
" gale,	Meadow fern,	Myrica gale.
" goldenrod,		Solidago odora.

COMMON.	ENGLISH.	BOTANICAL.
Sweet grass,	Sweet flag,	Acorus Calamus.
" gum,	the concrete. juice of	Liquidambar styraciflua.
" " Oriental,	Liquid styrax,	" orientale.
" horsemint,	Mountain dittany,	Cunila mariana.
" John,	the narrow leaved variety,	Dianthus barbatus.
" laurel,	Florida anise,	Illicium Floridanum.
" leaf,		Hopea tinctoria.
" lemon,		Citrus lumia.
" liquorice,	see Liquorice root.	
" lucerne,	Sweet clover,	Melilotus leucanthe.
" magnolia,	Magnolia,	Magnolia glauca.
" marjoram,		Origanum Marjorana.
" Meadow,		Spiræa ulmaria.
" "	Hardhack,	" tomentosa.
" oleander,		Nerium oleander.
" pea,		Lathyrus odoratus.
" pepperbush,	White alder,	Clethra alnifolia.
" pod,	Saint John's bread, .	Ceratonia siliqua.
" potato,		Convolvulus (Ipomæa) batatus.
" rocket,	Dames' violet,	Hesperis matronalis.
" root,	Sweet flagroot,	Acorus Calamus.
" rush,	" "	" "
" scabious,	Philadelphia fleabane,	Erigeron Philadelphicum.
" scabish,	.	Scabiosa atropurpurea.
" scented datura,		Datura Wrightii.
" " cedar,	Barbadoes cedar,	Cedrela odorata.
" " goldenrod,		Solidago odora.
" " life everlasting,		Gnaphalium polycephalum.
" " virgins' bower,		Clematis flammula.
" " water lily,	White pond lily,	Nymphæa odorata.
" " yellow Jasmine,		Jasminum odoratissimum.
" sedge,	Sweet flag,	Acorus Calamus.
" shrub,	Carolina allspice,	Calycanthus Floridus.
" sicily,	Sweet cicily,	Osmorrhiza longistylis.
" smelling trefoil,	Water maudlin,	Eupatorium Cannabinum.
" sop,		Anona squamosa.
" sultan,		Centaurea moschata.
" tree,	Maple sugar tree,	Acer saccharinum.
" verbena,	Sweet scented verbena,	Verbena triphylla.
" vernal grass,		Anthoxanthemum odoratum.
" viburnum,	Nannybush,	Viburnum lentago.
" violet,		Viola odorata.
" water,	a variety of White grape,	Vitis vinifera.
" william,		Dianthus barbatus.
" willow,	Meadow fern,	Myrica gale.
" wood,	Liquorice,	Glycyrrhiza glabra.
" " bark,	Cascarilla bark,	Croton Eluteria.
Swine bread,	Sowbread,	Cyclamen Europæum.
" cress,	Warteress,	Coronopus Ruellii.
" snout,	Dandelion,	Taraxacum Dens-leonis.
Swinesbane,		Chenopodium rubrum.
Swiss stone pine,	Stone pine,	Pinus cembra, and pinea.
Sword lily,	Round mandrake,	Gladiolus communis.
Sycamine,	White mulberry,	Morus alba.
Sycamore,	.	Acer pseudo platanus.
" of the ancients,		Ficus Sycamorus.
" False,	Buttonwood,	Platanus occidentalis.
Syrian herb mastich,	Cat thyme,	Teucrium marum.
" mallow,		Abelmoschus moschata.
" nard,	Indian nard,	.	Nardostachys Jatamansi.

COMMON.	ENGLISH.	BOTANICAL.
Syringia,	White lilac,	Lilaca vulgaris, var. alba.
"	Mock orange,	Philadelphus coronarius.
Syringe tree,	India rubber tree,	Siphonia elastica.
Tacamahac,	Balsam poplar,	Populus balsamifera.
"	a resinous product of	Fagara octandra.
" Curacoa,	" " "	Amyris tomentosa.
" Marituas,	" " "	Calophyllum iuophyllum.
" Orientale,	" " "	" "
" poplar,	Balsam poplar,	Populus balsamifera.
Tag alder,	Red alder,	Alnus rubra (serratula).
Tailed pepper,	Cubebs,	Piper Cubeba and other species.
Tall ambrosia,		Ambrosia trifida.
" boneset,	Joepye weed,	Eupatorium verticillatum.
" coneflower,	Thimble weed,	Rudbeckia laciniata.
" speedwell,	Black root,	Leptandra Virginica.
Tallow shrub,	Bayberry,	Myrica cerifera.
Tallow-tree,	Japan vegetable wax,	Stillingia sebifera.
"	Chinese sumach,	Ailanthus glandulosa.
Tallow, Vegetable,	Bayberry tallow (wax),	Myrica cerifera.
Tamarack,	American larch,	Larix Americana.
Tamarind,		Tamarindus Indica.
Tamarisk,		Tamarix Anglica.
" German,		Myricaria Germanica.
Tanners' bark,	Oak and Hemlock barks,	Quercus, and Abies.
" sumach,		Rhus coriaria.
Tansy,		Tanacetum vulgare.
" Double,	Double flowered tansy,	" crispum.
Tape grass,	Eel grass,	Vallisneria spiralis.
Tapioca,	the fecula of the root of	Janipha (Jatropha) manihot.
" Pearl,	prepared from grain,	
Tar, American,	an impure oleo resin by burning pine wood. Liquid pitch.	
" Barbadoes,	Liquid bitumen.	
" Beech,	obtained from beechwood,	Fagus sylvatica.
" Birch,	" " birchwood,	Betula alba.
" Juniper,	see oil of Cade.	
" Stockholm,	European tar,	Pinus silvestris and Larix Sibirica.
Taragon,		Artemisia dracunculus.
Tare,	Darnel,	Lolium tremulentum.
"	Spurrey,	Spergula arvensis.
"	Vetch,	Vicia sativa and other species.
Taro,		Caladium esculentum.
Tartarian moss,	yields Litmus,	Lecanora Tartarea.
" southernwood,		Artemisia santonica.
Tassel flower,		Senecio sonchifolia.
Tat grass,	Butterwort,	Pinguicula elatior.
Tawkin,	Golden club,	Arum (Orontium) aquaticum.
Tea, plant and species,		Thea sinensis.
" Abyssinian,	Cafta leaves,	Catha edulis.
" Appalachian,		Prinos glaber.
" Arabian,	Cafta leaves,	Catha edulis.
" Assam,		Thea Assamica.
" Australian,		Leptospermum lanigerum.
" berry,	Checkerberry,	Gaultheria procumbens.
" Black,		Thea Bohea.
" Botany bay,		Smilax glycyphylla.
" Bourbon,	Faham tea,	Angraecum fragrans.
" Brazilian,		Stachytocarpa Jamaicensis.
" Bush,	African tea plant,	Cyclopia genistoides.
" Camellia,		Camellia sesanqua.
" Canary,		Sida Canariensis.

COMMON.	ENGLISH.	BOTANICAL.
Tea, Carolina,		Ilex vomitoria.
" Chinese,		Thea Chinensis.
" Coffee,	leaves of the Coffee plant,	Coffea Arabica.
" Continental,	Labrador tea,	Ledum latifolium.
" Faham,		Angræcum fragrans.
" Green,		Thea viridis.
" Himalaya,	tea grown in North'n India,	" of several species.
" Jesuits',	Mate,	Ilex Paraguaiensis.
" Labrador,		Ledum latifolium.
" Lemon grass,		Andropogon Schœnanthus.
" Malay,		Psoralea corylifolia.
" Marsh,		Ledum palustre.
" Mexican,		Chenopodium ambrosioides.
" "		Ambrina ambrosioides.
" Mountain,	Checkerberry,	Gaultheria procumbens.
" New Jersey,		Ceanothus Americana.
" " Zealand,		Leptospermum scoparium.
" of Heaven,		Hydrangea Thunbergii.
" Oswego,		Monarda didyma.
" Paigle,	English cowslip,	Primula veris.
" Paraguay,	Mate,	Ilex Paraguaiensis.
" South sea,	Indian black drink,	" vomitoria.
" Swamp,	Marsh tea,	Ledum palustre.
" Sweet,	Botany bay tea,	Smilax glycyphylla.
" Theezan,		Sageretia theezan.
" West Indian,	Carib tea,	Capraria biflora.
" Wild,	Lead plant,	Amorpha canescens.
Teakwood,		Tectonia grandis.
" African,		Oldfieldia Africana.
" N. So. Wales,		Eudiandra glauca.
Tearel,	Boneset,	Eupatorium perfoliatum.
Tear thumb,	Sickle grass,	Polygonum arifolium.
Teasel,		Dipsacus sylvestris.
" Fullers',		" fullonum.
" Wild,		" sylvestris.
Tea-tree,		Thea.
" Ceylon,		Elœodendron glaucum.
" New Jersey,		Ceanothus Americana.
" White,		Melaleuca genistifolia.
Teazel,	see Teasel,	Dipsacus.
Teel,	Til seed,	Sesamum orientale.
Tellicherry bark,	Conessi bark,	Wrightia antidysenterica.
Ten bark,		Cinchona ovata vulgaris.
Ten o'clock,	Star of Bethlehem,	Ornithogalum umbellatum.
Tentwort,	Wall rue,	Asplenium ruta muraria.
Terra Firma wood,	Peach wood,	Cæsalpina bijuga.
" Japonica,	Gambir. Catechu,	Acacia Catechu.
" Merica,	Turmeric,	Curcuma longa.
Tetter berry,	White bryony,	Bryonia alba.
Tetterwort,	Garden celandine,	Chelidonium majus.
Texas sarsaparilla,	Yellow parilla,	Menispermum Canadense.
" snakeroot,		Aristolochia reticulata.
Teyl tree,	Linden tree,	Tilia Europæa.
Thalictroc,	Hedge mustard,	Erysimum Nasturtium.
Thick leaved pennywort,		Hydrocotyle Asiatica.
" weed,	Pennyroyal,	Hedeoma pulegioides.
Thimble berry,		Rubus occidentalis.
" weed,		Rudbeckia laciniata.
Thistle,		Cirsium.
" Bitter,	Blessed thistle,	Centaurea benedicta.

COMMON.	ENGLISH.	BOTANICAL.
Thistle, Blessed,		Centaurea benedicta.
" Canada,	Cursed thistle,	Cirsium (Cnicus) arvense.
" Common,		Cnicus lanceolatus.
" Cotton,	Scotch thistle,	Onopordon acanthium.
" Cursed,	Canada thistle,	Cirsium (Cnicus) arvense.
" Distaff,		" lanatum.
" Globe,		Echinops multiflora.
" Golden,		Scolymus.
" Holy,	Blessed thistle,	Centaurea benedictus.
" Horse,		Cirsium arvense.
" Jersey,		Centaurea Isnardi.
" Mary,		Carduus marianus.
" Mexican		Erythrolæna conspicua.
" Milk,		Carduus marianus.
" Musk,		" nutans.
" Saffron,		Carthamus tinctorius.
" Scottish,	Cotton thistle,	Onopordon acanthium.
" Sow,		Sonchus oleraceus, also Carduus.
" Star,	Knapweed,	Centaurea calcitrapa.
" Torch,		Cereus,
" Yellow,	Prickly poppy,	Argemone Mexicana.
Thorn-apple,		Datura Stramonium.
" tree,	May tree,	Cratægus oxyacanthus.
Thorn, Black,		Prunus spinosa.
" Buck,	Purging buckthorn,	Rhamnus catharticus.
" Buffalo,		Acacia latronum.
" Camels',		Alhagi camelorum.
" Christ's,		Paliurus aculeatus.
" Egyptian,	Gum Arabic tree,	Acacia vera.
" Elephant,		" tomentosa.
" Evergreen,		Cratægus pyracanthus.
" Goats',	yields Tragacanth,	Astragalus Tragacantha.
" Haw,	Hawthorn,	Cratægus oxyacanthus.
" Orange,		Citriobatus.
" poppy,	Prickly poppy,	Argemone Mexicana.
" Washington,		Cratægus cordata.
" White,		" oxyacanthus.
" "		" punctata.
" " W. I.,		Macromerium Jamaicense.
Thorough stem,	Boneset,	Eupatorium perfoliatum.
" wax,	"	" "
Thoroughwort,	"	" "
Thousand men wort,		Aristolochia cymbifera.
Three leaved arum,	Dragon root,	Arum triphyllum.
" thorned acacia,	Honey locust,	Gleditschia triacanthos.
" seeded leek,		Allium tricoccum.
Thrift, American,	Marsh rosemary,	Statice Caroliniana.
" European,		" Armeria.
Throat root,	White avens,	Geum Virginianum.
" wort,	Button snakeroot,	Liatris spicata.
Throw wort,	Motherwort,	Leonurus cardiaca.
Thunbergia,		Thunbergia alata.
Thunder plant,	Houseleek,	Sempervivum tectorum.
Thus, American,	Hard turpentine obtained from Pinus (Abies) sylvestris.	
"	Frankincense,	Abies excelsa.
Thyme,		Thymus vulgaris.
" Basil,	Calamint,	Calamintha officinalis.
" Cat,		Teucrium marum.
" Common,		Thymus vulgaris.
" Field,	Dogmint,	Clinopodium vulgare.

COMMON.	ENGLISH.	BOTANICAL.
Thyme, Garden,		Thymus vulgaris.
" Horse,		Calamintha officinalis and clinopo-
" Lemon,		Thymus citriodorus. [dium.
" Mother of,	Wild thyme,	" serpyllum.
" Mountain,		Acinos vulgaris.
" Virginia, Narrow		Pycnanthemum Virginicum.
" Water,		Anacharis alsinastrum.
" Wild,		Thymus serpyllum.
Tickle weed,	American Hellebore,	Veratrum viride.
Tick seed sunflower,		Coreopsis tripteris.
Tickweed,	Pennyroyal,	Hedeoma pulegioides.
"		Coreopsis.
Ticute,	Upas tree of Java,	Strychnos ticute.
Tiger flower		Tigridia parvonia.
" lily,		Lilium bulbiferum.
Tikor,	East Indian arrow root,	Curcuma angustifolia.
Tilseed,		Sesamum orientale and Indicum.
" Black,		Guizotia oleifera.
Tillseed,	Lentil,	Ervum lens.
Tillyseed,		Croton parvana.
Til-tree,	Linden tree,	Tilia Europæa.
Timothy grass,	Herds' grass,	Phleum pratense.
Tinder,	see Agaric.	
" German,	Amadou,	Polyporus fomentarius.
Tinker weed,	Fever root,	Triosteum perfoliatum.
Toad-flax,		Antirrhinum (Linaria) vulgaris.
Toad-lily,	White pond lily,	Nymphæa odorata.
Toad-mouth,	Snap dragon,	Antirrhinum majus.
Toad-root,	Red cohosh,	Actæa rubra.
Toadstool,	common name of	Agarics and Boleti.
Toa-tree,		Casuarina equisetifolia.
Tobacco,		Nicotiana Tabacum.
" Indian,	Lobelia,	Lobelia inflata.
" Mountain,	Arnica,	Arnica montana.
" root, Oregon,	Valerian,	Valeriana edulis.
" Wild,	Lobelia,	Lobelia inflata.
" wood,		Hamamelis Virginica.
Tolu balsam,	Balsam Tolu,	Myrospermum Toluiferum.
" " White,	" "	" "
Tomato,		Solanum lycopersicum.
Tonca-bean,	Tonka bean,	Dipterix odorata.
Tonga-beau,	" "	
Tongue-grass,	Peppergrass,	Lepidium sativum.
Tonka-bean,		Dipterix odorata.
" wood,	Scentwood,	Alyxia buxifolia.
Tonquin-bean,	Tonka bean,	Dipterix odorata.
Toothache bark,	Prickly ash,	Xanthoxylum fraxineum.
" bush,	" "	" "
" grass,		Aplocera maritima.
" tree,	Prickly ash,	Xanthoxylum fraxineum.
Tooth root,		Dentaria diphylla.
Toothwort,		" "
Toot-plant,	Toot poison,	Coriaria ruscifolia.
Torch wood,		Cereus heptogonus.
Tormentil,	Cranesbill root,	Geranium maculatum.
Tormentilla,		Potentilla Tormentilla.
Tory weed,	Hounds' tongue,	Cynoglossum officinale.
Touch-me-not,		Impatiens pallida.
" Pale,	Jewel weed, speckled,	" fulva.
Touch-weed,	Sensitive plant,	Mimosa pudica and other species.

COMMON.	ENGLISH.	BOTANICAL.
Touch-wood,	Agaric,	Polyporus ignarius.
Tous-les-mois,	St. Kitt's arrow root,	Canna edulis and other species.
Tow-cock,		Dolichos sinensis.
Tower mustard,		Turritis glabra.
Toywort,	Shepherds' purse,	Capsella bursa-pastoris.
Tragacanth,	the gummy exudation of	Astragalus verus and other species
" Senegal,		Sterculia Tragacantha.
Trailing arbutus,	Mayflower,	Epigæa repens.
" sumach,	Poison oak,	Rhus Toxicodendron.
Travellers' joy,	Vervain,	Verbena hastata.
" "		Clematis vitalba.
" tree,		Urania speciosa.
Tread-softly,		Cnidoscolus stimulans.
Treacle, Countryman's,	Rue,	Ruta graveolens.
Tree beard,	Florida long moss,	Tillandsia usneoides.
" of chastity,	Chaste tree,	Vitex Agnus-castus.
" of Heaven,	Chinese sumach,	Ailantus glandulosa.
" lungwort,	Maple lungwort,	Sticta pulmonaria.
" mallow,		Lavatera arborea.
" nightshade,	Jerusalem cherry,	Solanum pseudo-capsicum.
Trefoil,	Clover,	Trifolium.
"	Liverwort,	Hepatica Americana.
" Birdsfoot,	Lotus,	Lotus corniculatus.
" Hop,	Dutch clover,	Medicago lupulina.
" Marsh,	Buckbean,	Menyanthes trifoliata.
" Moon,		Medicago arborea.
" Shrubby,	Wafer ash,	Ptelea trifoliata.
" Tick,		Desmodium.
Trembling tree,	American poplar,	Populus tremuloides.
Trickle,	Tooth root,	Dentaria diphylla.
Tripoli powder,	the flinty integuments of several species of Diatomaceæ.	
True cinnamon,	Ceylon cinnamon,	Cinnamomum Zeylanicum.
Truelove,	Herb Paris,	Paris quadrifolia.
"	Beth root,	Trillium erectum.
Truffles,		Tuber æstivum and other species.
Trumpet flower,	see Bignonia, Tecoma, Solandra, etc.	
" leaf honeysuckle		Caprifolium sempervirens.
" leaf,	Water cup,	Sarracenia flava.
" lily,	Calla lily,	Richardia Africana.
" moss,		Cladonia pyridata.
" tree,		Cecropia peltata.
" weed,	Queen of the meadow,	Eupatorium purpureum.
" wood,	Trumpet tree,	Cecropia peltata.
Tub camphor,	Crude camphor,	Camphora officinarum.
Tubereuse,	Tuberose,	Polianthes tuberosa.
Tuber root,	Pleurisy root,	Asclepias tuberosa.
Tuberose,		Polianthes tuberosa.
Tulip,		Tulipa Gesneriana.
Tulip-tree,	Yellow poplar,	Liriodendron Tulipifera.
Tumeric,	Turmeric,	Curcuma longa.
Tun hoof,	Ground ivy,	Nepeta glechoma.
Tupelo-tree,		Nyssa multiflora.
Turkey bur seed,	Burdock seed,	Arctium lappa.
" berry,		Solanum mammosum.
" " tree,		Cordia callococca.
" claw,	Crawley root,	Corallorhiza odontorhiza.
" corn,		Corydalis formosa.
" pea,	Turkey corn,	" "
Turkish corn,	Indian corn,	Zea mays.
Turks'-cap,	Melon cactus,	Melocactus communis.

COMMON.	ENGLISH.	BOTANICAL.
Turks'-cap,	Martagon lily	Lilium martagon.
" lily,	" "	" "
Turmeric,		Curcuma longa.
Turnip,		Brassica rapa.
" Devil's,	Bryony root,	Bryonia dioica.
" Dragon,	Dragon root,	Arum triphyllum.
" Indian,	" "	" "
" Meadow,	" "	" "
" Pepper,	" "	" "
" Prairie,	the tubers of	Psoralea esculenta.
" St. Anthony's,	Crowfoot buttercup,	Ranunculus bulbosus.
" Swamp,	Dragon root,	Arum triphyllum.
" Swedish,	Rutabaga,	Brassica campestris Rutabaga.
" Wild,	Dragon root,	Arum triphyllum.
Turnsole,		Heliotropium Indicum.
Turpentine,	the resinous exudation from Pine trees.	
" American,	" " " "	Pinus palustris and tæda.
" Bordeaux,	" " " "	Abies picea.
" Canada,	Balsam of fir,	" balsamea.
" Chian,	the (oleo) resin obtained from Pistacia terebinthus.	
" Com. Amn.,		Pinus palustris and other species.
" " Europ'n,		" sylvestris.
" Cyprus,	Chian turpentine,	Pistacia terebinthus.
" Damarra,		Pinus damarra.
" Dombeya,		Dombeya excelsa.
" Gum,	Hard pine gum,	Pinus palustris and tæda.
" Scio,	Chian turpentine,	Pistacia terebinthus.
" Strasburg,	obtained from European silver fir, Abies taxifolia and picea.	
" Venice,	an oleo resin from	Abies (Pinus) Larix.
" White,	American turpentine,	Pinus palustris.
Turpentine sunflower,	Rosin weed,	Silphium gummiferum.
Turpeth root,		Convolvulus turpethum
Turtle-bloom,	Balmony,	Chelone glabra.
Turtle-head,	"	" "
Tutsan,	Park leaves,	Androsæmum officinalis.
Tway-blade,		Listera ovata and other species.
Twin-berry,		Xylosteum cliatum.
Twin-flower,		Linnæa borealis.
Twin-leaf,		Jeffersonia diphylla.
Twisted horn (stick),	Screw tree,	Helicteres isora.
Ule-tree,		Castilloa elastica.
Umbil,	American valerian,	Cypripedium.
Umbrella tree,	Magnolia, .	Magnolia tripetala.
Uncomocomo,		Aspidium athaninticum.
Uncum,	Female regulator,	Senecio gracilis.
Unicorn,		Aletris farinosa.
" False,		Helonias dioica.
" plant,	Double claw,	Martynia proboscidea.
" " Swt. sctd.		" fragrans.
" root,		Aletris farinosa.
Unicorns' horn,	False unicorn,	Helonias dioica.
Universe vine,	Bearberry,	Arctostaphylos Uva Ursi.
Unkum,	Female regulator,	Senecio gracilis.
Upas tree,	yields a powerful poison,	Strychnos tieute.
Upland cranberry,	Bearberry,	Arctostaphylos Uva Ursi.
" sumach,		Rhus glabra.
Upright birthwort,		Aristolochia clematis.
" virgins' bower,		Clematis erecta.
Upstart,	Colchicum,	Colchicum autumnale.
Urari,	Upas tree poison,	Strychnos toxifera (tieute).

COMMON.	ENGLISH.	BOTANICAL.
Uva ursi,	Bearberry,	Arctostaphylos Uva Ursi.
Valerian,	Officinal valerian,	Valeriana officinalis.
" American,		Cypripedium of many species.
" " Greek,	Abscess root,	Polemonium reptans.
" English,	Officinal valerian,	Valeriana officinalis.
" " Am'n,	Valerian grown in the U. S.,	" "
" False,	Life root plant,	Senecio aureus.
" German,	Valerian grown in Germany,	Valeriana officinalis.
" Great wild.	"	" "
" Greek,		Polemonium cœruleum.
" root,	Officinal valerian,	Valeriana officinalis.
" Swamp,		" dioica.
" Vermont,	American English valerian,	" officinalis.
Valerianworts,	the plants of the order Valerianaceœ.	
Vandal root,	Valerian,	Valeriana officinalis.
Vaniglia,	Vanilla,	Vanilla aromatica.
Vanilla,		" " (planifolia).
" bean,	Vanilla,	" "
" cactus,	Night blooming cereus,	Cactus (Cereus) grandiflorus.
" grass,	Seneca grass,	Hierochloa arborealis.
" leaf,	Carolina vanilla,	Liatris odoratissima.
Vanilloes,	a sort of bastard vanilla,	Vanilla pompoma.
Varnish-tree, Black,		Melanorrhœa usitatissima.
" Chinese,	Japan lacquer,	Rhus vernicifolia.
" False,	Chinese sumach,	Ailantus glandulosus.
" India,	Black varnish-tree,	Melanorrhœa usitatissima.
" Japan,		Rhus vernicifolia.
" of China,		Terminalia vernix.
Vegetable antimony,	Boneset,	Eupatorium perfoliatum.
" brimstone,	Lycopodium,	Lycopodium clavatum.
" hair,	Florida long moss,	Tillandsia usneoides.
" ivory,	Ivory nut,	Phytelephas macrocarpa.
" marrow,	Common squash,	Cucurbita verrucosa medullosa.
" musk,		Mimulus moschata.
" oyster,		Tragopogon porrifolius.
" powder,	Lycopodium,	Lycopodium clavatum.
" silk,		Chorisa speciosa.
" sulphur,	Lycopodium,	Lycopodium clavatum.
" tallow, Am'n,	Bayberry wax, "tallow,"	Myrica cerifera.
" " Chinese,	the product of	Stillingia sebifera.
" wax,	Japan wax,	Rhus succedanum.
" "	Carnauba palm wax,	Ceroxylon carnauba and other spe-
Veiny leaved hawkweed,	Hawkweed,	Hieracium venosum. [cies.
Velvet bark,	Hard Carthagena bark,	Cinchona cordifolia vera.
" leaf,	Pareira brava, false,	Cissampelos Pareira.
" "	Indian mallow,	Abutilon avicena and other species
" plant,	Mullein,	Verbascum Thapsus.
" sumach,		Rhus typhium.
Venice sumach,	Young fustic,	" cotinus.
" turpentine,	obtained from European larch,	Abies (Pinus) larix.
Venus' bath,		Dipsacus sylvestris.
" comb,	Shepherds' needle,	Scandix pecten.
" cup,	American valerian,	Cypripedium.
" flytrap,		Dionœa muscipula.
" looking-glass,		Specularia speculum.
" hair,	European maidenhair,	Adiantum capillus veneris.
" shoe,	American valerian,	Cypripedium.
Verbena,	Lemon scented verbena,	Aloysia citriodora.
Vernal-grass,		Anthoxanthum odoratum.
Vervain,	commonly applied to	Verbena hastata, and officinalis.

COMMON.	ENGLISH.	BOTANICAL.
Vervain, American,		Verbena hastata.
" Blue,		" spuria.
" European,		" officinalis.
" False,	American vervain,	" hastata.
" mallow,		Malva alcea.
" Nettle leaved,	White vervain,	Verbena urticifolia.
" White,		" "
Vervine,	Blue vervain,	" spuria.
Vetch,	Tare plant,	Vicia sativa.
" Bitter,	.	Orobus (Lathyrus) niger.
" Kidney,		Anthyllis vulneraria.
" Tare,		Vicia sativa.
Vetiver,	Vittivert,	Andropogon muricatus.
Vetivert,	"	" "
Victoria,	Amazon water lily,	Victoria Regina.
Vine,	the grape vine,	Vitis vinifera.
" Poison,		Rhus radicans.
" White,		Clematis vitalba.
" " wild,		Bryonia dioica.
" Wild,	White bryony,	" alba.
Vinegar plant,		Rhus typhina.
" "		Penicillium glaucum.
Violet,		Viola.
" Adders',	Net leaf plantain,	Goodyera pubescens.
" Birdsfoot,		Viola pedata.
" bloom,	Bittersweet,	Solanum Dulcamara.
" Cauker,	Beaked violet,	Viola rostrata.
" Common,	Blue violet,	" pedata, cuculata.
" Corn,		Specularia hybrida.
" Dames',	Garden rocket,	Hesperis matronalis.
" Dog,		Viola canina.
" Dogstooth,	Adders' tongue,	Erythronium Americanum.
" Garden,	Pansy,	Viola tricolor.
" Marsh,		" palustris.
" Rattlesnakes',		" ovata.
" "	Adders' tongue,	Erythronium Americanum.
" Sweet scented,		Viola odorata.
" Water,		Hottonia palustris.
" wood,	Kingwood,	Dalbergia nigra.
" "		Andira violacea.
Vipers' bugloss,	Adderswort,	Echium vulgare.
" grass,	Winter asparagus,	Scorzonera Hispanica.
Virginia aloes,		Agave Virginica.
" cowslip,	American lungwort,	Pulmonaria Virginica.
" creeper,	" ivy,	Ampelopsis quinquefolia.
" cypress,	Black cypress,	Cupressus disticha.
" dogwood,	Florida dogwood,	Cornus Florida.
" lungwort,		Mertensia (Pulmonaria) Virginica.
" mouse ear,		Cynoglossum Morrisoni.
" poke,	Garget,	Phytolacca decandra.
" sarsaparilla,		Smilax Sarsaparilla.
" snakeroot,		Aristolochia Serpentaria.
" speedwell,		Veronica officinalis.
" tobacco,		Nicotiana Tabacum.
" thorn,		Cratægus cordata.
Virgins'-bower,		Clematis vitalba.
" Common,		" Virginica.
" Swt. sctd.,		" flammula.
Virgin tree,	Oriental sassafras,	Sassafras Parthenoxylon.
Vittivert,	an East Indian perfumery root,	Andropogon muricatus.

COMMON.	ENGLISH.	BOTANICAL.
Vittievar,	Vittivert,	Andropogon muricatus.
Volkameria,		Volkameria inermis.
Vomit nut,		Strychnos Nux vomica.
" wort,	Lobelia herb,	Lobelia inflata.
Waahoo,	Wahoo,	Euonymus atropurpureus.
Wafer ash,	Ptelea,	Ptelea trifoliata.
Wahoo,		Euonymus atropurpureus.
Wake-pintle,	Dragon root,	Arum maculatum.
Wake-robin,	"　　"	" triphyllum.
"	Beth root,	Trillium pendulum.
Walewort,	English dwarf elder,	Sambucus ebulus.
Walking-leaf,		Asplenium rizophyllum.
Wall cress,		Arabis sagittata.
" flower,		Cheiranthus Cheiri.
"　" Western,	Bitter root,	Apocynum androsæmifolium.
" germander,		Teucrium chamædrys.
" moss,		Lichen parietinus.
" pellitory,		Parietaria officinalis.
" pennywort	Navelwort,	Umbilicus pendulinus.
" pepper	Small houseleek,	Sedum acre.
" rue,		Asplenium Ruta muraria.
" wort,	European dwarf elder,	Sambucus ebulus.
Walnut,		Juglans.
" Belgaum,	Spanish walnut, Candleberry tree, Aleurites triloba.	
" Black,		Juglans nigra.
" Common,	Hickory,	Carya (Hicorya) alba and sulcata.
" European,	English walnut,	Juglans regia.
" Indian,	Candleberry tree,	Aleurites triloba.
" Jamaica,		Pterodendron Juglans.
" Lemon,	Butternut,	Juglans cinerea.
" Otaheite,	Candleberry tree,	Aleurites triloba.
" White,	Butternut,	Juglans cinerea.
Walpole tea,	New Jersey tea,	Ceanothus Americana.
Wandering Jew,	Strawberry geranium,	Saxifraga sarmentosa.
" milkweed,	Bitter root,	Apocynum androsæmifolium.
Wartcress,		Coronopus Ruellii.
Wartwort,	Sunspurge,	Euphorbia helloscopia.
Water aloe,	Water soldier,	Stratiotes aloides.
" apple,	Alligator apple,	Anona palustris.
" avens,		Geum rivale.
" beech,	Iron wood,	Carpinus Americana.
" betony,	Water figwort,	Scrophularia aquatica.
" blob,	Marsh marigold,	Caltha palustris.
" bugle,	Bugle weed,	Lycopus Virginicus.
" cabbage,	White pond lily,	Nymphæa odorata.
" can,	European yellow lily,	Nuphar lutea.
" calamint,		Mentha arvensis.
" chickweed,	Water starwort,	Callitriche verna.
" chincapin,	Rattlenut,	Nelumbium luteum.
" cress,		Nasturtium officinale.
"　" American,	Pepper cress,	Lepidium Virginicum and sativum.
"　" English,		Nasturtium officinale.
"　" Fools',	Honewort,	Trinia vulgaris.
"　" Yellow,	Marsh water cress,	Nasturtium palustre.
" cup,		Sarracenia flava.
" crowfoot,		Ranunculus aquatica.
" dock,		Rumex Britannica.
"　" Great,		" aquaticus.
" dragon,	Marsh marigold,	Caltha palustris.
" dropwort,	Water fennel,	Œnanthe Phellandrium.

COMMON.	ENGLISH.	BOTANICAL.
Water eryngo,		Eryngium aquaticum.
" feathers,		Chara vulgaris.
" fennel,	Fine leaved water hemlock,	Œnanthe Phellandrium.
" figwort,		Scrophularia aquatica.
" flag,	Blue flag,	Iris versicolor.
" germander,		Teucrium scordium.
" hemlock,	see also Cicuta,	Œnanthe fistulosa.
" " American,		Cicuta (Conium) maculatum.
" " dropwort,		Œnanthe crocata.
" " fine leav'd,		" Phellandrium.
" hemp,		Acnida cannabina.
" horehound,	Bugle weed,	Lycopus Virginicus.
" horseradish,		Cochlearia aquatica.
" jessamine,		Gratiola Virginica.
" jelly,	Water shield,	Brasenia hydropeltis.
" leaf,	" "	" "
" "		Rhodymenia palmata.
" lily,	Common pond lily,	Nymphæa odorata.
" " Amazon,		Victoria Regia.
" " White,		Nymphæa odorata.
" " " Europ'n,		Nymphæa alba.
" " " Royal, Amazon water lily,		Victoria Regia.
" " Yellow,		Nuphar advena.
" lovage,	Water hemlock dropwort,	Œnanthe crocata.
" maize,	Amazon Water lily,	Victoria Regia.
" mallow,		Hibiscus Moscheutos.
" marigold,		Bidens Beckii.
" maudlin,	Sweet smelling trefoil,	Eupatorium Cannabinum.
" melon,		Cucumis (Cucurbita) citrullus.
" milfoil,	Myriad leaf,	Myriophyllum verticillatum.
" mint,		Mentha aquatica.
" navelwort,		Hydrocotyle umbellata.
" nerve root,	White Indian hemp,	Asclepias incarnata.
" nuts,		Trapa natans.
" "		Nelumbium luteum.
" nymph,		Najas Canadensis.
" oats,	Indian rice,	Zizania aquatica.
" parsley,	Poison hemlock,	Conium maculatum.
" parsnip,		Sium nodiflorum.
" " Common,		" latifolium.
" pepper,		Polygonum hydropiper.
" pimpernel,		Samolus valerandi.
" plantain,	Mad-dog weed,	Alisma Plantago.
" "	Heart leaved plantain,	Plantago cordata.
" poplar,	Cottonwood tree,	Populus angulata.
" purslain,	Brooklime,	Veronica beccabunga.
" radish,	Water cress,	Erysimum Nasturtium.
" rocket,	" "	" "
" rose,	White water lily, European,	Nymphæa alba.
" rush,	Bulrush,	Typha (Juncus) effusus.
" shamrock,	Buckbean,	Menyanthes trifoliata.
" shield,		Brasenia hydropeltis.
" soldier,		Stratiotes aloides.
" starwort,		Callitriche verna.
" thyme,	Water-weed,	Anacharis alsinastrum.
" torch,		Typha latifolia.
" weed,		Anacharis alsinastrum.
" wort,		Elatine hexandra and hydropiper.
Wattle,	Cassie,	Acacia Farnesiana.
" tree,	Australian name for Acacia.	

COMMON.	ENGLISH.	BOTANICAL.
Wattle tree gum,	Australian gum,	Acacia decurrens.
Wauhoo,	Wahoo,	Euonymus atropurpureus.
Waw weed,	Life root plant,	Senecio aureus.
Waxberry,	Bayberry,	Myrica cerifera.
" cornel,		Cornus alba.
Waxbush,		Cuphea viscosissima.
Wax, Carnauba,	Brazilian palm wax,	Corypha cerifera.
" cluster,	Checkerberry,	Gaultheria procumbens.
" myrtle,	Bayberry,	Myrica cerifera.
" palm,		Ceroxylon carnauba and other spe-
" "	Palm oil palm,	Elæis Guineensis. [cies.
" pink,	Garden purslane, var.,	Portulaca grandiflora.
" plant,		Hoya carnosa.
" tree,		Rhus succedanum.
" Vegetable,	see Bayberry wax, Japan wax, Carnauba wax, Vegetable tallow,	
Waxwork,	False bittersweet,	Celastrus scandens. [etc.
Waybread,	Plantain, Common,	Plantago major.
Wayfaring-tree,		Viburnum lantana and other species
Waythorn,	Buckthorn,	Rhamnus catharticus.
Weasel-snout,	Yellow archangel,	Galeobdolon luteum.
Weathercock,	Wild celandine,	Impatiens pallida.
Weatherglass,	Red chickweed,	Anagallis arvensis.
Weeping spruce,	Hemlock spruce,	Abies (Pinus) Canadensis.
" willow,		Salix Babylonica.
Weigelia,	Bush honeysuckle,	Diervilla rosea (Canadensis).
Weld,	a dye-stuff,	Reseda luteola.
Welch clover,		Trifolium arvense.
" sorrel,	Wood sorrel,	Oxalis acetosella.
Western dropwort,	Small flow'd Indian physic,	Gillenia stipulacea.
West Indian bead tree,	Necklace tree,	Ormosia dasycarpa.
Whahoo,		Euonymus atropurpureus.
"		Ulmus alata.
Wheat, Common,	the grain bearing Triticum vulgare of several varieties.	
" Egyptian,		Triticum compositum.
" Guinea,	Common Indian corn,	Zea mays.
" Spelt,		Triticum spelta.
" Summer,		" æstivum.
" Winter,		" hybernum.
Whig plant,	Garden chamomile,	Anthemis nobilis.
Whin,	Dyers' broom,	Genista tinctoria.
" berry,	Whortleberry,	Vaccinium resinosum (myrtillis).
Whip-poor-will's shoe,	American valerian,	Cypripedium acaule.
Whip-tongue,		Galium mollugo.
Whistle wood,	Red maple,	Acer rubrum.
White agaric,	Agaric,	Boletus laricis.
" alder,	Spiked alder,	Clethra alnifolia.
" apple,		Apios tuberosa.
" archangel,		Lamium album.
" ash,		Fraxinus Americana (acuminata).
" avens,		Geum Virginianum.
" ball,	Button bush,	Cephalanthus occidentalis.
" balsam,	Sweet scent'd life everlast'g,	Gnaphalium polycephalum.
" " of Peru,		Myrospermum Peruiferum.
" baneberry,	White cohosh,	Actæa alba.
" bay,	Magnolia,	Magnolia glauca.
" beads,	White cohosh,	Actæa alba.
" beam,		Pyrus aria.
" beech,		Fagus sylvatica.
" berry snakeroot,	White cohosh,	Actæa alba.
" birch,		Betula alba.

COMMON.	ENGLISH.	BOTANICAL.
White bryony,		Bryonia alba.
" bush,	Spiked alder,	Clethra alnifolia.
" "		Andromeda paniculata.
" cankerweed,		Prenanthes alba.
" cap,	Hardhack,	Spiræa tomentosa.
" cedar,		Cupressus Thyoides.
" " False,	Arbor vitæ,	Thuja occidentalis.
" chestnut oak,		Quercus prinus.
" cinnamon,	Canella bark,	Canella alba.
" clover,		Trifolium repens.
" cohosh,	White baneberry,	Actæa alba.
" cranberry,		Oxycocca hispidula.
" daisy,	White weed,	Leucanthemum vulgare.
" elm,	Common shade elm,	Ulmus Americana.
" fl'd ladies' slipper,	American valerian,	Cypripedium candidum.
" fraxinella,	Bastard dittany,	Dictamus alba.
" fringe,	Fringe tree,	Chionanthus Virginicus.
" galls,	see Galls,	Enblica officinalis.
" gentian,	Fever root,	Triosteum perfoliatum.
" "		Laserpitium latifolium.
" gum,	Sweet gum,	Liquidambar styraciflua.
" " tree,		Eucalyptus rostrata.
" hellebore,		Veratrum album.
" " Am'n,		" viride.
" henbane,		Hyosciamus alba.
" horehound,	Common horehound,	Marrubium vulgare.
" Indian hemp,		Asclepias incarnata.
" ipecac,	American ipecac,	Euphorbia Ipecacuanha.
" "	Undulated ipecac,	Richardsonia scabra.
" "		Ionidum Ipecacuanha.
" jassamine,	White jasmine,	Jasminum officinale.
" leaf,	Hardhack,	Spiræa tomentosa.
" lettuce,	Lionsfoot,	Nabalus albus.
" lilac,	Syringia,	Lilaca vulgaris, var. alba.
" lily,		Lilium candidum.
" " Pond,	White pond lily,	Nymphæa odorata.
" " Water,	" " "	" "
" linden,		Tilia alba.
" maple,	Silver maple,	Acer dasycarpum.
" melilot,		Melilotus albus.
" " clover,	Sweet clover,	" leucanthe.
" mullein,		Verbascum album.
" mustard,		Sinapis alba.
" nettle,	White archangel,	Lamium album.
" oak,		Quercus alba.
" osier,		Andromeda racemosa.
" Pareira brava,		Abuta rufescens.
" pepper,	Black pepper deprived of its husk or skin.	
" " bush,	White osier,	Andromeda racemosa.
" pine,	Soft deal pine,	Pinus strobus.
" plantain,		Antennaria plantagineum.
" "		Goodyera repens.
" poison vine,	White jessamine,	
" pollom,	" cranberry,	Oxycocca hispidula.
" pond lily,		Nymphæa odorata.
" poplar,	American poplar,	Populus tremuloides, and alba.
" "	Tulip tree,	Liriodendron Tulipifera.
" pursley,	Blooming spurge,	Euphorbia corollata.
" rhubarb,	Tartarian rhubarb,	Rheum leucorrhizum.
" root,	Pleurisy root,	Asclepias tuberosa.

COMMON.	ENGLISH.	BOTANICAL.
White rose,		Rosa alba.
" rosin,	Refined rosin,	Pinus palustris.
" sage,		Salvia argentea.
" sandal,		Santalum album and other species.
" sanicle,		Eupatorium ageratoides.
" saunders,	White sandal,	Santalum album.
" shamrock,	" clover,	Trifolium repens.
" snakeroot,		Eupatorium aromaticum.
" snowdrop,		Galanthus nivalis.
" soaproot,	White soapwort,	Lychnis dioica.
" soapwort,		" "
" sorrel,	Wood sorrel,	Oxalis acetosella.
" squills,		Scilla maritima.
" stonecrop,	Lesser houseleek,	Sedum album.
" swallow-wort,		Asclepias vincetoxicum.
" swamp sumach,	Varnish tree,	Rhus vernicifera.
" thorn,		Cratægus Crus-galli.
" top grass,		Agrostis alba.
" turpentine,	see Turpentine.	
" umbel,	White flow'd ladies' slipper,	Cypripedium candidum.
" vervain,	Nettle leaved vervain,	Verbena urticifolia.
" vine,	Travellers' joy,	Clematis vitalba.
" violet,	Snowflake,	Leucoium vernum and other species
" walnut,	Butternut,	Juglans cinerea.
" water lily,	White pond lily,	Nymphæa odorata.
" " " Europ'n,		" alba.
" weed,	Oxeye daisy,	Leucanthemum vulgare.
" willow,		Salix alba.
" wood,	Tulip tree,	Liriodendron Tulipifera.
" "		Tilia Americana (glabra).
Whitewood b'k, Bahama,	Canella bark,	Canella alba.
Whitewort,		Polygonatum officinalis.
Whitlavia,		Whitlavia grandiflora.
Whitlow-grass,	Nailwort,	Draba incana and other species.
Whitten-tree,	High cranberry bush,	Viburnum opulus.
Whorlywort,	Black root,	Leptandra Virginica.
Whortleberry,		Vaccinium.
" Black,		" myrtillis.
" Blue,		" frondosum.
" " black,		" Pennsylvanicum.
" Bush,		" dumosum.
" Giant,		" corymbosum.
Wicke,	Laurel,	Kalmia latifolia and other species.
Wickup,	Willow herb,	Epilobium angustifolium and other
Wicopy bark,	Leatherwood bark,	Dirca palustris. [species.
" herb,		Epilobium spicatum and other spec.
" Indian,		" " " " "
" root,	Wickopy,	" " " " "
Widow-wail,		Cneorum tricoccon.
Wiegelia,	see Weigelia.	
Wig-tree,	False fringe tree,	Rhus cotinus.
Wild allspice,	Spice wood,	Benzoin odoriferum.
" amaranth,	Allseed,	Amarantus blitum.
" angelica,		Angelica sylvestris.
" anise,	Florida anise,	Illicium Floridanum.
" arum, Water,	Swamp robin,	Calla palustris.
" balsam apple,	Elaterium cucumber,	Momordica Elaterium.
" basil,		Pycnanthemum aristatum.
" "		Cunila mariana.
" bergamot,		Monarda oblongata.

COMMON.	ENGLISH.	BOTANICAL.
Wild black currant,		Ribes Floridum.
" " cherry,		Prunus (Cerasus) Virginiana.
" brier,	Eglantine rose,	Rosa rubiginosa.
" bryony,		Bryonia alba.
" bugloss,	German madwort,	Asperugo procumbens.
" caraway,		Cacalia reniformis and other spec.
" cardamom,		Elettaria Cardamomum major.
" carrot,		Daucus Carota.
" " of Europe,	Spanish toothpicks,	" sylvestris.
" celandine,	Touch-me-not,	Impatiens pallida.
" celery,		Apium graveolens.
" chamomile,	Mayweed,	Anthemis Cotula.
" cherry,		Prunus (Cerasus) Virginiana.
" chervil,		Anthriscus sylvestris.
" cicily,	Cow weed,	Cicutaria vulgaris.
" cinnamon,	Canella bark,	Canella alba.
" clary,	Clear eye,	Salvia verbeniea.
" clove,		Eugenia acris.
" " bark,	Wild clove, Bay rum leaf tree,	Myreia acris.
" coffee,	Fever root,	Triosteum perfoliatum.
" comfrey		Cynoglossum amplexicaule.
" crab tree,	Crab apple tree,	Pyrus caronaria.
" cranberry,	Bearberry,	Arctostaphyllos Uva Ursi.
" cranesbill,		Geranium of several species.
" cucumber,	Elaterium cucumber,	Momordiea Elaterium.
" cummin,		Lagœeia euminoides.
" elder,	Dwarf elder,	Aralia hispida.
" endive,	Dandelion,	Taraxacum Dens Leonis.
" eryngo,		Eryngium campestre.
" fennel,		Fœniculum officinalis.
" flax,		Linum Virginianum.
" " False,		Camelina sativa.
" ginger, Southern,		Asarum Virginicum.
" grape,	Fox grape,	Vitis vulpina and other var.
" "	Frost grape,	" cordifolia and other var.
" hemlock,	Spotted hemlock,	Cicuta maculata.
" hemp,	Tall ambrosia,	Ambrosia trifida.
" honeysuckle,	Swamp pink,	Azalea nitida.
" hops,	White bryony,	Bryonia alba.
" horehound,		Eupatorium teucrifolium.
" hyacinth,		Scilla Fraseri.
" hydrangea,		Hydrangea arborescens.
" hyssop,	Vervain, American,	Verbena hastata.
" indigo plant,		Baptisia tinctoria.
" ipecac,	Fever root,	Triosteum perfoliatum.
" "	American ipecac,	Euphorbia Ipecacuanha.
" jalap,	Wild potato,	Convolvulus panduratus.
" jessamine,	Yellow jessamine,	Gelsemium sempervirens.
" Job's tears,	False gromwell,	Onosmodium Virginianum.
" lemon,	American mandrake,	Podophyllum peltatum.
" lettuce,		Lactuca elongata.
" "	Round leaved pyrola,	Pyrola rotundifolia.
" liquorice,		Astragalus glycyphyllos.
" "	American sarsaparilla,	Aralia nudicaulis.
" mandrake,	" mandrake,	Podophyllum peltatum.
" marjoram,	Origanum,	• Origanum vulgare.
" melon,		Citrullus amarus.
" mustard,	Charlock,	Sinapis arvensis.
" nard,	Asarabacca,	Asarum Europæum.
" nutmeg,	Male nutmeg,	Myristica tomentosa.

COMMON.	ENGLISH.	BOTANICAL.
Wild parsley,	Meadow parsnip,	Zizia aurea.
" parsnip,		Sium latifolium.
" patience,	Blunt leaved dock,	Rumex obtusifolius.
" pear,		Pyrus communis.
" pepper,	the berries (fruit) of	Daphne Mezereum.
" pine,		Pinus sylvestris.
" pink,	Catchfly,	Silene Virginica.
" plantain,	Indian shot plant,	Canna Indica.
" plum tree,	Sloe,	Prunus spinosa.
" pomegranate,		Punica granatum.
" poplar,	Whitewood,	Liriodendron Tulipifera. ,
" potato,	Wild jalap,	Convolvulus panduratus.
" radish,	Charlock,	Sinapis arvensis.
" rhubarb,	Wild potato,	Convolvulus panduratus.
" red cherry,		Prunus Pennsylvanica.
" rose,		Rosa parviflora, blanda, etc.
" " bay,	Mountain laurel,	Rhododendron maximum.
" rosemary,	Marsh tea,	Ledum palustre.
" " W. I.,		Croton lineare.
" rue,		Ruta sylvestris.
" sage,		Salvia lyrata.
" sarsaparilla,	American sarsaparilla,	Aralia nudicaulis.
" scammony,	Wild potato,	Convolvulus panduratus.
" senna,	American senna,	Cassia Marilandica.
" " European,		Globularia alypum.
" snowball,	Red root,	Ceanothus Americana.
" strawberry,		Fragaria Virginiana.
" succory,	Chicory,	Cichorium intybus.
" "	American centaury,	Sabbatia angularis.
" sunflower,		Helianthus giganteus.
" sweet potato,	Mendo,	
" tea,		Amorpha canescens.
" thyme,		Thymus serpyllus.
" "	Lemon thyme,	" citriodorus.
" tobacco,	Lobelia,	Lobelia inflata.
" turkey pea,		Corydalis formosa.
" turnip root,	Dragon root,	Arum triphyllum and other species
" vanilla,	Deer tongue,	Liatris odoratissima.
" valerian,		Valeriana officinalis.
" vine,	Bryony, Red,	Bryonia dioica.
" "	Travellers' joy,	Clematis vitalba.
" woodbine,	American ivy,	Ampelopsis quinquefolia.
" woodvine,	" "	" "
" wormseed,	Oak of Jerusalem,	Chenopodium anthelminticum.
" yam,		Dioscorea villosa.
" "	West India yam,	" sativa.
" zerumbet,		Zingiber Zerumbet.
Willow,		Salix.
" Black,		" nigra.
" catkins,	Pussy willow,	" "
" hemp,	Water hemp,	Acnida cannabina.
" herb,	Rose bay,	Epilobium angustifolium.
" " Hooded,	Scullcap,	Scutellaria lateriflora.
" " Purple,	Loose-strife,	Lythrum salicaria.
" " Swamp,		" verticillatum.
" " "	Wickop,	. Epilobium palustre.
" Purple, .		Salix purpurea.
" Pussy,	Black willow,	" nigra.
" Red,	Swamp dogwood,	Cornus sericea.
" Rose,	" "	" "

COMMON.	ENGLISH.	BOTANICAL.
Willow Sage,	Willow herb, Purple,	Lythrum salicaria.
" sponge,		Boletus suaveolens.
" weed,	Willow herb, Purple,	Lythrum salicaria.
" Weeping,		Salix Babylonica.
" White,		" alba.
" wort,	Willow herb, Purple,	Lythrum salicaria.
Will-o'-the-wisp,		Tremella nostoc.
Wind-bloom,		Anemone Virginica.
Wind-flower,	Wood anemone,	" nemorosa.
Wind-root,	Pleurisy root,	Asclepias tuberosa.
Wineberry,	Red currant,	Ribes rubrum.
"	Huckleberry,	Vaccinium myrtillus.
Wing-seed,	Wafer ash,	Ptelea trifoliata.
Winter asparagus,	Vipers' grass,	Scorzonera Hispanica.
" berry,	Black alder,	Prinos verticillatus.
" bloom,	Witch hazel,	Hamamelis Virginica.
" brake,	Rockbrake,	Pteris atropurpurea.
" cherry,		Physalis alkekengi.
" clover,	Squaw vine,	Mitchella repens.
" cress, Water,		Barbarea vulgaris and præcox.
" fern,	Rockbrake,	Pteris atropurpurea.
" hellebore,		Helleborus hyemalis.
Wintergreen,	Pipsissewa.	Chimaphila umbellata and maculata
. "	Checkerberry,	Gaultheria procumbens.
" Aromatic,	"	" "
" Cancer,		" hispidula..
" Chickweed,		Trientalis Europæa.
" False,	Round leaved pyrola,	Pyrola rotundifolia.
" Pear leaf,	" " "	" "
" Spicy,	Checkerberry,	Gaultheria procumbens.
" Spotted,		Chimaphila maculata.
" Spring,	Checkerberry,	Gaultheria procumbens.
Winter laurel,	Carolina laurel cherry,	Prunus Caroliniana.
Winterlien,	Flaxseed,	Linum usitatissimum.
Winter marjoram,	Wild marjoram,	Origanum vulgare.
" pink,	Trailing arbutus,	Epigæa repens.
" plum,	Persimmon,	Diospyros Virginiana.
" savory,		Satureja montana.
" sweet,	Wild marjoram,	Origanum vulgare.
" weed,		Veronica hederæfolia.
Winter's bark,		Wintera (Drimys) aromatica.
" cinnamon,	Winter's bark,	" " "
Wistaria,		Wistaria sinensis and frutescens.
Witch alder,		Fothergilla alnifolia.
" grass,	Dog grass,	Triticum repens.
" hazel,		Hamamelis Virginica.
" " Big leaf,		" macrophila.
Witches' thimble,		Silene maritima.
Withe,	see Willow,	Salix.
Withy,	" "	"
" Hoop,		Rivinia octandra.
Withwind,	Woodbine,	Caprifolium.
Woad,	Dyers' woad,	Isatis tinctoria.
" Waxen,	" broom,	Genista tinctoria.
" Wild,	Weld,	Reseda luteola.
Wolf-berry,		Symphoricarpus occidentalis.
Wolf's-bane,	Aconite,	Aconitum Napellus.
"	Arnica,	Arnica montana.
Wolf claw,	Lycopodium,	Lycopodium clavatum.
" foot,	Bugle weed,	Lycopus Virginicus.

COMMON.	ENGLISH.	BOTANICAL.
Wolf grape,	Woody nightshade,	Solanum Dulcamara.
" root,	Aconite root,	Aconitum Napellus.
" wort,		" luparia.
Wolveboon,	Hyæna poison,	Hyænanche capensis.
Wonder-of-the-world,	Chinese ginseng,	Panax schinseng.
Wood anemone,	Wind flower,	Anemone nemorosa.
" apple,	Elephant apple,	Feronia elephantum.
" betony,		Betonica officinalis.
" "		Pedicularis Canadensis. [other spec
Woodbine,	Honeysuckle, *	Caprifolium Periclymenum and
"	American ivy,	Ampelopsis quinquefolia.
"	Yellow jessamine,	Gelsemium sempervirens.
" Rough,	.	Lonicera hirsuta.
Wood boneset,		Eupatorium perfoliatum.
" broom,	Teasel,	Dipsacus sylvestris.
" fringe,	Climbing fumitory,	Adlumia cirrhosa.
" geranium,	Cranesbill,	Geranium dissectum.
" grass,		Sorghum nutans.
" nettle,		Laportea Canadensis.
" pea,	Tare,	Vicia sativa.
" "	Heath pea,	Orobus atropurpurea.
" pimpernel,		Lysimachia nemorum.
" roof,		Asperula odorata.
" row,	Brook liverwort,	Marchantia polymorpha.
" sage,	Germander,	Teucrium Canadensis.
" "		" scorodonia.
" sanicle,	European sanicle,	Sanicula Europæa.
" scabious,	Devil's bit root,	Scabiosa succissa.
" sorrel,		Oxalis acetosella.
" sour,	Wood sorrel,	" "
" strawberry,		Fragaria vesca.
" vetch,	Tare,	Vicia sativa.
" vine,	Bryony, White,	Bryonia alba.
" waxen,	Dyers' broom,	Genista tinctoria.
" of the holy cross,	the Oak mistletoe,	Loranthus Europæus.
Woody climber,	American ivy,	Ampelopsis quinquefolia.
" nightshade,		Solanum Dulcamara.
Woolen,	Mullein,	Verbascum Thapsus.
Woolybut,	Eucalyptus,	Eucalyptus.
Woorari,	a powerful poison supposed to be obtained from Strychnos	
Woorali,	Woorari,	[toxifera.
World's wonder,	False jalap,	Mirabilis Jalapa and other species.
Worm-bark,	Cabbage tree bark,	Andira inermis.
Worm-grass,	Carolina pink,	Spigelia Marilandica.
Worm moss,	Corsican moss,	Fucus helminthocorton.
Worm root, American,	Carolina pink,	Spigelia Marilandica.
Wormseed,		Chenopodium anthelminticum.
" Aleppo,	Levant wormseed,	Artemisia contra and other species
" Alexandria,	" "	" "
" American,	Oak of Jerusalem,	Chenopodium anthelminticum.
" Barbary,		Artemisia Judaica and other species
" European,	Levant wormseed,	" contra.
" Levant,	the unexpanded flowers of	" and other species
" plant,	Oak of Jerusalem,	Chenopodium anthelminticum.
Wormweed,	Clammy weed,	Polanisia viscosa.
" Corsican,	Corsican moss,	Fucus helminthocorton.
" Creeping,		Artemisia rupestris.
Wormwood,		" Absinthium.
" Roman,		" pontica.
" Am'n Roman,		Ambrosia artemisiæfolia.

COMMON.	ENGLISH.	BOTANICAL.
Woundweed,	Goldenrod,	Solidago odora.
Woundwort,		Anthyllis vulneraria.
" Clowns',		Stachys palustris.
" Hedge,		" sylvatica.
" Marsh,		" palustris.
Wrack,	Sea weeds thrown ashore,	
" Sea,	Fucus,	Zostera mariana and other species.
Wuckopy,	Lime tree,	Tilia glabra.
Wurrus,	Kameela,	Rottlera tinctoria.
Wych elm,		Ulmus montana.
Wymote,	Marsh mallow,	Althæa officinalis.
Yam,		Dioscorea.
" Chinese,		" Batatas.
" root, Cultivated,		" sativa.
" " Wild,		" villosa.
" " Tara,	W. I. arrow root,	Colocassia esculenta.
Yarrow,		Achillea Millefolium.
" Noble,		" nobilis.
Yasmyn,	Jasmine,	Jasminum officinale.
Yaupon,	South sea tea,	Ilex vomitoria.
Yaw root,	Queens' root,	Stillingia sylvatica.
" weed,		Morinda royoc.
Yellow adders' tongue,		Erythronium flavum.
" anemone,		Anemone vernalis.
" archangel,		Galeobdolon luteum.
" ash,	Fustic tree,	Cladrastis tinctoria.
" balm,		Lysimachia quadrifolia.
" balsam,		Impatiens Noli-tangere.
" bark,	Peruvian bark,	Cinchona Calisaya.
" bedstraw,	Cleavers, Yellow,	Galium verum.
" berries,	the dried unripe fruit of	Rhamnus infectorius.
" berry,	American Mandrake,	Podophyllum peltatum.
" broom,	Wild Indigo,	Baptisia tinctoria.
" cleavers,	Maids' hair,	Galium verum.
" dock,		Rumex crispus.
" dye berries,	Yellow berries,	Rhamnus infectorius.
" " tree,	Green heart tree,	Nectandra Rodiæi.
" erythronium,	Yellow adders' tongue,	Erythronium flavum.
" eye,	Goldenseal root,	Hydrastis Canadensis.
" figwort,		Scrophularia vernalis.
" foxglove,		Digitalis lutea.
" gentian,	American columbo,	Frasera Carolinensis.
" ginseng,	Blue cohosh,	Caulophyllum thalictroides.
" gowans,		Ranunculus.
" helmet flower,	variety of aconite,	Aconitum anthora.
" henbane,	Ground cherry,	Physalis viscosa.
" hercules,		Xanthoxylum clavus-Herculis.
" indigo,	Wild Indigo,	Baptisia tinctoria.
" jessamine,		Gelsemium sempervirens.
" jonquil,		Narcissus odora.
" ladies' slipper,	American valerian,	Cypripedium pubescens.
" lily root,	Yellow pond lily,	Nuphar advena.
" locust,		Robinia pseudo-Acacia.
" loose strife,	Yellow willow herb,	Lysimachia vulgaris.
" mallows,		Abutilon condalum.
" melilot clover,		Melilotus officinalis.
" moccasin,	American valerian,	Cypripedium pubescens.
" monkey flower,		Mimulus luteus.
" mustard,	White mustard,	Sinapis alba.
" nicker tree,	Bonduc nut tree,	Guilandina bonduc.

COMMON.	ENGLISH.	BOTANICAL.
Yellow paint root,	Goldenseal root,	Hydrastis Canadensis.
" parilla,		Menispermum Canadense.
" pitch pine,		Pinus palustris.
" plums,	Persimmon,	Diospyros Virginiana.
" pond lily,		Nuphar advena.
" poplar,	Tulip tree,	Liriodendron Tulipifera.
" puccoon,	Goldenseal root,	Hydrastis Canadensis.
" rattle,	Yellow cockscomb,	Rhinanthus Crista-galli.
" rhodendron,		Rhododendron Chrysanthemum.
" root,	Goldenseal root,	Hydrastis Canadensis.
" "	Shrub yellow root,	Xanthorrhiza apiifolia.
" "	Twin-leaf root,	Jeffersonia diphylla.
" "	Goldthread root,	Coptis trifolia.
" "	Bittersweet, false,	Celastrus scandens.
" root'd wat'r dock,	see Dock,	Rumex Britannica.
" sandal,		Santalum freycinetianum.
" sarsaparilla,	Yellow parilla,	Menispermum Canadense.
" saunders,	" sandal,	Santalum freycinetianum.
" snake leaf,	Adders' tongue,	Erythronium flavum.
" snow drop,	" "	" "
" star,		Helenium autumnale.
" star-root,		Aletris aurea.
" umbel,		Cypripedium pubescens.
" violet,	Gillyflower, stock,	Mathiola incana.
" water cress,		Nasturtium palustre.
" " lily,		Nuphar advena.
" weed,		Reseda luteola.
" "	Crowfoot buttercup,	Ranunculus acris.
" willow herb,		Lysimachia vulgaris.
" wolf'sbane		Aconitum lycotonum.
Yellow-wood,	Prickly ash, Southern,	Xanthoxylum Carolinianum.
"	Yellow ash, fustic,	Cladrastis tinctoria.
" E. Indian,		Chloroxylon Swietenia.
" Prickly,		Xanthoxylum clava-Herculis.
" W. Indian,		" " "
Yellow-wort,	Shrub yellow root,	Xanthorrhiza apiifolia.
"		Chlora perfoliata.
" Parsley l'vd,	Shrub yellow root,	Xanthorrhiza apiifolia.
Yellowish white gentian,	Marsh gentian,	Gentiana ochroleuca.
Yellows,		Cypripedium pubescens.
Yew,		Taxus baccata.
" Dwarf,		" Canadensis.
Yoke-elm,		Carpinus Betulus.
Ylang ylang,		Unona odoratissima.
Young fustic,	False fringe tree,	Rhus cotinus.
Youthwort,	Sundew,	Drosera rotundifolia.
"	Masterwort,	Heracleum lanatum.
Yucca,	the cassava plant,	Jatropha (Janipha) manihot.
Yulan,		Magnolia conspicua.
Zafran,	Saffron,	Crocus sativus.
Zante wood,	False fringe tree,	Rhus cotinus.
Zebra "		Omphalobium Lambertii.
Zedoary root,		Amomum Zerumbet.
" Round,		Curcuma Zedoary.
Zerumbet,		" Zerumbet.
Zinnia,		Zinnia elegans.

C. E. HOBBS'

BOTANICAL HAND-BOOK.

EXPLANATION OF ABBREVIATIONS USED IN THE BOTANICAL DEPARTMENT.

MEDICAL PROPERTIES, ETC.

Abo.	Abortive,	A medicine which claims the property of causing abortion.
Aci.	Acidulous,	Substances which possess a sourish taste.
Acr.	Acrid,	Hot, biting, irritating.
Adc.	Adenagic,	Relieving or arresting glandular pain.
Ale.	Alexipharmic,	Preventing the bad effects of poisons inwardly.
Alt.	Alterative,	Producing a salutary change without perceptible evacuation.
Ano.	Anodyne,	Relieving pain, or causing it to cease.
Ant.	Anthelmintic,	A remedy which destroys or expels worms.
Ape.	Aperient,	Gently laxative, without purging.
Aph.	Aphrodisiac,	Medicine believed capable of exciting the venereal appetite.
Aro.	Aromatic,	Odoriferous, stimulant, spicy, agreeable.
Ast.	Astringent,	Having the property of constringing the organic texture.
A-aph.	Anti-aphrodisiac,	A substance capable of blunting the venereal appetite.
A-bil.	Anti-bilious.	Opposed to biliousness; acting on the bile.
A-eme.	Anti-emetic,	A remedy for vomiting.
A-epi.	Anti-epileptic,	Opposed to epilepsy; relieving fits.
A-hys.	Anti-hysteric,	A remedy for hysteria.
A-lit.	Anthilitic,	Preventing the formation of calculi in the urinary organs.
A-per.	Anti-periodic,	Arresting morbid periodical movements.
A-phl.	Anti-phlogistic,	Opposed to inflammation.
A-rhe.	Anti-rheumatic,	Relieving, preventing, or curing rheumatism.
A-sco.	Anti-scorbutic,	Curing or preventing scurvy.
A-sep.	Anti-septic,	Opposed to putrefaction.
A-spa.	Anti-spasmodic,	Relieving or preventing spasm.
A-syp.	Anti-syphilitic,	Opposed to or curing venereal diseases.
A-ven.	Anti-venomous,	Used against bites of venomous insects or snakes, etc.
Bal.	Balsamic,	Mitigatory, healing, soothing to inflamed parts.
Bit.	Bitter,	Having a tonic effect.
Car.	Carminative,	Expelling wind from the bowels.
Cat.	Cathartic,	Increasing evacuations from the bowels.
Cau.	Caustic,	The property of burning or disorganizing animal substance.
Cep.	Cephalic,	Relating to diseases of the head.
Cho.	Cholagogue,	Increasing the flow of bile.
Con.	Condiment,	Improving the savor of food, as salt, pepper, salad, etc.
Cor.	Cordial,	A warm stomachic; exciting the heart.
Cos.	Cosmetic,	Used for improving the complexion or skin.
C-irr.	Counter-irritant,	Causing irritation in one part to relieve pain in another part.
Dem.	Demulcent,	Soothing, mucilaginous, relieving inflammation.
Deo.	Deobstruent,	Removing obstructions; aperient in a general sense.
Dep.	Depurative,	Purifying the blood.

Des.	Dessicative,	Drying the moisture of wounds and ulcers.
Det.	Detergent,	Cleansing to wounds, boils or ulcers.
Dia.	Diaphoretic,	Producing insensible perspiration.
Dis.	Discutient,	Dispelling or resolving tumors.
Diu.	Diuretic,	Increasing the secretion and flow of urine.
D-pil.	Depilatory,	Removing superfluous hair.
D-ter.	Detersive,	Detergent.
Dra.	Drastic,	Powerfully cathartic.
Eme.	Emetic,	Producing or causing vomiting.
Emm.	Emmenagogue,	Promoting menstruation.
Emo.	Emollient,	Softening to inflamed parts; soothing.
Esc.	Esculent,	Eatable as food.
E-sch.	Escharotic,	A substance which, applied to a living part, causes an eschar.
Exa.	Exanthematous,	Relating to eruption or skin diseases of an eruptive nature.
Exc.	Excitant,	Producing excitement; stimulant.
Exp.	Expectorant,	A medicine capable of facilitating expectoration.
Far.	Farinaceous,	Containing farina; mealy; employed as nutriment.
Feb.	Febrifuge,	Abating or driving away fever.
F-com.	Female complaints	Ailments peculiar to women, as dysmenorrhœa, amenorrhœa,
Fœt.	Fœtid,	Bad smelling, disgusting, nauseous, stinking. [etc.
For.	Forage,	Used as food for domestic cattle, sheep or horses.
Fum.	Fumigating,	Disinfecting by burning substances which counteract on
Gal.	Galactagogue,	Favoring the secretion of milk. [noxious odors.
Hep.	Hepatic,	Relating to diseases of the liver.
Her.	Herpatic,	Relating to or curing eruptions or skin diseases, as ringworm
Hyd.	Hydragogue,	Medicines that cause watery evacuations, and believed capa-
Hyp.	Hypnotic,	Producing or inducing sleep. [ble of expelling serum.
Ins.	Insecticide,	A substance that destroys insects.
Lax.	Laxative,	A medicine that acts gently on the bowels, without griping.
Len.	Lenitive,	Palliating or allaying irritation; also laxative.
Lit.	Lithontryptic,	Medicine believed to dissolve calculi in the urinary organs.
Mat.	Maturating,	Favoring the maturation or ripening of tumors, boils and
Muc.	Mucilaginous,	Gummy, glutinous, viscid, demulcent. [ulcers.
Nar.	Narcotic,	Stupefying, sedative, poisonous.
Nau.	Nauseant,	Causing inclination to vomit.
Nep.	Nephreticum,	Relating to or curing kidney complaints.
Ner.	Nervine,	Allaying nervous excitement; acting on the nervous system.
Nut.	Nutritious,	Having the quality of nourishing or sustaining life.
Opt.	Ophthalmicum.	A remedy for diseases of the eye.
Orn.	Ornamental,	Cultivated for ornament.
Par.	Parturient,	A medicine that induces or promotes labor or child-birth.
Pec.	Pectoral,	Medicines considered proper for relieving affections of the
Per.	Perfume,	A plant or substance used for its fragrance. [chest.
Poi.	Poisonous,	Producing death, if taken in improper doses.
Pun.	Pungent,	Biting, hot, acrid; prickly to the taste.
Pur.	Purgative,	A medicine that physics more powerfully than a cathartic.
Ref.	Refrigerant,	Depressing the morbid temperature of the body; cooling.
Res.	Resolvent,	Discutient; dispelling or resolving tumors.
Rub.	Rubifacient,	Producing or causing redness of the skin.
Sac.	Saccharine,	Containing sugar; sweetish.
Sad.	Salad,	Fresh herbs eaten as condiments or as food.
Sal.	Saline,	Containing or having the properties of a salt.
Sap.	Saponaceous,	Soapy; making a lather with water.
Sed.	Sedative,	Directly depressing to the vital forces.
Sia.	Sialagogue,	Provoking the secretion of saliva.
Ste.	Sternutatory,	A substance which provokes sneezing.
Sti.	Stimulant,	Exciting or inducing organic action of the animal economy.
Sto.	Stomachic,	Strengthening and giving tone to the stomach; tonic.
Sty.	Styptic,	Externally astringent; arresting hemorrhage or bleeding.
Sud.	Sudorific,	A medicine which provokes sweating; see Diaphoretic.

Ton.	Tonic,	Permanently strengthening; in a durable manner invigora-
Ver.	Vermifuge,	Anthelmintic; expelling worms. [ting.
Ves.	Vesicant,	Producing blisters.
Vis.	Viscid,	Having a glutinous or ropy consistency; tenacious.
Vul.	Vulnerary,	Healing to fresh cuts or wounds.

PARTS OF PLANTS AND SUBSTANCES.

Ba.	Balsam.	Hb.	Herb.	Po.	Powder.
Bd.	Buds.	Ju.	Juice.	Ps.	Pods.
Bk.	Bark.	Ke.	Kernels.	Re.	Resin.
Bu.	Bulb.	Ls.	Leaves.	Rh.	Rhizome.
Ca.	Capsules.	Nu.	Nuts.	Rt.	Root.
Ex.	Extract.	Ol.	Oil.	Sd.	Seed.
Fl.	Flowers.	Ol-re.	Oleo-resin.	Sh.	Shell.
Ft.	Fruit.	Pd.	Peduncle.	St.	Stem.
Fu.	Fungus.	Pe.	Peel.	Tp.	Tops.
Gu.	Gum.	Pi.	Pith.	Tw.	Twigs.
Gu-Re.	Gum resin.	Pl.	Plant, entire.	Wd.	Wood.

BOTANICAL HAND-BOOK.

BOTANICAL.

Botanical synonyms are marked with a *.

BOTANICAL.	COMMON.	PROPERTIES, PRODUCTIONS, USES, ETC.
Abelmoschus esculentus,	Musk leaf plant, Okra,	Ca. muc. dem. edi.
" moschatus,	Musk seed,	Sd. cor. sto. ner. pun.
Abies Americana,*	Tamarack,	see Larix Americana.
" balsamea,	Fir balsam tree,	Bk. sti. diu. ant. det. vul.
" " *	Balm of Gilead,	Bk. sti. diu. ant. det. vul.
" balsamifera,*	Fir balsam tree,	see Abies balsamea.
" Canadensis,	Hemlock tree,	Bk. ast. dia. ; Ls. abo. ; Gu. rub. sti.
" cedrus,	Cedar of Lebanon,	see Cedrus Libani. [diu.
" communis,	Common spruce,	Bk. diu. a-sco. emm.
" excelsa,	Norway pine,	yields Burgundy pitch; re. c-irr. sti.
" Larix,*	American larch,	see Larix Americana.
" Larix,	European larch,	yields Venice turpentine ; stl. c-irr.
" nigra,	Black spruce,	Bk. alt. diu. sti. ; Gu. sti. diu.
" pectinata,	Red spruce,	Bk. diu. a-sco. emm.
" pendula,*	Tamarack,	see Larix Americana.
" picea,	Silver fir tree,	yields Burgundy pitch : Gu. c-irr.sti.
" rubra,	Norway pine,	see Abies excelsa.
" taxifolia,	European silver fir tree,	yields Strasburg turpentine.
Abietis resina,	Burgundy pitch,	Gu. c-irr. sti. diu.
Abronia umbellata,	Abronia,	Pl. orn.
Abrus precatorius,	Love pea, Wild liquorice,	Rt. and Ls. pec. muc. dem. ; Sd. far.
Abuta Rufescens,	White Pareira brava,	Rt. diu.
Abutilon condalum,	Yellow mallows,	Pl. muc. emo. dem.
" Indicum,	Indian mallows,	Pl. dem. muc. diu.
" striatum,	Striped abutilon,	Pl. dem. muc.
Absinthium vulgare,*	Wormwood,	see Artemisia Absinthium.
Acacia Adansonia.	Thorn tree,	yields Gum Senegal.
" albida,	Brittle gum Senegambia,	Gu. muc. dem. bit.
" Arabica,	Sont. Babul,	yields Morocco gum Arabic.
" Bassora,	Bassora gum,	Gu. dem. muc.
" Bambolah,	Bablah pods,	Ast.
" Catechu,	Catechu gum,	Gu. ast.
" decurrens,	Black wattle tree,	yields Australian gum.
" Ehrenbergiana,	Gum Arabic,	Gu. nut. dem. muc. pec.
" floribunda,	Gum tree of N. Zealand,	Gu. nut. dem. muc. pec.
" Farnesiana,	Sponge tree,	yields the fragrant Cassie flowers.
" gummifera,	Barbary gum tree,	yields Gum Arabic.
" horrida,	Doornboom,	yields So. African gum Arabic.
" Jurema,	Adstringens bark,	Bk. ast.
" Karroo,	Cape gum Arabic,	Gu. dem. nut. muc. pec.

BOTANICAL.	COMMON.	PROPERTIES, PRODUCTIONS, USES, ETC.
Acacia Melanoxylon,	Mimosa bark extract,	Ex. ast.
" leucoplæa,	Bassora gum,	Gu. muc. dem.
" nebued,	Red gum Senegal,	Gu. nut. dem. pec. muc.
" Nilotica,*	Sont. Babul,	see Acacia Arabica.
" nostras,	a preparation of Prunus spinosa, Mild ast.	
" Senegal,	Gum Senegal,	Gu. muc. dem. pec.
" scyal,	Soffar,	yields Egyptian gum Arabic.
" vera,	Gum Arabic,	yields African gum Arabic.
" vereek,	White Gum Arabic,	Gu. nut. dem. pec. muc.
Acalypha Virginiana,	Mercury weed,	Pl. exp. diu.
Acanthus spinosus,	Bears' breech,	IIb. diu. ast.
Accrates viridiflora.	Green milkweed,	Pl. car. ton. diu. a-spa.
Acer negundo,	Ash maple,	Bk. ast. vul.
" nigrum,	Black maple,	Bk. bit. ast.
" Pennsylvanica,	Striped maple,	Bk. ast.; Ls. a-emc.
" pseudo-platinus,	Sycamore,	Bk. ast. opt. vul.
" rubrum,	Red maple,	Bk. ast. opt.
" saccharinum,	Maple sugar tree,	Bk. ton. ant. opt.
" striatum,	Striped maple,	Bk. opt. diu.
Acharas sapota,	Sapodilla plum,	Bk. feb.; Ft. diu.; Sd. diu. acr.
Achillea Ageratum,	Maudlin tansy,	Pl. cor. sto. cep.
" Millefolium,	Yarrow,	Pl. ast. alt. diu. ton. vul.
" nobilis,	Noble yarrow,	Pl. ast. alt. diu. ton. vul.
" Ptarmica,	German pellitory,	Rt. acr.; Ls. ste. pun.
Achyranthes repens,	Forty knot,	Pl. diu. ton.
Acinos vulgare,	Mountain thyme,	IIb. sti. ton. emm.
Acnida cannabina,	Water hemp,	
Aconitum anthora,	Yellow helmet flower,	Rt. poi. cat. ant.
" cammarum,	Var. Aconitum Napellus,	Rt. ano. nar. irr.
" Ferox,	Indian aconite; Bish,	Rt. poi. acr. ano. deo.
" heterophyllum,	Atis,	Rt. a-per. ton. aph.
" Luparia,	Wolfwort,	Pl. nar. irr. poi.
" luridum,	Var. Aconitum Ferox,	Rt. ano. deo. poi.
" Lycoctonum,	Great yellow wolfsbane,	Pl. nar.
" meloctonum,	Badgers' bane,	Pl. nar.
" Napellus,	Aconite root,	Pl. deo. ano. nar. sed. poi.
" Neubergeuse,	Styrian monkshood,	Pl. nar.
" palmatum,	Var. Aconitum Ferox,	Rt. ano. deo. poi.
" theriophonium,	Beastsbane,	Pl. nar.
" tragoctonum,	Goatsbane,	Pl. nar.
" uncinatum,	Am'n wild monkshood,	Pl. nar. ano. acr.
" vulparis,	Foxbane,	Pl. nar.
Acorus Calamus,	Sweet flag,	Rt. aro. car. ton. vul.
Acrostichum aureum,	Forked fern,	Pl. pec. ast.
Actæa alba,	White cohosh,	Rt. pur. emc.
" Americana,*	" "	Rt. pur. emc.
" Cimicifuga,*	Black cohosh,	Rt. a-per. nar. ner. diu. a-rhe.
" podocarpa,*	" "	Rt. a-per. nar. ner. diu. a-rhe.
" racemosa,*	" "	Rt. a-per. nar. ner. diu. a-rhe.
" rubra,	Red baneberry,	Rt. pur. emm.
" spicata,	Baneberry,	Rt. vul. ast.; Bs. dye black.
Adansonia digitata,	Baobab tree,	Bk. and Ls. emo. muc. feb. ton. sud.
" Gregorii,	Ethiopian sour gourd,	Ft. aci.; Bk. feb.
Adenanthera parvonia,	False red sandal wood,	Wd. dyes a deep red.
Adiantum capillus veneris,	European maidenhair,	Pl. pec. muc. exp. ref.
" pedatum,	American "	Pl. pec. muc. exp. ref. ton.
Adlumia autumnalis,	Climbing fumitory,	Orn.
Adonis autumnalis,	Pheasants' eye,	Pl. ast. bit.
" vernalis,	False hellebore,	Pl. nar. epi. cau. ves.
Ægle Marmelos,	Bengal quince,	Ft. nut. cat. ref. feb.

BOTANICAL.	COMMON.	PROPERTIES, PRODUCTIONS, USES, ETC.
Ægopodium podagaria,	Gout weed,	Rt. and Hb. a-sco. diu. sad.
Ægopogon pusillos,	Goats' beard grass,	
Æsculus glabra,	Buckeye tree,	Bk. ton. feb. nar.
" Hippocastanum,	Horse chestnut,	Bk. ton. feb. ast.; Sd. nar.
" pavia,	Small buckeye,	Bk. ton. feb. nar.; Sd. poi. far.
Æthusa cynapium,	Small hemlock leaves,	Ls. poi. nau. irr.
" divaricata,	Fools' parsley,	Pl. poi. nau.
Agallocha commiphora,	Bdellium, Aloe wood,	Wd. and Gu. aro.
Agapanthus minus,	African lily,	Fl. orn.
Agaricus campestris,	Mushroom,	Pl. edi.; see Boletus.
Agathophyllum aromaticum,	Madagascar nutmeg,	Sd. aro.
Agathosma pulchella,	Bucco,	Ls. diu. aro.
Agathis Damarra,*	Damarra pine,	see Pinus Damarra.
" lorauthifolia,*	" "	see Pinus Damarra.
Agathotes Chirayta,	Chiretta,	Pl. bit. a-per. a-bil. tou.
Agati grandiflora,		Bk. bit. ton.; Ps. edi.
Agave Americana,	Am'n Century plant,	Pl. diu. a-syp.
" Virginica,	False aloe,	Rt. bit. car.
Agdendron laurel,*	Sassafras nuts,	see Nectandra puchury.
Ageratum Mexicanum,	Ageratum,	Pl. and Fl. orn.
Agrimonia Eupatoria,	Agrimony,	Pl. ast. sto.
Agropyron repens,	Couch grass,	Rt. diu.
Agrostis alba,	Bent grass,	Tp. for.
" canina,	Dog bent grass,	Tp. for.
" dispar,	Herds' grass,	Tp. for.
" scabra,	Bent grass, White top,	Tp. for.
" stolonifera,	Knot grass,	Tp. for.
" vulgaris,	Red top grass,	Tp. for.
Agrostemma coronaria,	Rose campion,	Fl. orn.
" githago,*	Corn cockle,	see Lychnis githago.
Ailantus glandulosa,	Chinese sumach,	Bk. ant.; Fl. strong scented.
Ailanthus,*	" "	see Ailantus.
Aira,	Hair grass,	Tp. for.
" cæspitosa,	Hassock grass,	Tp. for.
Ajuga chamæpitys,	Ground pine,	Ls. aro. bit. ape. ton.
" pyrimidalis,	Upright bugloss,	Pl. sub-ast. bit.
" repens,	Bugle,	Pl. sub-ast. bit.
Albizzia anthelmintica,	Messena bark,	Bk. ant. for tapeworm.
" Julibrissin,	Silk tree,	Tree, orn.
Alchemilla alpina,	Ladies' mantle,	Pl. ast.
" arvensis,	Parsley.piert,	Pl. used in strangury.
" vulgaris,	Ladies' mantle,	Pl. ast. in hemorrhage.
Alchornea latifolia,	Alcornoque bark,	Bk. bit. ton. ast.; in asthma.
Aletris aurea,	Yellow star-root,	Rt. see Aletris farinosa.
" farinosa,	Unicorn root,	Rt. bit. ton. sto. f-com.
Aleurites laccifera,	Gum lac of Ceylon,	Gu. ton. ast.; in varnish.
" triloba.	Candlenut tree,	yields Spanish walnut oil.
Algarobia glandulosa,	Mexican gum Arabic,	yields Mesquit gum.
Alhagi camelorum,	Camel thorn,	yields Persian manna.
" Maurorum,	Hebrew manna,	Sac. nut. cat.
Alisma plantago,	Water plantain,	Ls. ves. diu. lit.
Alkanna tinctoria,	Alkanet root,	Rt. used for coloring.
Alliaria officinalis,	Hedge garlic,	Pl. a-sco. det.; Sd. acr. lit.
Allium ascolonicum,	Shallot,	Bu. edi. diu. con.
" Canadense,	Meadow garlic,	Bu. stl. diu. exp. dia. ant.
" cepa,	Common onion,	Bu. sti. diu. aut. emo. mat.
" porrum,	Leek,	Bu. sti. diu. ant. emo. mat.
" sativum,	Garlic,	Bu. sti. diu. ant. emo. mat.
" schænoprasum,	Cives, Chives,	Ls. used as salad.
" senescens,	False narcissus,	Pl. orn.

BOTANICAL.	COMMON.	PROPERTIES, PRODUCTIONS, USES, ETC.
Allium scodoprasum,	Rocambole,	Ls. used as salad.
" tricoccum,	Three seeded leek,	Pl. sti. diu. emo.
" ursinum,	Ransoms,	Pl. for gravel, etc.
" victorialis,	Allerman's-root,	Rt. a-spa.
" vineale,	Crow garlic,	Bu. diu. sti. emo.
Alnus glutinosa,	European alder,	Bk. and Ls. bit. ast. ton. int.
" incana,*	Tag alder,	see Alnus serratula.
" rubra,	Red alder,	Bk. alt. cmc. ast.
" serratula.*	American tag alder,	see Alnus rubra.
" serratida,*	" " "	see Alnus rubra.
Aloe, Africana,	Cape (Good Hope) aloes,	Ex. cat. cmm. ant. sto. bit.
" Americana,*	American century plant,	see Agave Americana.
" Arborescens,*	Cape aloes,	see Aloe spicata.
" Barbadensis,*	Barbadoes aloes,	see Aloe vulgaris.
" Bethelsdorf,*	Cape aloes of fine quality	see Aloe spicata.
" Commelyni,*	" "	" " "
" Ferox,	" "	" " "
" Guiniensis,	Fetid aloes,	Ex. cat. cmm. fœt. bit.
" Indica,*	Socotrine aloes,	see Aloe Socotrina.
" multiformis,*	Cape aloes,	see Aloe spicata.
" perfoliata,*	Barbadoes aloes,	see Aloe vulgaris.
" plicatillis,*	Cape aloes,	see Aloe spicata.
" purpurascens,*	Barbadoes aloes,	see Aloe vulgaris.
" Socotrina,	Socotrine aloes,	Ex. cat. sto aro. cmm. dra. aut.
" spicata,	Cape aloes,	Ex. sti. cat. cmm. ant. sto.
" vera,*	Socotrine aloes,	see Aloe Socotrina.
" vulgaris,	Barbadoes aloes,	Ex. cat. cmm. ant.
" " var. Indica,	Indian aloes,	Ex. cat. cmm. fœt. bit.
Aloepecurus agrostis,	Black grass,	Pl. for.
Aloexylon Agallochum,	Aloe wood,	Wd. fra. cor. sti.
Aloysia citriodora,	Sweet scented verbena,	Pl. fra. orn.
Alpina Cardamomum,*	Bastard cardamom,	see Amomum Cardamomum.
" officinarum,*	Galangal root,	see Alpina Galanga.
" Galanga,	" "	Rt. aro. sti. sto. cor.
" racemosa,		Sd. poi.
" repens,*	Small cardamom,	see Amomum repens.
Alstonia scholaris,	Devil's tree,	Bk. bit. ast.
" Theœformis,	Santa Fe tea,	Ls. used as tea.
Altrœmeria ligtu,	Talcahuana arrow root,	Rt. yields arrow starch.
Althæa officinalis,	Marsh mallow,	Rt. and Ls. muc. dem. diu.
" rosea,	Hollyhock,	Fl. emo. dem. diu.
Altinga excelsa,	Rasamala resin,	Re. allied to storax.
Alyssum maritimum,*	Sweet alyssum,	see Koniga maritima.
" saxatile,	Basket of gold,	Pl. orn.
Alyxia aromatica,	a Batavian tree,	Bk. allied to canella.
" buxifolium,	Scentwood,	Bk. aro. cor. feb.
" stellata,		Bk. aro. cor. feb.
Amanita aurantiaca,	Eatable mushroom,	Pl. edi.
" muscaria,	Poisonous mushroom,	Pl. poi. fœt. nau.
Amaranthus,*	.	see Amarantus.
Amarantus albus,	Pigweed,	Pl. vul. a-phl. det.
" blitum,	Wild amaranth, ·	Pl. emo. dem. sad.
" caudatus,	Flower gentle,	see A. hypochondriacus
" hypochondriacus	Pilewort, Princes' feath'r	Ls. ast. det. in piles.
" melancholicus,	Love lies bleeding,	Ls. ast. det.
" tricolor,	Joseph's coat,	a variety of A. melancholicus.
Amaryllis Atamasco,	Atamasco (daffodil) lily,	Pl. orn.; Rt. acr.
Ambrina ambrosioides,*	Mexican tea,	Ls. a-spa. ver. car.
Ambrosia artemisiæfolia,	Am'n Roman wormwood,	Pl. det. a-phl.
" elatior,	Roman wormwood,	Pl. see Artemisia Pontica.

BOTANICAL.	COMMON.	PROPERTIES, PRODUCTIONS, USES, ETC.
Ambrosia elatior,	Ragweed,	Pl. det.
" trifida,	Tall ambrosia,	Pl. sti. ast. a-sep. opt.
Amelanchier aronaria,*	Medlar,	see Mespilus Germanica.
" Canadensis,	Juneberry,	see Aronia.
Amianthium muscætoxicum	Fly poison,	Bu. with honey, a fly poison.
" erythrospermum,	" "	Bu. nar. poi.; Ls. poi.
Ammi copticum,	Bishops' weed,	Sd. aro. pun. car. diu. ton.
Amomum angustifolium,	Madagascar cardamom,	Sd. aro. pun. sto. car.
" aromaticum,*	Cardamom seed,	see Elettaria Cardamomum.
" Cardamomum,	" "	" " "
" Granum paradisi,	Grains of Paradise,	Sd. pun. sti. aro.
" maximum,	Java (Bengal) cardamom	Sd. aro. pun. car. cor.
" Melegueta,	Grains of Paradise,	see Amomum Granum paradisi.
" racemosum,	Round (clust.) cardamom	Sd. aro. pun. car. cor.
" repens,	Small cardamom,	Sd. aro. pun. car. cor.
" Xanthioides,*	Wild cardamom,	see Elettaria Cardamomum major.
" uva,*	Cardamom seed,	" " "
" Zerumbet,	Zedoary root,	Rt. car. aro. bit. ton. sti.
" Zingiber,	Ginger root,	see Zingiber officinale.
Amorpha canescens,	Lead plant,	Pl. alt. orn.
Ampelopsis quinquefolia,	American ivy,	Bk. and Tw. alt. ton. ast. exp.
Amygdalus communis,	Bitter and sw't almonds,	see A. amara, and dulcis.
" " var. amara	Bitter almond,	Ke. bit. sed.; yields prussic acid.
" " dulcis	Sweet "	Ke. sweet, soft, emo. dem. pec.
" glabra,	Nectarine,	Ft. edi.; Ls. and Ke. bit. poi.
" nana,*	Dwarf (flower'g) almond	Pl. orn.
" Persica,	Peach,	Ls. and Ke. sed. bit. aro. lax. poi.
" pumila,	Dwarf almond,	Pl. orn.
Amyris ambrosiaca,	So. Am'n gum elemi,	Re. vul.
" balsamifera,	Jamaica rosewood,	Wd. dia. aro. cep.
" caranna,	yields Caranna gum,	Re. bit. bal.
" commiphora,	Bdellium, Indian,	Gu. similar to myrrh.
" elemifera,	yields Gum elemi,	used in ointments and plasters.
" Gileadensis,	Balsam of Gilead,	Ba. bal. vul. cos. aro.
" hexandra,	Gum elemi,	Gu. in ointments and plasters.
" Niouttont,	Bdellium gum,	Re. similar to myrrh.
" opobalsamum,*	Balsam of Mecca,	Ba. see Amyris Gileadensis.
" tomentosa,*	Tacamahac,	see Fagara octandra.
" toxifera,	Poison ash (rosewood),	Wd. cep. aro. dia. bal.
Anabasis,	Berry bearing glasswort,	Pl. yields barilla.
Anacardium occidentale,*	Malacca (marking) nuts,	see Semecarpus Anacardium.
Anacharis alsinastrum,	Water thyme,	Pl. eaten by water fowl.
Anacyclus officinarum,	German pellitory,	Rt. sia. sti. acr.
" Pyrethrum,	Pellitory of Spain,	Rt. sia. sti. acr. hot. rub.
" radiatus,	Oxeye,	Pl. vul. ape.; dyes yellow.
Anagallis arvensis,	Scarlet pimpernel,	Pl. a-ven. poi. ner. exp. sti.
" coerulea,	Blue pimpernel,	Pl. var. of Anagallis arvensis.
" Phoenicea,*	Scarlet pimpernel,	see Anagallis arvensis.
Anamirta Cocculus,	Oriental berries,	Bs. acr. nar. poi. ins.; fish poison.
" paniculata,*	" "	see Anamirta cocculus.
Ananassa sativa,	Pineapple,	Ft. aci. edi. a-sco.
Anapodophyllum Canadensis	Mayapple,	see Podophyllum peltatum.
Anastatica hierochuntica,	Resurrection plant,	Pl. a hygrometric plant.
Anatherum muricatum,*	Vettivert,	see Andropogon muricatus.
" nardus,*	Ginger grass,	see Andropogon schœnanthus.
Anchusa Italica,	Italian bugloss,	Pl. muc. dem. ref. dia.
" officinalis,	Garden alkanet,	Hb. cor. in hypochondria.
" tinctoria,	Alkanet root,	Rt. used to color oils red.
Anda Brasiliensis,	Anda seed,	Sd. yields oil of anda.
" Gomesei,	" "	Ol. cat. eme.; Sh. ast.

BOTANICAL.	COMMON.	PROPERTIES, PRODUCTIONS, USES, ETC.
Andira anthelmintica,	Cabbage tree bark,	Bk. ant. ast.
" inermis,	Yellow cabbage tree b'k,	Bk. bit. ast. feb. ver.
" retusa,	Brown " " "	Bk. cat. ant. bit.
" violacea,	Violet wood,	Wd. fra. orn.
Andrographis paniculata,	Creyat,	Pl. bit. sto. feb.
Andromeda angustifolia,	Sorrel tree,	Ls. poi. ; see Kalmia.
" arborea,	" " elk wood,	Bk. and Ls. aci. ast. feb. ref.
" calyculata,	Leather leaf,	Pl. orn.
" hypnoides,	Moss bush,	Pl. orn.
" mariana,	Sorrel tree, Wicke,	Ls. used for itch; see Kalmia.
" nitida,	Pipe stem,	Ls. aer. poi. ; see Kalmia.
" ovalifera,	Oval leaved elk tree,	Ls. acr. poi. ; see Kalmia.
" paniculata,	White bush,	Pl. orn.
" pulverulenta,	Mealy bush,	Ls. err. stc.
" racemosa,	White ozier,	Wd. orn. ; useful.
" speciosa,	Large flow'd andromeda,	Ls. err. ; see Kalmia.
Andropogon bicornis,*	Ginger grass,	see Andropogon schœnanthus.
" Calamus,	Indian grass oil,	Ol. fra. aro. sti. per.
" citratum,	Lemon grass (verbena),	Ol. per. fra. aro. sti.
" furcastus,	Fork'd spike wood grass,	Pl. for.
" muricatus,	Vettivert, Khus khus,	Rt. used as a perfume.
" nardus,	yields Citronella oil,	Ol. used as a perfume ; aro. sti. car.
" nutans,*	Beard grass,	see Sorghum nutans.
" schœnanthus,	Ginger grass,	Ol. aro. sti. car.
Androsœmum officinale,	Park leaves,	Pl. vul.
Anemia,	Buckhorn,	see Osmunda.
Anemone cernua,	Droop'g flow'd Wind fl'r,	Pl. bit. ton.
" hepatica,*	Liverwort,	see Hepatica Americana.
" hortensis,	Garden anemone,	Pl. opt.
" nemorosa,	Wood anemone,	Fl. acr. corrosive, poi. rub.
" pratensis,	Meadow anemone,	Hb. a-syp. corrosive, poi. rub.
" Pulsatilla,	Pasque flower,	Hb. acr. corrosive, poi. rub.
" vernalis,	Yellow anemone,	Hb. cau. corrosive, rub.
" Virginica,	Wind bloom,	Hb. cau. ; in hollow teeth.
Anethum fœniculum,	Fennel seed,	Sd. aro. car. pec. diu.
" graveolens,	Dill seed,	Sd. sti. car. aro. sto.
" pastinaca,*	Garden parsnip,	see Pastinaca sativa.
" sowa,	Indian womum,	Sd. aro. car. edi. cor.
Angelica archangelica,	Angelica,	Rt. Ls. and Sd. aro. car. edi.
" atropurpurea,	Purple (high) angelica,	Rt. and Sd. car. sti. cinm.
" lucida,	Bellyache root (Angelica)	Rt. bit. sub-acr. fra. sto. ton.
" officinalis,	Europ'u garden angelica,	Rt. and Sd. aro. sti. res. ton. car.
" sylvestris,	" wild angelica,	Rt. bit. acr. aro. sti. ; Sd. ins.
Angostura cusparia,	Angustura bark,	see Galipea officinalis.
Angrœcum fragrans,	Faham tea,	Ls. substitute for Chinese tea.
Anisum vulgare,*	Anise seed,	see Pimpinella anisum.
Anona Myristica,	American nutmeg tree,	Ft. aro. pun. sti.
" muricatus,	Sour sop,	Ft. edi. aci. a-sco. ; Rt. a-ven.
" palustris,	Alligator apple,	Ft. nar. ; Bk. used as cork.
" reticulata,	Custard apple,	Ft. esc. ; Ls. aro. fra.
" squamosa,	Sweet sop,	Ft. esc ; preserved in syrup.
Antennaria dioica,	Life everlasting,	see Gnaphalium.
" margaritacea,*	Pearl flow'd everlasting,	" " margaritaceum.
" plantagineum,*	White plantain,	" " plantaginifolia.
Anthemis arvensis,	Corn chamomile,	Fl. bit. acr. eme. ton.
" Cotula,	Mayweed,	Ls. and Fl. pun. bit. a-hys.
" nobilis,	Garden chamomile,	Fl. bit. aro. ton. sti. eme. sud.
" "	Roman "	Fl. bit. aro. ton. sto.
" parthenoides,*	Feverfew,	see Pyrethrum parthenium.
" pyrethrum,*	Pellitory of Spain,	see Anacyclus Pyrethrum.

BOTANICAL.	COMMON.	PROPERTIES, PRODUCTIONS, USES, ETC.
Anthora vulgaris,*	Yellow helmet flower,	see Aconitum anthora.
Anthospermum Æthiopicum	Amber tree,	Ls. aro. fra.
Anthoxanthum odoratum,	Sweet vernal grass,	Pl. aro. fra. per.
Anthyllis vulneraria,	Woundwort,	Pl. sty. vul.; dyes yellow.
Anthriscus cerefolium,	Chervil,	Pl. deo. diu. cmm. lit.
" sylvestris,	Wild chervil,	Pl. acr. diu. poi.
Anthristira Australis,	Kangaroo grass.	
Antiaris toxicaria,	Upas tree of Java,	Ju. a violent poison.
Antirrhinum Linaria,	Toad flax snapdragon,	Pl. cat. diu. deo.
" apera spica-venti,	Wind grass,	Pl. orn.
" majus,	Snapdragon,	Pl. a-hys. opt.
Aphelandra cristata,		see Justicia.
Aphelanthes monspeliensis,	Lily pink,	Fl. orn.
Apios tuberosa,	Indian potato (pea),	Rt. far. edi.
Apium graveolens,	Wild celery,	Rt. and Sd. diu. acr. poi.
" " var. dulcis,	Garden celery,	Rt. and Sd. diu.; blanched tops a
" Petroselinum,*	Parsley,	see Petroselinum sativum. [salad.
Aplectrum hyemale,	Putty-root,	Rt. muc. pec.
Aplocera maritima.	Toothache grass,	Rt. acr. bit. sia.
Apocynum androsæmifolium	Bitter root, Dogsbane,	Rt. cmc. cat. diu. sud. exp. ant.
" cannabinum,	Black Indian hemp,	Rt. cmc. cat. sud. diu. ste.
" hypericifolium,	St. John's dogsbane,	Rt. cme. cat. sud. exp.
Aquilegia Canadensis,	Wild columbine,	Pl. diu. cmm. sud. ton.
" sylvestris,*	Garden columbine,	see Aquilegia vulgaris.
" vulgaris,	" "	Pl. diu. cmm. sud. ton. dis.
Arabis alpina,	Alpine rock cress,	Pl. orn. a-sco.
" Canadensis,	Sicklepod,	Rt. and Ls. diu. pun. a-sco.
" Chinensis,	Chinese cress,	Sd. sto. sti. abo. cmm.
" rhomboides,	Meadow cress,	Rt. and Ls. diu. pun. a-sco.
" sagittata,	Wall cress,	Pl. a-sco. ant. pun.
" stricta,	Bristol wall cress,	Pl. orn. a-sco.
" thatiana,	Mouse ear cress,	Pl. pun. a-sco.
Arachis hypogœa,	Peanut, Groundnut,	Sd. edi. nut. cat.; yield oil.
Aralia Canadensis,*	American ginseng,	see Panax quinquefolium.
" hispida,	Dwarf elder,	Ls. sud.; Rt. cmc. hyd. diu. alt.
" Japonica,*	Rice-paper plant,	see Fatsia papyrifera.
" nudicaulis,	American sarsaparilla,	Rt. aro. alt. pec. dep.
" quinquefolia,*	" ginseng,	see Panax quinquefolium.
" 'racemosa,	" spikenard,	Rt. sti. bal. dia. alt. pec.
" spinosa,	Prickly elder,	Bk. sti. sia. sud. cme. aro.
Araucaria Brasiliensis,	Norfolk Island pine, var.	Re. fra. aro. sti.
" Dombeyi,*	Dombeya turpentine tree	see Dombeya excelsa.
Arbor alba minor,*	Cajuputi tree,	see Melaleuca Cajuputi.
Arbutilon condalum,	Yellow mallow,	Pl. pec. muc. vul.
Arbutus Uva Ursi,*	Bearberry,	see Arctostaphylos Uva Ursl.
" unedo,	Strawberry tree, Europ'n	Ls. ast.; Ft. sac.
Archangelica officinalis,	Europ'n garden angelica,	Pl. and Rt. aro. fra. sto. car. dia.
" atropurpurea,*	Purple angelica,	see Angelica atropurpurea.
Arctium bardana,*	Burdock,	see Arctium Lappa.
" Lappa,	"	Rt. alt. diu. dep.; Sd. diu. alt.; Ls.
Arctostaphylos alpina,	Black bearberry,	Ls. ast.; Ft. esc. [mat.
" glauca,	Manzinita,	Ls. ast. a-syp.
" Uva Ursi,	Bearberry,	Ls. diu. ast. ton. nep. a-lit.
Arctotis,	Bears' ear,	Pl. orn.
Areca Catechu,	Areca palm,	Pl. yields catechu; ast.
" "	Catechu, Betel nut tree,	Ft. masticatory of the Orientals.
" oleracea,	Cabbage palm,	Ft.-Bd. edi. nut. a-sco.
Arenaria mariana,	Sandwort,	Pl. a-phl. sal. edi.
Arethusa cynapium,*	Small hemlock,	see Æthusa cynapium.
Argemone Mexicana,	Prickly poppy,	Pl. ano. diu. cat. cmc.; Sd. opt.

10*

BOTANICAL.	COMMON.	PROPERTIES, PRODUCTIONS, USES, ETC.
Argostemma githago,	Corn cockle, Lychnis,	Rt. vul. ast.; Sd. pur. acr.
Arisœma Dracontium,*	Green Dragon,	see Arum Dracontium.
" triphyllum,*	Three leaved arum,	see Arum triphyllum.
Aristida dichotoma,	Poverty grass,	Pl. forage.
Aristolochia,	Birthwort,	Rt. aro. cmm.
" anguicida,	So. American birthwort,	Ls. alc. nar.; Rt. cmm.
" bracteata,	Coramandel birthwort,	Ls. ant. cmm. bit. hep.
" clematitis,	Common birthwort,	Rt. feb. cmm. sti. acr. nar.
" cordifolia,		see Aristolochia anguicida.
" cymbifera,	Thousand men root,	Rt. aro. bit. sti. dia. diu. cxa.
" grandiflora,*	" " "	see Aristolochia cymbifera.
" hastata,*	Virginia snakeroot,	" " Serpentaria.
" hirsuta,	" "	" " "
" Indica,	Indian birthwort,	" " "
" longa,	Long birthwort,	Rt. acr. bit. sud. cmm. ton.
" odorata,	Jamaica contrayerva,	Rt. diu. pur. sto. cmm. aro.
" pistolochia,	French birthwort,	Rt. cmm. ton. sud.
" reticulata,	Texas snakeroot,	Rt. see Aristolochia Serpentaria.
" rotunda,	Round rooted birthwort,	Rt. aro. sti. ton. sud. cmm.
" sagittata,	Virginia snakeroot,	see Aristolochia Serpentaria.
" sempervirens,	Evergreen birthwort,	Rt. a counter poison; sti. ton.
" Serpentaria,	Virginia snakeroot,	Rt. sti. ton. dia. diu. feb. cxa. ano.
" Sipho,	Dutchman's pipe,	Rt. aro. sti. sud. cmm.
" tomentosa,	Climbing birthwort,	see Aristolochia Serpentaria.
" vulgaris,*	Common birthwort,	" " clematitis.
Armeria vulgaris,*	European thrift,	see Statice.
Armeniaca vulgaris,	Apricot,	Ft. edi. nut. lax.; Sd. bit. sap.
Armoracia rusticana,*	Horseradish,	see Cochlearia Armoracia.
Arnica angustifolia,*		see Arnica montana.
" montana,	Mountain arnica,	Pl. nar. sti. cmm. diu. vul. pol.
" scorpoides,	Creeping leopardsbane,	Rt. aro.; see Doronicum.
Aronia botrychium,	Shadbush,	Ft. edi. nut. a-sco.
" ovalis,	Juneberry, Medlar,	Ft. edi. nut.
" sanguinea,	Choke cherry,	see Cerasus Virginiana.
Arrenanthemum avenaceum,	Oat grass,	Pl. forage.
Artanthe adunca,	Matico, West Indian,	see Artanthe elongata.
" elongata,	"	Ls. sty. aro. bit. sti. vul. aph. a-syp.
Artemisia,	Wormwood,	Pl. aro. fra. bit ton.
" abrotanum,	Southernwood,	Tp. fra. bit. ton. deo. ant. ast.
" Absinthium,	Wormwood,	Pl. bit. ton. aro. feb. ant. a-sep. dis.
" afra,	a South African species,	Hb. ton. ant. a-spa. opt. [hep.
" alba,*	Levant wormseed,	see Artemisia Santonica.
" biennis,	Biennial wormwood,	" " pontica.
" balsamita,*	Roman wormwood,	" " "
" Botrys,*	" "	see Chenopodium ambrosioides.
" campestris,	Field southernwood,	see Artemisia abrotanum.
" Canadensis,	Canada wormwood,	" " Absinthium.
" Chinensis,	from this the Chinese form their moxas, Sty.	
" cina,	Levant wormseed,	Pe. ant. sti.
" Contra,	" "	Pe. ant. sti.
" dracunculus,	Taragon,	Ls. and Ol. con. to season food.
" glacialis,	Silky wormwood,	Pe. aro. bit. ton. sto.
" glomerata,*	Barbary wormseed,	Pe. and Sd. ant. sto.
" hirsuta,	Indian mugwort,	see Artemisia vulgaris.
" Indica,	" moxa,	" " Chinensis.
" Judaica,	Barbary wormseed,	Pe. ver. ant. sto.
" leptophylla,*	Roman wormwood,	see Artemisia Pontica.
" maritima,	Sea wormwood,	Pl. aro. bit. ton. feb. ant. hep.
" moxa,*	Moxa weed,	see Artemisia Chinensis.
" Pontica,	True Roman wormwood,	Pl. ton. dia. her. vul.

BOTANICAL.	COMMON.	PROPERTIES, PRODUCTIONS, USES, ETC.
Artemisia Romana,*	Roman wormwood,	see Artemisia Pontica.
" rubra,*	Levant wormseed,	" " Santonica.
" rupestris,	Creeping wormwood,	IIb. aro. feb. f-com.
" Santonica,	Levant wormseed,	Pe. ant. sti. aro.
" Sieberi,	Barbary wormseed,	Pe. aut. aro. bit. sto.
" suaveolens,*	Sea wormwood,	Pl. see Artemisia maritima.
" tenuifolia,*	Roman wormwood,	see Artemisia Pontica.
" valdiana,	Levant wormseed,	" " Santonica.
" vulgaris,	Mugwort,	IIb. cmm. a-cpi.
Arthanita Cyclamen,	Sowbread,	see Cyclamen Europæum.
Artocarpus incisa,	Bread fruit tree,	Ft. cdi. nut.
" integrifolia,	Jack fruit tree,	Ft. cdi. nut.
Arum Dracontium,	Green dragon,	see Arum triphyllum.
" dracunculus,	Common dragon,	see Arum maculatum.
" esculentum,	Indian kale,	Rt. acr. ast. cdi.
" Indicum,	" arum,	Rt. far. cdi. nut.
" maculatum,	Spotted arum,	Rt. acr. ves. far. Portland arrow rt.
" triphyllum,	Dragon root,	Rt. acr. sti. exp. car. dia.
" Virginicum,*	Arrow arum,	see Peltandra Virginica.
" vulgare,*	Common Indian turnip,	see Arum maculatum.
Arundo donax,	Spanish cane,	Rt. diu. cmm.
" phragmites,	Common reed,	Rt. diu. dep.
"	Bamboo,	Pl. sac.; Ju. cmm. par.
Asagræa officinalis,*	Cevadilla seed,	see Veratrum Sabadilla.
Asagraya officinalis,*	" "	see Asagræa officinalis.
Asarum Canadense,	Canada snakeroot,	Rt. aro. sti. dia. car. exp.
" Europæum,	European snakeroot,	Ls. err. cme. cat. ; Rt. pur. cmc. diu.
" Carolinianum,*	Canada snakeroot,	see Asarum Canadense.
" officinalis,*	European snakeroot,	" " Europæum.
" Virginicum,	Southern wild ginger,	" " Canadense.
Asclepias asthmatica,	Ceylon ipecac,	Rt. exp. dia. cme. sud.
" alba,	White swallow-wort,	see Asclepias vincetoxicum.
" Apocynum,	Common milkweed,	" " Syriaca.
" cornuti,	" "	" " "
" Curassavica,	Bloodweed,	Ls. and Rt. cme. irr. a-syp.
" contrayerva,	Mechoacan r't of Europe	Rt. cat. cme.
" decumbens,	Creeping milkweed,	Rt. esc. cat. sud. diu.
" fibrosa,	Meadow silkweed,	Rt. diu. alt. ape.
" gigantea,	Mudar bark,	Bk. alt. ton. diu. dia. pur. ver.
" incarnata,	White Indian hemp,	Rt. alt. ant. cme. cat.
" mataperro,	Cundurango,	Rt. diu. alt.
" obovata,*	Milkweed,	see Asclepias Syriaca.
" procera,	an Egyptian plant,	Ju. cau.
" pseudo-sarsa,*	East Indian sarsaparilla,	see Hemidesmus Indicus.
" pubescens,*	Milkweed,	see Asclepias Syriaca.
" pulchra,*	White Indian hemp,	" " incarnata.
" sullivantii,	Smooth milkweed,	" " tuberosa.
" Syriaca,	Common milkweed,	Rt. ton. diu. alt. emm. pur. cme.
" tomentosa,*	" "	see Asclepias Syriaca.
" tuberosa,	Pleurisy root,	Rt. exp. diu. dia. car. ton. [sects.
" verticillata,		Antidote to bites and stings of in-
" vincetoxicum,	White swallow-wort,	Rt. sud. diu. cmm. ton. pur. car.
Ascophoro mucedo,	Mould.	
Ascyrum crux-andræa,	St. Andrew's cross,	Pl. orn.
" hypericum,	St. Peter's wort,	see Hypericum; Sd. pur.
Asimina triloba,*	Papaw,	see Carica papaya.
Aspalanthus,	African broom,	Pl. diu. ; see Spartium.
Asparagus officinalis,	Asparagus,	Young shoots, cdi. diu. ape. deo.
Asperugo procumbens,	Wild bugloss,	Pl. sud. vul.
Asperula cynachica,	Quinsywort,	Pl. externally in quinsy.

BOTANICAL.	COMMON.	ROPERTIES, PRODUCTIONS, USES, ETC.
Asperula odorata,	Woodroof,	Pl. diu.·deo. ton. vul.
Asphodelus luteus,	Kings' spear, ·	Bu. acr. diu.
" ramosus,	Asphodel,	Bu. acr. far. edi. nut.; see Scilla.
Aspidium athamanticum,	a South African fern,	Po. ant. ver.; in tapeworm. ·
" coriaceum,*	Callahuala root,	see Polypodium calagualæ.
" depastum,*	Male fern root,	see Aspidium Filix Mas.
" discolor,*	Callahuala root,	see Polypodium calagualæ.
" ferrugineum,*	" "	" " "
" Filix Fœmina,	Female fern,	Rt. ant. ver.; in tapeworm.
" Filix Mas,	Male fern,	Rt. " " "
" Goldianum,*	" "	see Aspidium Filix Mas.
" marginale,*	" "	" " " "
Asplenium,	Spleenwort,	see the different species.
" Adiantum nigrum	Black spleenwort,	Ls. muc.; see Adiantum.
" angustifolium,	Swamp spleenwort,	Rt. ast. ant.
" aureum,* ·		see Asplenium Ceterach.
" Ceterach,*	Ceterach leaves,	Ls. sub-ast. muc. pec.
" Filix Fœmina.	Female fern,	Rt. ant. ast.
" Filix Mas,	Male fern,	Rt. ant. ast.
" latifolium,*	Ceterach leaves,	see Ceterach officinarum.
" murale,*	Wall rue,	see Asplenium ruta.
" obtusum,*	" "	" " "
" rhizophyllum,	Walking leaf,	Pl. ton. sub-ast. muc.
" ruta muraria,	Wall rue,	Rt. ant. ast.
". Scolopendrium,	Harts' tongue spleenw't,	Ls. ast. pec. vul.
" trichomanes,	Common maidenhair,	see Adiantum.
Assafœtida disgunensis,*	Assafœtida,	see Narthex Assafœtida.
Aster æstivus,	Sampson snakeroot,	Rt. a-spa. alt. a-rhe. f-com.
" amellus,	Starwort,	Pl. said to relieve Rhus poisoning.
" cordifolius,	Heart leaved aster,	Rt. aro. ton. ner. dia.
" dysentericus,*	Small fleabane,	see Inula dysenterica.
" Helenium,*	Elecampane,	see Inula Helenium.
" inguinalis,*	Wild eryngo,	see Eryngium campestre.
" officinalis,*	Elecampane,	see Inula Helenium.
" puniceus,	Cocash,	Rt. dia. a-rhe. ton. aro. dia. ner. ·
" tradescantia,	Michaelmas daisy,	Pl. orn. aro.·dia.
" undulatus,*	Small fleabane,	see Inula dysenterica.
Astragalus aculeatus,*	Goats' thorn,	see Astragalus Tragacantha.
" aristatus,	Tragacanth of Greece,	" " "
" caryocarpus,	Ground plum.	Ft. edi. ast.
" Creticus,	Tragacanth of Crete,	see Astragalus Tragacantha.
" exscapus,	" root,	Rt. a-syp. ast.
" glaux,	Milk vetch,	see Astragalus Tragacantha.
" glycyphyllos,	Wild liquorice,	Rt. sac. diu.; Ls. diu.
" gummifer,	Syrian Tragacanth,	see Astragalus Tragacantha. .
" massiliensis,*	Tragacanth,	" " "
" Tragacantha,	Gum Tragacanth,	Gu. dem. vis.; in emulsions.
" verus,	" "	Gu. " "
Astrantia major,*	Imperial masterwort,	see Imperatoria.
Athamanta annua,*	Candy carrot,	see Athamanta Cretensis.
" aureoselinum,*	Speedwell root,	see Athamauta oreoselinum.
" cervaria,*	Black gentian,	see Libanotis vulgaris.
" Cretica,*	Candy carrot,	see Athamanta Cretensis.
" Cretensis,	" "	Sd. acr. aro. car. diu.
" libanotis,*	Black gentian,	see Libanotis vulgaris.
" meum,*	Spignel, Bearswort,	see Meum Athamanticum.
" oreoselinum,	Speedwell root,	Pl. Sd. and Rt. aro. ape. deo. lit.
Antherosperma moschata,	Australian sassafras,	Ol. dia sed. aro.
Athyrium Filix Fœmina,*	Female Fern,	see Aspidium Filix Fœmina.
Atractylis gummifera,	Piney thistle,	Ju. milky viscid gum; see Mastich.

BOTANICAL.	COMMON.	PROPERTIES, PRODUCTIONS, USES, ETC.
Atriplex halimoides,	Sea orache,	Ilb. aro. a-sco. con. cdi.
" hortensis,	Garden orache,	Ilb. emo. ; Sd. eme.
" laciniata,*	Spreading orache,	see Atriplex patula.
" mummularia,	Salt bush,	Ls. pickled as condiment; Ju. cos.
" patula,	Spreading orache,	Ju. cat. dra.
" Purshiana,*	" "	see Atriplex patula.
" Mexicana,*	Mexican tea,	see Chenopodium ambrosioides.
" odorata,*	Oak of Jerusalem,	" " Botrys.
" olidum,*	Stinking orache,	" " vulvaria.
Atropa Belladonna,	Deadly nightshade,	Pl. nar. diu. dia. opt. pol.
" Mandragora,	European mandrake,	Rt. a narcotic poison.
Avena elatior,	Yellow oat grass,	Tp. and Sd. for.
" flavescens,	Tall oat grass,	Tp. and Sd. for.
" sativa,	Common oats,	Sd. edl. nut. far. emo.
" " nigra,	Black oats,	Sd. for.
" sterilis,	Animated oats,	Pl. orn. ; Sd. for.
Averrhoa bilimbi,	Blimbing tree,	Ft. aci. edl. ; Ls. dis.
" carambola,	Caramba tree,	Ft. " "
Avicennia tomentosa,*	Malacca bean,	see Semecarpus anacardium.
" nitida,	Courida,	Bk. ast.
Avoira elias,	Palm fruit,	see Elais Guinensis.
Aydendron laurel,*	Pichurim beans,	see Nectandra puchury.
Azadirachta Indica,	Nim bark,	see Melia Indica.
Azalea Indica,	Garden azalea,	Pl. ast. orn.
" nitida,	Swamp pink,	Pl. ast. used in conserve.
" nudiflora,	Pinxter flower,	Pl. ast. orn.
" procumbens,*	Creeping azalea,	Pl. ast.; see Loiseleuria.
" viscosa,	Swamp honeysuckle,	Pl. ast. orn.
Azedaracha amœna,*	Pride of India,	see Melia azedarach.
Baccharis genistelloides,	Plowmans' spikenard,	Ls. in horse diseases.
" halimifolia,	Groundsel tree,	Bk. dem. pec. aro.
Bactyrilobium fistula,*	Purging cassia,	see Cassia fistula.
Ballota fœtida,	Stink'g (bl'k) horehound	Pl. a-spa. res. det.
" lanata,	Wolfsballote,	Pl. a-rhe. diu. ; in dropsy.
" nigra,*	Black horehound,	see Ballota fœtida.
" suaveolens,	Jamaica spikenard,	Pl. emm. a-hys. a-cpl. exp. ver.
Balsamadendron,*		see Balsamodendron.
Balsamaria Inophyllum,*	Tacamahac, ⚊	see Fagara octandra.
Balsamita,*	Costmary,	see Pyrethrum Tanacetum.
Balsamodendron Myrrha,	Gum myrrh,	Gu. aro. stl. ton. vul. emm. exp.
" Gileadensis,*	Balsam of Gilead,	see Amyris Gileadensis.
· Mukul,	Bdellium,	Gu. cor. sti.
" pubescens,	Bayce balsam,	Ba. fra. diu.
Bambusa Arundinacea,	Bamboo,	Pi. sac. ; Ju. emm. par.
Bamia moschata,*	Okra,	see Abelmoschus esculentus.
Banisteria angulosa,		Pl. sud. ; antidote to poison.
Banksia Abyssinica,	Koosso,	see Brayera anthelmintica.
Baphia nitida,	Camwood,	Wd. dyestuff.
Baptisia alba,	Prairie indigo,	Pl. alt. a-sep.
" Australis,	False indigo,	Pl. alt. eme.
" leucantha,*	Wild indigo,	see Baptisia tinctoria.
" tinctoria,	" " weed,	Pl. alt. a-sep. emm. eme. dis.
Barbarea præcox,*	Winter watercress,	see Erysimum barbarea.
" stricta,*	" "	" " "
" vulgaris,*	" "	" " "
Bardana,	Burdock,	see Arctium Lappa.
" minor,	Sea burdock,	see Xanthium.
Barosma betulina,*	Buchu,	see Barosma crenata.
" crenata,	" short leaved,	Ls. sti. diu. dia. a-rhe ton.
" crenulata,	" medium leaved,	Ls. " " "

BOTANICAL.	COMMON.	PROPERTIES, PRODUCTIONS, USES, ETC.
Barosma serratifolia,	Buchu, long leaved,	Ls. sti. diu. dia. a-rhe. ton.
Barringtonia speciosa,	Seed used like cocculus,	see Anamirta Cocculus.
Basella,	Malabar nightshade,	Ls. edi. con.
Bassia latifolia,	Mahwah,	Sd. yield Galam butter.
" Parkii,	Butter tree,	Sd. yield Shea butter.
Batschia canescens,	Alconet,	Rt. ver.; Indian paint.
Bauhinia tomentosa,	St. Thomas' tree,	Bd. and Fl. muc.; in dysentery.
" variegata,	Mountain ebony,	Ls. muc.; in dysentery.
Battata Virginiana,*	the potato plant,	see Solanum tuberosum.
Begonia grandiflora,	Elephants' ear,	Rt. ast. a-sco. feb. aci.
" tomentosa,		Rt. " "
Behen abiad,*	White behen,	see Centaurea behen.
" album,*	" "	" " "
" officinarum,*	Bladder campion,	see Cucubalus behen.
Belladonna baccifera,*	Deadly nightshade,	see Atropa Belladonna.
Bellis perennis,	Common garden daisy,	Rt. a-sco.; Ls. vul.
" hortensis,*	" " "	Rt. " "
" major,	" " "	Rt. " "
Benzoe amygdaloides,	Gum benzoin, white,	Gu.-re. sti. aro. exp. pec. vul. per.
Benzoin odoriferum,*	Feverbush,	see Laurus Benzoin.
" officinalis,*	Gum benzoin,	see Styrax Benzoin.
Berardia,*	Burdock,	see Arctium.
Berberis,	Barberry of Europe,	Oxyacantha Galeni.
" aquifolium,*	Holly leaved barberry,	see Mahonia.
" aristata,	India barberry, Ruswut,	Ex. of Bk. opt. ton. feb.
" Canadensis,	American barberry,	Bk. ton. lax. bit. hep.; Bs. aci. ref.
" ' irratibilis,*	" "	see Berberis Canadensis. [opt. ast.
" lycium,	Barberry of India,	Ex. of Bk. opt. ton. feb.
" vulgaris,	Com'n barb'y of Europe,	Rt. and Bk. bit. ast. hep.; Bs. aci.
Berchemia volubilis,	Supple Jack.	[ref.
Bergera Konigii,	Curry leaf tree,	Bk. and Rt. sti. a-ven. a-eme.
Bertholletia excelsa,	Castana (Brazil) nut,	Sd. edi.; yield castana oil.
Beta hybrida,	Root of scarcity,	Rt. edi. nut. sac. sad.
" macrorhiza,	Mangel wurzel,	Rt. " "
" vulgaris alba,	White garden beet,	Rt. " "
" " rubra,	Red garden beet,	Rt. " " emm.
Betonica aquatica,*	Water figwort,	see Scrophularia aquatica.
" officinalis,	Wood betony,	Ls. ape. cor.; Rt. eme.
" Pauli,*	Paul's betony (Bugle),	see Veronica officinalis.
Betula alba,	Common white birch,	Ls. ast. bit.; Ju. sac.; Ol. fra.
" " var. populifolia,	American white birch,	Ls. ast. bit.; Ju. a-sco. ant. vul.
" emarginata,*	European alder,	see Alnus glutinosa.
" glutinosa,	" "	" " "
" lenta,	Black birch,	Ls. and Bk. aro. exe. dia. sti. ast.
" nigra,	" "	an oil like Gaultheria.
" papyracea,	Paper birch,	see Betula alba.
" rubra,	Red birch,	" " "
Bidens Beckii,	Water marigold,	Pl. her. vul. sia.
" bipinnati,	Spanish needles,	Hb. exp. emm. pec. f-com.
" connata,	Cuckold, Harvest lice,	Hb. exp.; palpitation of the heart.
" frondosa,	Beggars' tick,	Hb. in croup; exp. emm.
" tripartita,	Swamp beggars' tick,	Hb. and Sd. emm. exp.
Bigaradia myrtifolia,	Mandarin orange,	Ft. edi. aci. ref. a-sco. fra.
Bigonia alliacea,	Garlic shrub,	Pl. tastes and smells like garlic.
" capreola,		Rt. used like sarsaparilla.
" Catalpa,*	Catawba tree,	see Catalpa cordifolia.
" Indica,	Indian bignonia,	Ls. emo. to ulcers.
" leucoxylon, ·	Trumpet tree,	Sap antidote to manchineel.
" ophthalmica,	Guiana eyevine,	Ju. of Rt. opt.
" radicans,	Trumpet flower,	Rt. vul. sud. a-ven.

BOTANICAL.	COMMON.	PROPERTIES, PRODUCTIONS, USES, ETC.
Bignonia sempervirens,*	Yellow Jessamine,	see Gelseminium sempervirens.
Bittera febrifuga,	Quassia,	see Simaruba excelsa.
Bixa Americana,*	Annatto,	see Bixa Orellana.
" Orellana,	yields Annatto,	Pulp, a yellow dye; ast. sto.
" orleana,*	Manihot antidote,	see Bixa Orellana.
Blechnum borealis,	Roman fern,	Rt. ape. diu.
" lignifolium,*	Harts tongue spleenwort	see Asplenium Scolopendrium.
" squamosum,*	Ceterach leaves,	see Asplenium Ceterach.
" Virginicum,	Blechnum,	Ls. ast.
Blighia sapida,	Akee tree,	Ft. edi. sub-aci. a-sco. nut.
Blephilia hirsuta,	Ohio horsemint,	Hb. aro. ton. sti. car.
Blitum Americanum,	Garget,	see Phytolacca decandra.
" Capitatum,	Indian strawberry,	Ls. lax.; as spinach.
Bocconia cordata,	Celandine tree,	Rt. dyes red; Ju. acr. cau.
Bœhmeria nivea,	Ramie (cloth) plant,	Ls. ȧu decoction for hemorrhoides.
Boletus Agaricus,	Larch agaric,	Pl. pur. eme. sty.
" albus,*	" "	see Boletus Agaricus.
" cervinus,*	Puff ball,	see Lycoperdon proteus.
" esculentus,	an eatable mushroom,	Pl. aph. edi. nut.
" fulvus,*	Oak agaric,	see Boletus ignarius.
" fomentarius,	Spunk,	" " "
" ignarius,	Oak agaric,	Pl. sty. ast. bit.
" laricis,	Larch agaric, European,	Pl. cat. eme. sty.
" " Canadensis,	Canadian agaric,	Pl. a-rhe.
" officinalis,*	Larch agaric,	see Boletus laricis.
" suaveolens,	Willow sponge,	Fu. in phthisis pulmonalis.
" ungulatus,*	Oak agaric,	see Boletus Ignarius.
Boldoa,	Boldo,	Ls. ton. cor. sti. feb.; Bk. ast.
"	"	see Peumus Boldoa.
Bomarea salsalla,	Climbing Bomarea,	Pl. sud. her. vul.
Bombax ceiba,	Silk cotton tree,	yields a fibre; cloth as bandages.
" pentandrum,	Cotton tree gum,	Rt. and Gu. nut. dem.
Bonplandia trifoliata,*	Angustura,	see Galipea officinalis.
Bontia Germinans,*	Malacca bean,	see Avicennia tomentosa.
Borago officinalis,	Borage,	Pl. cor. pec. ape.
Borbonia ruscifolia,	Borbonia,	Ls. diu.; in asthma.
Borassus flabelliformis,	Fan palm,	yields Bdellium; Ju. suc.
Boswellia fereana,	African elemi,	Gu-re. similar to the turpentines.
" floribunda,*	Arabian olibanum,	see Juniperus lycia.
" papyrifera,	Olibanum,	" " "
" serrata,	Indian olibanum,	Re. fra. aro. bit. ast. fum.
" thurifera,*	Arabian frankincense,	see Juniperus lycia.
Botrophis racemosa,*	Black cohosh,	see Cimicifuga racemosa.
" Serpentaria,*	" "	" " "
Botrychium fumaroides,	Rattlesnake fern,	Ls. ast. vul.
" lunaria,	Moonwort,	Ls. ast. vul.
" Virginianum,*	Rattlesnake fern,	see Botrychium fumaroides.
Botrys ambrosioides,*	Mexican tea,	see Chenopodium ambrosioides.
" Americana,*	" "	" " "
" anthelminticum,*	Oak of Jerusalem,	" " anthelminticum.
" Mexicana,*	Mexican tea,	" " ambrosioides.
Boussingaultia baselloides,	Madeira vine,	Pl. orn.
Bouvardia triphylla,	Cape jessamine,	Pl. orn.
" versicolor,	Red cape jessamine,	Pl. orn.
Bovista nigrescens,*	Puff balls,	see Lycoperdon.
Bowdichia virgilioides,	American alcornoque,	Bk. feb.; see Alchornea latifolia.
Brabejum stellatum,	African almond,	Sd. edi.; roasted as coffee.
Brachycome iberidifolia,	Swan daisy,	Pl. orn.
Branca Germanica,*	Cow parsley,	see Heracleum spondylium.
" ursina,*	Bears' breech,	see Acanthus spinosus.

BOTANICAL.	COMMON.	PROPERTIES, PRODUCTIONS, USES, ETC.
Branca vera,*	Bears' breech,	see Acanthus spinosus.
Brasenia hydropeltis,	Water shield,	Ls. muc. pec. ast.; see Cetraria.
Brassica canina,	Dogs' mercury,	see Mercurialis perennis.
" capitata,	Headed cabbage,	Ls. edi. nut. car.; Sd. oil.
" campestris,	Cole (rape) seed,	seed yield rape oil.
" " v. rutabaga	Swedish turnip,	Rt. edi.; Tp. forage.
" eruca,	Garden rocket,	Sd. aph. aro. sti. pun.
" Florida,	Cauliflower,	Ls. edi. nut. a-sco. diu.
" hispida,*	Garden rocket,	Ls. aph. a-sco.; Sd. aph.
" hydropeltis,*	Water shield,	see Brasenia hydropeltis.
" marina,	Sea bindweed,	see Convolvulus Soldanella.
" napus,	Rape, Navette,	Sd. yield rape oil; bird seed.
" nigra,*	Black mustard,	see Sinapis nigra.
" oleracea,	Garden cabbage,	Ls. edi. nut. car.; Sd yield oil.
" " cauliflora,	Cauliflower, var.,	Ls. edi. nut. pec. diu.
" rapa,	Turnip,	Rt. edi. nut.; Sd. oil.
" sabellica,	Borecole,	a variety of B. Florida.
Brayera anthelmintica,	Koosso,	Fl. and unripe Ft. ant. ver.
Briza maxima,	Quaking grass,	Pl. orn.
" media,	Maidenhair grass,	Pl. orn. for.
Brizopyrum boreale,	Quaking spike grass,	Pl. for.
" spicatum,	Spike grass,	a salt marsh grass.
Bromelia ananas,	Pineapple,	Ft. aci. edi. a-sco. nut.
" pinguin	Pinguin fruit,	Ft. ref. austere, acl.
Bromus ciliatus,	Brome-grass,	Pl. eme. ant. cat. diu.
" glaber,*	Dog grass,	see Triticum repens.
" mollis,	Soft brome-grass,	Sd. nar.; said to be fatal to fowl.
" pubescens,	Brome-grass,	see Bromus ciliatus.
" purgans,	"	" " "
" secalinus,	Chess (cheat) grass,	Sd. said to be pur.
" tremulentus,	Darnel, Tare,	see Lolium tremulentum.
Brosimum alicastrum,	Bread nut,	Sd. edi. nut.; Wd. like mahogany.
Broussonetia tinctoria,*	Old Fustic,	see Morus tinctoria.
Browallia elata,	Blue amethyst,	Pl. orn.
Brucea antidysenterica,*	Bk. of Strychnos Nux Vomica, see Strychnos Nux Vomica.	
Bryonia Africana,	African bryony,	Pl. eme. cat. diu. her. a-syp.
" alba,	White bryony,	Rt. and Bs. acr. bit. dra.-cat. far.
" dioica,	Red bryony,	Rt. and Bs. acr. bit. dra.-cat. eme.
" mechoacanna,	Man root,	see Convolvulus panduratus.
" nigra,*	Black bryony, Tamus,	see Tamus communis.
" Peruviana,	Jalap root,	see Convolvulus Jalapa.
Bubon Galbanifera,	South African Galbanum,	Gu-re. a-spa. exp.
" gummiferum,*	Gum Ammoniac,	see Dorema Ammoniacum.
" Macedonicum,	Macedonian parsley,	similar to common parsley.
Bucida buceras,	Olive bark tree,	Bk. ast.; used for tanning.
Buena acuminata,	False Peruvian bark,	Bk. feb.; see Cinchona grandiflora.
" hexandra,	China bark,	Brazilian Quinquinia bark.
" obtusifolia,*	False Peruvian bark,	see Cinchona grandiflora.
Bulbocapnos cavus,	Round birthwort,	see Aristolochia.
" digitatus,	Thick "	" "
Bunias Americana,	Sea rocket,	Pl. diu. a-sco. edi.
Bunium bulbocastanum,	Kipper nut, (Earth nut),	Bu. eaten roasted; diu. a-lit.
" flexuosum,	Ar nut, (Earth nut),	Bu. eaten roasted; diu. a-lit.
Bupleurum rotundifolium,	Hares' ear,	Hb. and Sd. aro.; used in ruptures
Bursera acuminata,*	Jamaica birch,	see Bursera gummifera.
" gummifera,	Caranna gum,	Gu. like balsams and turpentines.
Butea frondosa,	Dhak tree, Bengal kino,	Ex. ast.; used for tanning.
Buxus sempervirens,	Box tree,	Bk. dia. pur. sud. a-per. alt.
" suffructicosa,	Dwarf box,	Bk. sud. alt. ant.
Byrsonima crassifolia,	Alcornoque bark,	Bk. ast. feb.; dyes red.

BOTANICAL.	COMMON.	PROPERTIES, PRODUCTIONS, USES, ETC.
Byttera febrifuga,*	Quassia,	see Simaruba excelsa.
Cacalia anteuphorbium,	Kleinia, Pl. said to temper the caustic property of Euphor-	
" atriplicifolia,	Wild caraway,	Pl. emo.; like the mallow. [bium.
" reniformis,	" "	Pl. emo.; like the mallow.
" suaveolens,	" "	Pl. emo.; like the mallow.
" tuberosa,	Indian plantain,	Ls. emo. pec.; Bu. edi.
Cacao Theobroma,*	Cacao (Chocolate) nut,	see Theobroma Cacao.
Cacoucia coccinea,	Tikimma,	Pl. and Ft. eme. cat.
Cactus grandiflorus,	Night blooming cereus,	Fl. and St. sed. diu. cardiac.
" Opuntia,*	Prickly pear cactus,	see Opuntia vulgaris.
Cæsalpina bijuga,	Peach wood,	Wd. a dyestuff.
" Bahamausis,	Braziletto wood,	Wd. a dyestuff.
" Bonducella,	Indian hazelnut,	Ft. feb. ton.
" Brasiliensis,	Braziletto wood,	Wd. a red dyestuff.
" coriaria,	Divi divi,	Ps. ast.; used in tanning.
" crista,	Nicaragua wood,	Wd. a red dyestuff.
" echinata,	Brazil wood,	Wd. ast.; a dyestuff.
" pipai,	Pipi pods,	Ps. ast.
" pulcherrima,	Spanish carnation,	Ls. and Sd. abo. pur. feb.
" sappan,	Sappan wood,	Wd. a red dyestuff; sty.
Caffea Arabica,	Coffee,	see Coffea Arabica.
Caladium esculentum,	Taro, Indian kale,	see Arum esculentum.
" palustris,*	Water dragon,	" " "
" seguinum,	Dumb cane,	Ju. aph. a-rhe. acr.
Calamagrostis arenaria,	Beach grass,	Pl. on beaches prevents drifting.
" brevipilis,	Purple bent grass,	Pl. for.
" Canadensis,	Blue joint grass,	Pl. for.
Calamintha acinos,	Basil thyme,	Hb. dia. dis. exp. aro.
" alpina,	Mountain thyme,	Hb. aro. sto. cor. feb. emm.
" clinopodium,	Horse thyme,	Hb. aro. sti. dia. dis. exp.
" Nepeta,*	Field calamint,	see Melissa nepeta.
" officinalis,	Calamint,	Hb. sto. cor. diu. feb. emm. aro.
" pulegii,	" Lesser,	Hb. aro. cor. feb. emm.
Calamus Alexandrinus,*	yields Citronnella oil,	see Andropogon nardus.
" aromaticus,*	Sweet flag,	see Acorus Calamus.
" Draco,	Dragons' blood,	Ex. ast.; in plasters for color.
" Indicus,*	Sugar cane,	see Saccharum.
" odoratus,*	Sweet flag,	see Acorus Calamus.
" petræus,	Dragons' blood,	Ex. ast.; to impart a red color.
" Rotang,	Rattan, Dragons' blood,	Ex. (Gu-re.) to impart a red color.
" rudentum,	Dragons' blood,	" " " " " "
" verus,	" "	" " " " " "
" vulgaris,*	Sweet flag,	see Acorus Calamus.
Calandrina discolor,	Calandrina,	Fl. cos.
Calatropus gigantea,*	Mudar bark,	Bk. excites powerful uterine con-
Calceolaria arachnoidea,	Relbun,	yields a yellow dyestuff. [tractions.
" primata,	Slipperwort, Peruvian,	Hb. pur. eme.
" trifida,	Tumpu,	Hb. feb. ton.
Calceum Equinum,	Coltsfoot,	see Tussilago.
Calendulá alpina,*	Arnica,	see Arnica montana.
" arvensis,	Wild marigold,	Fl. and Ls. ape. dia. ton.
" officinalis,	Garden marigold,	Fl. and Ls. vul. dis. sti. dia.
Calla palustris,	Wild water arum,	Rt. sti. cau. muc.; see Arum.
" Virginica,	Arrow arum,	see Peltandra Virginica.
Callicarpa Americana,	French mulb'y, Sourbush	Bs. acl. ast.; Ls. for dropsy.
Callicocca Ipecacuanha,	Ipecac,	see Cephælis Ipecacuanha.
Calliopsis tinctoria,*		see Coreopsis.
Callirrhœ digitata,	Mallow-wort,	see Malva.
Callistephus Chinensis,	China aster,	Pl. sti. dia. ner. orn.
Callitriche verna,	Water starwort,	Pl. diu. in dropsy.

BOTANICAL.	COMMON.	PROPERTIES, PRODUCTIONS, USES, ETC.
Callitris articulata,	So. African sandarach,	Rc. fum. in gout, rheumatism, etc.
" quadrivalis,*	Sandarach, see Thuja,	Rc. to make pounce.
Calls arum,*	Skunk cabbage,	see Ictodes fœtida.
Calluna vulgaris,	Heather,	Pl. orn.
Calophyllum Calaba,	Calaba balsam,	Rc. fra. aro. ; as gum thus.
" inophyllum,	Oriental tacamahac,	Rc. fra. aro. bit. ; as gum thus.
" spuria,	yields Oil pootingee,	Ol. used in itch; lamp oil.
" tacamahaca,	Curacoa tacamahac,	see Fagara octandra.
Calotropis gigantea,*	Mudar bark,	see Asclepias gigantea.
" mudarii,*	" "	" " "
Caltha alpina,*	Arnica,	see Arnica montana.
" arvensis,*	Wild marigold,	see Calendula arvensis.
" officinalis,*	" "	" " "
" palustris,	Marsh marigold,	Pl. exp. pec. ; in coughs.
" vulgaris,*	Garden marigold,	see Calendula officinalis.
Calycanthus Floridus,	Florida allspice,	Fl. aro. fra. ; Rt. cmc. ; Bk. aro. sti.
Calyptranthes Caryophyllata	Clove bark,	see Myrcia acris.
Calystegia sepium,	Bracted bindweed,	see Convulvulus sepium.
Camassia esculenta,	Prairie turnip, Quamash,	Rt. edi. ; as food by Indians.
Cambogia gutta,	Gamboge,	see Garcinia Cambogia.
Camellia Japonica,	Common camellia,	Pl. orn.; Ls. as tea flavor.
" sesanqua,	Lady Banks' camellia,	Ls. used to flavor tea leaf.
Camelina sativa,	False wild flax,	Ls. ver.; Sd. oil; in palsy.
Cameraria latifolia,	Bastard manchineel,	Ju. an arrow poison; Caoutchouc.
Campanula glomerata,	Clustered bellflower,	Fl. orn. cmc. pec.
" latifolia,	Haskwort,	Fl. orn. cmc. pec.
" media,	Canterbury bell,	Fl. orn.; Pl. used as a potherb.
" rotundifolia,	Harebell,	Pl. orn. cmc. pec.
" trachelium,	Canterbury bell,	Pl. relaxant, pec. cmc.
Camphora officinarum,	Camphor,	Gu. uar. dia. sed. aph. sti. ano. a-spa.
Camphorosma Monspelica,	Stinking ground pine,	Pl. aro. diu. dia. cep. a-spa.
Campomanesia Lineatifolia,	Palillo,	Ft. aro. per.
Canarium commune,*	Gum elemi,	see Amyris elemifera.
Canella alba,	White canella,	Bk. aro. sti. ton. cor. a-sco.
" axillaris,	Paratoda bark,	Bk. feb. ton. sto.
" caryophyllata,	Clove bark,	see Myrtus Caryophyllata.
" Cubana,	Canella bark,	see Canella alba.
" giroffe,	Clove bark,	see Myrtus Caryophyllata.
" malabarica,	China cinnamon	see Laurus Cassia.
" Winterana,	Canella bark,	see Canella alba.
Canna achiras,*	Canna starch,	Fecula of Rt. as arrow root.
" coccinea,*	" "	" " " " "
" edulis,	St. Kitts' arrow root,	" edi. nut.
" Indica,	Indian shot plant,	St. Kitt's arrow root.
" gigantea,	Great canna,	Pl. orn.
" solutiva,	Purging cassia,	see Cassia Fistula.
" speciosa,	African turmeric,	Canna starch; Rt. dyes yellow.
Cannabina aquatica,*	Water maudlin,	see Eupatorium Cannabinum.
Cannabis Americana,	American hemp,	see Cannabis Indica; Sd. bird seed.
" Indica,	Indian hemp, Bangue,	Ex. ner. ano. sud. a-spa. a-syp.
" sativa,	Common hemp, ·	as above; Sd. yield oil.
Capillus veneris,*	European maidenhair,	see Adiantum Capillus veneris.
" " Canadensis,*	American "	" " pedatum.
Capparis baducca,	Baducca,	Fl. orn. pur.; Ju. sti. vul.
" spinosa,	Caper bush,	Bk. of Rt. ast. diu. ; Bd. as con.
Capraria biflora,	Carib tea,	Fl. ast.; used as tea.
Caprifolium perfoliatum,	Honeysuckle,	Ls. det. ; Fl. orn.
" sempervirens,	Trumpet honeysuckle,	see Lonicera.
Capsella bursa-pastoris,	Shepherds' purse,	Hb. acr. det. ast.
Capsicum annuum,	Cayenne pepper,	Ft. pun. sti. dia. ton. sia. alt. a-rhe.

BOTANICAL.	COMMON.	PROPERTIES, PRODUCTIONS, USES, ETC.
Capsicum baccatum,	Bird pepper,	Ft. pun. sti. dia. ton. sia. alt. a-rhe.
" Brasilicum,*	Cayenne pepper,	see Capsicum annuum.
" cerasiforme,	Cherry pepper,	Ft. see Capsicum grossum.
" fastigiatum,	African pepper,	Ft. sti. a-spa. rub. ton. pun. sia.
" frutescens,	Bird (goats') pepper,	Ft. acr. sti. ton. car.
" Hispanicum,*	Cayenne pepper,	see Capsicum annuum.
" grossum,	Bell pepper, (garden),	Ft. used in pickles.
" micranthium,*	Chili "	see Capsicum annuum.
" minimum,	Bird "	Ft. acr. sti. ton. car.
" tetragonium,	Bonnet pepper,	Ft. as Capsicum grossum.
Carapa Guianensis,	Kundoo nut tree,	Bk. ant. pur.; Coondi oil.
" Touloucouna,	" "	Bk. ant. pur.
Cardamine amara,	Bitter cress,	Pl. sal.; Rt. pur.; Fl. a-spa.
" fontana,*	Water "	see Sisymbrium Nasturtium.
" Nasturtium,*	" "	" " "
" pratensis,	Cuckoo flower,	Pl. sti. dia. diu. a-sco. dep. a-epi.
" rhomboides,	Spring cress,	Pl. a-sco. diu. sti. con.
Cardaminum minus,*	Nasturtium,	see Tropæolum majus.
Cardamomum majus,*	Grains of Paradise,	see Amomum Granum paradisi.
" malabarense,*	Malabar cardamom,	see Elettaria Cardamomum.
" minus,*	Lesser cardamom,	" " "
" piperatum,*	Grains of Paradise,	see Amomum Granum paradisi.
" rotundum,*	Round cardamom,	see Elettaria Cardamomum.
Cardiaca crispa,*	Motherwort,	see Leonurus cardiaca.
Cardiospermum halicacabum	Balloon vine,	Pl. orn.
Carduus altilis,*	Garden artichoke,	see Cynara Scolymus.
" benedictus,*	Blessed thistle,	see Centaurea benedicta.
" Brazilianus,*	Pineapple,	see Bromelia ananas.
" marianus,*	Mary thistle,	Hb. bit. ton.; Sd. oleaginous.
" nutans,	Musk "	Hb. bit. tou. aro.
" pineus,*	Piney "	see Atractylis gummifera.
" sativus,*	American saffron,	see Carthamus tinctorius.
" veneris,*	Fullers' teazel,	see Dipsacus fullonum.
Carex arenaria,	German sarsaparilla,	St. dia. dem. alt.
" flava,	Marsh hedgehog grass,	Pl. coarse marsh grass.
" glauca,	Carnation grass,	" " " "
" pulicaris,	Flea grass,	" " " "
Carica papaya,	Papaw, (Custard apple),	Ft. edi.; Ju. ver.
Carlina acaulis,	Carline thistle,	Rt. ton. sud. emm.
" sub-acaulis,	" "	Rt. ton. sud. emm.
Carludovica palmata,	Panama hat palm,	Ls. for manufactur'g Panama hats.
Carpinus Americana,	Hornbeam, Ironwood,	see Ostrya Virginica.
" betulus,	Blue beech,	Ls. a mild ast.
Carthamus tinctorius,	Dyers' (Am'n) saffron,	Fl. dia. diu. sud. cos.; Sd. aro. cat.
Carum carui,	Caraway seed,	Sd. car. aro. fra. sto. [diu.
Carya alba,	Walnut, " Shagbark,"	Bk. cat.; Ls. mild ast.; Ft. edi.
" amara,	Bitter hickory,	Bk. acr. cat.; Ft. edi.
" olivæformis,	Illinois pecan,	Ft. edi.; Sh. ast.
" porcina,	Hog nut hickory,	Ft. edi.; Sh. ast.; Bk. cat.
" sulcata,	Shellbark hickory,	Ft. edi.; Sh. ast.; Hu. ast.
" tomentosa, ·	Mocker nut,	Ft. edi.; Sh. ast.; Bk. ast.
Caryocar nuciferum,	Guiana butternuts,	· Sd. esc. con.; oily.
" tomentosum,	" almonds,	Sd. esc. con.; oily.
Caryophyllata aquatica,*	Water avens,	see Geum rivale.
" nutans,*	" "	" " "
" urbana,*	European avens,	" " urbanum.
" vulgaris,*	" "	" " "
Caryophyllus Americanus,	Allspice,	see Eugenia Pimenta.
" aromaticus,	Cloves,	Fl.-Bd. aro. sti. irr. a-cme.
" hortensis,*	Clove pink,	see Dianthus caryophyllus.

BOTANICAL.	COMMON.	PROPERTIES, PRODUCTIONS, USES, ETC.
Caryophyllus Pimenta,*	Pimento, Allspice,	see Eugenia Pimenta.
" vulgaris,*	European avens,	see Geum urbanum.
Caryota urens,	Jaggery palm,	yields sugar, sago and fibre.
Cascarilla,	Cascarilla bark,	see Croton eluteria.
Cascarilla, a sub-genus of Cinchonaceæ.		
" acutifolia,	mixed with true barks,	see Cinchona acutifolia.
" aborquillado,	" " " "	" " dichotoma.
" amarillo,	False red cinchona,	" " oblongifolia.
" bora,	of Bolivia,	" " caduciflora.
" Carabaya,	Carabaya bark,	" " ovata Rufinervis.
" delgadilla,	Wiry crown bark,	" " hirsuta. ·
" echinique,	False Calisaya,	" " amygdalifolia.
" Ichu,	" "	" " Calisaya Josephiana.
" macrocarpa,	Guiana bark,	" " macrocarpa.
" magnifolia,	Spurious red bark,	" " caduciflora.
" negrilla,	Huanuco bark,	" " glandulifera.
" ovata,	Carabaya bark,	" " ovata.
" provinciana,	Coarse gray bark,	" " micrantha.
" quepo,	False Calisaya,	" " amygdalifolia.
" riveroana,	" Cinchona bark,	" " oblongifolia.
" sebiferum,*	Tallow tree,	see Stillingia sebifera.
" tinctorium,*	Blue patch,	see Croton tinctorium.
" verde,	Coarse gray bark,	see Cinchona micrantha.
Cassia absus,	Absus seed,	Sd. opt. ; a dry collyrium.
" acutifolia,	Alexandrian senna,	Ls. a valuable cathartic.
" Æthiopica,	Tripoli senna,	Ls. " " [a-syp.
" alata,	Ringworm bush,	Ls. for ringworm, tetter and itch ;
" angustifolia,	Wild India senna,	cultivated var. is Tinnevelly senna.
" Bouplandiana,*	Purging cassia,	see Cassia Fistula.
" Brasiliana,	Horse cassia of W. I.,	Pu. bit. pur. ; as C. Fistula.
" caryophyllata,*	Wild clove bark,	see Myrtus Caryophyllata.
" chamæcrista,	Prairie senna,	Ls. as Cassia Marilandica.
" cinnamomea,*	Chinese cinnamon,	see Cinnamomum aromaticum.
" elongata,*	India (Tinnevelly) senna	Ls. hyd.-cat.
" emarginata,	West Indian senna,	Ls. pur. ; used as C. Senna.
" excelsa,*	Purging cassia,	see Cassia Fistula.
" Fistula,	" "	Pu. mildly cathartic.
" Forskalii,*	Mecca senna,	see Cassia lanceolata.
" lanceolata,	" "	" " Senna.
" lenitiva,*	Alexandrian senna,	" " acutifolia.
" lignea,*	Chinese cinnamon,	see Cinnamomum aromaticum.
" Marilandica,	American senna,	Ls. cat. ver. a-syp. diu.
" Moschata,	Purging cassia,	see Cassia Fistula.
" nictitans,	Sensitive plant,	Ls. cat. ver.
" nigra,	Purging cassia,	see Cassia Fistula.
" obovata,	Aleppo senna,	" " Senna.
" obtusata,*	Alexandrian senna,	" " acutifolia.
" occidentalis,	Styptic weed,	Ju. exa. ; Rt. diu. deo.
" officinalis,*	the senna plant,	see Cassia Senna.
" orientalis,*	" " "	" " "
" ovata,	Tripoli senna,	" " "
" Senna,	Senna plant,	Ls. Hydragogue-cathartic.
" Tora,	Sickle senna,	" " "
Cassie Farnesiana,*	Cassie flower tree,	see Acacia Farnesiana.
Cassine Caroliniana,*	Paraguay tea,	see Ilex Paraguaiensis.
" colpoon,	Ladle wood,	Wd. orn.
Cassuvium pomiferum,*	Marking (Cashew) nut,	see Anacardium occidentale.
Cassytha filiformis,	So. African cassytha,	Pl. wash in scald head ; ins.
Casuarina stricta,	He oak,	Bk. ast. sty.
" torulsa	Botany Bay oak,	Bk. ast. sty.

BOTANICAL.	COMMON.	PROPERTIES, PRODUCTIONS, USES, ETC.
Casuarina equisetifolia,	Toa-tree,	Bk. ast. vul. sty.
" quadrivalvia,	She oak,	Bk. ast. sty. vul.
Castalia speciosa,*	Europ'n white water lily	see Nymphæa alba.
Castana,	Chestnut,	see also Fagus.
" Americana,	" American,	Bk. ast. ton. fcb.; Ft. cdi.; Ls. ast.
" Equinæ,*	Horse chestnut,	see Æsculus Hippocastanum.
" pumila,	Chinquapin,	Bk. ast. ton. fcb.; Ft. cdi.
" vesca,	European chestnut,	Bk, ast. ton.; Ft. cdi. nut. [cough
" " Americana,	American chestnut,	Bk. ast. ton.; Ls. in whooping
" vulgaris,	Spanish chestnut,	Bk. ast. ton. fcb.; Ft. cdi. nut.
Castela Nicolsonia,	Goats' bush.	
Castiglionea lobata,	Pinoncillo tree,	Ft. cdi.; Ju. powerful caustic.
Castilloa elastica,	Ule tree,	yields Caoutchouc. [ano.
Catalpa arborea,	Catawba tree,	Ps. in nervous asthma; ton. ver.
" cordifolia,	" "	Ps. in nervous asthma; ton. ver.
Cataputia major,*	Castor oil plant,	see Ricinus communis.
" minor,*	Caper plant,	see Euphorbia lathyris.
Catha edulis,	Cafta leaves, Khat,	Ls. exc. sti.; as Coca ls.
Cathartocarpus Fistula,*	Purging cassia,	see Cassia Fistula.
Caucaulis carota,*	Garden carrot,	see Daucus Carota.
" daucoides,	Hensfoot, Bastard parsley, Pl. diu. aro.	
Caulophyllum thalictroides,	Blue cohosh,	Rt. sti. cmm. sud. a-spa. par. diu.
Causarra Chirayta,*	Chiretta,	see Agathotes Chirayta.
Ceanothus Americana,	Red-root-bark-tree,	Bk. of Rt. ast. exp. sed. a-syp. sti.
" colubrinus,	Redwood tree, Bahama,	see Soymoida febrifuga.
Cecropia peltata,	Trumpet tree, Guarumo,	Bk. ast.; yields Caoutchouc.
Cedrela febrifuga,*	Febrifuge bark,	see Swietenia febrifuga.
" odorata,	Barbadoes cedar,	Bk. aro. fcb.; Wd. fra.
" surena,*	Febrifuge bark,	see Swietenia febrifuga.
Cedrus baccifera,*	Savin,	see Juniperus Sabina.
" deodora,	Fountain tree, Indian cedar, Wd. fra. res. orn.	
" Libani,	Cedar of Lebanon,	Wd. fra. resinous.
" Mahogani,*	Mahogany tree,	see Swietenia Mahogani.
Celastrus edulis,*	Cafta leaves,	see Catha edulis.
" scandens,	False bittersweet,	Bk. of Rt. alt. diu. her. nar. sti.
Celosia cristata,	Cockscomb,	Ls. ast. apt. vul. [a-syp.
Celtis occidentalis,	Nettle tree, Sugarberry,	Bk. ano. ref.; Bs. sweet, ast.
Cenchrus,	Bur grass.	
Centaurea Americana,	American star thistle,	Rt. ast. bit. ton.
" behen,	White behen,	Rt. ast. bit. ton.
" benedicta,	Blessed thistle,	Pl. ton. dia. bit. feb. sto.
" calcitrapa,	Knapweed,	Rt. ton. feb. bit.; Sd. dia.
" centaurium,	Greater centaury,	Rt. bit. ton. feb. vul.
" cineraria,	Dusty miller,	Pl. orn.
" cyanus,	Blue centaury,	Fi. cor. ton. opt. orn.
" Isnardi,	Jersey thistle,	Rt. ton. feb.
" maritima,	Dusty miller,	Pl. orn.
moschata,	Sweet sultana,	Pl. orn.
" niger,	Horse knob thistle,	Rt. bit. deo. ton.
" stellata,*	Knapweed,	see Centaurea calcitrapa.
Centaurium magnum,*	Greater centaury,	" " centaurium.
" officinalis,*	" "	" " "
Centunculus echinatus,	Hedge-hog grass,	on salt marshes.
" minimus,	Chaff weed, Bastard pimpernel, see Anagallis.	
Cepa,*	Onion,	see Allium.
" maritima,*	Squill,	see Scilla.
Cephaelis emetica,*	Peruvian black ipecac,	see Psychotria emetica.
" Ipecacuanha,	Ipecac root,	Rt. eme. exp. nau. nar. sed.
" muscosa,	West Indian ipecac,	Rt. eme.; as Ipecacuanha.
" punicea,	Jamaica ipecac,	" " " "

BOTANICAL.	COMMON.	PROPERTIES, PRODUCTIONS, USES, ETC.
Cephaelis-reniformis,	Peruvian ipecac,	Rt. eme.; as Ipecacuanha.
Cephalanthus occidentalis,	Button bush,	Bk. ton. lax. feb. ape. diu.
Cerastium vulgatum,*	Mouse ear chickweed,	Hb. ref. nut.; as salad, a-sco.
Cerasus acida,	Wild sour cherry,	Bk. ast. pec. feb.; Ft. aci.
" avium,	Sweet cherry,	Bk. ast. pec. feb.; Ft. ast.
" hiemalis,	Wild black cherry,	Bk. acr. ast.; Ft. ast.　　[edi.
" hortensis,	Garden cherry,	Bk. and Ls. ast. pec. feb. ton.; Ft.
" Lauro-cerasus,	Cherry laurel,	Ls. aro. frn. poi. con.; see Prunus.
" padus,*	Service or bird cherry,	see Prunus padus.
" rubra,	Red cherry,	"　　" cerasus.
" Serotiana,	Wild cherry,	"　" Serotina.
" Virginiana,*	"　" Choke cherry	"　" Virginiana.
" vulgaris,	Common cherry,	Bk. ast. Feb.; Ft. ref. aci. edi.
Cerasus mahaleb,	St. Lucie wood,	Wd. aro. sud.; Ke. Macanet grains
Ceratonia siliqua,	St. John's bread,	Po. sac. edi. nut. pec. cat.
Cerbera Tanghin,	Madagascar Tanghinia,	Ke. deadly paralyzing poison; Ls.
Ceratochloa pendula,	Horn grass,	see Bromus.　　　　　　[cat.
Ceratophyllum demersum,	Hornwort,	Pl. dem. cmo.
Cercus grandiflorus,*	Night blooming cereus,	see Cactus grandiflorus.
" senilis,	Old man cactus,	Pl. curious ornament.
Cercis Canadensis,	Judas tree,	Po. a-sco.; Ft. edi.
Ceroxylon andicola,	Vegetable wax palm,	Wax in ointments and for tapers.
" carnauba,	Carnauba wax palm,	"　　"　　"
Ceterach officinarum,	Ceterach leaves,	Ls. sub-ast. muc. pec.
Cetraria Islandica,	Iceland moss,	Pl. dem. ton. nut. pec. muc.
Chærophyllum angulatum,	Wild chervil,	see Anthriscus.
" odoratum,*	Sweet cicily,	see Myrrhis odorata.
" sativum,*	Chervil,	see Scandix cerefolium.
" sylvestre,	Wild chervil,	Pl. fœt. aro. acr. a-syp.
Chamædrys incana,*	Syrian herb-mastich,	see Teucrium marum.
" marum,*	"　　"	"　　"　　"
" palustris,*	Water germander,	"　　" scordium.
" scordium,*	"　　"	"　　"　　"
" vulgaris,*	Germander,	"　　" chamædrys.
Chamæleon album,*	Southernwood root,	see Carlina acaulis.
Chamælirium luteum,*	False unicorn root,	see Helonias dioica.
Chamæmelum,*	Chamomile,	see Anthemis nobilis.
" fœtidum,*	Mayweed,	"　　" Cotula.
" odoratum,*	Chamomile,	"　　" nobilis.
" vulgaris,*	German chamomile,	see Matricaria Chamomilla.
Chamærops hystrix,	Blue palmetto,	Pl. yields farina.
" Palmetto,*	Palm tree of So'n States	see Sabal Palmetto.
" serratula,*	Saw palmetto,	Rt. yields Florida arrow root.
Chamomilla fœtida,*	Mayweed,	see Anthemis Cotula.
" nostros,*	German chamomile,	see Matricaria Chamomilla.
" Romana,*	Chamomile,	see Anthemis nobilis.
" spuria,*	Mayweed,	"　　" Cotula.
Chara vulgaris,	Water feathers,	Pl. a-spa. ver.
Chaulmoogra odorata,*	Chaulmoogra seed,	see Gynocardia odorata.
Chavica Betel,*	Betel nuts,	see Piper Betel.
" officinarum,	Java long pepper,	spikes, analogous to black pepper.
" Populoides,	India long pepper,	"　　"　　"　　"
" Roxburghii,*	Long pepper,	see Piper longum.
Cheiranthus annuus,	July flower,	Pl. nar. deo. orn.
" Cheiri,	Wall flower,	Fl. ner. nar. deo.
" incana,	Brompton queen,	Fl. cmo. det.
Chelidonium glaucium,	Horn poppy.	
" majus,	Celandine,	Pl. hyd. ape. diu.; Ju. exa. sti. diu.
" minus,	Lesser celandine,	Rt. acr. sty. in piles; Ls. cau.
Chelone barbata,	Bearded penstemon,	Pl. cultivated for ornament.

BOTANICAL.	COMMON.	PROPERTIES, PRODUCTIONS, USES, ETC.
Chelone elatior,*	Balmony,	see Chelone glabra.
" glabra,	"	Ls. bit. ton. cat. ant. hep.
" lancolata,	"	" " " " " "
" maculata,	"	" " " " " "
" purpurca,	"	" " " " " "
Chenopodium album,	Lambs' quarter,	Ls. a-sco.
" ambrosioides,	Mexican (Jerusalem) tea	Hb. ant. as wormsced.
" anthelminticum,	Am'n wormseed plant,	Ft. ant. ; yield oil wormsced.
" Bonus Henricus,	English mercury plant,	Ls. emo. ref. lax.
" Botrys,	Jerusalem oak,	Pl. ant. pec.
" fœtidum,	Stinking goosefoot,	see Chenopodium vulvaria.
" olidum,	" orache,	" " "
" quinoa,	Quinua, Peruvian,	Ls. edi. nut. ; Sd. as food.
" rubrum,	Swinesbane,	Ls. ant. nar.
" sagittatum,*	English mercury,	see Chenopodium Bonus Henricus.
" vulvaria,	Stinking goosefoot (orache),	Pl. a-spa. ner. fœt.
Chilochloa Bœhmeri,	Fodder grass,	see Phalaris.
Chimaphila maculata,	Spotted wintergreen,	Pl. diu. ton. alt. a-syp. nar. a-cpi.
" umbellata,	Wintergreen, Pipsissewa	Pl. diu. ton. alt. ast. exa. a-syp.
Chinchona,*	Peruvian bark,	see Cinchona.
Chiococca anguifuga,*	Cahinca,	see Chiococca racemosa.
" densifolia,*	"	' " "
" racemosa,	" Snowberry,	Bk. cat. eme. ton, diu. dia.
" scandens,*	"	see Chiococca racemosa.
Chionanthus Virginica,	Fringe tree,	Bk. ape. alt. diu. ton. feb.
Chironia augularis,*	Am'n Red centaury,	see Sabbatia augularis.
" centaurium,	European centaury,	Tp. and Rt. aro. bit. ton. feb. ins.
" chilensis,	Peruvian (Chili) gentian	Rt. bit. ton. ; as gentian.
Chlora perfoliata,	Yellow-wort,	Pl. bit. ton. ; see Erythræa.
Chloroxylon Swietenia,	E. I. yellow wood, Satin wood, said to yield E. I. wood oil.	
Chondrus crispus,	Irish moss,	Pl. nut. muc. dem. pec.
" mamilosus,	" "	see Chondrus crispus.
" polymorphus,	" "	" " "
Chondrilla juncea,	Succory Gum,	Ls. diu. ; used in dropsy ; Gu. nar.
Chondrodendron tomentosum	True Pareira brava,	Bk. and Rt. diu. feb. bit. ton.
Chondrodendron "	" " "	see Chondrodendron tomentosum.
Chrysanthemum carneum,*	Persian feverfew,	see Pyrethrum roseum.
" leucanthemum,	Oxeye daisy,	Fl. and Ls. acr. ; see Leucanthe-
" parthenium,*	Feverfew,	see Pyrethrum Parthenium. [mum.
Chrysobalanus Galeni,	Nutmeg,	see Myristica fragrans.
" Icaco,	Cocoa plum,	Ft. edi. ; Ls. pec.
Chrysopsis argentea,	Silver aster,	Fl. orn. ; Ls. pec.
" graminifolia,	" "	Fl. orn.
Chrysophyllum Buranheim,	Monesia bark,	Bk. ast. bit. ; Ex. called Monesia.
" glycyphlœum,	" "	Bk. ast. bit. ; Ex. called Monesia.
" Jamaicensc,	Star apple,	Ft. esc. nut. a-sco.
" cainito,	Broad-leaved star apple,	Ju. ast.
Cicer arietinum,	Chick pea,	Sd. res. edi. nut. ast.
" lens,	Lentil,	see Ervum lens.
Cichorium endiva,	Garden chicory,	Rt. ton. lax. diu.
" intybus,	Wild chicory,	Rt. apc. deo. bit.
Cicuta aquatica,	Water hemlock,	Pl. acr. nar. poi. eme. res. ano.
" maculata,	Am'n water hemlock,	Pl. acr. nar. poi. eme. res. ano.
" major,*	Common poison hemlock	see Conium maculatum.
" Stœrkii,	" " "	" " "
" virosa,	Water hemlock,	see Cicuta aquatica.
" vulgaris,	Poison hemlock,	see Conium maculatum.
Cicutaria vulgaris,	Cow weed,	see Cicuta aquatica.
Cimicifuga racemosa,	Black cohosh,	Rt. alt. ner. exp. emm. dia. a-rhe.
" Serpentaria,*	" "	see Cimicifuga racemosa. [nar. ins.

BOTANICAL.	COMMON.	PROPERTIES, PRODUCTIONS, USES, ETC.
Cimicifuga fœtida,	Black cohosh,	see Cimicifuga racemosa.
Cinchonas.	These are arranged according to Weddel's system.	
Cinchona acutifolia,	mixed with other barks,	Cascarilla acutifolia.
" amygdalifolia,	False Calisaya bark,	not imported separately.
" angustifolia,	var. lancifolia,	see Cinchona Condaminea.
" asperifolia.	Bolivia bark,	not imported separately.
" Australis,	" "	inferior bark mixed with Calisaya.
" Boliviana,	False Calisaya bark,	"flimsy," mixed with Calisaya bark.
" brachycarpa,	Jamaica bark,	see Exostemma brachycarpum.
" caduciflora,	Spurious Red bark,	see Cinchona oblongifolia.
" Calisaya,	Yellow bk. of U. S. and Br. Ph. Bk., a-per. ton. feb. ast. a-sep.	
" " Josephiana,	False Calisaya bark,	Bk. a-per. ton. feb. ast. bit.
" " vera,	Royal yellow bark,	Bk. a-per. ton. bit. feb. ast. a-sep.
" " morada,	Flimsy Calisaya bark,	see Cinchona Boliviana.
" Candollii,	Black Cinchona,	" " Condaminea.
" Carabayensis,		not collected.
" Caribœa,	Sea-side beech,	see Exostemma Caribœa.
" Caroliniana,	Florida bark,	see Pinckneya pubens.
" chomeliana,	Carabaya bark,	see Cinchona ovata.
" cinerea,	Lima bark,	" " micrantha.
" Condaminea,	Pale bk., Crown bk. of Loxa, rich in alkaloids.	
" " candollii,	Black Cinchona,	mixed with Loxa barks.
" " lancifolia,	Fibrous Carthagena bark yields alkaloids.	
" " lucumæfolia,	White crown bark,	used in manufacture of Quinine.
" " Pitayensis,	Pitaya bark, "hard,"	used in manufacturing Quinine.
" " vera,	Loxa bark, Pale Cinchona of U. S. P., Bk. a-per. ton. feb.	
" cordata,	Yellow bark of Santa Fe see Cinchona cordifolia. [a-sep. bit.	
" cordifolia,	Royal yellow bark, Carthagena, yields cinchonia and quinia.	
" "	Columbia bark,	used in Quinine manufacture.
" " rotundifolia,	Ashy crown bark,	' " "
" " vera,	Hard Carthagena bark,	Bk. bit. nau. ton. feb. a-sep.
" coriacea,		see Exostemma coriaceum.
" dichotoma,	False red Cinchona,	Cascarilla aborquillado.
" discolor,	Olive leaved Cinchona,	not known in commerce.
" elliptica,	Carabaya bark,	scarcely known in commerce.
" erythroderma,	Red bark of commerce,	see Cinchona ovata.
" excelsa,	of India,	Bk. bit. ast. feb.
" ferruginea,	Brazilian bark,	substituted for Cinchona bark.
" floribunda,	St. Lucia bark,	see Exostemma floribunda.
" glandulifera,	Huanuco bark,	Bk. not used as a distinct sort.
" grandiflora,	False Peruvian bark,	Bk. bit. feb. ton.
" grandifolia,	New Granada red bark,	Cascarilla amarillo.
" hexandra,	China bark,	see Buena hexandra.
" hirsuta,	Wiry crown bark,	Cascarilla delgadilla.
" Humboldtiana,	Jaen bark of Loxa,	see Cinchona villosa.
" Jamaicensis,	Sea-side beech,	Bk. feb. cmc. ton. bit. nau. aro.
" Josephiana,	False Calisaya bark,	see Cinchona Calisaya.
" lanceolata,	var. Pitayensis,	" " Condaminea.
" lancifolia,	Fibrous Carthagena bark	" "
" lucumæfolia,	White crown bark,	" " " var.
" Luziana,	St. Lucia bark,	see Exostemma floribunda.
" macrocalyx,	var. lucumæfolia,	a var. of Cinchona Condaminea.
" macrocarpa,	Guaiana bark,	Bk. bit. scentless.
" magnifolia,	False Cinchona,	Cascarilla aborquillado.
" micrantha,	Pale Lima bark of U. S. P., Bk. a-per. ton. feb. ast. a.sep.	
" " oblongifolia,	Coarse grey bark,	mixed with Calisaya.
" " rotundifolia,	" " "	Bk. bit. pun. ast., see Calisaya.
" microphylla,	Loxa bark (in part),	see Cinchona Mutisii microphylla.
" montana,	St. Lucia bark,	see Exostemma floribunda.
" Mutisii,	yields aricine,	Bark not used in commerce.

BOTANICAL.	COMMON.	PROPERTIES, PRODUCTIONS, USES, ETC.
Cinchona Mutisii crispa,	Fine crown bark,	Bark not used in commerce.
" " microphylla,	Loxa bark (in part),	" " "
" nitida,	Lima bark, Huanuco, fine gray, see Cinchona Condaminea.	
" oblongifola,	False red Cinchona,	Cascarilla aborquillado.
" officinalis,	Loxa or Pale crown bark see Cinchona Condaminea.	
" ovalifolia,	Jaen bark of Loxa,	" " villosa.
" ovata rufinervis,	False Calisaya bark, Carabaya, yields Quinia; bit. ton. feb.	
" " Erythroderma,	Red bark of commerce, see Cinchona succirubra.	
" " vulgaris,	Ash, Jaen or Carabaya.	
" Pahudiana,	of India,	cultivation stopped.
" Palton,	sub species of C. macrocalyx, see Cinchona Condaminea.	
" pelalba,	Peruvian bark,	not used.
"' Peruviana,	a gray bark,	see Exostemma Peruviana.
" Pitayensis,	Pitaya bark,	see Cinchona Condaminea.
" pubescens,	Arica bark,	mixed with yellow barks.
" " Pelletieriana,	Yellow Cusco bark,	yields cinchonia.
" " purpurea,	Cusco bark, Booby bark, mixed with yellow barks.	
" purpurascens,	White Loxa bark,	scarcely known in commerce.
" purpurea,	Cusco bark,	see Cinchona pubescens.
" quercifolia,	Loxa bark, false,	" " Mutisii.
" rosea,		Bk. ast. acr. nau. feb.
" rotundifolia,	Ashy crown bark,	see Cinchona cordifolia.
" scorbiculata,	Red bark of Cusco,	yield Sulph. cinchonia and quinia.
" " Delondriana,	St. Ann's or Red Cusco bark, yield Su. cinchonia and quinia.	
" " genuina,	Peruvian Calisaya bark, yield Sulph. cinchonia and quinia.	
" succirubra,	Red bk. of U. S. and Br. Ph., Bk. a-per. feb. ton. ast. a-sep.	
" vera,	Royal yellow bark,	see Cinchona Calisaya. [bit.
" villosa,	Jaen bark of Loxa,	not imported separately.
" vulgaris,	Carabaya bark,	see Cinchona ovata vulgaris.
" triflora,	Jamaica bark,	see Exostemma triflora.
Cineraria heterophylla,	Ashwort,	Pl. ant.; see Senecio.
" maritima,*	Dusty-miller,	Pl. orn. vul.; see Senecio.
Cinna arundinaceæ,	Indian reed-grass.	
Cinnamomum,		Laurus Cinnamomum.
" albiflorum,	Tamala mother cinnam'n see Cinnamomum nitidum.	
" album,*	Canella bark,	see Canella alba.
" aromaticum,	Chinese cinnamon,	yields Cassia buds.
" Burmanni,*	Massoy bark,	see Cinnamomum Kiamis.
" Camphora,	Japan camphor,	see Camphora officinarum.
" Cassia,*	Bastard cinnamon,	Bk. aro. sti. ast.; see Laurus.
" Culilawan,	Culilawan (Clove) bark, Bk. aro. sti. ast. pun. car. sto.	
" Indicum,	Cassia bark,	see Laurus Cassia.
" Kiamis,	Massoy bark,	yields oil of Massoy.
" Lourcirii,*	Chinese cinnamon,	Cassia buds; aro. pun. sto. sti.
" Magellanicum,	Winters' bark,	see Wintera aromatica.
" Malabaricum,	Cassia bark,	see Laurus Cassia.
" nitidum,	Tamala cinnamon leaves, Ls. aro.; substitute for cinnamon.	
" rubrum,*	Clove bark,	see Cinnamomum Culilawan.
" sintoc,	Sintoc bark of Java,	Bk. aro. bit.
C. sylvestre Americanum,*	Santa Fe cinnamon,	see Nectandra Cinnamomoides.
Cinnamomum tamala,*	Cassia bark,	see Cinnamomum nitidum.
" Xanthoneurum	Papuan bark,	Bk. fra. pun. sti.
" Zeylanicum,	Ceylon true cinnamon,	Bk. aro. cor. sto. car. sti. ast.
Circæa Canadensis,	Enchanters' nightshade,	Pl. res. vul.
" lutetiana,	" "	Pl. res. vul.
Cirsium arvensis,	Canada thistle,	Rt. ton. diu. hep. a-bil. a-phl.
" lanatum,	Distaff thistle,	Rt. dep. hep.; Ls. a-phl.
" maculatum,*	Mary thistle,	see Carduus marianus.
Cissampelos Capensis,	Cape ivy vine,	Rt. eme. cat.

BOTANICAL.	COMMON.	PROPERTIES, PRODUCTIONS, USES, ETC.
Cissampelos glaberrima,*	Pareira brava,	see Cissampelos Pareira.
" Mauritiana,	" "	" " "
" Pareira,	White Pareira brava,	Rt. diu.; see Chondodendron.
Cistus Canadensis,	Frostwort,	see Helianthemum Canadense.
" Creticus,	Rose of Crete,	Labdanum resin; sti. emm.
" ladaniferus,	Gum cistus,	Gu-re. sti. exp. emm.
" laurifolius,	Labdanum,	Gu-re. sti. exp. emm.
" salvifolius,	Holly rose,	Ls. ast.
Citrullus amarus,	African Wild Watermelon,	Pulp drastic cathartic.
" colocynthis,	Colocynth,	Pulp hyd. cat. dra. diu.; in dropsy.
" vulgaris,*	Watermelon,	see Cucumis Citrullus.
" " var.,	Citron melon,	Ft. edl. con.
Citrus acida,	Lime tree,	Ft. aci. acr. a-sco. ref.
" acris,	" "	Ft. aci. acr. a-sco. ref.
" Aurantium, \	Common sweet orange,	Ft. swt. edl. nut. ref.; Fl. fra. per.
" Bergamia,*	Bergamot lemon,	see Citrus Limetta.
" Bigaradia,	Seville orange,	unripe Ft. issue peas; Ft. fra.
" " Sinensis,	Mandarin orange,	Ft. edl.; oil of Orange peel.
" " myrtifolia,	" "	Ft. edl. sweet; Fl. yield oil Neroli.
" decumana,	Shaddock,	Ft. esc. aci. a-sco.
" lumia,	Sweet lemon,	see Citrus Limonum.
" Limetta,	Bergamot lemon,	Oil Bergamot; fra. per. odo.
" Limonum,	Lemon,	Ft. aci. ref. a-sco.; Pe. oil Lemon.
" medica,	Citron, (Cedrat),	Pe. candied; oil Cedrat.
" " var. limon,	Lemon,	Ft. aci. a-sco.; Pe. bit. aro. ton.
" mella-rosa,	Bergamot lemon,	see Citrus Limetta.
" Paradisi,	Forbidden fruit,	" " aurantium.
" vulgaris,	Common bitter orange,	Ft. sour, bit. feb.; Pe. bit. aro. ton.
" "	Seville orange,	Ls. and Fl. a-spa. cor.; Pe. bit. sto.
Cladonia Islandica,*	Iceland moss,	see Cetraria Islandica. [aro.
" pyxidata,	Trumpet moss,	Pl. ast. feb. eme.; whooping cough.
" rangiferina,*	Reindeer moss,	Pl. nut. muc. aro. per. sto.
Cladrastis tinctoria,	Yellow ash, Fustic,	Bk. cat.; Wd. yellow.
Claviceps purpurea,	Ergot fungus in general,	see Ergot.
Clavus secalinus,*	" of rye	see Sclerotium clavus.
Claytonia lanceolata,	Pigeon root,	Ls. sad. a-sco.
" Virginica,	Spring beauty,	Pl. orn.
Clematis crispa,	Curled flowered clematis	Ls. and Fl. ves. acr. irr. diu.
" erecta,	Upright virgins' bower,	" " ves. acr. irr. a-syp. diu.
" flammula,	Swt. sctd. " "	" " cau. acr.; Sd. dra. [dia.
" viorna,	Leather flower,	" " acr. ves. sti. diu. sud.
" Virginica,	Common virgins' bower,	" " acr. ves. sti. diu. sud.
" vitalba,	Travellers' joy,	" " used for itch; cau. opt.
Clethra alnifolia,	White alder,	" " dia. exc.
" tinifolia,	Soapwood,	" " dia. sap. det.
Clianthus Dampieri,	Glory pea,	Pl. and Fl. orn.
" puniceus,	Parrots' bill,	Pl. and Fl. orn.
Cliffortia Ilicifolia.	African tea plant,	Ls. emo. exp.; in catarrh.
Clinopodium arvense,*	Dogmint,	see Clinopodium vulgare.
" vulgare,	Field thyme, Dogmint,	Ls. a-ven. ast. diu.
Clusia,	Card leaf tree,	Re. a-phl. to sore breasts.
" galactodendron,	Cow tree,	Ju. said to be vegetable milk.
Clutia cascarilla,*	Cascarilla,	see Croton Eluteria.
" Eluteria,*	" "	" " "
Cneorum tricoccum,	Widow wail, (see Daphne),	Pl. acr. irr.; Ls. pur. det.
Cnicus arvensis,*	Canada thistle,	see Cirsium arveuse.
" benedictus,*	Blessed thistle,	see Centaurea benedicta.
" lanceolata,	Common thistle,	Rt. ton. alt. hep.; Ls. a-phl.
" marianus,*	Mary thistle,	see Carduus marianus.
Cnidum palustre,*	Marsh smallage,	see Selinum palustre.

BOTANICAL.	COMMON.	PROPERTIES, PRODUCTIONS, USES, ETC.
Cobæa scandens,	Mexican cobæa,	Pl. orn.
Coccoloba uvifera,	Seaside grape,	yields Jamaica kino.
Cocculus Bakii,	Senegal root,	Pl. bit. ton. diu. feb. a-per.
" Carolinus,	Carolina cocculus,	Sd. used as fish poison.
" chondodendron,*	True Pareira brava,	see Chondodendron tomentosum.
" cordifolius,	. Gulancha,	Pl. ton. a-per. diu. a-syp. a-rhe.
" crispus,		see Menispermum tuberculatum.
" Indicus,	Oriental berries,	see Anamirta Cocculus.
" Inda aromaticus,	Pimento,	see Eugenia Pimenta.
" lacunosus,*	Cocculus Indicus,	see Anamirta Cocculus.
" Levanticus,*	" "	" " "
" palmatus,	Columbo root,	Rt. a pure bitter tonic; feb. a-eme.
" Plukenetil,*	Cocculus Indicus,	see Anamirta Cocculus.
" suberosus,*	" "	" " "
Cochlearia aquatica,	Water horse radish,	Rt. a-sco. sti. ton. diu.; Ls. sad.
" Armoracia,	Horse radish,	Rt. sti. ton. diu. abo. rub. con.; Ls.
" coronopus,	Wild scurvy grass,	Pl. a-sco. diu. [sad.
" hortensis,*	Scurvy grass,	see Cochlearia officinalis.
" officinalis,	" "	Ls. a-sco. dia. rub.; Ls. sad.
" Pyrenaica,	" "	see Cochlearia officinalis.
" vulgaris,	" "	" " "
Cocos butyracea,	Oil palm, (see also Elais)	Palm oil; emo.; in toilet soaps.
" fusiformis,	Black-ebony tree,	yields Macaw fat.
" nucifera,	Cocoanut,	Kc. edi.; Milk dep. lax.; Ol. pec.
Cœlocline polycarpa,*	Yellow dye-tree, Soudan,	Bk. topically in obstinate ulcers.
Coffea Arabica,	Coffee,	Roast'd Sd. sti. exc. a-eme. lit. fum
" vulgaris,	"	Sd. f-com.; Ls. as a beverage.
Coix lachryma,	Job's tears,	Sd. diu. ton.; strung for children
Cola acuminata,	Cola nut, (Caffein nut),	Nu. stl. ner. [teething.
Colchicum autumnale,	Colchicum,	Rt. and Sd. acr. nar. pol. sed. a-rhe.
" illicium,*	Chequer flower,	see Colchicum variegatum. [diu. cat
" variegatum,	" "	said to yield Hermodactyle.
Coleus Blumei,	Coleus,	Pl. orn.
" Verschaffeltii,	German coleus,	Pl. orn.
Collinsonia Canadensis,	Stone root,	Rt. diu. irr. a-spa. exp.; Ls. vul.
Colocasia esculenta,	Yam root, Tara,	yields W. I. arrow root.
Colocynthis officinalis,*	Colocynth,	see Citrullus Colocynthis.
Columnea longiflora,	Bahel,	Ls. vul. mat.; in tumors.
Coluteа arborascens,	Bladder senna,	Ls. pur.; often mixed with senna.
" vesicaria,	Senna herb,	Ls. cat.; used as senna.
Comandra umbellata,	Bastard toad-flax,	Ls. feb. ton.
Comarum palustre,	Marsh five-finger,	Hb. ast. feb. ton.
Commelina angustifolia,	Day flower, .	Pl. exp. emo. ano.; Rt. feb.
Comocladia dentata,	Bastard Brazil wood,	Wd. a dyewood; see Brazil-wood.
" ilicifolia,	St. Domingo Braziletto,	" " " "
Comptonia asplenifolia,	Sweet fern,	Ls. ast. dia. ton. exp. sed. feb.
" dulci-fllix,*	" "	see Comptonia asplenifolia.
Conferva helminthocortos,*	Corsican moss, (Wormweed),	see Laurencia and Fucus.
" rivalis,	River weed,	Pl. in spasmodic asthma; phthisis.
Coniclinium cœlestinum,	Mist flower,	Pl. dia. a-spa. exp.
Conio-selenium Canadense,*	Marsh or hemlock parsley,	Pl. anti-dysenteric.
Conium maculatum,	Poison hemlock,	Ls. and Sd. nar. poi. sed. ano. deo.
" moschatum,	Aracacha,	Rh. edi. nut. con. [a-spa. diu.
Consolida major,*	Comfrey,	see Symphytum.
" media,*	Oxeye daisy	see Chrysanthemum leucanthemum
" minor,*	Healall,	see Prunella vulgaris.
" regalis,*	Branching larkspur,	see Delphinium consolida.
" rubra,*	Tormentilla root,	see Potentilla Tormentilla.
" sarracenica,*	European goldenrod,	see Solidago virgaurea.
Contrayerva Virginiana,*	Virginia snakeroot,	see Aristolochia Serpentaria.

BOTANICAL.	COMMON.	PROPERTIES, PRODUCTIONS, USES, ETC.
Convallaria biflora,	Dwarf Solomon seal,	Rt. sti. cep. ste.
" borealis,	Dragoness plant,	Rt. sti. cep. ste, exp. muc. pun.
" majalis,	Lily of the valley,	Rt. muc. sweet; Fl. fra. ste.
" multiflora,	Giant Solomon seal,	Rt. ton. muc. pec. emc. ast. f-com.
" polygonatum,	Solomon seal,	Rt. ast. ton. muc. exp. f-com.
" racemosa,	Small Solomon seal,	Rt. ast. ton. muc. exp. f-com.
Convolvulus arvensis,	Small bearbind,	Rt. pur. diu.
" Batatus,	Sweet potato,	Bu. edl. esc. nut.
" cantabrica,	Lavender bindweed,	Rt. cat. ant.; Ilb. ver.
" Floridus,*	Canary rosewood,	Rt. err.; see Convolvulus Scoparius
" Indicus,*	Sweet potato,	see Convolvulus Batatus.
" Jalapa,*	Jalap root, .	see Ipomœa Jalapa.
" major albus,*	German scammony plant	see Convolvulus sepium.
" mechoacanna,	Mechoacan root, .	Rt. ape. diu.; as Jalap.
" megalorrhizus,*	Wild Jalap,	see Convolvulus panduratus.
" nil,	Blue morning-glory,	Rt. and Sd. cathartic.
" operculatus,	Brazilian purge,	Rt. pur. yields a gum resin.
" Orizabensis,*	Male Jalap,	see Convolvulus Jalapa.
" panduratus,	Wild potato, (Jalap),	Rt. cat. diu. her. lit. a-ven.
" perennis,*	Hops,	see Humulus lupulus.
" purga,*	Jalap root,	see Ipomœa Jalapa.
" purpurea,	Purple morning-glory,	Pl. and Fl. orn.
" repens,*	Hedge bindweed,	see Convolvulus sepium.
" sagittæfolius,	Small bindweed, '	" " arvensis.
" Scammonia,	Aleppo scammony plant,	Gu.-re. dra.-cat.; griping.
" Scoparius,*	Canary rosewood,	yields oil Rhodium.
" sepium,	German scammony plant	Gu.-re. a powerful cat.; diu.
" sibthorpii,*	Small bindweed,	see Convolvulus sagittæfolius.
" soldanella,	Sea bindweed,	Ls. dra.-cat. diu.; Re. pur.
" spithameus,	Dwarf morning-glory,	Pl. and Fl. orn.
" Syriacus,*	Scammony plant,	see Convolvulus Scammonia.
" Turpethum,	Turpeth root,	Rt. cat.
Conyza,	Small fleabane,	see Inula dysenterica.
" bifoliata,	Silk seed aster,	Pl. diu.
" major,*	Marsh fleabane,	see Conyza squarrosa.
" Marilandica,	Plowmanswort,	Pl. sti. ner. a-spa. pec. bal. a-hys.
" media,	Small fleabane,	see Inula dysenterica.
" pycnostachyum,	Black root,	Rt. alt. nar.
" squarrosa,	Marsh fleabane,	Pl. ins. emm. a-epi.
Copaifera Beyrichii,	Balsam copaiba tree,	Ol.-re. sti. diu. cat. ast. pec. a-syp.
" bijuga,	Rio Negro copaiba,	" " " " " " "
" cordifolia,*	Brazilian copaiba,	see Copaifera coriacea.
" coriacea,	" "	Ol.-re. sti. diu. cat. pec.
" Guianensis,*	Para copaiba,	see Copaifera bijuga.
" Jacquini,*	West Indian copaiba,	" " officinalis.
" Jussieui,*	Brazilian copaiba,	" " Langsdorffii.
" Langsdorffii,	" "	Ol.-re. sti. diu. cat. ast. pec.
" laxa,	" "	see Copaifera Langsdorffii, var.
" Martii,	Peruvian copaiba,	Ol.-re. see Copaifera officinalis.
" multijuga,	Para copaiba,	Ol.-re. sti. diu. cat. ast. pec. a-syp.
" nitida,*	Brazilian copaiba,	see Copaifera Langsdorffii.
" officinalis,	Balsam copaiba, W. I.,	Ol.-re. sti. diu. cat. ast. pec. a-syp.
" Sellowii,*	So. American copaiba,	see Copaifera Langsdorffii.
Coptis tecta,	Chinese goldthread,	Rt. bit. ton. sto. opt.
" trifolia,	Goldthread root,	Rt. a pure bitter tonic, for canker.
Corallina Corsicana,	Corsican moss,	see Fucus helminthocorton.
" officinalis,	" worm moss,	" " "
Corallorhiza odontorhiza,	Crawley root,	Rt. dia. sud. sed. feb.; Po. fever po.
Corchorus capsularis,	Jute plant,	Pl. yields the fibre known as jute.
" olitorius,	Jews' mallow,	Ls. emo. sad. feb.

BOTANICAL.	COMMON.	PROPERTIES, PRODUCTIONS, USES, ETC.
Cordia Africana,	Assyrian plum,	Ft. ape. edi. pec. ; Ju. vis.
" Boisseri,	Anacahuite wood,	Wd. formerly used in phthisic.
" domestica,	Sebesten plum,	Ft. ape. edi. muc. vis. ; Wd. fra.
" myxa,	" "	Ft. muc. esc. lax.; Bk. ton. ast. ;
" Gerascanthus,	Jamaica rosewood,	Wd. fra. ; yields oil. [bird lime.
Coreopsis trichosperma,	Tickweed sunflower,	Ls. and Fl. alt. exp.
" tripteris,	" "	Ls. and Fl. alt. exp.
" tinctoria,	Nutall's weed,	Pl. orn.
Coriandrum Cicuta,*	Water hemlock,	see Cicuta aquatica.
" maculatum,*	Poison hemlock,	see Conium maculatum.
" sativum,	Coriander seed,	Sd. aro. pun. car. cor. sto.
Coriaria myrtifolia,	Tanners' sumach, (Redoul)	Ls. acr. nar. poi.
" · ruscifolia,	Toot poison,	Tw. and Ls. poisonous to animals.
' sarmentosa,*	" "	see Coriaria ruscifolia.
" thymifolia,	Ink plant of New Granada,	Ju. used as ink; chanchi.
Cornucopia cucculata,	Horn-of-plenty grass,	Pl. cultivated orn.
Cornus alba,	Waxberry cornel,	Bs. waxy white ; Fl. orn.
" alternifolia,	Alternate leaved cornel,	Bk. dia. ast. feb.
" amomus,*	Swamp dogwood,	see Cornus sericea.
" Canadensis,	Bunchberry cornel,	Bk. ast. ton. dia. feb.
" circinata,	Green osier,	Bk. ast. ton. bit.
" Fœmina,*	Swamp dogwood,	see Cornus sericea.
" Florida,	Florida dogwood (boxwood),	Bk. ast. ton. feb. a-per. ; Fl.
" mas,*	Cornelian cherry,	see Cornus mascula. [ton. f-com.
" mascula,	" "	Ft. aci. a-sco. ; Fl. ast.
" mas odorata,*	Sassafras,	see Sassafras officinalis.
" paniculata,	White cornel (dogwood)	Bk. as Cornus Florida.
" rubiginosa,*	Swamp dogwood,	see Cornus sericea.
" sanguinea,	Dogberry tree,	Bk. wash for mangy dogs.
" sericea,	Red osier, Swamp dogwood,	Bk. ast. ton. feb. bit.
" tomentosa,	Green osier,	see Cornus circinata.
Coronopus,*	Wild scurvy grass,	see Cochlearia coronopus.
" Ruellii,	Wart cress, Herb ivy,	Fl. fœt. dem.
Coronilla emerus,	Scorpion senna,	Ls. cat. ; less active than Senna.
Corvisartia Helenium,*	Elecampane,	see Inula Helenium.
Corydalis bulbosa,	Round birthwort,	Rt. dep. dis. cos. ver. emm.
" Canadensis,	Squirrel corn,	Rt. a-syp. diu. alt.
" cucullaria,	Colic weed,	Rt. a-syp. diu. alt. dia.
" digitata,	Thick birthwort,	Rt. ver. emm. dep.
" formosa,	Turkey corn,	Rt. ton. diu. alt. a-syp.
" nobilis,	Noble fumitory,	Hb. bit. dia. ape. ton.
" solida,*	Thick birthwort,	Rt. see Corydalis bulbosa.
" tuberosa,	Birthwort,	Rt. bit. acr. ; Ju. det. dis.
Corylus Americana,	American filbert,	Ft. edi. diu. ano. ant.
• " avellana,	European filbert,	Ft. edi. nut.
" rostrata,	Beaked hazel,	Spicula ant. ; see Mucuna pruriens.
Corypha cerifera,*	Carnauba wax palm,	see Ceroxylon carnauba.
Costus Arabicus,	Sweet costus,	Rt. aro. ton. car. diu. emm.
" nigra,*	Garden artichoke,	see Cynara Scolymus.
Coscinium fenestratum,	Colombo wood,	Wd. sto. ant. ; dyes yellow.
Cotoneaster vulgaris,	Common cotoneaster,	Ft. ast.; Pl. orn.
Cotula,	Mayweed,	see Anthemis Cotula.
" fœtida,	" "	" " " "
" multifida,	African cotula,	Pl. a-rhe. exa. ; for burns.
Cotyledon orbiculata,		Ju. a-epi. ; Ls. used for "corns."
" umbilicus,	Navelwort, Kidneywort,	Hb. ton. diu. a-epi.; in asthma.
Coumarouna odorata,*	Tonka bean,	see Dipterix odorata.
Coutarea speciosa,	a false Peruvian bark,	Bk. similar to Cinchona.
Coutoubea alba,		Pl. bit. emm. ant. anti-dysenteric.
Cowania Stansburiana,	a Salt Lake plant,	Pl. ast. sty.

BOTANICAL.	COMMON.	PROPERTIES, PRODUCTIONS, USES, ETC.
Crassula arborescens,		see Cotyledon orbiculata.
" tetragona,	So. African crassula,	Pl. ast.; in diarrhœa.
Cratægus aria,	White beam tree,	Ft. ast. edl.; Ls. pec.
" coccinea,	Hawthorn,	Bk. and Ls. ton. exp.; Ft. edi. ton.
" cordata,	Washington thorn,	cultivated for hedge rows.
" crus-galli,	White cockspur thorn,	Fl. fra.; Ls. pec.
" Oxyacanthus,	English hawthorn,	Fl. fra.; Ft. called "haws;" ast.
" pyracanthus,	Evergreen thorn,	Ft. ast.
Cratæva gynandra,	Garlic pear,	Bk. of Rt. said to blister.
" marmelos,	Bengal quince,	see Ægle marmelos.
Crepis virens,	Hawk beard,	see Cichorum intybus.
Crinium Americanum,	Jonquil,	Bu. exp. sti.; as squill.
Crithmum Maritimum,	Samphire,	Pl. spicy aro.; con. when pickled.
Crocus odorus,	Sicilian saffron,	see Crocus sativus.
" officinalis,*	Saffron,	" " "
" orientalis,*	Ceylon saffron,	" " "
" Indicus,	Long turmeric root,	see Curcuma longa.
" Germanica,	Dyers' saffron,	see Carthamus tinctorius.
" sativus,	Spanish saffron,	Fl. sti. dia. exa. car. emm.
" vernus,	Spring crocus,	Pl. orn.
Crotalaria Juncea,	Madras hemp fibre,	yields a strong fibre.
" sagittalis,	Rattlebox,	
Croton balsamiferum,	Seaside balsam,	Bk. aro. ton. emm. f-com.
" benzœ,*		see Terminalia Benzoin.
" Cascarilla,*	Cascarilla bark,	see Croton Eluteria.
" Draco,	Mexican dragons' blood,	Gu.-re. vul. ast.; resembles kino.
" Eleutheria,*	Cascarilla bark,	see Croton Eluteria.
" Eluteria,	· " "	Bk. bit. aro. ton. car. sto. fum.
" Gossypifolium,	Blood tree,	Red juice; as dragon's blood.
" humilis,	Pepper rod,	Bk. in baths for nervous weakness.
" Jamalgota,*	Croton oil plant,	see Croton Tiglium.
" lacciferum,*	Seed lac, Ceylon,	see Aleurites laccifera.
" lineare,	Wild rosemary of W. I.,	specific in colic.
" lucidus,	Spurious cascarilla,	Bk. ast.; mixed with cascarilla.
" malambo,	Matias bark,	Bk. ton. feb.; in intermittents.
" niveus,*	Copalchi bark,	see Croton pseudo-china.
" pavana,	Tilly seed,	Sd. yield croton oil in part.
" pedicipes,	Alcamphora,	Bk. diu. a-syp.
" Philippinensis,*	Kamcela,	see Rottlera tinctoria.
" pseudo-china,	Copalchi bark of Mexico	Bk. bit. ton. feb.; analogous to cascarilla.
" Sloanei,*	Cascarilla bark,	see Croton Eluteria.
" suberosum,*	Copalchi bark,	see Croton pseudo-china. [cat.
" tiglium,	Croton oil plant,	Rt. cat.; Sd. yield Croton oil; dra.
" tinctorium,	Blue or red patch,	Pl. acr. emc. dra. cat.; dyes purple
Crozophora tinctoria,*	Turnsol,	see Croton tinctorium.
Cruccata,	Crosswort,	Hb. pec. exp.
Cryptocarya moschata,	Brazilian nutmeg,	Nu. aro.; Bk. aro. sti.
Ctenium Americanum,	Toothache grass,	
Cubeba Clusii,	African black pepper,	see Piper afzelii.
" officinalis,*	Cubebs,	see Piper Cubeba.
Cucubalus behen,	Spatling poppy,	Pl. alc. cor.
Cucumis anguria,*	Watermelon,	Ft. edl. ref. sac.
" agrestis,*	Squirting cucumber,	see Momordica Elaterium.
" Citrullus,*	Watermelon,	see Cucurbita Citrullus.
" colocynthis,	Colocynth apple,	Ft.-pulp, bit. dra.-cat.
" melo,	Muskmelon,	Ft. edi. a-sco.; Sd. muc. diu.; oily.
" sativus,	Cucumber,	Ft. edl. pickled; Sd. muc. diu. oily.
" sylvestris,*	Squirting cucumber,	see Momordica Elaterium.
Cucurbita anguria,	Watermelon,	see Cucurbita Citrullus.
" aurantia,	Orange gourd,	" " ovifera.

BOTANICAL.	COMMON.	PROPERTIES, PRODUCTIONS, USES, ETC.
Cucurbita Citrullus,	Watermelon,	Ft. edi. ref.; Sd. diu. ref.
" leucantha,*	Pumpkin,	see Cucurbita Pepo.
" lagenaria,	Gourd,	Sd. ref. diu.; Rt.-pulp, dra.-cat.
" maxima,	Winter squash,	Ft. edi. nut. diu.; Sd. diu.
" melopepo,	Round squash,	see Cucurbita Pepo.
" ovifera,	Egg squash,	Ft. edi. nut. diu.
" Pepo,	Pumpkin,	Sd. ver. oil; in tapeworm; diu.
" pinnatifida,	Watermelon,	see Cucurbita Citrullus.
" verrucosa,	Crookneck squash,	Ft. edi. nut.; Sd. muc. diu.
" " medullosa,	Marrow squash,	Ft. edi. nut.; Sd. muc. diu.
Cuminum Cyminum,	Cummin plant,	Sd. aro. sti. car.; veterinary.
" minutum,*	" "	see Cuminum Cyminum.
" nigrum,*	Black cummin,	see Nigella sativa.
" pratense,*	Caraway,	see Carum carui.
" Romanum,*	Cummin seed,	see Cuminum Cyminum.
Cunila mariana,	Mountain dittany, Am'n,	IIb. sti. ner. sud. ton. vul. cep.
" mascula,*	Small fleabane,	see Inula dysenterica.
" pulegioides,*	Pennyroyal,	see Hedeoma pulegioides.
" sativa,*	Summer savory,	see Satureja hortensis.
Cuphea viscosissima,	Waxbush,	Pl. vis.
Cupressus disticha,	Virginia cypress,	Ls. dye brown.
" sempervirens,	Cypress tree,	Bs. Ls. and Wd. ast. feb. bit.
" thyoides,	White cedar,	Ls. sti. aro. dia. sto.
Curcas multifidus,	French physic nuts,	see Jatropha multifida.
" purgans,	Barbadoes nuts,	Nu. pur. eme.; Ol. pur. dra. her.
Curcuma amada,	Mango (amada) ginger,	Rt. gently sti.; in curry po. [pol.
" angustifolia,	yields E. I. arrow root,	Rt. and Fecula edi. nut.
" aromatica,*	Zerumbet,	see Curcuma Zerumbet.
" leucorrhiza,*	"Tikor" arrow root,	" " angustifolia.
" longa,	Turmeric root, long,	Rt. colors yellow; in curry po.
" rotunda,	" " round,	see Curcuma longa.
" viridifolia,*	" "	" " "
" Zedoaria,	Round zedoary root,	Rt. fra. sti. sto.; as ginger.
" Zerumbet,	Zerumbet root,	Rt. sti. a-spa. aro.; as ginger.
Cuscuta Americana,	Dodder,	Pl. ton. ast. alt. a-per.
" Europæa,	Greater dodder, Hell weed,	Ju. pur. deo.; for itch.
" epithymum,	Dodder of thyme,	Pl. pun. fœt. cat.; in melancholia.
" filiformis,*	"	see Cuscuta Europæa.
" glomerata,	American dodder,	Pl. bit. sub-ast. ton. a-per.
" major,*	Greater dodder,	see Cuscuta Europæa.
" minor,*	Dodder of thyme,	" " epithymum.
" tetrandra,*	Greater dodder,	" " Europæa.
" vulgaris,	" "	" " "
Cusparia febrifuga,	Angustura bark,	see Galipea officinalis.
Cyanus Ægyptiacus,	Lotus,	see Nelumbium luteum.
" segetum,	Blue bottle,	Fl. ast. ref. opt.; colors blue.
Cycas circinalis,	Broad leaved cycas,	St. yields Japan sago.
" revoluta,	Narrow leaved cycas,	St. yields Japan sago.
Cyclamen Europæum,	Sow bread,	Rt. acr. bit. dra. ant. pol.
Cyclopia genistoides,	So. African tea plant,	Ls. exp. pec.
Cydonia Japonica,	Japan quince,	an ornamental shrub.
" maliformis,*	Quince,	see Cydonia vulgaris.
" vulgaris,	"	Ft. ast. edl.; Sd. vis. muc. opt.
Cynanchum argel,	Argel leaves,	Ls. mixed with Alexandrian Senna.
" Ipecacuanha,*	Ceylon ipecac,	see Asclepias asthmatica.
" Monspeliacum,	Montpellier Scammony,	Gu.-re. pur. diu.
" oleæfolium,*	Argel leaves,	see Cynanchum argel.
" tomentosum,*	Ceylon ipecac,	see Asclepias asthmatica.
" · vincetoxicum,*	White swallow-wort,	" " vincetoxicum.
" viridiflorum,*	Ceylon ipecac,	" " asthmatica.

BOTANICAL.	COMMON.	PROPERTIES, PRODUCTIONS, USES, ETC.
Cynanchum vomitorium,*	Ceylon ipecac,	see Asclepias asthmatica.
Cynara cardunculus,	Cardoon,	Rt. ape. diu. aph.; Ls. sad.
" scolymus,	Garden artichoke,	Rt. edi. nut. diu.; Ju. of Ls. in
Cynodon dactylon,	Bermuda grass,	Rt. as sarsaparilla. [dropsy.
Cynoglossum amplexicaule,	Wild comfrey,	Rt. muc. dem. ast. pec. vul.
" Morrisoni,	Virginia mouse-ear,	Rt. muc. ton. ast. diu.
" officinale,	Hounds' tongue,	Pl. ast. aro. ano. muc. nar.
" Virginicum,*	Wild comfrey,	Rt. see Cynoglossum amplexicaule.
Cynometra Agallocha,*	Aloe wood,	see Aloexylon agallochum.
Cynomorium coccineum,	Scarlet mushroom,	Fu. ast. sty.; Fungus melitensis.
Cynosurus cristata,	Leghorn-straw grass	Tp. for.; straw for hats.
" echinatus,	Cockscomb grass,	Tp. for.
Cyperus antiquorum,*	Henne plant,	see Lawsonia inermis.
" esculentus,	Rush nut,	Rt. edi.; roasted; as coffee.
" hydra,	Nut grass,	Rt. edi.; roasted; as coffee.
" Indicus,	Long rooted turmeric,	see Curcuma longa.
" longus,	Galingale,	Rt. aro. bit.; see Dorstenia Contra-
" officinalis,	English galingale,	Rt. aro. ton. bit. dia. diu. [yerva.
" odorus,	Lisbon contrayerva,	Rt. aro. sto. dia.; see Aristolochia.
" Romanus,	Long rooted cyperus,	see Cyperus longus.
" rotundus,	Round rooted cyperus,	" " " in cholera.
Cypripedium acaule,	Ladies' slipper root, stemless,	Rt. ton. sti. dia. ner. a-spa.
" arietinum,	Rams' head,	Rt. ton. sti. ner. dia. a-spa. [nar.
" calceolus,*	Showy ladies' slipper,	see Cypripedium spectabile.
" Canadensis,*	" " "	" " "
" caudidum,	White fl'd ladies' slipper,	Rt. analogous to C. pubescens.
" flavum,	Nerve rt., Moccasin rt.,	Rt. " " "
" humile,*	" "	see Cypripedium flavum.
" luteum,	American valerian,	Rt. ner. ton. asp. dia. hyp.
" parviflorum,	" "	see Cyripedium luteum.
" pubescens,	Yellow ladies' slipper,	Rt. ton. sti. ner. dia. a-spa. nar.
" spectabile,	Showy ladies' slipper,	Rt. ton. sti. dia. nar. ner.
" venustum,	an East Indian variety,	Pl. orn.
Cytinus hypocistus,	parasite on Woody cistus	Ju. of Ft. acl. ast. sty.
Cytisus cajan,	Pigeon pea, Angola pea,	Rt. aro. pec.; Sd. acr. edi.
" Laburnum,	Laburnum tree,	Ls. diu. res.; Sd. and Bk. poi.
" Scoparius,	Broom herb,	Tp. diu. cat.; Sd. eme. cat.
Dacrydium cupressimum,	New Zealand thus,	Re. used in varnish.
Dactylis cæspitosa,	Tussock grass,	Pl. for.
" glomerata,	Orchard grass,	Pl. for.
Dactyloctenium Ægypticum,	Egyptian grass,	Pl. for.
" rudulans,	Comb fringe grass,	Pl. for.
Dædalea suaveolens,*	Willow sponge,	see Boletus suaveolens.
Dæmonorops Draco,*	Dragons' blood resin,	see Calamus Draco.
Dahlia variabilis,	Common dahlia,	Pl. orn.; Rt. acr.
Dalbergia nigra,	Violet (rose) wood, E. I.	Wd. orn.
Damarra Australis,	Damarra turpentine tree	see Pinus Damarra.
Daniellia thurifera,	African frankincense,	Gu.-re. aro. fra.
Danthouia,	Wild oat grass,	Pl. and Sd. for.
Daphne alpina,	Widow wail, Dwarf olive	Bk. acr. ves. poi.
" garou,	Garou bark,	see Daphne Gnidium.
" Gnidium,	Spurge flax, Mezereon,	Bk. acr.-poi. ves.; Bs. acr.-poi.
" Laurcola,	" laurel,	Bk. " " " "
" major,*	" "	see Daphne laureola.
" Mezercum,	Mezereon bark,	Bk. aer. poi. sti. alt. diu. dia. cat.
" odora,	Sweet scented mezereon,	Pl. and Fl. orn. [a-syp.
" paniculata,*	Spurge flax,	see Daphne Gnidium.
Darlingtonia Californica,	Side flow'g saddle flow'r,	Pl. curious orn.; ins.
Dasystoma pedicularia,*	Bushy gerardia,	see Gerardia pedicularis.
Datisca cannabina,	False hemp,	Pl. cat.

BOTANICAL.	COMMON.	PROPERTIES, PRODUCTIONS, USES, ETC.
Datisca hirta,	False hemp,	Pl. cat.
Datura alba (fatuosa*)	Indian datura (white),	Sd. scd. nar.; Ls. ano.-poultice.
" ferox,	East Indian datura,	Rt. and Ls. poi.; smok'd for asthma
" meteloides,*	Sweet scented datura,	see Datura Wrightii.
" Stramonium,	Thorn apple plant,	Pl. poi. sti. scd,; fumes for asthma.
" sanguinea,	Red th'n apple, Grave pl.	Peruvians make a nar. drink; Tonga
" tatula,	Purple stramonium,	Pl. analogous to D. Stramonium.
" Wrightii,	Sweet scented datura,	Pl. cultivated for ornament.
Daucus candianus,*	Candy carrot,	see Athamanta Cretensis.
" Carota,	Wild, and Garden carrot,	Rt. and Sd. sti. diu. car.; Rt. poul-
" gingidium,	Chervil,	Rt. bit. bal.; see Opoponax. [tices.
" sativa, var.,	Garden carrot,	Rt. edi.; Sd. 'see Daucus Carota.
" sylvestris,	European wild carrot,	Ft. diu.; rays of umbels, Spanish
" vulgaris,*	Common carrot,	see Daucus Carota. [toothpicks.
Davilla Brasiliana,	Davilla,	Pl. ast.; in glandular swellings.
" rugosa,*	"	see Davilla Brasiliana.
Decodon verticillatum,*	Slink weed,	see Lythrum verticillatum.
Delabechea rupestris,	Bottle tree of Australia,	Gu. resembles Tragacanth.
Delphinium ajacis,	Rocket larkspur,	a garden variety; orn. poi.
" consolida,	Branch'g larksp'r, com'n,	Rt. and Fl. poi. diu. cmm. vul. ver,
" exaltum,	Tall wild larkspur,	see D. consolida. [Sd. poi. ins.
" officinalis,	Stavesacre,	see Delphinium Staphisagria.
" Staphisagria,	"	Sd. irr.-poi. cme. cat. nar. ins.
· " versicolor,	Larkspur,	see Delphinium consolida.
Dendropogon usneoides,*	Spanish moss,	'see Tillandsia usneoides.
Dentaria diphylla.	Tooth root,	Rt. used as mustard.
Desmodium gyrans,	Moving plant,	Pl. orn.; curious.
Deutzia gracilis,	Graceful Deutzia,	Pl. orn.
" scabra,	Rough Deutzia,	Pl. orn.
Diagraphis arundinacea,	Ribbon grass,	see Phalaris.
Dianthus Armeria,	Deptford pink,	Pl. see Dianthus Caryophyllus.
" barbatus,	Sweet William,	" " "
" Caryophyllus,	Clove pink,	Fl. ccp. a-spa. ner. cor. fra.
" Chinensis,	China pink,	Pl. see Dianthus Caryophyllus.
" hortensis,	Cob (Garden) pink,	" " "
" plumarius,	Plumed (Single) pink,	" " "
Diatomaceæ,	Brittleworts,	microscopic moving plants.
Diccutra Canadensis,	Squirrel corn, (Colle weed),	see Corydalis Canadensis.
" cucullaria,	Dutchman's breeches,	· " " cucullaria.
" spectabilis,*	Bleeding heart,	" " formosa.
Dicksonia pilosiuscula,	Fine haired fern,	Pl. orn.
Dictamnus albus,	White fraxinella,	Rt. bit. ton. sto. cmm. feb.
" Creticus,	Dittany of Crete, .	see Origanum dictamnus.
" Fraxinella,	Bastard dittany,	see Dictamnus alba.
Dicyphellium caryophyllatum,*	Clove bark, Brazilian,	see Laurus Caryophyllatum.
Didiscus cœrulea,	Sky blue didiscus,	Pl. orn.
Dielytra Canadensis,*	Squirrel corn,	see Corydalis.
" cucullaria,*	Dutchman's breeches,	" " cucullaria.
" formosa,*	Turkey corn,	" " formosa.
Diervilla Canadensis,	Bush honeysuckle,	Rt. Ls. and Tw. diu. ast. alt.; in
" rosea,	Weigela rosea, var.	see Diervilla Canadensis. [ivy poi.
" trifida,*	Bush honeysuckle,	" " "
Digitalis lutea,	Yellow foxglove,	used as Digitalis purpurea.
" purpurea,	Foxglove,	Ls. irr. nar.-poi. scd. diu. a-epi.
Digitaria sanguinalis,	Crab grass,	Pl. for.
Dignetia quitarvensis,	Lance wood,	Wd. useful.
Digraphis, .	Ribbon grass,	see Phalaris.
Dilatris tinctoria,*	Red root, Spirit weed,	see Lacnanthes tinctoria.
Diodia Virginica,	Button weed,	Fl. orn.
Djonæa muscipula,	Venus' flytrap,	Pl. ins.

BOTANICAL.	COMMON.	PROPERTIES, PRODUCTIONS, USES, ETC.
Dioscorea batatas,	Chinese yam,	Rt. edi. nut. far.; as potatoes.
" sativa,	Cultivated yam,	Rt. edi. far. nut.; yields arrow rt.
" villosa,	Wild yam, Colic root.	Rt. a-spa. dia. f-com.; in bilious
Diosma crenata,*	Short leaf Buchu,	see Barosma crenata. [colic.
" crenulata,*	Medium (round) Buchu,	" " crenulata.
" serratifolia,*	Long leaf Buchu,	" " serratifolia.
Diospyros ebenum,	Ebony tree, Ceylon,	Wd. orn.; ebony wood.
" Embryopteris,	Indian persimmon,	Ju. vis. ast.; in dysentery.
" Virginiana,	Persimmon,	Bk. ast. feb. a-per.; Ft. ast. edi.
" Lotus,	Indian date plum,	Ft. ast. acerb.
Dipsacus fullonum,	Fullers' teasel,	Rt. diu. sud. sto. opt.
" pilosus,	Shepherds' rod,	Rt. " "
" sylvestris,	Teasel, cultivated,	Rt. " "
Dipterix odorata,	Tonka bean,	Ft. aro. bit. fra ste.; in snuff.
Dipterocarpus turbinatus,	Gurgun copaiba,	Ba. analogous to Copaiba.
Dirca palustris,	Leatherwood,	Bk. acr. rub. ves. sud. exp.; Bs.
Discopleura capillacea,*	Bishop weed, Am'n,	see Ammi. [poi.
Diserneston gummiferum,*	Gum ammoniac,	see Dorema Ammoniacum.
Distylium racemosum,	Chinese, velvet Galls,	Galls ast.; for dyes.
Ditoplaxis muralis,*	Flixweed, Sinapis,	see Sisymbrium muralis.
Dodecatheon media,	American cowslip,	Fl. orn.; Pl. sad
Dodonæa Thunbergiana,		Rt. a gentle cathartic in fevers.
Dolichos bulbosus,	Bulbous Dolichos,	Sd. eaten.
" filiformis,	Cats' claw,	Sd. ast. din.
" Lablab,	Purple hyacinth bean,	used as food in Egypt. [mintic.
" prurieus,*	Cowhage,	Hairs of pods a mechanical anthel-
" prurita,	East India cowhage,	" " " "
" purpureus,	Wild cowhage,	" " " "
" Sinensis,	Chinese bean,	Sd. eatable.
" Soja,	Sooja,	Sd. used in Japanese soy.
" tuberosa,	Martinique potato,	Rt. eatable.
Dombeya excelsa,	Dombeya (Chili) turpentine, Re. sti. diu. ant. c-irr.
Dorema Ammoniacum,	Gum resin ammoniac,	Gu.-re. exp. a-spa. dis. res.; plasters
" Aucheri,	" " "	" " " "
" robustum,		Gu.-re. resembles ammoniac.
Doronicum Arnica,	Arnica,	see Arnica montana.
" cordatum,*	Great leopardsbane,	see Doronicum pardalianches.
" Germanicum,*	Arnica,	see Arnica montana.
" montanum,*	"	" " "
" officinalis,*	Great leopardsbane,	see Doronicum pardalianches.
" pardalianches,	Gt. (Roman) leop'dsbane	Rt. and Tp. analogous to Arnica.
" Romanum,*	Gt.	see Doronicum pardalianches.
Dorstenia Brasiliensis,	Brazil contrayerva,	Rt. dia. aro. bit. ton. eme.
" contrayerva,	Lisbon contrayerva,	Rt. aro. bit. pun. muc. ton. dia.
" Drakena,*	Vera Cruz contrayerva,	see Dorstenia contrayerva.
" Houstonia,	Campeachy contrayerva,	Rt. sti. sud. dia. ton. alc.
Draba acaulis,	English pepper,	Sd. pun.; detersive.
" incana,	Whitlow grass,	" " "
" verna,	" "	" " "
Dracæna Draco,	Dragons' blood, Canary Isles,	Re. ton. ast.
Draco sylvestris,*	German pellitory,	see Achillea Ptarmica.
Dracocephalum Canariense,	Sweet balm,	Hb. aro. sti. sud. dia. cep. ton.
" "	Balm of Gilead herb,	see Dracocephalum Canariense.
" moldavica,	Moldavian balm,	" " " "
Dracontium fœtidum,*	Skunk cabbage,	see Symplocarpus fœtidus.
" polyphyllum,	Labaria plant,	Pl. nau. fœt. a-ven. sti.
Dracunculus vulgaris,	Common dragon plant,	see Arum Dracunculus.
Drimys aromatica,*	Winters' bark,	see Drimys Winteri.
" Chillensis,	" " of Chili,	analogous to Drimys Winteri.
" Granatensis,	" " N. Granada	" " "

BOTANICAL.	COMMON.	PROPERTIES, PRODUCTIONS, USES, ETC.
Drimys Mexicana,	Mexican Drimys,	analogous to Drimys Winteri.
" Winteri,	Winters' bark,	Bk. aro. fra. sti. pun. a-sco.
Drosera longifera,	Long leaved sundew,	Pl. acr. det. rub. pec. aph.
" rotundifolia,	Round leaved sundew,	" " "
Dryobalanops aromatica,*	Sumatra camphor, also an oil called Camphor oil.	
" Camphora,	Borneo camphor,	see also Camphora officinarum.
Drymis,*	Winters' bark,	see Drimys.
Duvana dependens,		Re. pur. a-spa.; in plasters.
Dyospyros,*	Ebony tree,	see Diospyros.
Dysophylla avicularia,	Earwort, Ceylonian plant, Pl. sti.; for deafness.	
Ecballium agreste,*	Squirting cucumber,	see Momordica Elaterium.
" elaterium,*	" "	" " "
" officinalis,*	" "	" " "
Ecbolium elaterium,*	" "	see Momordica.
Echinacea purpurea,*	Purple coneflower,	see Rudbeckia purpurea.
Echinocactus Texensis,	Hedgehog cactus,	a curious orn.
Echinochloa echinata,	" grass,	referred to Panicum.
Echinops multiflorus,	Globe thistle,	Rt. in nosebleed; Sd. diu.
" strigosus,	Spanish tinder,	Down and Ls. as amadou.
Echinus Phillipinensis,*	Kameela,	see Rottlera tinctoria.
" scandens,	Allamanda, (Guiana)	Ls. used in painters' colic.
Echinum vulgare,	Vipers' bugloss.	Pl. orn.; Rt. ast. dep.
Echites Chiliensis,	Chilinac,	Rt. eme. ste.
" suaveolens,	Chili Jessamine,	Pl. orn.
Elæocarpus copalliferus,	Copal tree,	Re. copal; in varnish.
" granitrus,	Necklace seed,	Sd. as rosaries.
" Hinau,	Olive nut tree, N. Zealand, Bk. dyes brown or black.	
Elais Guiniensis,	Palm oil tree,	Ol. emo.; in toilet soap.
" occidentalis,*	Oil palm,	see Cocos butyracea.
Elaphomyces granulatus,*	Deer balls,	see Lycoperdon cervinum.
Elaphrium elemiferum,	Mexican elemi,	Re. similar to the turpentines.
" excelsum,*	Tacamahac,	see Fagara octandra.
" tomentosum,*	"	" " "
Elaterium cordifolium,	Squirting cucumber,	see Momordica Elaterium.
Elatine hexandra,	Water-wort,	Pl. on pond banks; orn.
" hydropiper,	"	" " "
Elettaria Cardamomum,	Malabar cardamom,	Sd. aro. sti. car. con.
" major,	Wild cardamom,	Sd. " "
Elleborus,	Hellebore,	see Helleborus, and Veratrum.
Elæodendron glaucum,	Tea shrub of Ceylon,	Ls. substituted for tea.
Eleusine Indica,	Crab grass,	Pl. for.
Elymus arenarius,	Marram sea grass,	Pl. on beaches prevents drifting.
" geniculatus,*	Lime-grass,	Pl. for.
" villosus,	"	Pl. for.
" Virginicus,	" Wild rye,	Pl. and Sd. for.
Elytropappus Rhinocerotis,		Pl. bitter resinous; ton.
Emblica officinalis,*	White galls,	see Phyllanthus.
Embryopteris glutinifera,*	Indian persimmon,	see Diospyros Embryopteris.
Empetrum nigrum,	Crowberry,	Bs. edi.; dye black.
Entada scandens,	Sea bean, W. I.,	Sd. polished as ornaments.
Enula campana,*	Elecampane,	see Inula Helenium.
Ephedra Americana,	a fibrous plant, "pingo pingo," Ls. sty.	
Epidendron,	Vanilla,	see Vanilla aromatica.
Epidendrum vanilla,*	"	" " "
Epideudum conopseum,	Magnolia air plant,	Pl. orn.; on magnolia trees.
Epifagus Americanus,*	Beech drops, Cancer root, see Orobanche Virginiana.	
Epigæa repens,	May flower, Trailing arbutus, Ls. diu. ast.; as Uva Ursi.	
Epilobium angustifolium,	Willow herb,	Ls. and Rt. ton. ast. dem. emo. f-com
" colaratum,	Purple veined willow hb. Ls. and Rt. see Epilobium angusti-	
" hirsutum,	Codlins and cream,	Ls. ast. sty. det. [folium.

BOTANICAL.	COMMON.	PROPERTIES, PRODUCTIONS, USES, ETC.
Epilobium palustre,	Marsh epilobium, Wickop,	Ls. and Rt. ton. ast. a-spa. dem.
" spicatum,*	Willow herb, Rose bay,	see Epilobium angustifolium.
Epimedium alpinum,	Barrenwort,	Pl. ast.
Epipactis lateriflora,	Bastard hellebore,	Rt. far.
Epiphegus Americanus,*	Beech drops,	see Orobanche Virginiana.
" Virginiana,*	" "	" " "
Equisetum arvense,	Horsetail rush,	Pl. diu. ast.
" hyemale (majus*),	Scouring rush,	Pl. diu. ast. ton.; contains silex.
" minus,*	Mares' tail.	see Equisetium arvense.
Eragrostis,	Love grass,	
Eranthis hyemalis,	Winter hellebore,	see Helleborus hyemalis.
Erechthites hieracifolius,	Fireweed,	Pl. eme. ton. ast. alt.; in piles.
Ergotaetia abortifaciens,	Ergot of rye fungus,	see Sclerotium clavus.
Erica carnea,	Heath,	Pl. orn.
" vulgaris,	"	Pl. a-rhe. sud. ast. diu.
Erigeron annuum,	Scabious,	see Erigeron heterophyllum.
" bellidifolium,	Robert's plantain,	Pl. see Erigeron Philadelphicum.
" Canadense,	Canada fleabane,	Pl. ast. ton. diu. sty.
" heterophyllum,	Scabious,	Pl. diu. ast. ton. dia. emm.
" Philadelphicum,	Philadelphia fleabane,	Fl. " " "
" pusilum,*	a variety of	Erigeron Canadense.
" strigosum,*	Scabious,	see Erigeron Philadelphicum.
Eriodendron aufrectuosum,*	God tree, Cotton tree gum,	see Bombax pentandrum.
Eriocephalus umbellatus,	So. Af'n wild rosemary,	Pl. diu.; in dropsy.
Eriophorum polystachyron,	Cotton grass.	
" Virginicum,	Moss crop.	
Eriospermum latifolium,		Bu. mat. f-com.
Erodium cicutarium,	Hemlock storksbill,	Rt. ast. det. diu.; in poultices.
" moschatum,*	Musky storksbill,	see Geranium moschatum.
Eruca barbarea,	Winter water cress,	see Erysimum barbarea.
Ervum lens,	Lentils,	Sd. edi. nut. ast.
Eryngium aquaticum,	Water eryngo,	Rt. diu. sti. dia. exp. eme. a-syp.
" campestre,	Wild eryngo,	Rt. aph. diu. sud. ape.
" maritimum,	Sea holly,	Rt. aro. diu. exp.
" ovalifolium,	" " .	" " "
" yuccefolium,	Corn snake root,	Rt. diu. sud.
Erysimum alliaria,*	Hedge garlic,	see Alliaria officinalis.
" barbarea,	Winter watercress,	Pl. a-sco. pec. det.; Sd. acr. lit.
" cheiranthoides,	Treacle mustard,	Hb. ver. sto.
" Nasturtium,*	English watercress,	see Nasturtium officinale.
" officinale,*	Hedge mustard,	see Sisymbrium officinale.
Erythrolœna conspicua,	Mexican thistle,	Pl. orn.
Erythoxylon Coca,*	Coca leaves,	see Erythroxylon Coca.
Erythræa acaulis,	Rejagnou,	Rt. dyes yellow.
" centaurium,*	Com'n Europ'n centaury,	see Chironia centaurium.
" Chilensis,	Centaury of Chili,	Pl. a mild tonic.
Erythrina Abyssinica,	Carat seed,	Sd. orn.
" crista galli,	Coral tree,	Pl. orn.
" herbacea,	Coral bloom,	Rt. exp. sud. feb.
" monosperma,*	Dhak tree, Bengal kino,	yields Gum lac; see Butea.
Erythronium Americanum,	Dogs'-tooth violet,	Pl. eme emo.; anti-scrofulous.
" dens Canis,*	" "	see Erythronium Americanum.
" flavum,*	Yellow adders'-tongue,	Pl. anti-scrofulous; eme.
" Indicum,	East Indian squills,	Bu. used as squills; a-lit.
" lanceolatum,*	Dogs'-tooth violet,	see Erythronium Americanum.
Erythrophleum Judiciale,	Ordeal (sassy) bk. tree,	Bk. poi. ast. emc. nau.
" Guineense,*	" (Doom) bk. tree,	Bk. " "
Erythroxylon Coca,	Coca leaves,	Ls. ner. sti.; as tea, coffee, etc.
Eschscholtzia Californica,	California poppy,	Pl. and Fl. orn.
Esenbeckia febrifuga,	Esenbeckia,	Bk. used as Cinchona.

BOTANICAL.	COMMON.	PROPERTIES, PRODUCTIONS, USES, ETC.
Eucalyptus citriodora,	Lemon scent'd gum tree,	yields kino.
" dumosa,	Manna of New Holland,	Manna; as food.
" globosus,	Fever tree of Australia,	Ls. a-per. feb.; Tree anti-malarial.
" globulus,	" " "	Ls. " " "
" mannifera,	Manna of N. S. Wales,	Manna; edi. ape. sac.
" piperita,	Peppermint tree,	Ol. resembles Cajeput oil.
" resinifera,	Red (brown) gum tree,	yields Botany Bay kino.
" robusta,	Brown gum tree,	" " " "
" rostrata,	White gum tree,	" " " "
Euchroma coccinea,	Painted-cup plant,	Fl. orn.
Eudiandra glauca,	Teak wood, N. S. Wales.	
Eugenia acris,	Wild clove, Bay rum l'f,	see Myrcia acris.
" caryophyllata,*	Cloves,	see Caryophyllus aromaticus.
" cauliflora,*	Jabuticaba fruit,	see Myrtus cauliflora.
" Jambos,	Rose apple,	Ft. edi. aro.
" Luma,	W. I. myrtle,	Pl. orn.
" Pimenta,	Pimento, Allspice,	unripe Ft. aro. car. sto. sti. con.
Euonymus Americanus,	Burning bush,	Bk. ton. lax. alt. diu. exp.; Sd. eat.
" atropurpureus,	Wahoo bark,	" " " "
" Europæus,	Burning bush, Spindle tree,	Bk. ton. lax. alt.; Sd. eme. pur.
" Japonicus,	Chinese box,	Pl. orn. [ins.
" tingens,	Tika dye,	Pl. opt.; dyes yellow.
Eupatorium ageratoides,	White sanicle,	Rt. aro. ton. dia. diu.
" aromaticum,	" snakeroot,	Rt. dia. a-spa. exp. aro. diu.
" aya-pana,	Brazilian ayapana,	see Eupatorium perfoliatum.
" cannabinum,	Hemp agrimony,	Ju. eme. pur. diu. eat.
" connatum,*	Boneset,	see Eupatorium perfoliatum.
" hyssopifolium,	Justice weed,	Pl. a-ven.
" incarnatum,	Mata,	smoked with tobacco for its odor.
" Guaco,*	Guaco,	see Mikania Guaco.
" Japonicum,*	Hemp agrimony,	see Eupatorium cannabinum.
" leucolepsis,	Justice weed,	Hb. a-ven.
" maculatum,	Spotted boneset,	used as Eupatorium purpureum.
" nervosum,	Jamaica bitter bush,	Pl. bit. ton. feb.
" perfoliatum,	Boneset, Thoroughwort,	Hb. tou. ape. dia. eme. feb. a-per.
" pilosum,*	Wild horehound,	see Eupatorium teucrifolium.
" purpureum,	Queen-of-the-meadow,	Rt. diu. sti. ast. ton.; Hb. diu. ton.
" rotundifolium,*	Wild horehound,	used as Eupatorium perfoliatum.
" sessilifolium,	Upland (bastard) boneset,	" " purpureum.
" teucrifolium,	Wild horehound,	see Eupatorium perfoliatum.
" trifoliatum,	Wood boneset,	" " cannabinum.
" urticifolium,	Deerwort boneset,	" " ageratoides.
" verbenæfolium,*	Wild horehound,	" " teucrifolium.
" verticillatum,	Tall boneset, Joepye,	used as Eupatorium purpureum.
" violaceum,	Violet boneset,	" " "
" Virginicum,*	Boneset,	see Eupatorium perfoliatum.
Euphorbia antiquorum,	Euphorbium, Egyptian,	see Euphorbia tirucalli.
" canariensis,	" Canary Isles,	Gu. eme. cat. err. c-irr. ves.
" capitata,	Brazilian Caacica,	Pl. ast. a-ven.
" caput medusæ,	Medusa's head,	Ju. an arrow poison.
" corollata,	Blooming spurge,	Rt. eme. dia. exp.; in dropsy.
" cotinifolia,	a Brazilian species,	Ju. an arrow poison.
" cyparissias,	Cypress spurge,	Ju. acrid cathartic; poi.
" helioscopia,	Sun spurge, Churnstaff,	Ju. cau. to warts.; eat. a-syp.
" hypericifolia,	Large spotted spurge,	Ls. ast. ton. nar. [dropsy.
" Ipecacuanha,	American ipecac,	Bk. of Rt. eme. dia. exp. diu.; in
" lathyris,	Mole (Caper) plant,	Sd. eat. emm.; Rt. cat.
" maculata,	Spotted spurge,	Ls. ast. ton. nar.
" officinarum,	Euphorbium,	Gu.-re. err. c-irr. ves.
" palustris,	Greater spurge, Russian,	Ju. pur.; cau. to warts; ves. .

BOTANICAL.	COMMON.	PROPERTIES, PRODUCTIONS, USES, ETC.
Euphorbia paniculata,	Greater spurge,	see Euphorbia palustris.
" paralias,	Sea spurge,	Ju. a violent cathartic; irr.
" peplus,	Petty spurge,	Rt. cat. diu.; in dropsy.
" pilosa,*	Greater spurge,	see Euphorbia palustris.
" prostrata,	Gollindrinera,	Ju. a-ven. emc. cat.
" protylacoides,*	American Ipecac,	see Euphorbia Ipecacuanha.
" pulcherrima,	Poinsettia,	Pl. and Fl. orn.
" resinifera,	Euphorbium, Morocco,	Gu. violent ste. emc. cat. ves.
" tirucalli,	Milk hedge of India, Gu.	said to induce uterine contractions.
" tithymaloides,	Jew bush,	see Pedilanthus tithymaloides.
" tortilis,	of the East Indies,	Ju. see Euphorbia tirucalli.
" villosa,		see Euphorbia palustris.
" virosa,	Ethiopian arrow poison,	" " caput-medusæ.
Euphrasia officinalis,	Eyebright,	Ls. ton. ast. a-epi. opt.
Euryangium Sumbul,	Musk root, Sumbul,	Rt. aro. bal. sti. ton.
Eurybia argophylla,	" tree,	Ls. have a musky odor.
Exidia auricula,	Jews' ear, a fungus,	Fu. in vinegar for a gargle in quinsy
Exodia aromatica,	Clove nutmeg,	see Agathophyllum aromaticum.
Exogonium purga,*	Jalap root,	see Ipomœa Jalapa.
Exostemma angustifolium,*		see Cinchona angustifolia.
" brachycarpum,	Jamaica bark,	Bk. emc. feb.
" Caribæa,	Seaside beech,	Bk. feb. emc. bit. aro.
" coriaceum,	St. Domingo bark,	Bk. as Cinchona.
" floribunda,	St. Lucia bark,	Bk. dra. emc. pur.
" Peruviana,	Bicolorata bark,	Bk. bit. nau. ton.
" triflora,	Jamaica bark,	Bk. as Cinchona.
" Souzanum,	Quinquinia de piantri,	Bk. bitter feb.
Faba vulgaris,	Common bean,	Sd. edl. nut. car.; Ls. emo. det.
Fagara octandra,	Tacamahac,	Re. fra. bal.; in plasters.
" microphylla,	Raingoat,	Re. analogous to turpentine.
" piperita,*	Japanese pepper,	see Xanthoxylum piperitum.
Fagarastrum Capensis,	Wild cardamom,	Ft. aro. car. sti.
Fagopyrum esculentum,*	Common buckwheat,	see Polygonum Fagopyrum.
" tartaricum,	Indian buckwheat,	Sd. edi. nut.; for flour.
Fagus castanea,	Spanish chestnut,	see Castana vulgaris.
" " pumila,	Chestnut, Chinquapin,	" " pumila.
" ferruginea,	American beech,	Bk. ast. ton. a-sep.; Ft. edl.
" purpurea,*	White beech,	see Fagus sylvatica.
" sylvatica,	European white beech,	Bk. ast. ton.; Ft. diu.; yield oil.
" sylvestris,*	" " "	see Fagus sylvatica.
Fatsia papyrifera,	Rice-paper plant,	Pi. is made into paper.
Fedia cornucopia,	Horn-of-plenty grass,	Pl. orn.
" olitoria,	Lambs' lettuce,	Ls. used as salad.
" radiata,	Corn salad,	" "
Feronia elephantum,	Elephant (wood) apple,	Ls. and Fl. aro. fra. sto. sti.; Gu.
Ferula Africana,*	So. African Galbanum,	see Bubon Galbanifera. [pec.
" ammoniacum,*	Gum ammoniac,	see Dorema Ammoniaca.
" ammonifera,*	" " "	" " "
" Asafœtida,*	Assafœtida,	see Narthex Assafœtida.
" crubescens,	Persian Galbanum,	Gu.-re. sti. exp.; in plasters.
" ferulgo,*	Gum Galbanum,	see Ferula Galbanifera.
" Galbanifera,	Russian Galbanum,	Gu.-re. sti. exp. a-spa. mat. res.
" galbaniflua,	Gum Galbanum,	Gu.-re. sti. exp.; in plasters.
" Narthex,	Assafœtida,	see Narthex Assafœtida.
" Persica,	Sagapenum,	Gu.-re. sti. emm. a-spa. dis. f-com.
" rubicaulis,	Persian Galbanum,	see Ferula crubescens.
" Sumbul,*	Sumbul root,	see Euryangium Sumbul.
" tingitana,	Ammoniac of the ancients,	Gu.-re. sti. exp. a-spa.
Festuca elatior,	Fescue grass,	Tp. nut. for.
" fluitans,	Russia grass,	Sd. edi. nut.; as grain.

BOTANICAL.	COMMON.	PROPERTIES, PRODUCTIONS, USES, ETC.
Festuca Myurus,	Capons' tail grass.	Pl. orn.; Ls. for.
" ovina,	Sheeps' fescue grass,	Tp. for.
Ficaria ranunculoides,	Lesser celandine,	see Chelidonium minus.
Ficus carica,	Fig tree,	Ft. nut. lax. dem. mat.; Ls. useful.
" communis,*	" "	see Ficus carica.
" elastica,	E. I. india rubber tree,	Ju. forms caoutchouc.
" Indica,	Banyan tree,	yields Lac; also India rubber.
" Sycamorus,	Sycamore fig tree,	Ft. inferior to Ficus carica.
" religiosa,	Sacred fig,	yields Lac; Sd. ref. alt.; Bk. ton.
Filago Germanica,	Cotton rose, Cudweed,	Pl. ast. vul. dis.
Fillasa suaveolens,*	Sassy bark,	see Erythrophleum.
Flammula Jovis,	Upright virgins' bower,	see Clematis erecta.
Fœniculum aquaticum,*	Fine leav'd water heml'k	see Œnanthe Phellandrium.
" dulce,	Sweet fennel,	Sd. aro. car. pec. diu. sto.
" officinale,	Wild fennel, Large,	" " "
" vulgare,	Common fennel,	see Anethum.
Forsythia suspensa,	Forsythia,	Shrub orn.
Fothergilla alnifolia,	Witch alder,	Bk. and Ls. ast. ton.
Fragaria anserina,*	Silver weed,	see Potentilla anserina.
" Canadensis,	Mountain strawberry,	Ft. edi. diu. a-sco.; Ls. ast. diu.
" elatior,	Hautbois "	Ft. edi. nut. diu.
" grandiflora,	Pineapple "	" " [ton. diu.
" vesca,	Common "	Ft. nut. diu. ref. a-sco.; Ls. ast.
" Virginiana,	Wild "	Ft. edi. diu. ref.; Ls. ast. diu.
" Indica,	Indian "	Ft. edi.
Frasera Carolinensis,*	American columbo,	see Frasera Walteri.
" officinalis,*	" "	" " "
" verticillata,*	" "	" " "
" Walteri,	" "	Rt. bit. ton. a-sep. feb.
Fraxinella Dictamnus,	White fraxinella,	see Dictamnus alba.
Fraxinus acuminatus,*	American white ash,	see Fraxinus Americana.
" Americana,	" " "	Bk. ton cat. diu.; Sd. in obesity.
" apetala,*	Manna ash,	see Fraxinus excelsior.
" Chinensis,	yields China wax,	Vegetable wax.
" excelsior,	European (Manna) ash,	Bk. feb. diu.; Ls. cat.; Sd. acr.
" florifera,*	Manna tree,	see Fraxinus ornus.
" ornus,	" "	yields manna; lax.
" paniculata,*	" ash,	see Fraxinus ornus.
" parviflora,*	Flowering (Manna) ash,	" " "
" polygamic,	Common ash,	Ls. used in gout.
" pubescens,	Red ash,	Bk. ast. ton. a-per.
" quadrangulata,	Blue ash,	Ls. used in gout; Bk. ast. a-per.
" rotundifolia,	Manna tree,	see Fraxinus ornus.
" sambucifolius,	Black ash,	Bk. dis. exa. ast. ton.
Fritillaria imperialis,	Crown imperial,	Fl. orn.; Bu. acr. poi.
" meleagris,	Guinea-hen flower,	Fl. feb.; Bu. acr.
" Persica,	Persian lily,	Fl. orn.
Fuchsia coccinea,	Eardrop,	Pl. and Fl. orn.
" Magellanica,	"	" " "
Fucus Amylaceus,	Jaffna (Ceylon) moss,	see Gracilaria lichenoides.
" crispus,*	Irish moss,	see Chondrus crispus.
" digitatus,	Bladder wrack,	yields kelp; see Fucus versiculosus
" helminthocorton,	Corsican moss,	Pl. ver. ant.
" palmatus,*	Dulse,	see Rhodymenia palmata.
" versiculosus,	Sea (bladder) wrack,	yields kelp; in obesity.
Fumaria bulbosa,*	Round birthwort,	see Corydalis bulbosa.
" cava,*	" "	" " "
" media,*	Fumitory,	see Fumaria officinalis.
" officinalis,	"	Ls. sal. bit. ton. alt. diu. lax. exa.
Fungus,	see Fungus in Pharmacopœial section.	

BOTANICAL.	COMMON.	PROPERTIES, PRODUCTIONS, USES, ETC.
Fungus Rosarum,	Bedeguar, on rose bushes,	Fu. lit. ver. ast.
Funkia ovata,	Blue day-lily,	Pl. and Fl. orn.
" subcordata,	White day-lily,	" "
Gaillardia picta,	Painted Gaillardia,	" "
Galactodendron utile,	Cow-tree,	see Clusia Galactodendron.
Galanthus nivalis,	White snowdrop,	Pl. and Fl. orn.
Galax rotundifolia,	Beetle weed, Carpenters' leaf,	Rt. ast. ; Ls. vul.
Galbanum officinalis,*	Galbanum,	see Bubon Galbaniferum.
Galega apollinea,	Silver leaved senna,	see Tephrosia apollinea.
" officinalis,	Goats' rue,	Pl. aro. sud. ale. feb. gal.
" Persica,*	" "	see Galega officinalis.
" tinctoria,	Galega,	said to yield Indigo.
" toxicaria,*	Fish poison,	see Tephrosia toxicaria.
" Virginiana,*	Hoary pea,	" " Virginiana.
Galeobdolon, luteum,	Yellow archangel,	used as Ballota fœtida.
Galeopsis angustifolia,*	see also Sideritis,	see Galeopsis grandiflora.
" grandiflora,	Hollow tooth herb,	IIb. vul. pec. bit. res.
" ladanum,*	Ironwort,	see Galeopsis grandiflora.
" ochroleuca,	Hemp nettle,	Pl. exp. ; in phthisic.
" tetrahit,	" "	Pl. a-spa. res. det.
" versicolor,	Bees' nettle,	Pl. analogous to G. grandiflora.
Galipea cusparia,*	Angustura,	see Galipea officinalis.
" febrifuga,*	" "	" " "
" officinalis,	" bark,	Bk. stl. ton. cat. feb. a-bil.
Galium album,	Wild madder,	see Galium mollugo.
" aparine,	Cleavers, Goose-grass,	IIb. diu. ape. ref.
" asprellum,	Pointed cleavers,	" "
" circæzans,	Wild liquorice,	IIb. dem. diu.
" crucciatum,	Maywort, Crosswort,	IIb. diu. ; Rt. dyes red.
" luteum,	Yellow cleavers,	see Galium verum.
" mollugo,	Wild madder,	Pl. in epilepsy; Rt. dyes red.
" odoratum,*	Woodroof,	see Asperula odorata.
" palustre,	Marsh cleavers,	IIb. in epilepsy.
" tinctorium,	Wild madder,	a var. of Galium trifidum.
" trifidum,	Small cleavers,	IIb. ner. diu. exp. dia.
" verum,	Yellow cleavers,	IIb. in epilepsy.
Garcinia Cambogia,	Ceylon Gamboge,	Gu.-re. irr. dra. hyd. cat. diu.
" Cochin Chinensis,	Siam Gamboge,	" " "
" mangostana,	Wild mangosteen,	Ft. edi. nut. ; Bk. ast. ; in sore thr't
" morella,	E. I. Gamboge (Ceylon),	Gu.-re. irr. dra. hyd. cat. diu.
" Pictoria,	Mysore Gamboge,	" " "
" purpurea,	Mangosteen,	Kocum butter; for chaps.
Gardenia campanulata,	an East Indian tree,	Ft. cat. ant.
" dumetorum,	Malabar Ipecac,	see Randia.
" florida,	Cape jasmine,	Fl. orn.
" grandiflora,	a Chinese tree,	Ft. yields a yellow dye.
" gummifera,	E. I. species,	yields a fragrant resin like elemi.
Garuleum bipinnatum,	So. African snakeroot,	Rt. pec. diu. dia.
Gastridium australe,	Nit grass,	Pl. orn.
Gaultheria hispidula,	Creeping wintergreen,	Pl. in cancer; scrofula.
" procumbens,	Checkerberry,	Ls. sti. aro. ast. diu. emm.
" repens,*	" Spicy wintergr'n,	see Gaultheria procumbens.
Gaylussacia resinosa,	Huckleberry,	see Vaccinium resinosum.
Gazania splendens,	Gazania,	Pl. and Fl. orn.
Gelidium corneum,	a seaweed yielding Japan isinglass.	
Gelseminum,*	Yellow jessamine,	see Gelsemium.
Gelsemium nitidum,*	" "	" " sempervirens.
" sempervirens,	" "	Rt. feb. a-spa. ner. alt. emm. nar.
Genipa oblongifolia,	Huito,	Ju. a-ven.
Genista canariensis,	Rhodium wd., Rosew'd,	Wd. fra. cor. ; Ol. attractive to fish.

BOTANICAL.	COMMON.	PROPERTIES, PRODUCTIONS, USES, ETC.
Genista scoparia,	Broom herb,	see Cytisus Scoparius.
" scorpius,	Scorpion plant,	Pl. and Fl. orn.
" spinosus,	Furze,	Pl. nau. diu. dia. exa.
" tinctoria,	Dyers' broom,	Tp. and Sd. diu.; dyes yellow.
Gentiana acaulis,	Dwarf gentian,	Rt. bit.; used as a tonic.
" alba,	White gentian,	see Laserpitium latifolium.
" amarella,	Bastard gentian,	Rt. bit. ton. sto.
" Catesbæi,	Blue gentian,	Rt. as Gentiana lutea.
" centaurium,*	European centaury,	see Chironia centaurium.
" Chirayta,*	Chiretta,	see Agathotes Chirayta.
" crinata,	Fringed gentian,	Rt. as Gentiana lutea.
" lutea,	Gentian (officinal),	Rt. ton. bit. sto. eme. ant. a-bil.
" major,*	Gentian,	Rt. see Gentiana lutea.
" ochroleuca,	Sampson snakeroot,	Rt. bit. ton. ant. ast. a-bil. feb.
" quinquefolia,	Gall weed,	Pl. and Rt. bit. ton. cho.
" rubra,*	Gentian,	see Gentiana lutea.
" saponaria,	Soapwort gentian,	Rt. bit.,ton. dia. a-bil.
" veterum,*	Gentian,	see Gentiana lutea.
Geoffræa inermis,*	Cabbage tree, yellow,	see Audira inermis.
" Jamaicensis,*	Bilge water tree,	" " "
" Surinamensis,*	Brown cabbage tree bk.,	" " retusa.
" vermifuga,	Arriba,	Ft. ant.
Geoffroya,	Cabbage tree bark,	see Andira.
Geranium Carolinianum,	Carolina geranium,	see Geranium Robertianum.
" dissectum,	Wood geranium,	Rt. ast. ton. sty. a-sep.
" maculatum,	Spotted cranesbill,	Rt. a powerful ast.; sty. a-sep.
" moschatum,	Musky storksbill,	Pl. ast. det.
" novaboracense,	Spotted cranesbill,	see Geranium maculatum.
" purpureum,*	Herb Robert,	" " Robertianum.
" pusillum,	Small cranesbill,	Rt. ast. ton. sty. det.
" Robertianum,	Herb Robert,	Hb. bit. ast. feb. nep.
" sanguineum,	Blood geranium,	Rt. ast. ton. sty.; in canker.
" sylvaticum,	Doves' foot,	Pl. ast. det.
Gerardia pedicularia,	Bushy gerardia,	Hb. dia. a-sep. sed. feb.
" quercifolia,	Golden oak,	Hb. sti. dia. sed.
Geum caryophyllatum,*	European aveus,	see Geum urbanum.
" nutans,*	Water (Purple) avens,	" " rivale.
" odoratissimum,	Spice root,	" " urbanum.
" rivale,	Water avens, Chocolate root, Rt. ast. ton. feb. sed. sto.
" strictum,*	Herb bennet, Upright avens, see Geum urbanum.
" urbanum,	European avens, Bennet, Rt. ast. ton. sty.
" vernum,	Early water avens,	analogous to Geum rivale.
" Virginianum,	White avens, Chocolate root, used as " "
Gigartina helminthocorton,* Corsican moss,	see Fucus helminthocorton.
" lichenoides,	Ceylon moss,	see Gracilaria lichenoides.
Gigarum Serpentaria,	Common dragon root,	see Arum dracunculus.
Gilia coronopifolia,	Standing cypress,	Pl. and Fl. orn.
Gillenia occidentalis,*	Indian physic,	see Gillenia trifoliata.
" stipalacea,	Bowmans' root,	Bk. of Rt. resembles ipecac.
" trifoliata,	Indian physic,	Bk. of Rt. eme. cat. sud. exp. ton.
Gladiolus communis,	Sword lily, (Round mandrake), Rt. aph. dis.; Fl. orn. [dia.
" luteus,*	False sweet flag,	see Iris pseudo-acorus.
" vulgaris,*	Sword lily, Corn flag,	see Gladiolus communis.
Glaucium flavum,*	Horned poppy,	see Chelidonium glaucum.
" luteum,*	" " "	" " "
Glaux maritima,	Sea milkwort,	Pl. sal.
Glechoma hederacea,*	Ground ivy, Gillrun,	see Nepeta glechoma.
" hirsuta,*	" " "	" " "
Gleditschia triacanthos,	Honey locust,	Sd. and Ls. for.; Sap sac. [gum.
Globularia alypum,	Wild Senna of Europe,	Ls. bit. cat.; Re. called Turbith

BOTANICAL.	COMMON.	PROPERTIES, PRODUCTIONS, USES, ETC.
Gloxinia,	Gloxinia,	Pl. and Fl. orn.
Glyceria Canadensis,	Rattlesnake grass,	Tp. dried; orn.
" distans,	Sea spur grass,	Tp. and Sd. nut. for.
" fluitans,	Manna croup, Russia sd.	Tp. for.; Sd. nut.; birdseed.
" maritima,	Sea meadow grass,	Tp. for. sal.
" nervata,	Meadow spear grass,	Tp. and Sd. for.
" pallida,	Pale manna grass,	" "
Glycyrrhiza echinata,	Wild liquorice of Sicily,	Rt. and Ex. sac. dem. pec.
" glabra,	Common liquorice,	" " "
" glandulifera, var.	Russian liquorice,	" " "
" lepidota,	American liquorice,	similar to Glycyrrhiza glabra.
" typica, var.,	Spanish liquorice,	Rt. and Ex. sac. dem. pec.
Gnaphalium arenarium,	German golden locks,	Tp. sti.; in palsy.
" dioicum,*	Life everlasting,	Pl. and Fl. ast. sty.; in hemorrhage
" luteum album,	Jersey livelong,	Pl. in obstructions and colds.
" Margaritaceum,	Pearl fl'd life everlasting	Fl. ano. pec. ast. ver.
" orientale,	Immortelles,	Fl. orn.
" plantaginifolia,	Mouse ear,	Fl. and Ls. diu. pec.
" polycephalum,	Swt. sctd. life everlast'g	Pl. ast. dia. pec. feb.; Ju. a-aph.
" stœchas,	Eternal flower,	Fl. in obstructions and colds.
" sylvaticum,	Golden motherwort,	Fl. ast. dia.; catarrh colds.
" uliginosum,	Cudweed, Mouse ear,	Pl. sud. sto. muc. diu.
Godetia,*		see Œnothera.
Gomphia Guianensis,	Candle wood.	Bk. ton. sto. a-eme.
Gomphosia chlorantha,	False calisaya bark,	sometimes mixed with Calisaya.
Gomphrena globosa,	Globe amaranth,	Pl. and Fl. ornamental.
Gonolobus hirsutus,	Negro vine,	Rt. dra.-cat.; Ju. nar.
" macrophyllus,	Angle pod,	Rt. eat.; Ju. pol.
Goodyera pubescens,	Adders' violet,	Pl. a-scrofulous; dem. opt. f-com.
" repens,	Net leaf plantain,	a var. of Goodyera pubescens.
Gordonia Hæmatoxylon,	Blood wood,	Ju. yields red extract; ast.
" lasianthus,	Holly-bay,	Bk. ast.; dyes red.
Gossypium album,	Sea Island cotton, (long)	see Gossypium herbaceum.
" Barbadense,	the principle species grown in North America. [olly.
" herbaceum,	Cotton plant (root),	Bk. of Rt. emm. par. diu.; Sd. muc.
" nigrum,	Short staple cotton,	" " " "
" Peruvianum,	the principle species grown in South America.
Gracilaria lichenoides,	Ceylon moss,	Pl. nut. muc.; see Cetraria.
Gratiola aurea,	Hedge hyssop,	Pl. acr. ver. diu. nar.
" officinalis,	" "	Pl. acr. dra. ver. diu. nar.
" Virginica,	Water Jessamine,	see Gratiola officinalis.
Gramen Crucis cyperloidis,	Egyptian cocksf't grass,	Rt. used as Triticum repens.
Gramitis aurea,*	Ceterach leaves,	see Asplenium Ceterach.
Grevillea robusta,	Silk bark oak,	Bk. ast.
Grias cauliflora,	Anchovy pear,	Ft. edl. acl.; foliage ornamental.
Grindelia hirsutula,	Grindelia,	Pl. antidote to Rhus poison.
" robusta,	"	Pl. used in asthma; dem. vul.
Grossularia nigra,	Black currant,	see Ribes nigrum.
" rubra,	Red currant,	" " rubrum.
Guaiacum arboreum,	Lignum vitæ tree,	used as Guaiacum officinalis.
" officinalis,	Guaiac, Lignum vitæ,	Wd. sti.- dia. diu. alt.; Gu.-re. sti.
" sanctum,	" " "	Wd. sti. dia.; Gu.-re. a-rhe.; Bk. acr
Guazuma tomentosa,	Bastard cedar,	Bk. sud. muc.; to clarify sugar.
Guibourtia copallifera,	African pan copal,	Gu.-re. in varnish.
Guilandina Bonducella,*	Indian hazel nut,	see Cæsalpina Bonducella.
" bonduc,	Bonduc nut,	Sd. ton. ast. a-syp. a-per.
" conducella,	Beazor nut,	see Guilandina bonduc.
" Moringa,	Behen nut,	see Moringa pterygosperma.
Guizotia oleifera,	Niger seed,	Sd. pressed for Huts' yellow oil.
Gunnera perpensa,	Marsh marigold lv'd gunnera,	Ls. ton. diu. dem. pec. vul.

BOTANICAL.	COMMON.	PROPERTIES, PRODUCTIONS, USES, ETC.
Gymneura lactiferum,	Ceylon cow-tree,	Ju. milky; nut. edi.
Gymnocladus Canadensis,	Kentucky coffee tree,	Sd. and Ps. eme. ner.; Ls. cat.
Gymnopogon racemosus,	Naked beard grass,	Tp. for.
Gymnopteris Ceterach,*	Ceterach leaves,	see Asplenium Ceterach.
Gynandropsis pentaphylla,	Cleome,	Ju. in earache; dia.
Gynerium argenteum,	Pampas grass,	Pl. ornamental.
Gynocardia odorata,	Chaulmugra,	Sd. alt. ton. her. a-rhe.
Gypsophila struthium.	Spanish soapwort,	Pl. and Rt. det. sap. lit.
Gyromia Virginica,*	Indian cucumber,	see Medeola Virginica.
Habilla Carthagena,	Carthagena bean,	Sd. a-ven.
Habzelia Æthiopica,*	Ethiopian pepper,	see Unona Æthiopica.
Hæmanthus coccineus,	Salmon leav'd blood fl'r,	Bu. diu.; in asthma; Ls. a-sep.
Hæmatoxylon Campechianum,	Logwood,	Wd. a dyestuff; Ex. ast. ton.
Hagenia Abyssynica,*	Koosso,	see Brayera anthelmintica.
Halesia tetraptera,	Snowdrop tree, .	Pl. and Fl. ornamental.
Hamamelis macrophylla,	Big leaf witch hazel,	Bk. and Ls. ast. ton. sed. dis.
" Virginica,	Witch hazel,	" " " f-com. a-phl
Hardenbergia monophylla,	Australian sarsaparilla,	Rt. substituted for sarsaparilla.
Hardwickia pinnata,		Ba. as copaiba; a-syp. vul.
Hebradendron Cambogioides,	Ceylon gamboge,	see Garcinia morella.
Hecatonia palustris,*	Marsh crowfoot,	see Ranunculus sceleratus.
Hedeoma pulegioides,	American pennyroyal,	Hb. sti. dia. emm. car. sud. abo.
Hedera arborea,	English ivy,	see Hedera Helix.
" Helix,	" "	Ls. sti. vul. exa. ins.; Bs. eme. cat.
Hedychium,	Garland flower,	Fl. orn. fra. [sud.
Hedysarum alhagi,	Hebrew manna,	see Alhagi maurorum.
Helenium autumnale,	False sunflower,	Hb. ton. dia. err. feb.; Fl. err.
" parviflorum,	Sneezeweed,	Hb. deleterious to animals.
" tennuiflorum,	"	Hb. " ".
Helianthemum Canadense,	Frostwort,	Hb. alt. aro. ton.; in scrofula.
" corymbosum	Rock rose,	Pl. as Helianthemum Canadense.
" roseum,	Holly rose,	Fl. orn.
" vulgare,*	Com'n garden sunflower,	see Helianthus annuus.
Helianthus annuus,	Garden sunflower,	Ls. and Sd. diu. exp. pec.
" divaricatus,	Rough sunflower,	used as Helianthus annuus.
" giganteus,	Wild sunflower,	" " "
" tuberosus,	Jerusalem artichoke,	Bu. edi. nut. diu.
Helichrysum arenarium,*	German golden locks,	see Gnaphalium arenarium.
" bracteatum,	Immortal flower,	Pl. and Fl. orn.
" nudifolium,	Caffer tea,	Pl. dem. pec.; in catarrh.
" stœchas,*	Eternal flower,	see Gnaphalium stœchas.
Helicteres isora,	Screw tree,	Rt. a-syp.
Heliotropium grandiflorum,	Heliotrope,	Pl. and Fl. orn. fra. per.
" Indicum,	Turnsole,	Pl. bit. vul. a-ven. exa.
" Peruvianum,	Garden heliotrope,	Pl. and Fl. orn. fra. per.
Helleboraster,*	Fetid hellebore,	see Helleborus fœtidus.
Helleborus albus,	White "	see Veratrum album.
" fœtidus,	Stink'g " Bearsfoot,	Ls. eme. pur. ver. poi.
" grandiflorus,*	Black "	see Helleborus niger.
" hyemalis,	Winter "	Rt. as " "
" niger,	Black "	Rt. poi. irr. rub. cat. diu. emm.
" officinalis,*	" "	see Helleborus niger.
" orientalis,	Levant "	properties as Helleborus niger.
" trifolius,*	Goldthread,	see Coptis trifolia.
" viridis,*	Green hellebore,	see Helleborus hyemalis.
Hellenia grandiflora,*	Sweet costus,	see Costus Arabicus.
Helminthochortos officinalis,*	Worm moss, (Corsican),	see Fucus helminthocorton.
Helonias dioica,	False unicorn,	Rt. alt. ton. diu. emm. ver. f-com.
" lutea,*	" "	see Helonias dioica; [uterine tonic.
" officinalis,*	Cevadilla seed,	see Veratrum Sabadilla.

BOTANICAL.	COMMON.	PROPERTIES, PRODUCTIONS, USES, ETC.
Hematoxylon Campechianum,*	Logwood,	see Hœmatoxylon.
Hemerocallis flava,	Day lily,	Rt. cat. ; Ls. ref.; Fl. orn.
Hemidesmus Indicus,	E. I. sarsaparilla,	Rt. a-syp. ton. diu.
Hepatica Americana,	Kidney liverleaf, (wort)	Pl. a mild muc.-ast. hep. pec.
" acutiloba,	Heart "	" " " "
" obtusa,	Kidney "	var. of Hepatica Americana.
" triloba.*	Kidney liverwort,	see Hepatica Americana.
Heptallon graveolens,	Hogwort,	Pl. fœt. sud. cat. a-spa.
Heracleum gummiferum,		said to yield Ammoniac gum.
" lanatum,	Cow parsnip, (Masterwort),	Rt. and Sd. stl. a-spa. car. nar.
" spondylium,	Cow parsley,	Rt. and Sd. ton. sto. car. diu.
Hermodactylus,	Chequer flower,	see Colchicum variegatum.
Hernandia sonora,	Jack-in-a-box,	Ju. of Ls. d-pil.
Herniaria glabra,	Rupturewort,	Pl. sal. ast. diu. opt.
Hesperis alliaria,	-	see Alliaria.
" matronalis,	Dames' violet,	Pl. diu. ; in strangury.
Heuchera Americana,	Alum root,	Rt. a powerful astringent.
" acerifolia,	American sanicle,	Rt. ast. ton. a-sep.
" cortusa,*	Alum root,	see Heuchera Americana.
" Richardsonii,	Cree Indian sanicle,	Rt. ast. vul. sty.
" viscida,*	Alum root,	see Heuchera Americana.
Heudelotia Africana,	African bdellium,	Re. similar to myrrh.
Hevea Guianensis,	India rubber tree,	yields Caoutchouc.
Hibiscus abelmoschus,*	Musk seed,	see Abelmoschus moschatus.
" esculentus,*	Okra, Gombo,	" " esculentus.
" moscheutus,*	Musk seed,	" " moschatus.
" palustris,	Marsh hibiscus,	analogous to Althœa officinalis.
" populeus,	Balimbago,	Ju. as gamboge; Rt. eme.
" rosa-sinensis,	Shoe black plant,	Fl. ast. orn. ; Ju. dyes black.
" sabdariffa,	Guinea sorrel,	Hb. aci. diu. ref.
" Syriacus,	Rose of Sharon,	Pl. and Fl. orn.
" trionum,	Bladder Ketmia,	Pl. muc. dem.
Hicorya alba,	Shagbark, Walnut,	see Carya.
" olivæformis,	Pecan nut,	" " olivæformis.
" porcina,	Pignut, Walnut,	" " porcina.
" sulcata,	Shellbark hickory,	" " sulcata.
" tomentosa,	Mocker nut,	" " tomentosa.
Hieracium lachenalli,	Hawkweed,	see Hieracium murorum.
" murorum,	French lungwort,	Pl. ton.
" oleraceum,	Sow thistle,	see Sonchus oleraceus.
" pilosella,	Mouse bloodwort,	Pl. bit. ast.
" venosum,	Hawkweed, Bloodwort,	Pl. ton. muc. ast.
Hierochloa borealis,	Seneca grass,	Pl. aro. stl. per.
Hippomane mancinella,	Manchineel,	Ft. cau. poi. ; Ju. ves. poi. ; arrow
Hippuris vulgaris,	Mares' tail,	see Equisetum. [poison.
Holchus lanatus,	Velvet grass,	Tp. for.
" saccharatus*	Imphee, Sugar cane,	see Saccharum officinarum.
" serghum,*	Italian millet, Panic grass,	see Setaria Italica.
Homeria collina,	Cape tulip,	Bu. acro-narcotic poison.
Homerocallis flava,*	Day lily,	see Hemerocallis.
Hopea tinctoria,	Sweet leaf,	Bk. of Rt. ton. ; sweetish.
Hordeum causticum,*	Cevadilla seed,	see Veratrum Sabadilla.
" distichon,	Common barley,	Sd. nut. dem. esc. ; malt.
" jubatum,	Squirrel tail grass,	Sd. and Tp. for.
" vulgare,	Scotch barley,	Sd. nut. dem. esc.
Hoteia Japonica,	Hoteia, Astilbe,	Fl. orn. fra.
Hottonia palustris,	Water violet,	Pl. and Fl. orn.
Hoya carnosa,	Wax plant,	Pl. orn.
Humiria floribundum,	Balsam of Umiri,	Ba. said to resemble Ba. Peru.
Humulus Lupulus,	Hop plant, Hops,	Fl. bit. ner. ton. ano. hyp. diu. feb.

BOTANICAL.	COMMON.	PROPERTIES, PRODUCTIONS, USES, ETC.
Hura Brasiliensis,	Assacou,	Ft. acr.-cmc. cat. nar. ant. dia.
" crepitaus,	Sand-box tree,	Ft. acr.-cmc. cat. rub. poi.
Hyacinthus muscari,	Musk grape flower,	Bu. diu. cmc.
" orientalis,	Hyacinth,	Bu. cmc. diu.; Fl. orn.
Hymenanche capensis,	Wolveboon,	sce Hymenanche globosa.
" globosa,	Hyæna poison,	Ft. lu po. used to poison hyænas.
Hydnora Africana,	Jackals' kost,	Pl. fœt.; caten by natives.
Hydrangea arborescens,	Hydrangea, Seven barks,	Rt. diu. lit.; Ls. ton. sia. cat. diu.
" hortensia,	Garden hydrangea,	Pl. orn.
" Japonica,	" "	"
" Thunbergii,	Tea of Heaven,	"
" vulgaris,*	Seven barks,	sce Hydrangea arborescens.
Hydrastis Canadensis,	Goldenseal root,	Rt. ton. a-per. lax. dct. alt. opt.
Hydrocotyle Asiatica,	Indian pennywort,	Pl. aro. nar. diu.; in leprosy.
" centella,	So. African pennywort,	Pl. aut.; in dysentery.
" umbellata,	Water navelwort,	Hb. aro. alc. cmc.
" vulgaris,	Thick leav'd pennywort,	Pl. diu. sud. aph.
Hydropeltis purpurea,*	Water shield,	sce Brasenia hydropelta.
Hydrophyllum verum,*	Goldenseal,	sce Hydrastis Canadensis.
" Virginicum,	Burr flower,	Pl. ast. diu.
Hymenæa courbaril,	So. Am'n locust tree,	yiclds Gum anime; pec.; incense.
" Mozambicensis,*	African copal,	yiclds Copal; in varnish.
" verrucosa,	Madagascar copal,	" " "
Hyosciamus albus,	White henbane,	properties analogous to H. niger.
" niger,	Black "	Hb. nar. ano. a-spa. stl. poi.
" scopolia,	Night "	Hb. and Rt. as Belladonna.
Hyperanthera Moringa,*	Behen nut,	sce Moringa aptera.
Hypericum baccatum,	So. Am'n Gamboge,	sce Gamboge.
" bacciferum,	the bark yields a juice which resembles Gamboge.	
" connatum,	a Brazilian species,	Ls. ast.; in throat complaints.
" coris,	Bastard St. Johuswort,	Ls. diu. a-spa.
" Guianense,	Brazilian Gamboge,	sce Hypericum bacciferum.
" laxiusculum,	Allerin brabo,	Ls. a-ven.
" officinale,*	St. Johnswort,	sce Hypericum perforatum.
" officinarum,*	" "	" " "
" parviflorum,	Low centaury,	Pl. bit. ton. ast.
" perforatum,	St. Johns— "t,	Pl. aro. ast. res. ncr.
" sarothra,	Orange grass, Pine weed	Hb. vul.; in sprains and contusions
" Virginicum,*	St. Johnswort,	sce Hypericum perforatum.
" vulgare,*	" "	" " "
Hypæne thebaica,	Gingerbread tree, Doum palm,	Ft. edi. insipid, nut. feb. ref.
Hypogon anisatum,	Anise root,	Rt. aro. fra. diu. car. feb.; Ls. as tea
Hypopitys lanuginosa,	False beech drops,	Rt. ncr. aph.
Hypoxis erecta,	Star grass,	Rt. edi. dct. vul.; in ague.
Hyssopus officinalis,	Hyssop,	Hb. aro. sti. sud. pec.
Iberis amara,	Bitter candytuft,	Pl. bit. a-sco. a-rhe.; Sd. acr. nar.
" bursa-pastoris,*	Shepherds' purse,	sce Capsella bursa-pastoris.
" coronaria,	Rocket candytuft,	Pl. and Fl. orn. a-sco.
" sophia,*	Cuckoo flower,	sce Cardamine pratensis.
" umbellata,	Purple candytuft,	Pl. and Fl. orn.
Icica altissima,	Curanna wood,	Wd. fra. bit.
" caranna,	Carauna,	" " sce Amyris.
" Guianensis,*	Spanish cedar,	" " Gu. fra. vul.
" heptaphylla,*	So. Am'n clemi (cedar),	sce Amyris ambrosiaca.
" icicariba,	Brazilian clemi,	" " clemifera.
Ictodes fœtidus,	Skunk cabbage,	Pl. fœt. acr.; Rt. exp. sud. a-spa.
" "	" "	Ft. and Sd. pec.; in cough. [pec.
Ignatia amara,	St. Ignatius' bean,	sce Strychnos Ignatia.
Ilex Aquifolium,	Common European holly	Ls. ton. a-sep. ast. cmc. feb.
" Canadense,	Canadian holly,	Bk. Ls. and Ls. ton. exp. lax. feb.

BOTANICAL.	COMMON.	PROPERTIES, PRODUCTIONS, USES, ETC.
Ilex cassina,*	Yaupou holly,	see Ilex vomitoria.
" dahoon,*	" "	" " "
" glaber,	Ink berry,	see Prinos glaber.
" major,	Spanish bellotas,	Bs. ast. muc. pec.
" mate,*	Paraguay tea,	see Ilex Paraguaiensis.
" opaca,	American holly,	Ls. ton. a-sep. ast. emc. feb.
" Paraguensis,	Paraguay tea,	see Ilex Paraguaiensis.
" Paraguaiensis,	" "	Ls. used as tea; ast. ton. bit.
" verticillatus,	Black alder,	see Prinos verticillatus.
" vomitoria,	South sea tea,	Ls. emc. aro. sto. sti. exp. diu.
Illicium anisatum,	Star anise seed,	Sd. yields oil anise; aro. sti. car.
" Floridanum,	Florida anise tree,	Bk. bit. pun. ; Sd. aro. ; Ls. aro. poi.
" parviflorum,	Small flow'd anise tree,	resembles Sassafras.
" religiosum,	Sacred anise tree,	Bk. aro. fra. ; as incense.
Imbricaria saxatilis,	Skull moss,	see Lichen saxatilis.
Impatiens balsamina,	Balsam weed,	Pl. diu. emc. alt. cat.
" fulva,	Speckled Jewel weed,	Pl. diu. emc. alt. ; Ju. acr.
" nolime tangere,	Europ'n Yellow balsam,	Pl. eme. cat. diu. ; Ju. acr.
" pallida,	Touch-me-not,	Pl. diu. emc. alt.
Imperatoria ostruthium,	Masterwort,	Rt. and Sd. aro. sti. cor. emm. dia.
Indigofera anil,	Indigo plant,	Rt. and Ls. hep. ; yields indigo.
" argentea,	" "	yields Egyptian indigo; poi.
" Caroliniana,	" "	Pl. ins. vul. a-epi.
" tinctoria,	E. I. indigo plant,	yields indigo; poi. ner. ton. a-spa.
Inga feuillei,	Peruvian Pacay,	Ps. pulp eaten; Ju. ast.
" spectabilis,	Guavo real,	" "
" vera,	Igna bark tree,	" cat. sac. ; Ju. ast.
Inula Britannica,*	Small fleabane,	see Inula dysenterica.
" conyza,*	Plowmans' spikenard,	" " "
" dysenterica,	Small fleabane,	Rt. ton. aro. acr. ; in dysentery.
" Helenium,	Elecampane,	Rt. exp. emo. ton. diu. dia.
" oculus Christi,*	Christ's eye,	Sd. dem. muc. opt.
" squarrosa,*	Marsh fleabane,	see Conyza squarrosa.
Ioannesia principis,*	Anda seed,	see Anda Brasiliensis.
Ionidium Ipecacuanha,	White Ipecac,	Rt. emc. dia. exp. sto.
" marcucci,	Cuichunchulli	Rt. emc. cat. ; in leprosy.
" microphyllum,*	"	see Ionidium marcucci.
" parviflorum,	"	Rt. dia. diu. emc. sia.
Ipomœa Batatas,	Sweet potato,	see also Convolvulus Batatus.
" Batatoides,	Jalap root,	see Ipomœa orizabensis.
" Convolvulus,	Morning glory,	Pl. and Fl. orn.
" Jalapa,	Jalap root,	Rt. hyd.-cat. diu. ; in dropsy.
" macrorhiza,*	Mechoacan root,	see Convolvulus mechoacanna.
" Nil,*	Blue morning glory,	Pl. and Fl. orn. ; Rt. cat.
" operculata,*	Brazilian purge,	see Convolvulus operculatus.
" orizabensis,	Male (false) Jalap,	Rt. cat. diu.
" purga,*	Jalap root,	see Ipomœa Jalapa.
" purpurea,	Morning glory,	Pl. and Fl. orn. ; Rt. cat.
" quamoclit,	Cypress vine,	Pl. and Fl. oru. pur. sti. eep.
" Schiedeana,*	Jalap root,	see Ipomœa Jalapa.
" simulans,	Tampico Jalap,	Rt. dra. diu.
" turpethum,*	Turpeth root,	see Convolvulus turpethum.
Iridœa edulis,	Red dulse,	a seaweed; edi. ant. sal.
Iris Florentina,	Florentine orris,	Rt. (Rh.) acr. fra. exp. ; in denti-
" fœtidissima,	Stinking gladwine,	Rt. ste. hyd. a-spa. nar. [frice.
" Germanica,	Common orris,	Rt. acr. nau. dra. ; in dropsy.
" hexagona,*	Blue flag,	see Iris versicolor.
" lacustris,	Dwarf lake iris,	analogous to Iris versicolor.
" lutea,*	False sweet flag,	see Iris pseudo-acorus.
" nostras,*	Common orris,	" " Germanica.

BOTANICAL.	COMMON.	PROPERTIES, PRODUCTIONS, USES, ETC.
Iris pallida,	Florentine orris (pale),	Rt. as Iris Florentina.
" palustris,*	Yellow water flag,	see Iris pseudo-acorus.
" pseudo-acorus,	False sweet flag,	Rt. acr. sty. err. sia. ast. diu.
" tuberosa,	European orris,	Rt. incisive; pur.
" versicolor,	Blue flag,	Rt. alt. res. sia. lax. diu. a-syp. ver.
" verna,	Slender blue flag,	analogous to Iris versicolor.
" Virginica,	" " "	" " "
" vulgaris,*	Common orris,	see Iris Germanica.
Isatis tinctoria,	Woad,	yields a blue dyestuff; ast. des.
Isœtea lacustris,	Quillwort,	Pl. and Sd. curious; orn.
Isnardia palustris,	Phthisic weed,	see Ludwigia palustris.
Isonandra gutta,	Gutta percha tree,	Ju. gutta percha; as bandages.
Ixia Chinensis,	Blackberry lily,	Fl. orn.; Bu. dra. diu.
Jacaranda caroba,	Caroba bark,	Wd. green ebony; in dyeing.
Jacea tricolor,*	Pansy, ·	see Viola tricolor.
Jacobœa obovata,*	Old Robert's herb,	see Senecio Jacobœa.
Jalappa alba,	Mechoacan root,	see Convolvulus mechoacanna.
Jambosa Malaccensis,	Malay apple,	Ft. esc. nut. ref.
" vulgaris,*	Rose apple,	see Eugenia Jambos.
Janipha Manihot,	Cassava plant,	yields Tapioca; see Jatropha.
" stimulans,	Sand nettle,	Ju. acr. irr.; Sd. pur.; Rt. pol.
Jasminum Arabicum,	Coffee,	see Coffea Arabica.
" fruticans,	Shrubby jasmine,	Pl. and Fl. fra. orn.
" grandiflorum,	Large flowered jasmine,	yields oil jasmine; fra. per.
" odoratissimum,	Sweet yellow jasmine,	Fl. orn. per.
" offlcinale,	Common white jasmine,	Ol. fra. bit. a-rhe.; anti-paralytic.
" sambac,	Jasmine,	" " "
Jateorrhiza Calumba,*	Columbo root,	see Cocculus palmatus.
" palmata,*	" "	" " "
Jatropha curcas,	Physic nuts,	Nu. and Ol. pur. eme.
" elastica,*	yields Surinam caoutchouc, see Hevea Guianensis.	
" gossypifolia,	Wild cassada,	Ls. cat. err.; Sd. eaten by poultry.
" manihot,	Cassava plant,	Ju. pol. acr.; Fecula of Rt. tapioca.
" multifida,	French physic nuts,	Nu. and Ol. dra.
" urens,	Stinging cassada,	Pl. irritant-pol.; stinging.
Jeffersonia Bartoni,*	Twin-leaf root,	see Jeffersonia diphylla.
" diphylla,	" "	Rt. diu. alt. a-rhe. a-syp.
" diphyllum,*	" "	see Jeffersonia diphylla.
" lobata,	" "	analogous to Jeffersonia diphylla.
" odorata,	" "	" " "
Jenkinsonia anti-dysenterica,* a So. African plant,		see Pelargonium anti-dysenterica.
Juglans alba,*	Shagbark, Walnut,	see Carya alba.
" cathartica,*	Butternut, ·	see Juglans cinerea. [esc.
" cinerea,	"	Bk. cat. alt. ton. ant. cho. ast.; Nu.
" nigra,	Black walnut,	Bk. ast.; Ju. her.: Nu. edi. oily.
" fraxinea,	Ash walnut,	Bk. cat. alt.; Nu. esc.
" rigia,	English (French) walnut	Nu. edi.; Ft.-rind a-syp.; Ol. a-syp.
" tomentosa,*	Mocker nut,	see Carya tomentosa.
Juncus bufonis,	Toad grass,	Pl. cat. diu.
" effusus,	Bulrush,	Pl. cat.; mildly-ast.
" odoratus,*	Ginger grass, Sw't rush,	see Andropogon schœnanthus.
Juniperus Bermudiana,	Jamaica red cedar,	Bk. fra. aro.
" communis,	Juniper,	Bk. and Ls. aro.; Bs. diu.; in gin as
" depressa,	"	see Juniperus communis. [flavor.
" lycia,	Arabian Olibanum,	Gu.-re. fra. aro. bit. ast. fum.
" oxycedrus,	Berry bearing cedar,	yields oil of cade; exa. aut.
" prostrata,	Dwarf red cedar,	analogous to Juniperus Virginiana.
" Sabina,	Savin,	Ls. sti. diu. emm. dia. pol. irr.
" Virginiana,	Red (pencil) cedar,	" " Excrescences ant.
Justicia adhatoda,	Malabar nut tree,	Ls. abo. pur.; Rt. a-spa. bit.

BOTANICAL.	COMMON.	PROPERTIES, PRODUCTIONS, USES, ETC.
Justicia cebolium,		Rt. and Ls. lit. diu.
" nasuta,	Braid root,	Rt. diu. aph. her. ale. vul.
" hyssopifolia,	Snap tree,	Pl. diu. vul.
" pectoralis,	Garden balsam,	Pl. ast. pec. vul. fra.
" pedunculosa,	Water willow,	Pl. bit. ton. sto.
Kæmpferia Galanga,	Greater galangal,	see Alpina Galanga.
" rotunda,*	Zerumbet,	see Curcuma Zerumbet.
Kalmia angustifolia,	Nar'w lv'd (sheep) laurel	Ls. nar.-poi. alt. a-syp. sed. ast. err.
" " var. ovata,	Dwarf (sheep) laurel,	" " " poison to sheep
" glauca,	Swamp laurel,	" " err. a-syp. her.
" " v. rosmarinifolia	" "	" " " "
" latifolia,	Br'd lv'd (Mount.) laurel	" " " sed. ast.
Kœleria cristata,	Crested hair grass,	Pl. for.
Knowltonia vesicatoria,	So. African buttercup,	Ls. blister; in rheumatism; ves.
Koniga maritima,	Sweet alyssum,	Pl. orn.
Krameria argentea,	Para rhatany,	Rt. ast. ton. bit. sty.
" ixina,	Savanilla (W. I.) rhatany,	Rt. " "
" lanceolata,	American rhatany,	Rt. ast.-ton.
" triandra,	Rhatany,	Rt. ton. ast. sty. diu. det.
Krigia Virginica,	Dwarf dandelion,	see Taraxacum Dens-leonis.
Kuhnia eupatorides,	False boneset,	Pl. bit. ton. dia.
Lacnanthes tinctoria,	Spirit weed, (Red root),	Hb. stl. exc.; see Belladonna.
Lactuca Canadensis,*	Wild lettuce,	see Lactuca elongata.
" elongata,	Wild opium lettuce,	Ls. and Ju. nar. sed. ano. hyp. diu.
" graveolens,*	Strong scented lettuce,	see Lactuca virosa. [dia.
" marina,*	Sea (Bladder) wrack,	see Fucus vesiculosus.
" sanguinea,	Wood (red) lettuce,	Ls. nar. ano. hyp.
" sativa,	Garden lettuce,	Ls. sad.; Ju. ano. hyp.
" scariola,*	Wild lettuce,	see Lactuca elongata.
" sylvestris,*	" "	" " "
" virosa,	Acrid lettuce,	Ju. is German lactucarium.
Lagenaria vulgaris,	Bottle gourd,	see Cucurbita Lagenaria.
Lagetta lintearia,	Lace bark tree,	Bk. properties like mezereon.
Lagœcia cuminoides,	Wild cummin,	Ft. car. sto. aro.
Lagurus ovata,	Hares'-tail grass,	Pl. for.
Laminaria bulbosa,	Bulbous laminaria,	Pl. yields iodine.
" esculenta,	Bladder locks,	Pl. eaten as food.
" digitata,	Sea tangles,	Pl. sac. edi.; yields kelp.
" saccharina,	Devil's apron,	Pl. sac. edi. sal.; yields kelp.
Lamium amplexicaule,	Dead nettle, Henbit,	Pl. stl. sud. lax. a-rhe. cep.
" album,	Blind nettle,	Pl. f-com. sty.
" foliosum,*	" "	see Laminum album.
" gurganicum,	Dead nettle,	" " amplexicaule.
Lansium domestica,	Ayer ayer,	Ft. esc. ref.; agreeable.
Lantana camara,	Bahama tea, Sage tea,	Ls. used as tea; diu. sed.
" pseudo-thera,	Brazil tea,	" "
Lapathum acutum,*	variety of Yellow dock,	see Rumex acutus.
" aquaticum,*	Great water dock,	" " aquaticus.
" hortense,*	Garden patience, Dock,	" " patientia.
" orientale,*	Rhubarb,	see Rheum.
" pratense,*	Common English sorrel,	see Rumex acetosa.
" sanguineum,*	Bloody dock (Olcott) rt.	" " sanguineus.
" sylvestre,*	Sharp dock,	" " acutus.
Lappa bardana,*	Burdock plant,	see Arctium Lappa.
" major,*	" "	" " "
" minor,	Burdock plant,	see Arctium Lappa.
" officinalis,	" "	" " "
" racemosa,*	" grass.	
Lapsana communis,	Nipplewort,	Pl. bit.; externally to sore nipples.
Larix Americana,	Am'n larch, Tamarack,	Bk. lax. ton. diu. alt.; Gu.-re. c-irr.

BOTANICAL.	COMMON.	PROPERTIES, PRODUCTIONS, USES, ETC.
Larix communis,*	European larch,	see Larix Europæa.
" Cedrus,*	Cedar of Lebanon,	see Cedrus Libani.
" decidua,*	European larch,	yields Venice turpentine ; see Abies
" Europæa,	" "	see Abies Larix.
" pyramidalis,	" "	" " "
Larrea Mexicana,	Creosote plant,	Pl. resinous ; a-rhe. a-syp.
Laserpitium asperum,*	White gentian,	see Laserpitum latifolium.
glabrum,*	" "	" " "
" latifolium,	" "	Rt. bit. ton.
" montanum,*		see Laserpitium siler.
" siler,	Heartwort,	Sd. and Rt. aro.
" trifoliatum,	"	see Laserpitium siler.
Lasiagrostis,*	Wooly grass,	see Stipa.
Lathyrus odoratus,	Sweet pea,	Pl. and Fl. orn.
" maritimus,	Beach pea,	Pl. and Sd. as poultices.
Laurelia sempervirens,	Peruvian nutmeg,	Sd. used as spice.
Laurencia obtusa,	Corsican moss,	Pl. Algæ ; anat.
" pinnatifida,	Pepper dulse,	Seaweed, aro. aer. sad.
Lauro cerasus,*		see Prunus laurus Cerasus.
Laurus Æstivalis,*	Spice (Fever) bush,	see Laurus Benzoin.
" alba,	White sassafras,	Bk. resembles sassafras.
" Benzoin,	Spice (Fever) bush,	Tw. aro. ver. feb. a-per. ; Ft. spicy
" Camphora,	Camphor tree, (Laurel),	see Camphora officinarum.
" Caryophyllata,	Clove bark,	Bk. anal. to Cinnamomum culilawan
" Cassia,	Cassia bark,	Bk. and Ls. as cinnamon ; Fl. cassia
" Cinnamomoides,*	Santa Fe cinnamon,	see Nectandra Cinnamomoides. [bd
" Cinnamomum,	Chinese cinnamon, (bd.)	see Cinnamomum Loureirii.
" Cubeba,	Cubebs,	see Piper Cubeba.
" culilawan,	Culilawan (Clove) bark,	see Cinnamomum culilawan.
" cupularis,	Isle of France cinnamon	Bk. aro. ast. ; Wd. strong scented.
" Indicus,	Indian bay,	Bk. aro. sto.
" Malabathrum,	Java cinnamon,	Bk. bit. aro. a-spa. par. ; Ls. yield
" myrrh,	Chinese cinnamon, (bd.)	see Cinnamomum aromaticum. [oil
" nobilis,	Bay tree,	Ls. and Ft. aro. fra. ast. sto. car.
" persea,	Avocado pear,	see Persea gratissima.
" pichurim,	Pichurim (Sassafras) nuts,	see Nectandra puchury.
" Quixos,	Peruvian cinnamon,	Bk. aro. ast. fra.
" Sassafras,*	Sassafras,	see Sassafras officinale.
" Winterana,*	Winter's bark,	see Wintera (Drimys) aromatica.
Lavatera arborea,	Tree mallow,	Pl. muc. ; yields fibre.
Lavandula angustifolia,*	Lavender,	see Lavandula vera.
" latifolia,*	Spike lavender,	" " spica.
" officinalis,*	Lavender,	" " vera.
" spica,	Spike lavender,	Hb. yields Oil spike ; rub. ; in var-
" Stœchas,	Arabian lavender,	Ls. and Fl. pec. [nish.
" vera,	Com'n garden lavender,	Pl. fra. aro. err. sti. ; Ol. fra. per.
Lawsonia alba,	Gopher wood,	see Lawsonia inermis.
" inermis,	Henne plant,	Rt. ast. ; Ju. orange dyestuff.
" spinosa,	East Indian plant,	Rt. in lepra ; exa.
Lecanora parella,	Archel, Litmus,	used in dyeing purple.
" tartarea,	Tartarean moss, (Cudbear),	" " "
Lechea major,	Greater Pinweed,	Pl. ton. feb. a-per.
Lecontia Virginica,*	Arrow arum,	see Peltandra Virginica.
Lecythis zabucajo,	Zabucajo Brazil nuts,	Ft. edible, nut. ; oily.
Ledum latifolium,	Labrador tea,	Ls. pec. ton. ; as tea.
" palustre,	Marsh tea,	Ls. bit. sub-ast. vul. ins.
" Rosmarinus,*	" "	see Ledum palustre.
Leersia oryzoides,	Rice cut-grass,	Pl. for. ; Sd. nut.
Leiophyllum buxifolium,	Sleek leaf,	
Lemna minor,	Duckweed,	Pl. dem. ; used in poultices.

BOTANICAL.	COMMON.	PROPERTIES, PRODUCTIONS, USES, ETC.
Leontice leontopetalum,	Black turnip,	Rt. sto. sap.
" thalictroides,*	Blue cohosh,	see Caulophyllum thalictroides.
Leonotis Leonurus,	a So. African plant,	Ls. nau. nar. cat. emm.
" ovata,	Male crow parsnip,	Ls. smoked as tobacco; see L. Leo-
Leontodon autumnale,	Fall dandelion,	Pl. lax. diu. hep. opt. [nurus.
" palustris,	Marsh dandelion,	Rt. " "
" Taraxacum,*	Dandelion,	see Taraxacum Dens-leonis.
Lepidium campestre,	Bastard cress,	Pl. sad.; Sd. acr. ast.
" piscidium,	Fish poison,	Pl. pun.; Sd. said to poison fish.
" sativum,	Pepper cress, Pepper grass,	Pl. sad.; Sd. a-sco. sti. ape.
" Virginica,	Wild pepper cress,	Ls. pun. diu. alt. a-sco.
Lepidotis clavata,*	Club moss,	see Lycopodium clavatum.
Leptandra purpurea,	Culver's root, Black root	see Leptandra Virginica.
" villosa,	" " " "	" " "
" Virginica,	" " " "	Rt. cat. alt. cho. ton. hep.
Leptanthus palustris,	Water star-grass,	Pl.
Leptochloa,	Slender grass,	Tp. for.
Leptospermum laulgerum,	Australian tea,	Ls. used as tea.
" scoparium,	New Zealand tea,	" " aro. fragrant.
Lespedeza sessilifolia,	Bush clover,	Ls. alt. diu.
Leucanthemum,	Chamomile,	see Anthemis.
" vulgare,	White wd., Oxeye daisy,	Ls. and Fl. acr.
Leucoium luteum,*	Wall flower,	see Cheiranthus Cheiri.
Levisticum officinalis,*	Lovage,	see Ligusticum levisticum.
Leysera gnaphaloides,	a So. African plant,	Pl. emo. pec.; in catarrh.
Liatris odoratissima,	Vanilla leaf,	Ls. aro. per. ton. sti. dia.
" scariosa,	Rattlesnakes' master,	Rt. diu. stl. ton. a-ven.
" spicata,	Button snakeroot,	Rt. " sti. dia. ton. emm.; Bright's
" squarrosa,	Blazing star,	Rt. " ton. alt. a-syp. [disease.
Libanotis annua,	Candy carrot,	see Athamanta Cretensis.
" coronaria,*	Rosemary,	see Rosmarinus.
" Cretensis,*	Candy carrot,	see Athamanta Cretensis.
" hirsuta,*	" "	" " "
" vulgaris,	Black gentian,	Rt. dia. diu. lit.
Licaria Guianensis,*	Brazilian clove bark,	see Laurus caryophyllata.
Lichen arboreum,*	Lung moss,	see Sticta pulmonaria.
" caninus,	Ash col'd ground liverwort,	Pl. in mania and asthma.
" carragheen,*	Irish moss,	see Chondrus crispus.
" cocciferus,	Scarlet-cup lichen,	Pl. ast. pec. feb.; in whoop'g cough
" Floridus hirtus,*	Lapland moss,	see Lichen plicatus.
" Islandicus,*	Iceland moss,	see Cetraria Islandica.
" laciniatus,*	Skull moss,	see Lichen saxatilis.
" parietinus,*	Trumpet moss,	see Cladonia pyxidata.
" plicatus,	Lapland moss,	Pl. ast. sty. vul.
" pulmonarius,*	Lung moss, Tree lungwort,	see Sticta pulmonaria.
" pyxidatus,	Trumpet moss,	Pl. ast. pec. feb.; in whoop'g cough
" rangiferinus,*	Reindeer moss,	Pl. nut. fra.; in scent bags.
" reticulatus,*	Tree lungwort (moss),	see Sticta pulmonaria.
" roccella,	Archel, Litmus plant,	yields a blue test dye; pec.
" saxatilis,	Skull moss,	in affections of the head.
" spurious,*	Ash col'd ground liverwort,	see Lichen caninus.
" stellatus,*	Woodrow, Liverwort,	see Marchantia polymorpha.
" tartareus,	Tartarean moss, Cudbear,	see Lecanora tartarea.
" tinctorius,*	Skull moss,	see Lichen saxatalis.
Ligusticum actæifolium,	American lovage,	Rt. aro. car. sti. sto.
" capillaceum,*	Baldmoney, Spignel,	Rt. aro. car. sto.; gummy.
" cornubiense,	Cornish lovage,	Rt. exudes a sti. resin.
" fœniculum,	Fennel plant,	see Fœniculum and Anethum.
" levisticum,	Lovage,	Pl. sti. car. emm. sto. aro. f-com.
" meum,*	Baldmoney, Spignel,	see Meum athamanticum.

BOTANICAL.	COMMON.	PROPERTIES, PRODUCTIONS, USES, ETC.
Ligusticum phellandrium*	Fine lv'd water hemlock,	see Œnanthe Phellandrium.
" podograria,	Gout weed,	Pl. used in gout; Ls. sad.
Ligustrum Ægyptiacum,*	Henne plant,	see Lawsonia inermis.
" vulgare, ᴠ	Privet, Prim,	Ls. ast. bit.; Fl. fra.; Ft. cat.
Lilaca vulgaris,	Lilac,	see Syringia vulgaris.
" " var. alba,	Syringia,	" " "
Lilium bulbiferum,	Tiger lily,	Rt. cat.; Ls. ref.; Fl. orn.
" Canadensis,	Nodding lily,	Rt. cat. mat.; Fl. fra. orn.
" candidum,	Common white lily,	Fl. emo.; in oil; Bu. mat.
" convallium,	Lily of the valley,	see Convallaria majalis.
" lancifolium,	Japan lily,	Fl. orn.; fragrant.
" martigon,	Turks' cap lily,	" Rt. diu. cmm.
" tigrinum,	Spotted tiger lily,	" Rt. mat.
Limonium,*	Thrift,	see Statice.
" malum,*	Citron, Cedrat,	see Citrus medica.
Limosella aquatica,	Mudwort,	Pl. and Fl. ref. mat.
" subularia,	"	" " " "
Linaria cymbalaria,	Ivy leaved toad flax,	Pl. a-sco.; in diabetes.
" vulgaris,*	Toad flax, Snap dragon,	see Antirrhinum Linaria.
Lindera Benzoin,*	Spice bush, Fever bush,	see Laurus Benzoin.
Linnæa borealis, ·	Twin flower,	Pl. bit. sub-ast. a-rhe.
Linosyrus vulgaris,	Goldy locks,	Pl. and Fl. orn. ant. deo.
Linum arvense,*	Common flax, Linseed,	see Linum usitatissimum.
" catharticum,	Purging flax,	Pl. bit. cat.
" perenne,	Perennial flax,	Pl. as Linum usitatissimum.
" usitatissimum,·	Common flax, Flaxseed,	Sd. pec. dem. muc. mat.
" "	" "	Pl. yields linen fibre; Sd. yield oil
" Virginicum,	Wild flax,	Pl. and Sd. as common flax. [lins'd.
Lippia citriodora,*	Sweet verbena,	see Verbena triphylla.
Liquidambar altinga,*	Rasamala,	see Altinga excelsa.
" asplenifolia,*	Sweet fern,	see Comptonia asplenifolia.
" imberbis,*	" gum,	see Liquidambar styraciflua.
" officinalis,*	Styrax,	see Styrax officinalis.
" orientale,	Oriental sweet gum,	see Styrax.
" styraciflua,	Styrax sweet gum,	Bk. ner.; Fl. aro.; Gu. in ointm'ts.
Liriodendron Tulipifera,	White wood, Tulip tree,	Bk. bit. aro. sti.-ton. feb. a-per. ver.
Lithocarpus Javensis,	Stone oak,	Bk. ast. sty.
Lithospermum arvense,	" seed,	Sd. diu. lit.; Pl. dyes red.
" officinale,	Gromwell,	Sd. supposed to be lit. diu.
" tinctorium,*	Alkanet root,	see Anchusa tinctoria.
" villosum,*	" "	" " " "
Litsæa Cubeba,	Cubeb,	see Piper Cubeba.
Littorella lacustris,	Shore weed,	Pl. ref. mat.
Loasa lateritia,	Stinging loasa,	Glands sting as nettles.
Lobaria Islandica,*	Iceland moss,	see Cetraria Islandica.
" pulmonaria,*	Lung moss,	see Sticta pulmonaria.
Lobelia cardinalis,	Red lobelia, Red cardinal	Pl. orn. ant. a-spa. ner.
" coccinea,*	" " "	see Lobelia cardinalis.
" fulgens,	Fulgent lobelia,	Pl. and Fl. orn.
" inflata, ·	Indian tobacco, Lobelia,	Pl. emc. dia. exp. a-spa. ner. diu.
" pinifolia,	So. African lobelia,	Pl. exc. dia. a-rhe. exa. [res.
" spicata,	Spiked lobelia,	Pl. diu.; Fl. orn.
" syphilitica,	Blue cardinal (lobelia),	Pl. emc. a-syp. diu. cat.
" urens,	Acrid lobelia,	Pl. acr.; reputed poisonous.
Loiseleuria procumbens,	Creeping azalea,	Pl. ast.
Lolium annuum,	Darnel,	see Lolium tremulentum.
" Italicum,	Italian rye grass,	Pl. cultivated forage plant.
" perenne,	Perennial rye grass,	Pl. " " "
" tremulentum,	Darnel, Tare,	Sd. poi. emc. nau.
Lonicera brachypoda,	a Japanese plant,	Ls. diu.; in infusion.

BOTANICAL.	COMMON.	PROPERTIES, PRODUCTIONS, USES, ETC.
Lonicera caprifolium,	Honeysuckle,	Fl. orn.; anti-asthmatic.
" cœrulea,	Blue honeysuckle,	" "
" Diervilla,*	Bush honeysuckle,	see Diervilla Canadense.
" Germanica,*	Woodbine, ·	see Lonicera periclymenum.
" hirsuta,	Rough woodbine,	Pl. and Fl. bit. ast. muc.
" Marilandica,*	Pink root,	see Spigelia Marilandica.
" periclymenum,	Woodbine, Honeysuckle,	Pl. ast. ton.; in gargles.
" sempervirens,	Trumpet honeysuckle,	Ls. in asthma.
Loranthus Europæus,	Oak misletoe,	Ft. pur.; yields bird-lime; Ls.
Lotus corniculata,	Birdsfoot trefoil,	Pl. and Fl. orn. [a-cpi.
" hirsutus,	Pile lotus,	Sd. used in piles.
" sylvestris,*	Blue melilot,	see Melilotus cœrulea.
" Virginiana,*	Persimmon,	see Diospyros Virginiana.
Lucuma mammosum,	Marmalade apple,	Sd. bit.; Ft.-pulp edible, ref.
Ludwigia alternifolia,	Seed box,	Pl. pec. '
" palustris,	Phthisic weed,	Pl. pec.; in asthma and cough.
Luffa acutangula,	of India,	Sd. yield an oil; cmc. cat.
Lunaria biennis,	Money flower,	Pl. and Fl. orn. diu.
" rediviva,	Satin flower, Honesty,	Rt. detersive; Ls. diu.; Sd. acr.
Lupinus albus,	White lupine,	Sd. edible, ant.; as a cataplasm.
" luteus,	Yellow "	" " " "
" perennis,	Wild "	Sd. bit. nut.
Lupulus communis,*	Hop plant,	see Humulus Lupulus.
Luzula campestris,	Cuckoo grass,	Pl. orn.
Lychnis chalcedonica,	Maltese cross,	Pl. and Fl. orn.
" coronaria,	Mullein pink,	- " " ·
" dioica,	White soapwort,	Rt. and Ls. sap. det.
" diurna,*	" "	see Lychnis dioica.
" flos-Jovis,	Flower of Jove,	Pl. and Fl. orn.
" flos-cuculi,	Ragged robin,	" "
" githago,	Corn cockle,	Sd. acr. dep. pur.; Rt. vul. ast.
" officinalis,*	Soapwort,	see Saponaria.
" vespertina,*	White soapwort,	see Lychnis dioica.
Lycium barbarum,	Matrimony vine,	Pl. and Fl. orn.; Rt. diu.; Ls. sad.
Lycoperdon bovistis,	Puff-ball,	Fu. a mechanical styptic.
. " cervinum,	"	Fu. sty. nar.; fumes stupefying.
" globosum,	"	" " " "
" proteus,	"	Fu. fumes anæsthetic.
" solidum,	Indian head, Tuckaho,	see Sclerotium giganteum.
" tuber,	Truffle,	Fu. aph.; an aliment.
Lycopersicum esculentum,*		see Solanum lycopersicum.
Lycopodium clavatum,	Club moss, Lycopodium,	Pollen in matted hair.
" "	" " "	Pollen in excoriations in infants.
" complanatum,	Ground pine,	Pl. cmc. pur. abo. ins.
" lucidulum,	Moon fruit pine,	Pl. " "
" officinale,*	Lycopodium moss,	see Lycopodium clavatum.
" rupestre,	Festoon pine,	Pl. used for decoration.
" selago,	Fir club moss,	Pl. cmc. cat. abo. ins. nar.
Lycopsis arvensis,	Wild bugloss,	· Pl. pec.
" vesicularia,	Creeping bugloss,	"
Lycopus Europæus,	Bitter bugle weed,	Pl. pec.; in hemorrhage of lungs.
" sinuatus,	Paul's betony, Bitter bugle,	Pl. pec.; in hemorrhage.
" Virginicus,	Sweet bugle weed,	Pl. ton. sed. ast.
Lygodium palmatum,	Hartford (Climbing) fern,	Pl. ornamental.
Lyperia crocea,	African saffron,	similar to saffron, Crocus.
Lysimachia nemorum,	Wood pimpernel,	Hb. vul. ast.
" Nummularia,	Moneywort,	Hb. vul. a-sco. ast. f-com.
" quadrifolia,	Yellow balm, Crosswort,	Hb. ast. sto. a-per. exp.
" vulgaris,	Loose strife, Ycl. willow	Hb. ast. exp.
Lythrum alatum,	" " Milk willow	Hb. ast. muc. a-syp. f-com.

BOTANICAL.	COMMON.	PROPERTIES, PRODUCTIONS, USES, ETC.
Lythrum Hyssopifolia,	Hyssop leaved Lythrum,	IIb. ast. muc. f-com. a-syp.
" salicaria,	Loose strife,	IIb. muc. ast. dem. a-syp.
" verticillatum,	Slink weed, Sw'p willow	IIb. abo. to cattle; muc. ast. dem.
Maclura aurantica,	Osage orange,	Pl. a hedge plant; Wd. dyes yellow
" tinctoria,	Fustic wood (old),	yields a yellow dyewood.
Macrochloa tenicrissima,	Long grass,	Tp. dried; ornamental.
Macropiper methysticum,*	Ava kava,	see Piper methysticum.
Macrotrys racemosa,*	Black cohosh,	see Cimicifuga racemosa.
Madia sativa,	Madia seed,	yields Madia oil; as olive oil.
Magnolia acuminata,	Cucumber tree,	Bk. as Magnolia glauca.
" fragrans,	Sweet bay tree,	see Magnolia glauca.
" glauca,	Swamp sassafras,	Bk. ton. ast. sti. bit. feb. exa.
" grandiflora,	Big laurel,	Bk. as Magnolia glauca.
" macrophylla,	Big leaved magnolia,	" " "
" tripetala,	Umbrella tree,	" " "
" yulans,	Chinese magnolia,	Fl. orn.
Mahernia verticillata,	Mahernia,	Fl. orn.; fragrant.
Mahonia,	Holly leaved barberry,	Shrub orn.
Mallotus Philippinensis,*	Monkey face tree, Kameela, see Rottlera tinctoria.	
Malpighia crassifolia,*	Alcornoque bark,	see Byrsonima crassifolia.
" glabra,	Barbadoes cherry,	Ft. sub-aci. car. sto.
" urens,	Cowhage cherry,	Ft. " "
" mourella,	a Cayenne shrub,	Bk. feb.; in diarrhœa.
Malus,		see Pyrus.
Malva arborea,*	Marsh mallow,	see Althæa officinalis.
" alcea,	Vervain mallow,	Rt. as Althæa officinalis.
" Mauritiana,	Tree mallow,	Pl. and Fl. orn.
" moschata,	Musk plant,	" " emo.; Fl. pec.
" rotundifolia,	Low mallow (Cheeses),	Pl. muc. diu. emo.; Fl. pec.
" sylvestris,	Common mallow, (high)	" " "
" vulgaris,	" "	" " "
Mammea Americana,	Mammea apple,	Tree cultivated for its fruit.
Mandevilla suaveolens,	Chili Jasmine,	Pl. and Fl. orn.
Mandragora officinalis,*	European mandrake,	see Atropa Mandragora.
Mangifera Indica,	Mango tree,	Ft. nut. edible; Ke. far.
Mangostana Cambogia,*	Ceylon Gamboge,	see Garcinia Cambogia.
" Garcinia,*	Wild mangosteen,	" " Mangostana.
" mangifer,	Gamboge,	" "
" morella,	"	" " morella.
Manihot utilissima,*	Cassava (Tapioca) plant see Janipha Manihot.	
Maranta allouya,	Arrow root,	yields W. I. arrow root starch.
" arundinacea,	" "	" " " " "
" Galanga,*	Galangal root, small,	see Alpina Galanga.
" Indica,* var.	Arrow root,	yields W. I. arrow root starch.
" nobilis,	" "	" " " " "
Marchantia polymorpha,	Woodrow, Liverwort,	Pl. acr. ast. hep.
Marrubium album,*	White horehound,	see Marrubium vulgare.
" Germanicum,*	Horehound,	" " "
" nigrum,	Black horehound,	see Ballota nigra (fœtida).
" vulgare,	Common horehound,	Pl. bit. aro. ton. dia. exp. pec. cmm.
Martynia fragrans,	Sweet scented unicorn,	Fl. orn.; fragrant.
" proboscidea,	Double claw, Martynia,	Ft. used as pickles.
Maruta Cotula,	Mayweed,	see Anthemis Cotula.
Matonia Cardamomum,*	Cardamom seed,	see Elettaria Cardamomum.
Matricaria Chamomilla,	German chamomile,	Fl. ton. emm. sto. car. nep. ver.
" glabrata,	S. Af'n wild chamomile,	properties anal. to M. Chamomilla.
" leucanthemum,	Oxeye daisy	see Chrysanthemum Leucanthe-
" parthenium,	Feverfew,	see Pyrethrum parthenium. [mum.
" parthenoides,		resembles Anthemis nobilis.
Maurandia Barclayana,	Maurandia,	Pl. climbing; orn.

BOTANICAL.	COMMON.	PROPERTIES, PRODUCTIONS, USES, ETC.
Mays Americana,*	Indian corn,	see Zea Mays.
Mechoacanna nigra,*	Mechoacan root,	see Convolvulus Mechoacanna.
Medeola Virginica,	Indian cucumber,	Rt. diu. hyd.-cat.
" verticillifolia,*	" "	see Medeola Virginica.
Medicago arborea,	Moon trefoil,	Shrub orn.
" circinata,	Caterpillar plant,	Ps. curious; orn.; Hb. dysury.
" intertexta,	Hedgehogs,	Pl. and Ps. curious; orn.
" lupulina,	Dutch clover, Nonsuch,	Pl. len. for.; Wd. hard.
" sativa,	Lucerne,	Pl. cultivated for.; Sd. dyes yell'w.
" scutellata,	Beehive, Snail plant,	Sd.-Ps. curious; orn.
Melaleuca Cajuputi,	Cajuput tree,	yields oil Cajeput; aro. diu. sto.
" genistifolia,	White tea tree,	Ls. ast. ton. {cmm.
" hypericifolia,	Cajuput tree,	Ol. as Melaleuca Cajuputi.
" latifolia,	New Caledonia cajuput,	Ol. aro. sti. a-spa. sto. a-rhe.
" Leucadendron,	White cajuput tree,	see Melaleuca Cajuputi.
" minor,*	Cajuput tree,	" " "
" viridifolia,	Green cajuput,	Oil anal. to Melaleuca Cajuputi.
Melampyrum arvense,	Cow wheat,	Sd. aph.; Ls. fatten cattle.
" sylvaticum,	Horse flower,	Rt. diu.; Sd. aph.
Melampodium,*	Black hellebore,	see Helleborus niger.
Melanorrhœa usitatissima,	Varnish tree of India,	Ju. irr.; as varnish.
Melanthium hybridum,	Bunch flower,	Pl. and Fl. poi.
" Virginicum,	Quafodil, Black flower,	Rt. a crow and fly poison; in itch.
Melia Azadarachta,	Nim bark, (Margosa),	Bk. ton. a-per. feb.; in small-pox.
" Azedarach,	Pride of India (China),	Bk. of Rt. cat. eme. ant. nar.; Ls.dis
" Indica,	Nim bark, (Margosa),	Bk. ton. a-per. feb.; in small-pox.
Melianthus major,	a So. African plant,	Ls. in Tinea capitis; det.; in sore
Melica nutans,	Melic grass,	Tp. cultivated for. [gums.
Melicocca paniculata,	Honeyberry,	Ft. edible, nut. sac.
Melilotus alba,	White melilot,	Fl. aro. emo. pec. dis.
" cœrulea,	Blue melilot,	properties anal. to M. officinalis.
" leucanthe,	Sweet clover,	Pl. and Fl. per. aro.
" officinalis,	White melilot,	Hb. pec. emo. exp. diu. aro.
" segetalis,	Sword grass.	
" vulgaris,	Yellow millet,	Pl. and Fl. exp. diu. emo. aro.
Melissa Calamintha,	Calamint, Mountain balm,	Hb. aro.-sti. dia. dis. exp.
" Canariensis,	Sweet balm,	see Dracocephalum Canariense.
" citrata,*	Lemon balm,	see Melissa officinalis.
" citrina,*	" "	" " "
" grandiflora,	Mountain calamint,	Hb. aro.-sti. car. dia. sti.
" Fuchsii,	Bastard balm,	Pl. diu. dia.
" Nepeta,	Field calamint,	Pl. aro. car. sto.
" officinalis,	Lemon balm,	Hb. and Fl. aro. cep. dia. emm.
Melittis melisophyllum,	Bastard balm,	Pl. aro. sti.; uterine obstructions.
Melocactus communis,	Melon cactus,	Pl. curious ornament.
Melothria pendula,	Creeping cucumber,	Ft. pickled; diu. ver.
Melothrum,*	White bryony,	see Bryonia alba.
Menispermum angulatum,*	Yellow parilla,	see Menispermum Canadense.
" Canadense,	" "	Rt. exc. ton. bit. lax. alt. diu.
" Cocculus,*	Cocculus Indicus,	see Anamirta Cocculus.
" fenestratum,*	False colombo,	see Coscinium fenestratum.
" palmatum,*	Colombo root,	see Cocculus palmatus.
" paniculatum,*	Oriental berries,	see Anamirta Cocculus.
" tuberculatum,	Cocculus crispus,	Pl. ast. ton.; in bowel complaints.
Mentha aquatica,	Water mint,	Pl. bit. pun.; as Mentha viridis.
" arvensis,	Water calamint,	Pl. bit. pun. a-spa. a-rhe.
" balsamea,*	Peppermint,	see Mentha piperita.
" Capensis,	Cape mint,	Pl. has the gen'l properties of mint
" cervina,	Hart's pennyroyal,	Hb. aro. nau.; as Hedeoma.
" citrata,	Bergamot mint,	Hb. see Melissa officinalis.

BOTANICAL.	COMMON.	PROPERTIES, PRODUCTIONS, USES, ETC.
Mentha crispa,	Balm mint (curled),	similar to Mentha piperita.
" gentilis,*	Spearmint,	see Mentha viridis.
" gratissima,	European horsemint,	Pl. aro. sud. pun. a-rhe.
" hirsuta,	Water mint,	see Mentha aquatica.
" lœvigata,*	Spearmint,	" " viridis.
" Langii,*	Peppermint,	" " piperita.
" officinalis,*	"	" " "
" palustris,*	Water mint,	" " aquatica. [a-spa.
" piperita,	Peppermint,	Pl. aro. sti. sto. car.; Ol. rub. sti.
" pulegium,	European pennyroyal,	Pl. aro. a-spa. cmm.; Ol. sti. cmm.
" Romanœ,*	Costmary,	see Pyrethrum Tanacetum.
" rotundifolium,	Patagonia mint,	Ls. fragrant, sto. sti. car.
" sativa,*	Spearmint,	see Mentha viridis.
" sylvestris,*	European horsemint,	" " gratissima.
" spicata,*	Spearmint,	. " " viridis,
" undulata,*	Balm mint,	" " crispa.
" villosa,*	European horsemint,	" " gratissima.
" viridis,	Spearmint,	Pl. car. a-spa. sti. aro. diu.; Ol. sti.
" vulgaris,*	"	see Mentha viridis. [rub.
Menyanthes nymphœoides,	Fringed buckbean,	St. and Rt. bit. ton. feb.
" trifoliata,	Buckbean,	Pl. bit. ton. ant. diu. cat. emc.
" verna,	American buckbean,	Pl. bit. ton. a-per. pur. dep.
Mercurialis annua,	Mercury herb,	Ls. det. pur. res. emm. poi.
" perennis,	Dogs' mercury,	Hb. poi. emc. pur. acr.-nar.
Mertensia Virginica,	Virginia lungwort (cowslip), Pl. muc. dem.	
Mesembryanthemum,		
" crystallinum,	European ice plant,	Pl. diu.; in dysuria.
" edule,	a So. African plant,	Ju. ast. diu. a-sep. vul.
" spectabile,	Fig marigold,	Pl. and Fl. orn.
" tortuosum,		Pl. possesses narcotic properties.
Mespilodaphne Sassafras,	Brazilian sassafras,	resembles Sassafras officinale.
Mespilus aria,	White beam tree,	see Cratægus aria.
" Germanica,	English (Dutch) medlar,	Ft. ast.; Ls. and Sd. in det. gargles
" Oxyacantha,	White hawthorn,	Fl. pec.; see Cratægus Oxyacanthus
Methystophyllum glaucum,	Bushman's tea,	Ls. as tea; in thoracic diseases.
Metroxylon,*	Sago plant,	see Sagus.
Meum Athamanticum,	Bearswort, Baldmoney,	Rt. gummy, car. aro. sto.
Michella champaca,	Vishnu tree,	Tree orn. frn.; Fl. in headache.
Mikania Guaco,	Guaco,	Ju. a-ven.; in cholera; feb. ant.
" scandens,	Climbing hempweed,	Pl. orn.
Milium effusum,	Millet grass,	Pl. orn.; Sd. as birdseed.
" solis,	Gromwell,	see Lithospermum officinale.
Mimosa abstergens,	Soap nuts,	Nu. det. sap.
" Catechu,	Catechu gum,	see Acacia Catechu.
" cineraria,*	Bablah pods,	" " Arabica.
" Nilotica,*	Morocco gum tree,	" "
" pudica,	Humble plant,	the sensitive plant of the garden.
" scandens,	Cachang Parang,	Pl. and Sd. in pleurisy.
" sensitiva,	Sensitive plant,	Pl. sensitive.
Mimulus luteus,	Yellow monkey-flower,	Pl. and Fl. orn.
" moschatus,	Musk plant,	Pl. smells musky; sti.
" ringens,	Monkey flower,	Pl. and Fl. orn.
Mirabilis Jalapa,	False jalap, Marvel of Peru,	Rt. cat.; Fl. orn.
" longiflora,	Metalista root,	Rt. cat.; Fl. orn. per.
Mirostylis ophioglossioides,	Adders' mouth,	Fl. orn.
Mitchella repens,	Squaw vine, Partridge berry,	Pl. diu. par. ast. ton. alt. f-com.
Mitella cordifolia,	Coolwort,	Ls. diu.; see Tiarella.
" diphylla,	Currant leaf,	Ls. diu. ast.
Mohria thurifera,	a So. African fern,	Ls. used in ointment for burns.
Mollugo verticillata,	Carpet weed,	Pl. diu. ref.

BOTANICAL.	COMMON.	PROPERTIES, PRODUCTIONS, USES, ETC.
Molucella lævis,	Molluca balm,	Pl. aro. sti.; as verbena.
Momordica aspera,	Squirting cucumber,	see Momordica Elaterium.
" balsamina,	Balsam apple,	Ft. nar. vul.; Rt. pur. diu. poi.
" charantia,	" pear,	Ft. bitter, ver.
" Elaterium,	Wild squirt'g cucumber,	Ju. yields Elaterium; hyd. ver. poi.
Monarda coccinea,	Mountain rosebalm,	Pl. dia. diu. ton. car. a-per.
" didyma,	Oswego tea, Mt. mint,	Pl. and Ol. rub. sti. car. aro.
" fistulosa,	Wild bergamot, Horsemint,	Ls. aro.-bit. ner. sto. deo.
" Kalmiana,*	Mountain mint,	see Monarda didyma.
" oblongata,	Wild bergamot,	" " fistulosa.
" punctata,	Horsemint,	Hb. sti. car. sud. diu. emm. a-eme.
Monodora Myristica,	American nutmeg,	less pungent than nutmegs.
Monnina polystachia,	So. American Polygala,	Pl. orn; powerfully astringent.
Monotropa hypopitys,	False beech drops,	Rt. bitter, eme. nau. diu.
" uniflora,	Fit root, Indian pipe,	Rt. ton. sed. ner. a-spa. a-epi.
Morœa sisyrinchium,	Spanish nuts,	Bu. nar. acr.
Morinda citrifolia,	Indian mulberry,	Rt. a red dyestuff.
" muscosa,*	West Indian ipecac,	see Cephaelis muscosa.
" royac,	Yaw weed,	Hb. in yaws. (Framboesia).
" tinctoria,	Ach root,	Rt. dyes red and yellow.
Moringa aptera,	Bonduc tree, Behen nuts	Ol. much used by perfumers.
" oleifera,*	Behen nuts,	Ol. ape.; Rt. Sd. and Ls. irr. sad.
" pterygosperma,	" Ben nut,	yield oil of Ben.; Ls. and Rt. acr.
Moronobea coccinea,	Hog gum Tragacanth,	a var. of Bassora gum.
Morus alba,	White mulberry,	Ft. esc. sac. lax. ref.; silkworm
" Indica,	Lopez root,	see Toddalia aculeata. [tree.
" nigra,	Black mulberry,	Ft. esc. aci.; Bk. ver. cat.
" rubra,	Red mulberry,	" " " "
" tinctoria,	Fustic tree (old),	Wd. dyes yellow.
" papyrifera,	Paper mulberry,	Bk. made into paper and cloth.
Mucuna pruriens,	Cowhage,	see Dolichos pruriens.
" prurita,*	East India cowhage,	" " prurita.
Mulgedium acuminatum,	Blue lettuce,	Ls. a-ven.
" Floridanum,	False lettuce,	
Muhlenbergia capillaris,	Awned hair grass,	Pl. for.
" diffusa,	Drop seed, Nimble Will,	"
Musa Paradisiaca,	Plantain,	Ft. esc. nut.; used as food.
" sapientum,	Banana,	a luscious fruit used as food.
Muscari botryoides,	Globe hyacinth,	Bu. acr. cat. diu. eme.; Fl. orn.
" moschatum,	Grape hyacinth,	Bu. " " " per.
Myopyrum platycarpum,	Sugar tree,	Ju. sac.
Myosotis arvensis,	Myosotis, Forget-me-not	Fl. orn. pec.
" palustris,	Scorpion weed,	Fl. "
Myrcia acris,	Bay rum leaf,	Ls. aro. fragrant; used in bay rum.
" "	Jamaica bayberry (clove),	Ft. aro. pun.; as spice.
Myrica cerifera,	Bayberry bush,	Bk. ast. ton. err. sti.; Ls. aro. sti.
" ,	Wax myrtle,	Berries yield Vegetable wax.
" Gale,	Meadow fern, Sw't gale,	Ls. and Bd. ton. alt. dep. vul. exa.
Myricaria Germanica,	German tamarisk,	Bk. bit. ast.; yield galls.
Myristica aromatica,*	Nutmeg, Mace,	see Myristica fragrans.
" fatua,	Male, long or wild nutmeg,	yield inferior nutmegs.
" fragrans,	Nutmeg, Mace,	Kc. of Ft. aro. sti. sto. a-eme. con.
" Moschata,*	" "	Arillus of Ft. called mace.
" officinalis,*	Female nutmeg,	see Myristica fragrans.
" otoba,	Santa Fe nutmeg,	a coarse variety of nutmeg.
" tomentosa,*	Male, long or wild nutmeg,	see Myristica fatua.
Myrobalanus chebula,	White myrobalan (galls)	Dried Ft. see Terminalia.
" citrina,	Yellow myrobalan,	" " "
" emblica,	Emblic, White galls,	see Phyllanthus.
" Indica,	Black myrobalan,	see Terminalia chebula.

BOTANICAL.	COMMON.	PROPERTIES, PRODUCTIONS, USES, ETC.
Myrocarpus frondosus,		allied to Myroxylon.
Myrospermum frutescens,	Balsam tree,	Ps. car. a-rhe. sti. vul.
" Pereiræ,*	Peruvian balsam tree,	see Myrospermum Peruiferum.
" Peruiferum,	" " "	Ba. aro. stl. pec. exp. vul.
" pubescens,	Myrrh seed,	Sd. aro. sti. ; yield balsam.
" sonsonata,*	Balsam of Peru,	see Myrospermum Peruiferum.
" Toluiferum,	Balsam of Tolu,	Ba. stl. aro. ton. pec. vul. per.
Myroxylon balsamiferum,	Trinidad balsam tree,	yields Balsam of Peru.
" Pereiræ,*	Balsam of Peru,	see Myrospermum Peruiferum.
" Peruiferum,	" "	" " "
" sonsonatense,*	" " (San Salvador),	" "
" Toluifera,*	" of Tolu,	Ba. see Myrospermum Toluiferum.
" Toluiferum,	" "	" " "
Myrrhis annua,*	Candy carrot,	see Athamanta Cretensis.
" Claytoni,	European sweet cicily,	Rt. aro. sti. car. pec. exp.
" odorata,	Sweet cicily,	" " "
Myrsiphyllum asparagoides,	Cape smilax,	Pl. orn. ; decorative.
Myrtus acris,*	Bay rum leaf,	see Myrcia acris.
" anglica,*	Meadow fern,	see Myrica Gale.
" Caryophyllata,	Clove bark,	Bk. aro. stl. cor. ; used as cloves.
" Caryophyllus,*	Cloves,	see Caryophyllus aromaticus.
" cauliflora,	Jabuticaba fruit,	Bk. aro. ast. ; in sore throat.
" communis,	Myrtle,	Ft. aro. ast. ; in alvine fluxes.
" leucodendron,*	Cajuput tree,	see Melaleuca Cajuputi.
" Pimenta,*	Allspice, Pimento,	see Eugenia Pimenta.
Nabalus albus,	White lettuce, Lions' ft.,	Pl. bit. a-ven.
" Serpentaria,	a variety of	Nabalus albus.
Najas Canadensis,	Water nymph,	Ls. ast. mat.
Napellus verus,*	Aconite,	see Aconitum Napellus.
Narcissus aurauticus.	Butter and eggs,	Fl. orn. ; Rt. eme.
" bulbocodium,	Medusa's trumpet,	" "
" jonquilla,	Jonquil flower,	" "
" odora,	Yellow jonquil,	" "
" poeticus,	Poets' narcissus,	Fl. orn. ; Rt. mat. ; whoop'g cough.
" pseudo-narcissus,	Daffodil,	Rt. eme. cat. ; Fl. orn. eme. a-spa.
Nardostachys Jatamansi,	Spikenard of the ancients,	Rt. sti. aro. per. pun. det.
Nardus Americana,*	American sarsaparilla,	see Aralia nudicaulis.
" Celtica,*	Celtic nard (valerian),	Rt. aro. sti. per. cos.
" Indica,	Indian (Syrian) spikenard	see Nardostachys Jatamansi.
" Jatamansi,	" " "	see also Valeriana; aro. sti. a-hys.
" montana,*	Mountain valerian,	see Valeriana montana.
" stricta,	Mat grass, Com'n nard,	Rt. aro. sto. sti. vul.
Narthecium Americanum,	Asphodel false,	Rt. diu. ; see Asphodelus.
Narthex Asafœtida,*	Assafœtida,	see Narthex Assafœtida. [cmm.
" Assafœtida,	Assafœtida,	Gu.-re. fœt. stl. a-spa. exp. lax. ant.
Nasturtium amphibium,	Water radish,	Pl. pun. acr. a-sco. sad. cmm.
" Armoracia,*	Horse radish,	see Cochlearia Armoracia.
" officinale,	Water cress,	Pl. dep. a-sco. emm. sad.
" palustris,	Marsh watercress,	Pl. sti. diu. a-sco. deo. hep. sad.
" Peruvianum,	Nasturtium, Indian cress	see Tropæolum majus.
" sativum,*	Pepper grass (cress),	see Lepidium sativum.
Nauclea Brunonis,	Pegu catechu,	Ex. ast. ton. sty. vul.
" Gambir,	Gambir, Terra Japonica,	Ex. " "
" longiflora,	Catechu,	Ex. " "
Nectandra cinnamomoides,	Santa Fe cinnamon,	Bk. used as cinnamon.
" cymbarum,	Orinoco sassafras,	Bk. aro.-bit. sto.
" puchury,	Pichurim (Sassafras) nut	Ft. aro. stl. ; Bk. feb. ton.
" Rodiæi,	Greenheart (Bebeeru) tree,	Bk. ton. a-per. feb.
Negretia pruriens,*	Cowhage,	see Mucuna.
Nelumbium codophyllum,	Napoleon plant,	Rt. Ls. and Sd. esc. ref. lax. diu. emo

13*

BOTANICAL.	COMMON.	PROPERTIES, PRODUCTIONS, USES, ETC.
Nelumbium luteum,	Yel. water lily, Wat'r nut	Rt. Ls. and Sd. esc. ref. lax. diu. emo
" specicosum,	Lotus rattlenut,	Nu. called chincapins; esc.
Nepentha distillatoria,	Bandura,	Rt. ast.
Nepenthe,*	Pitcher plant,	Pl. curious ornaments.
Nepeta cataria,	Catmint, Catnep,	Ls. aro. dia. car. a-spa. ano. emm.
" citriodora,*	Balm lemon,	see Melissa officinalis. [deo.
" glechoma,	Ground ivy, Gillrun,	Ls. sti. ton. pec. dlu. cep.; in lead
" vulgaris,*	Catmint,	see Nepeta cataria. [colic.
Nephelium longanum,	Dragons' eye, Longan	ft. Tree cultivated for its esc. fruit.
Nephrodium Filix Mas,*	Male fern,	see Aspidium Filix Mas.
Nerine Sarniensis,	Guernsey lily,	Pl. and Fl. orn.
Nerium anti-dysentericum,	Conessi bark,	see Wrightia anti-dysenterica.
" odorum,	Sweet oleander,	Bk. poi.; sweet smelling.
" oleander,	Oleander, South Sea rose	Ls. nar. her.; in itch; stc.
Nervosperma balsamita,*	Balsam apple,	see Momordica balsamina.
Nicotiana fruticosa,	a native of China,	also cultivated in Cuba.
" latissima,	variety of tobacco	see Nicotiana Tabacum.
" Loxensis,	" "	" " "
" macrophylla,	" "	" " "
" minor,*	Wild English tobacco,	" " rustica. ·
" paniculata,	Small flowered tobacco,	" " Tabacum.
" quadrivalvis	var. tobacco com'n in Mo. analogous to Nicotiana Tabacum.	
" repanda,	Havana (Cuba) tobacco,	Ls. fra.; see " "
" rustica, ·	Wild tobacco,	Ls. milder than cultivat'd tobacco.
" Tabacum,	Virginia tobacco,	Ls. nar. sed. eme. dlu. err. poi. ins.
Nigella Damascena,	Black cummin, Fennel fl.	Pl. and Fl. orn.; Sd. pun. car. gal.
" sativa,	Bl'k caraway, Nutm'g fl.	Pl. exp. deo. sia. err. emm.
Nuphar advena,	Yellow pond lily,	Rt. ast. cmo. dis. dem. f-com.
" lutea,*	European pond lily,	see Nymphæa lutea.
Nuphæa advena,	Yellow pond lily,	see Nuphar advena.
Nyctantes sambac,*	Jasmine,	see Jasminum Sambac.
Nyctanthes arbortristis,	Night jasmine,	Fl. orn.; fragrant.
Nymphæa advena,	Yellow pond lily,	see Nuphar advena.
" alba,	Europ'n white water lily	Rt. and Ls. dem. a-aph. ast. emo.
" Indica,*	Egyptian bean,	see Nymphæa nelumbo. [ano.
" lutea,	Europ'n yellow pond lily	Rt. and Ls. nar. sed. a-aph. ast.
" nelumbo,	Egyptian bean,	Sd. eaten; ton. ast.
" odorata,	Wh. pond lily, swt. sctd.	Fl. orn. per.; Rt. ast. vul. dis.; Ls.
" umbilicalis,*	Europ'n yellow pond lily	see Nymphæa lutea [vul.
Nyssa coccinea,	Ogee chee tree, Lime tree	Ft. aci. bit.; Wd. fine cross grain'd
" capitata,	" " "	" ". " "
" multiflora,	Pepperidge, Black gum,	Ft. aci.; Wd. fine cross grained.
Obione canescens,	Greasewood plant,	Pl. aro.; oily; as wormwood.
Ocotea pichurim,	Pichurim beans,	see Nectandra puchury.
Ocimum basilicum,	Common basil, (sweet),	Hb. aro. stl. ner. con.
" Caryophyllatum,	Small (Bush) basil,	Hb. aro. stl. ner. cor. err.
" citratum,*	Common basil,	see Ocimum basilicum.
" pilosum,*	" "	" " "
" racemosum,*	" "	" " "
Ocymum basilicum,	Sweet basil,	" " "
Œnanthe aquatica,*	Fine lv'd water hemlock,	see Œnanthe Phellandrium.
" crocata,	Hemlock water dropw't,	Pl. acr.-nar. poi.; in fomentation.
" fistulosa,	Water hemlock,	Pl. " " "
" Phellandrium,	Fine lv'd water hemlock,	Pl. nar. sti. exp. alt. diu.
Œnothera biennis,	Ev'g primrose, Scabish,	Pl. muc. acr. exa. orn.
" gauroides,*	" " "	see Œnothera biennis.
" glauca,	Sundrops,	Pl. exa. cmo. det.
Oidium abortifaciens,	Ergot fungus,	see Sclerotium Clavus.
Oldfieldia Africana,	Teak wood tree,	Wd. heavy and useful.
Oldenlandia,	Ind'n madder, Chay root	Rt. in dyeing; Ls. exp.

BOTANICAL.	COMMON.	PROPERTIES, PRODUCTIONS, USES, ETC.
Olea Americana,	Am'n olive, Devil wood.	
" Europæa,	Olive tree,	Bk. and Ls. feb.; Ft. y'lds Olive oil
" fragrans,	Fragrant olive,	see Osmanthus fragrans.
" Gallica,*	French olive tree,	Ft. pickled; esc. con.
" latifolia,	Spanish olive,	" " Olive oil.
" longifolia,	Italian olive,	" " "
" polymorpha,*	Olive tree,	see Olea Europæa.
" sativa,*	" "	" " " Ol. nut. dem. lax.
Omphalea diandra,	West India cobnut,	Nu. yield Oil ouabe. [emo.
Ompholobium Lambertii,	Zebra wood,	Wd. for ornamental purposes.
Onoclea sensibilis,	Sensitive fern,	Pl. orn.; see Aspidium.
Onobrychis sativa,	Sainfoin grass,	Pl. cultivated forage plant.
Ononis antiquarum,*	Rest harrow,	see Ononis spinosa.
" arvensis,	Cammock,	analogous to Ononis spinosa.
" natrix,	Goats' root,	" " "
" repens,	Rest harrow,	Rt. diu. det. ape.
" spinosa,	" "	" "
Onopordon acanthium,	Cotton (Musk) thistle,	Ju. externally in cancer.
Onopordum acanthium,*	" " "	see Onopordon acanthium.
Onosmodium Carolinianum,	False gromwell,	see Onosmodium Virginianum.
" hispidum,*	Gravel weed,	" " "
" strigosum,	Wild Job's tears,	anal. to Onosmodium Virginianum.
" Virginianum,	" " "	Rt. and Sd. diu. ton. lit.
Ophelia Chirayta,*	Chiretta,	see Agathotes· Chirayta.
Ophiocaryon paradoxum,	Snake nuts,	Nu. curious; a-ven.
Ophioglossum ovatum,*	Adders' tongue fern,	see Ophioglossum vulgatum.
" vulgatum·	" " "	Pl. vul.; in ointments.
Ophiorhiza mitreola,	Pink snakeroot,	Rt. ant.; reputed a-ven.
Ophiurus,	Hard sea grass.	
Ophrys apifera,	Bee orchis,	Fl. orn.
Opoidia Galbanifera,	Persian (African) galbanum, see Bubon Galbanifera. .	
Opopanax chironium,*	Opopanax,	see Pastinaca opopanax.
Opuntia cochinillifera,	Common prickly pear,	food of the cochineal insect.
" Ficus Indica,	Indian fig,	Ft. sac. diu. ref.
" vulgaris,	Com'n prickly pear cactus, Ls. ref.; Ju. colors red.	
Orchis bracteata,	Vegetable satyr,	Rt. muc. emo.
" latifolia,	Marsh orchis,	Rt. yields salep.
" mascula,	Male fool stones (orchis)	" "
" morio,	Fools' stones,	" "
" spectabilis,	Showy orchis	" "
Oreoselinum Africanum,	Galbanum,	see Bubon Galbaniferum.
" legitimum,	Speedwell,	see Athamanta oreoselinum.
" nigrum,	"	" " "
Origanum aquaticum,*	Hemp agrimony,	see Eupatorium Cannabinum.
" Creticum,	Dittany of Crete,	see Origanum dictamnus.
" dictamnus,	" . "	Pl. aro. ton. alc.
" Marjorana,	Sweet marjoram,	Pl. aro. tou. emm. cep. con.
" marjoranoides,*	" "	see Origanum Marjorana.
" Smyrnaceum,	Spanish marjoram,	Hb. aro. bit. sti. emm.
" Syriacum,*	Syrian herb mastich,	see Teucrium marum.
" vulgare,	Wild marjoram,	Hb. aro. pun. ton. sti. emm.
Ornithogalum altissima,	So. African squill,	Bu. resembles squills; see Scilla.
" maritimum,*	Squill,	see Scilla maritima.
" pyrenuicum,	French sparrowgrass,	young shoots as asparagus.
" umbellatum,	Star of Bethlehem,	Rt. eaten; Sd. aro. con.
Ormosia coccinea,	Guinea pea, (Necklace seed), Sd. ornamental.	
Ornithopus puppusillus,	Birdsfoot,	a fodder plant.
Ornus Europæa,	Manna ash,	see Fraxinus Ornus.
" mannifera,	" "	" " "
" rotundifolia,	" "	" " "

BOTANICAL.	COMMON.	PROPERTIES, PRODUCTIONS, USES, ETC.
Ornus sylvestris,*	European (Manna) ash,	see Fraxinus excelsior.
Orobanche Americana,	Beech drops, Earth club,	Rt. nau. bit. ast. dep. vul.
" uniflora,	One flow'd broom rape,	see Orobanche Americana.
" Virginiana,	Cancer rt., Beech drops,	Rt. nau. bit. ast. dep. vul.
Orobus atropurpureus,	Heath pea,	Sd. contain farina; dis.
" niger,	Black bitter vetch,	" " "
" sylvaticus,	Bastard vetch,	" " "
Orontium aquaticum,	Golden club,	Pl. orn.; Sd. acr. edible.
Oryza sativa,	Rice plant,	Sd. the rice of commerce.
Osmanthus fragrans,	Fragrant olive,	Ls. tea flavor.
Osmorrhiza Claytoni,	Sweet cicily root,	Rt. car. exp. dem. aro. sto.
" dulcis,	" " "	Rt. see Osmorrhiza longistylis.
" longistylis,	" " "	Rt. car. exp. dem. aro. sto.
Osmunda cinnamomea,	Cinnamon colored fern,	Rt. ast. muc. ton. sty. f-com.
" lunaria,	Moonwort,	see Botrychium lunaria.
" regalis,	Buckhorn, Male fern,	Rt. muc. ton. ast. sty. f-com.
Ostrya Virginica,	Hop horn-beam, Iron wd.	Bk. bit. ton. a-per. alt.
Oxalis acetosa,	Garden sorrel,	Pl. aci. ref. diu. a-sco. irr.
" acetosella,	Wood sorrel,	" " "
" Americana,	" "	see Oxalis acetosella.
" corniculata,	Ladies' sorrel,	" " "
" crassicaulis,	Peruvian sorrel,	Pl. aci. ast.
" stricta,	Yellow wood sorrel,	analogous to Oxalis acetosella.
" tuberosa,	Peruvian Oca,	Tuber used as food; Ju. ast. aci.
" violacea,	Violet sorrel,	Pl. ref. a-sco. aci.
Oxyacantha Galeni,*	European barberry,	see Berberis vulgaris.
Oxycoccos hispidula,	White cranberry,	Ft. aci. ref.; agreeable.
" macrocarpus,	Cranberry,	see Vaccinium oxycoccos.
" palustris,	Moorberry, Marshwort,	" " "
Oxyria reniformis,	Boreal sour dock,	properties as Oxalis acetosella.
Pæderota Virginica,*	Black (Culver's) root,	see Leptandra Virginica.
Pæonia albiflora,	Fragrant white peony,	Fl. fragrant; orn.
" anomala,	Peony,	Rt. see Pæonia officinalis.
" communis,*	Common peony,	" " "
" corallina,	Peony,	" " "
" moutan,	Tree peony,	cultivated variety.
" officinalis,	Double peony,	Rt. ast. a-epi. a-hys. sti. emm. a-spa
" peregrina,	Peony,	see Pæonia officinalis.
Palca Cabotii,	Java fern,	a mechanical styptic.
Pallicourea Crocea,*		see Psychotrea crocea.
" maregraavii,	Ervado rato,	Pl. poi.; used to kill rats.
" officinalis,	Palicourea,	Pl. reputed a powerful diuretic.
" speciosa,	Gold shrub,	Ls. a-syp. poi.
Pallurus aculeatus,	Christ's thorn,	Sd. diu.; Rt. and Ls. ast. det.
Palma Christa,*	Castor oil plant,	see Ricinus communis.
" cocos,*	Cocoanut palm,	see Cocos nucifera.
" Dactylifera,*	Date palm,	see Phœnix Dactylifera.
" humilis,*	Plantain fruit,	see Musa Paradisiaca.
Panax pseudo-ginseng,	Chinese ginseng,	Rt. dem.; a gentle stimulant.
" quinquefolia,	American ginseng,	" " "
" schinseng,	Asiatic ginseng,	Rt. dem. sti.
" trifolia,	Dwarf ground nut,	Rt. esc. dem. sac.
Pancratium,	Squillily,	Rt. emc. diu.; as Scilla.
Pandauus odoratissimus,		Sd. oil Keora; Fl. fragrant.
Panicum capillare,	Old witch grass,	Pl. for.
" crus-galli,	Barn grass,	Pl. nutritive forage.
" dactylon,	Bermuda grass,	see Cynodon dactylon.
" Germanicum,	Millet seed (Hungarian)	Pl. and Sd. cultivated forage.
" junctorum,	Panic grass.	
" miliaceum,	Common millet,	Sd. used as food.

BOTANICAL.	COMMON.	PROPERTIES, PRODUCTIONS, USES, ETC.
Panicum molle,	Scotch grass.	
" sanguinale,	Finger grass,	Pl. wild weed; for.
Papas Americanus,*	Potato plant,	see Solanum tuberosum.
Papaver Argemone,*	Prickly poppy,	see Argemone Mexicana.
" cœruleum,	Black seeded poppy,	Sd. maw seed; fed to birds.
" nigrum,*	Black poppy,	var. of Papaver somniferum.
" orientale,	Opium poppy,	concrete Jn. called opium.
" rhœas,	Wild, red or corn poppy,	Fl. ano. pec.
" somniferum,	Opium poppy,	yields opium; nar. ano. sti. hyp.
" " var. alba,	White var. Opium poppy	Sd. yields oil.
" " " nigrum	Black " " "	Sd. maw seed.
Pappea Capensis,	Caffer wild plum,	Ft. aci.; Sd. yields oil; cat. exa.
Pappus Americanus,*	Potato plant,	see Solanum tuberosum.
Pardanthus Chinensis,*	Blackberry lily,	see Ixia Chinensis.
Pareira medica,* .	Calamba wood,	see Coscinium fenestratum.
Parietaria diffusa,	Wall pellitory,	see Parietaria officinalis.
" erecta,*	" "	" " "
" officinalis,	" "	Pl. diu. emo.; contains Nit potash.
" Pennsylvanica,	American pellitory,	Pl. diu. deo. emm.
Paris quadrifolia,	Herb Paris, Fox grape,	Pl. nar. eme. ano. res.
Parmelia caperata,	Stone crottles,	Lichen dyes wool orange color.
" Islandica,*	Iceland moss,	see Cetraria Islandica.
" parietina,*	Trumpet (Wall) moss,	see Cladonia pyxidata.
" plicata,*	Lapland moss,	see Lichen plicatus.
" pulmonacea,*	Maple lungwort,	see Sticta pulmonaria.
" rocella,*	Archel, Litmus plant,	·see Lichen rocella.
" saxatilis,	Skull moss,	see Lichen saxatilis.
• Parmentiera cerifera,	Candle fruit tree,	Ft. edible to cattle.
Parnassia Caroliniana,	Grass of Parnassus,	Ju. opt.; Sd. diu. ape.
Parthenium febrifugum,	Feverfew,	see Pyrethrum Parthenium.
" hysterophorus,	Bastard feverfew,	Pl. ton. car. emm. ver.
" integrifolium,	Prairie dock, Cutting almond,	Rt. aro. bit. sti. diu. lit. nep.
Paspalum scrobiculatum,	Ghohona grass,	reputed poisonous.
Passiflora cœrulea,	Passion flower,	Pl. and Fl. orn. diu.
" edulis,	Granadilla,	Pl. eme.; Fl. orn.; Ft. edible.
" incarnata,	May pops,	Pl. and Fl. orn.
" lauriflora,	Bay lv'd passion flower,	Ft. aro. fra. sto. ton.
" maliformis,	Apple shaped granadilla,	Ft. a delicacy; eaten.
Pastinaca altissima,*	Opopanax,	see Pastinaca Opopanax.
" anethum,*	Dill plant,	see Anethum graveolens.
" graveolens,*	" "	" " "
" Opopanax,	Rough parsnip,	Gu.-re. similar to Assafœtida.
" pratensis,	Parsnip,	see Pastinaca sativa.
" sativa,	Garden parsnip,	Rt. esc. nut.; Sd. and Tp. diu.
" sylvestris,	Parsnip,	see Pastinaca sativa.
Patrinia Jatamansi,*	Spikenard of the ancients,	see Nardostachys Jatamansi.
Paullinia cupana,	Guarana,	Ex. ast. sti. exc.; in sick headache.
" sorbilis,	"	" " " "
Pavia rubra,	Small buckeye,	see Æsculus pavia.
Pedicularis Canadensis,	Wood betony,	Pl. ton. sed. ast. vul.
" gladiata,	High healall,	Pl. vul. ast. ton.
Pedilanthus tithymaloides,	Jew bush,	Pl. a-syp. f-com. emm. emc.
Pelargonium anceps,		Pl. abo. par. f-com.
" anti-dysentericum	a So. African species,	Pl. boiled in milk for dysentery.
" capitatum,	Rose-scented geranium,	Pl. and Fl. orn. per. sti.
" cucullatum,	Cape G. H. Herba althææ,	Pl. eneme in colic; nep. diu. ant.
" odoratissimum,	Rose geranium,	Pl. and Fl. orn.; Ol. per. sti.
" roseum,	Rose-scented storksbill,	yields oil Rose geranium.
" thiste,		Rt. ast. ver.
Peltandra Virginica,	Arrow arum,	see Arum; acr. sti. irr.

BOTANICAL.	COMMON.	PROPERTIES, PRODUCTIONS, USES, ETC.
Peltidea,*		see Lichen.
Penæa mucronata,	Sarcocolla,	Re. acrid; pur. fra. det.
" sarcocolla,	"	" " "
" squamosa,	"	" " "
Penicillium hyphomycetes,	Vinegar plant,	Fu. a thread mould.
Penstemon pubescens,	Beard tongue,	Pl. and Fl. orn.
Periandra dulcis,	Brazilian liquorice,	Rt. analogous to Glycyrrhiza.
Perilla Nankinensis,	Purple leaved perilla,	a foliage plant.
Periploca græca,	Scammony senna,	Ls. cat.; mixed with Senna.
" Indica,*	E. I. sarsaparilla,	see Hemidesmus Indicus.
" secamone,	Smyrna scammony,	Rt. yields a drastic resin.
Persea caryophyllata,*	Clove bark,	see Laurus caryophyllata.
" gratissima,	Avocado pear,	Ls. bal. pec. vul.; Sd. ast.; Ft. esc.
" Sassafras,*	Sassafras,	see Sassafras officinale.
" tamala,	Tamala cinnamon,	see Cinnamomum nitida.
Persica vulgaris,	Peach,	see Amygdalus Persica.
Persicaria mitis,*	Heartscase,	see Polygonum Persicaria.
" urens,*	Water pepper,	" " hydropiper.
Petroselinum Macedonicum,	Macedonian parsley,	see Bubon Macedonicum.
" sativum,	Garden parsley,	Rt. ape. diu. nep.; Sd. feb.; Ju.
Petunia violacea,	Petunia,	Pl. and Fl. orn. [a-per.
Peucedanum alsaticum,*	Sow fennel,	see Peucedanum officinale.
" Austriacum,*	Swamp sow fennel,	" " palustre.
" cervaria,	Black gentian,	see Libanotis vulgaris.
" officinale,	Sow fennel, Sulphurwort	Rt. diu. exp. ape.; Ju. diu. a-spa.
" oreoselinum,*	Speedwell,	see Athamanta oreoselinum.
" montana,*	Marsh parsley,	see Selenium palustre.
" ostruthium,	Masterwort,	see Imperatoria ostruthium.
" palustre,	Swamp sow fennel,	see Peucedanum sylvestre.
" paniculatum,*	Sow fennel,	" " officinale.
" pratense,*	Meadow saxifrage,	" " Silaus.
" Silaus,	" "	Pl. ape. diu. car.
" sylvestre,	Milk parsley,	Rt. alc. a-epi. [oily; Ft. sac.
Peumus Boldoa,	Boldoa, Boldo,	Bk. ast.; Ls. ton. sti. as tea; Sd.
" fragrans,	" "	Ls. ton. cor. sti. feb.; Bk. ast.
Peziza auricula,*	Jews' ear, a fungus,	see Exidia auricula.
Phacelia viscida,	Phacelia,	Border plants; orn.
Phalagium pomeridianum,	California soapwort,	Bu. sap. det.; used as soap.
Phalaris arundinacea,	Reed canary grass,	cultivated as fodder.
" Americana,	Ribbon grass,	Pl. orn.
" canariense,	Canary (seed) grass,	Sd. diu. nut.; birdseed.
" colorata,	Striped grass,	Ls. orn.; striped.
" Zizanoides,*	Vettivert,	see Andropogon muricatus.
Pharbitis nil,*	Blue morning glory,	see Convolvulus nil.
Phaseolus caracolla,	Snail flower, Caracol,	Fl. orn.; sweet scented.
" multiflorus,	Scarlet runner (bean),	Pl. and Fl. orn.; Ft. esc.
" vulgaris,	Common bean, Faba,	Ft. esc. nut. car. nep. cos.
Phellandrium aquaticum,	Fine lv'd water hemlock,	see Œnanthe Phellandrium. [det.
Philadelphus coronarius,	Mock orange, Syringia,	Fl. orn.; powerfully odorous; Ls.
Phillyrea latifolia,	Stone linden,	Ls. ast. diu. cep.; in sore mouth.
" media,*	" "	see Phillyrea latifolia.
Phleum pratensis,	N. E. Herds (Timothy) grass, cultivated as forage.
Phlomis tuberosa,	Jerusalem sage,	Pl. and Fl. orn.
Phlox Drummondi,	Garden phlox,	" "
" subulata,	Moss pink,	" "
Phœnix Dactylifera,	Date palm,	Ft. nut. lax. emo. pec. sac.
" excelsa,*	" "	see Phœnix Dactylifera.
" farinifera,	Sago palm,	Pl. yields sago; nut. far.
" sylvestris,	Sugar palm, Jaggery,	yields Palm sugar.
Phoradendron flavescens,*	Mistletoe on elm trees,	see Viscum flavescens.

BOTANICAL.	COMMON.	PROPERTIES, PRODUCTIONS, USES, ETC.
Phormium tenax,	New Zealand flax,	Fibre used for cloth and paper.
Phragmites communis,*	Common reed,	see Arundo phragmites.
Phryma leptostachys,	Lop seed.	
Phyllanthus emblica,	White galls,	unripe Ft. ast. acl.
" urinaria,		Pl. diu. feb. ast.
Phyllitis,	Spleenwort,	see Asplenium.
Physalis alkekengi,	Winter cherry,	Ft. diu. feb.
" obscura,*	Ground cherry,	see Physalis viscosa.
" Pennsylvanica,*	" "	" " "
" Stramonium,	a plant of Nepal,	acts as Belladonna.
" viscosa,	Ground cherry,	Ft. diu. sed.
Physolobium,	Bladder pod,	Pl. orn.
Physostegia Virginiana,*		see Dracocephalum.
Physostigma venosum,	Ordeal bean of Calabar,	Sd. poison, sed.; myositic.
Phytelephas macrocarpa,	Vegetable ivory,	Sd. substituted for ivory in small
Phyteuma obiculare,	Rampion,	Pl. and Fl. orn. sad. [articles.
Phytolacca decandra,	Poke, Garget,	Rt. alt. res. deo. det. a-syp. n-sco.
" vulgaris,	" "	Rt. and Ft. a-rhe.; see P. decandra.
Picræna excelsa,*	Quassia, Bitterwood,	see Simaruba excelsa.
Picrodendron Juglans,	Jamaica walnut,	Ft. ese.; yield oil.
Pillea pumila,	Clearweed, Coolweed,	see Urtica pumila.
" serpyllifolia,	Artillery plant,	Pl. orn.
Pilocarpus pennatifolius,	Jaborandi,	see also Piper; Rt. sti. sia. a-ven.
Pilularia globulifera,	Pillwort, Peppergrass,	Pl. acr. sad.
Pimenta acris,	Black cinnamon,	Bk. see Myrcia acris.
" officinalis,	Allspice, Pimento,	see Eugenia Pimenta.
Pimpinella Anisum,	Anise plant (seed),	Sd. aro. fragrant, car. ton. stl. sto.
" alba,*		see Pimpinella Saxifraga.
" magna,	Greater (False) Pimpinella,	Rt. diu. res. ast. a-spa. cos.
" major,*	" " "	see Pimpinella magna.
" nigra,*	" " "	" " "
" Saxifraga,	Burnet saxifrage,	Rt. aro. nau. pun. stl. sto.
" " var. nigra,	Black Burnet saxifrage,	" " "
" umbellifera,		see Pimpinella Saxifraga.
Pinckneya pubens,	Florida (Georgia) bark,	Bk. ton. a-per. ast. feb.
Pineus purgans,	Physic nuts,	see Jatropha cureas.
Pinguicula elatior,	Butterwort,	Fl. pur. vul. det.
" vulgaris,	" Rot grass,	" "
Pinus Abies,*	Norway spruce (pine),	see Abies excelsa.
" alba,*	White (Deal) pine,	see Pinus strobus.
" Australis,*	Wild pine, Scotch fir,	" " sylvestris.
" balsamea,*	Balsam fir tree,	see Abies balsamœ.
" Canadensis,*	Hemlock tree,	" " Canadensis.
" caudicans,*	Silver fir tree,	" " picea.
" Cembra,	Siberian stone pine,	yields Carpathian (Riga) balsam.
" communis,*	Spruce,	see Abies communis.
" Damarra,	Damarra pine,	yields a fine hard resin.
" excelsa,	Norway pine (spruce),	yields Burgundy pitch; see Abies.
" Fraseri,	Double, Balsam fir tree,	yields balsam similar to Fir.
" Gallica,	Silver fir,	see Abies picea.
" Lambertiana,	Sugar pine,	y'lds a substance resembl'g manna.
" Larix,	European larch,	yields Venice turpentine and Oren-
" maritima,*	Cluster pine,	see Pinus pinaster. [burg gum.
" mughos,	Mountain (Mugho) pine,	yields Hungarian balsam.
" microcarpa,	American larch,	see Larix Americana.
" nigra,	Black spruce, (Gum tree)	see Abies nigra.
" palustris,	Yellow pitch pine,	y'lds Gum Thus, turpentine and tar.
" pendula,*	American larch,	see Larix Americana.
" picea,*	Silver fir tree, Red spruce,	see Abies picea.
" pinaster,	Cluster pine,	yields tar, pitch and turpentine.

BOTANICAL.	COMMON.	PROPERTIES, PRODUCTIONS, USES, ETC.
Pinus pinea,	Stone pine, Pine nuts,	Nu. used in emulsions; esc.
" pumilo,	Mountain (Mugho) pine,	yields Hungarian balsam.
" resinosa,	Red pine, Norway pine,	yields a hard resin.
" rigida,	Common pitch pine,	yields tar, pitch and turpentine.
" strobus,	White (Deal) pine,	Wd. soft; yields Gum Thus.
" sylvestris,	Wild Scotch fir (pine),	y'lds Thus, oil, tar, resin and pitch.
" tæda,	Loblolly pine, Old field pine,	yields common pine resin.
" taxifolia,	Silver fir tree,	see Abies picea.
" vulgaris,	"　　"	"　　"　　"
Piper album,	White pepper,	Black pepper freed from its cuticle.
" Afzelii,	African black pepper,	Ft. analogous to Piper nigrum.
" angustifolium,*	Matico plant,	see Artanthe elongata.
" aromaticum,*	Black pepper,	see Piper nigrum.
" anisatum,	Anise pepper (Cubeb),	Ls. and Ft. smell and taste as anise
" Betel, (Betle*),	Betel,	Ls. bit. sto. ton. aph. sia. dia.
" Brazilianum,*	Capsicum,	see Capsicum annuum.
" Calecuticum,*	"	"　　"　　"
" caninum,	Cubebs,	see Piper Cubeba.
" Capense,	a So. African species,	Ft. resembles Cubebs.
" Caryophyllatum,*	Pimento, Allspice,	see Eugenia Pimenta.
" caudatum,*	Cubeb, Tailed pepper,	see Piper Cubeba.
" Chiapæ,*	Pimento, Allspice,	see Eugenia Pimenta.
" Clusia,*	African black pepper,	see Piper Afzelii.
" Cubeba,	Cubeb,	unripe Ft. sti. diu. a-syp. pur. aro.
" elongatum,*	Matico,	see Artanthe elongata.
" Guineense,*	Cayenne (Guinea) pepper	see Capsicum annuum.
" Hispanicum,*	Capsicum,	"　　"　　"
" Indicum,*	"	"　　"　　"
" Jamaicense,*	Pimento (Jamaica) pepper,	see Eugenia Pimenta.
" longum,	Long pepper,	see Chavica; sti. car. ton.
" lusitanicum,*	Cayenne pepper,	see Capsicum annuum.
" Melegueta,*	Grains Paradise,	see Amomum Melegueta.
" methysticum,	Ava kava, Kawa,	Rt. in gout; a-syp
" murale,*	Biting stonecrop	see Sedum acre.
" nigrum,	Black pepper,	unripe Ft. aro. pun. sti. feb. con.
" odoratum Jamaicense,*	Pimento,	see Eugenia Pimenta.
" officinarum,*	Long pepper,	see Piper longum.
" reticulatum,	Jaborandi root, Brazil,	Rt. and Catkins sia. sti. exp. a-ven.
" Tabascum,*	Pimento,	see Eugenia Pimenta.
" Turcicum,*	Cayenne pepper,	see Capsicum annuum.
Piptostegia operculata,*	Brazilian purge,	see Convolvulus operculatus.
Piscidia erythrinia,	Jamaica dogwood,	Bk. acr.-nar. sed.; fish poison.
Pistacio Lentiscus,	Mastich tree,	yields Mastic gum; Ol. a-spa.
" terebinthus,	Turpentine tree,	yields Chian turpentine.
" vera,	Pistachio nut tree,	Nu. resemble sweet almonds.
Pisum sativum,	Garden or field pea,	Sd. far. nut. esc.
Pitaya Condaminea,*	Hard Pitaya bark,	see Cinchona Condaminea.
Pitosporum Tobira,	Pitch seed plant,	Pl. and Fl. orn.; Sd. sticky.
Plauanthus fastigatus,	Fir club moss,	see Lycopodium selago.
Plantago Alisma,*	Water plantain,	see Alisma Plantago.
" aquatica,*	"　　"	"　　"　　"
" arcisma,	Fleawort,	see Plantago Psyllium.
" arenaria,*	"	"　　"　　"
" cordata,	Heart leaved plantain,	Ls. ref. diu. sub-sty. ast. ano.a-eme.
" coronopus,	Bucks' horn plantain,	Ls. as poultice in agues.
" cynops,*	Fleawort,	see Plantago Psyllium.
" decumbens,	Spogel seed, Ispaghul,	Sd. muc. pur. det.
" Genevensis,*	Fleawort,	see Plantago Psyllium.
" Ispaghula,*	Spogel seed,	"　　" decumbens.
" lanceolata,	Snake plantain, (Ribwort),	Rt. and Ls. ast. vul. alt.

BOTANICAL.	COMMON.	PROPERTIES, PRODUCTIONS, USES, ETC.
Plantago lancifolia,	Rib grass,	Rt. and Ls. ast. vul. alt.
" major,	Plantain, Waybread,	Pl. alt. diu. a-sep. a-syp. f-com.
" media,	Rib grass, Lambs' lettuce	Pl. ast. vul. det. feb.
" nitida,*	Fleawort, Fleaseed,	see Plantago Psyllium.
" psyllium,	" Branch'g plant'n,	Sd. muc. pur. det. dem.
" Virginica,	White (Dwarf) plantain,	Pl. vul. diu. a-veu. f-com.
" vulgaris,*	Plantain,	see Plantago major.
Platanthera orbiculata,	Large round lv'd orchis,	Ls. for dressing blisters. [ambar.
Platanus occidentalis,	False sycamore, Plane tree,	Ls. opt.; Bk. a-sco.; see Liquid-
" orientalis,	Oriental plane tree,	Bk. and Wd. feb.; y'lds turpentine.
Plossica floribunda,	Frankincense,	African Olibanum.
Plumbago Europæa,	Leadwort,	Ju. cau. corrosive; in toothache.
" rosea,	Blister root,	Pl. ves.; Radix vesicatoria.
" scandens,	Herbe au Diable,	Rt. an energetic blister.
" Zeylanica,		Bk. of Rt. emm. par.; in itch.
Plumiera alba,	W. I. white jasmine,	Ju. milky; cathartic.
" drastica,	a Brazilian species,	in jaundice and chronic obstruct'ns
" rubra,	Red jasmine, Frangipanni,	Fl. deliciously fragrant.
Plumifera lancifolia,	Agonia bark,	Bk. feb. ton.
Poa annua,	Low spear grass,	Pl. a common fodder plant.
" compressa,	Blue grass,	a valuable pasture grass.
" pratensis,	Kentucky blue grass,	" . "
" trivialis,	Round stalked meadow grass, a valuable pasture grass.	
Podalyria tinctoria,*	Wild indigo plant,	see Baptisia tinctoria.
Podophyllum callicarpum,	a variety of	Podophyllum peltatum.
" diphyllum,*	Twin leaf plant,	see Jeffersonia diphylla.
" montanum,	Mountain mayapple,	Rt. anal. to Podophyllum peltatum.
" peltatum,	Mayapple, Am. mandrake	Rt. cat. cho. res. alt. eme. dia. ver.
Pogostemon Patchouly,	Patchouly, Pucha pat,	Pl. a favorite perfume; ius. [emm.
Pohon antlar,*	Upas tree of Java,	see Antiaris toxifera.
Poinciana pulcherrima,	Spanish carnation,	see Cæsalpina pulcherrima.
Polanisia graveolens,	Wormweed,	Pl. ves. c-irr. ver.
" icosandra,	Bastard mustard,	" " pun.
Polemonium cæruleum,	Greek valerian,	Rt. ast. vul.
" reptans,	Am'n Greek valerian,	Rt. alt. dia. pec. diu. exp. a-ven.
Polianthes tuberosa,	Tuberose,	Fl. orn. fragrant; Rt. a-ven. eme.
Polygala amara,	Bitter polygala, Europe,	Pl. bit. dia. exp. pec.; in pleurisy.
" amarella,*	" "	see Polygala amara.
" incarnata,	Milkwort,	" " Senega.
" Nuttallii,	Ground centaury,	Pl. ton. alt. diu.; in boils and cry-
" paucifolia,	Flowering wintergreen,	anal. to Polygala amara. [sipelas.
" polygama,*	Bitter polygala,	see Polygala rubella.
" polymorpha,	Common milkwort,	" " Senega.
" rubella,	Bitter polygala, Am'n,	Rt. and Hb. bit. ton. sti. exp.
" Senega,	Senega snakeroot,	Rt. sti. sia. exp. diu. dia. emm.
" Serpentaria,	a shrub of Caffraria,	Rt. a-ven.
" Virginiana,*	Senega root,	see Polygala Senega.
" vulgaris,	Bitter polygala,	" " polymorpha.
Polygonatum articulatum,	Joint weed,	Pl. in dysentery.
" multiflorum,*	Great Solomon seal,	see Convallaria multiflora.
" officinale,	White wort,	" " "
" uniflorum,	Solomon seal,	" " polygonatum.
" vulgare,	" "	" " "
Polygonum acre,	Water pepper,	see Polygonum hydropiper.
" amphibium,	Water persicaria,	Rt. used as sarsaparilla; dep.
" arifolium,	Sickle grass,	see Polygonum articulatum.
" articulatum,	Joint weed,	Rt. and Pl. ast. vul.
" aviculare,	Knot grass,	Rt. ast. vul.; Sd. aro. pur. eme.
" Bistorta,	Bistort, Patience dock,	Rt. ast. diu. alt. sad.
" coccineum,*	Water persicaria,	see Polygonum amphibium.

BOTANICAL.	COMMON.	PROPERTIES, PRODUCTIONS, USES, ETC.
Polygonum convolvulus,	Black-bird (Bind) weed,	Sd. used as buckwheat.
" divaricatum,	Eastern buckwheat,	Rt. as meal used as food.
" erectum,*	Knot grass,	see Polygonum aviculare.
" Fagopyrum,	Common buckwheat,	Sd. esc. nut. for.; as food.
" hydropiper,	Water pepper,	Ls. aer. pun. a-sep ape. diu.
" hydropiperoides	" " mild,	Ls. a-sep. ape. diu.
" orientale,	Princes' feather,	Pl. and Fl. orn.
" Persicaria,	Heartsease,	Pl. vul. a-sep.
" punctatum,	Smart weed, Am'n Water pepper,	Pl. aer. ast. dia. ton.
" tinctorium,	Indigo plant,	Pl. said to yield indigo.
" Virginianum,	Virginia bistort,	Pl. ast. ton. diu.
Polymnia Canadensis,	Leaf cup,	Pl. vul. ton. a-spa.
" uvedalia,	Yel. leaf cup. (Bearsfoot)	" "
Polypodium Adiantiforme,*	Callahuala root,	see Polypodium Calaguala.
" ammifolium,*	" "	" " "
" Calaguala,	" "	Rt. diu.; in dropsy, pleurisy, etc.
" Filix Fœmina,*	Female fern,	see Aspidium Filix Fœmina.
" Filix Mas,*	Male fern,	" " " Mas.
" fragile,	Cup fern,	Pl. used as Adiantum pedatum.
" incanum,	Rock brake, (Bracken),	Rt. dem. cat. ant.
" incisum,*	Female fern,	see Aspidium Filix Fœmina.
" molle,*	" "	" " " "
" trifidam,*	" "	" " " "
" Virginianum,*	Polypody, Rockbrake,	see Polypodium vulgare.
" vulgare,	" "	Pl. diu. cho. pur. sac.
Polyporus,*	Agaries,	see Boletus, each species.
Polystichum Filix Mas,*	Male fern root,	see Aspidium Filix Mas.
Polytrichum commune,*	Common maidenhair,	see Adiantum.
" vulgare,	Golden maidenhair,	Pl. diu. ast.; in lung complaints.
" juniperum,	Hair cap moss,	Pl. diu.; in dropsy.
Pombalia itubu,*	White ipecac,	see Ionidium Ipecacuanha.
Ponticum Absinthium,*	True Roman wormwood,	see Artemisia Pontica.
Populus albus,	Abele tree,	Bk. in strangury.
" angulata,	Cottonwood tree,	Bk. ast. ton. bit.
" balsamifera,*	Tacamahac,	see Fagara octandra.
" candicans,	Balm of Gilead (bud) tree,	Bk. sti. ton. diu. a-sco.; Bd. vul.
" dilatata,	Italian poplar,	Tree cultivated for ornament. [bal.
" grandidentata,	Poplar tree, large,	Bk. ton. sto. feb. a-per.
" lævigata,	Cottonwood tree,	Bk. ast. ton. bit.
" nigra,	Black poplar,	Bd. resinous, vul.; in ointments.
" pendula,*	European aspen,	see Populus tremula.
" tremula,	" "	Bk. used in strangury.
" tremuloides,	American aspen, Poplar,	Bk. ton. sto. feb. a-per.
Porcelia triloba,*	Papaw, Custard apple.	see Carica papaya.
Portulaca grandiflora,	Showy (Garden) portulaca,	Pl. and Fl. orn.
" oleracea,	Gard'n purslane, Pursley	Pl. a-sep. diu. ape. vul.
Potentilla anserina,	Silver weed, Wild tansy,	Ls. ast. ton.
" argentina,*	" " " "	see Potentilla anserina.
" Canadensis,	Common fivefinger,	Pl. ton. ast.; in night sweats.
" fragraria,	Barren strawberry,	Rt. ast.; in summer complaints.
" Norvegica,	Norway Cinquefoil,	" " "
" palustris,	Marsh fivefinger,	see Comarum palustre.
" reptans,	Cinquefoil,	Pl. ast. feb.
" Tormentilla,	Tormentilla,	Rt. ast. feb. ton.; in hemorrhage.
Poterium sanguisorba,	Small bibernel,	Pl. in salads; cor. sto.
Pothos fœtida,*	Skunk cabbage,	see Ictodes fœtidus.
Pothomorpha peltata,	Caapeba,	Pl. diu.; in strangury.
Premna integrifolia,	Headache tree,	Ls. bit. ast. cep.
Prenanthes alba,	Canker weed, Lions' foot	Pl. see Nabalus albus.
" Fraseri,	Gall of the earth,	Rt. bit. ast. ton.

BOTANICAL.	COMMON.	PROPERTIES, PRODUCTIONS, USES, ETC.
Prenanthes Serpentaria,*	White lettuce,	see Nabalus albus.
Primula auricula,	Bears' ear,	Hb. vul. exp.
" elatior,	Oxlip,	Rt. eme.; Hb. ano. c-irr.
" officinalis,	Tree primrose,	see Primula veris.
" sinensis,	Chinese primrose,	Pl. and Fl. orn.
" veris,	Eng. Cowslip primrose,	Fl. ton. a-spa. ano.
" vulgaris,	Common primrose,	Rt. and Ls. ste. ano. eme.
Prinos glaber,	Appalachian tea,	Ls. ast.; used as tea.
" glabra,	Ink berry,	Ls. ast.; Ft. eme. ton.
" verticillatus,	Black alder,	Bk. ton. bit. alt. feb. ast.; Ft. cat.
Prosopsis dulcis,	Gum mesquite tree,	Ft.-pulp eaten; Gu. as Arabic. [ver.
Protococcus vulgaris,	a fungus moss,	Fu. yields Phycite sugar.
Prunella officinalis,*	Healall, Selfheal,	see Prunella vulgaris.
" vulgaris,	"	Pl. ast.; in sore throat.
Prunus acacia,	Bird cherry,	see Prunus padus.
" amygdalus,*	Almonds,	see Amygdalus communis
" " var. amara,*	Bitter almond,	" " "
" " var. dulcis,*	Sweet almond,	" " "
" Armeniaca,*	Apricot,	see Armeniaca vulgaris.
" avium,*	Sweet cherry, Gaskins,	see Cerasus avium.
" Caroliniana,	Winter laurel,	Ls. give almond flavor; poi.
" capricidia,	Goatsbane,	Ls. yield Prussic acid.
" Cerasus,	Garden red cherry,	Ft. esc. aci.; Bk. ton. sed.
" Communis,	Bullace plum,	Ft. esc. aci.; in preserves.
" Damascena,*	Damson plum,	see Prunus domestica.
" domestica,	" "	Ft. esc. lax. con.
" hortensis,	Garden red cherry,	see Prunus cerasus.
" insitia,	Bullace plum,	Ft. esc. lax. con.
" Lauro-cerasus,	Cherry laurel,	Ls. aro. fragrant, poi. con.
" macrophylla,*	Sweet cherry, Gaskins,	see Cerasus avium.
" maritima,	Beach plum,	Ft. aci. esc. a-sco. con.
" nana,*	Flowering almond,	see Amygdalus pumila.
" padus,	Bird cherry,	Bk. yields oil like almonds.
" Pennsylvanica,	Wild red cherry,	Bk. bit. acr. ton. ast.
" pygmæa,	American sloe,	Ft. in gargles.
" Sebestena,	Sebesten plum,	see Cordia domestica.
" serotina,	Wild black cherry,	see Prunus Virginiana.
" spinosa,	Sloe tree, Wild plum,	Ft. aci. ast.; used in hemorrhage.
" sylvestris,*	" "	see Prunus spinosa.
" Virginiana,	Wild choke (black) cherry,	Bk. bit. ton. sti. sed. pec. poi.
" vulgaris,	Wild cluster or bird cherry, see Prunus padus.	
Psidium pyriferum,	Guava, Bay plum,	Ft. yields Guava jelly; Ls. ast.
Psoralea corylifolia,	Malay tea, Bawchang sd.	Ps. aro.; Sd. yields oil.
" esculenta,	Bread root, Ind'n turnip,	Rt. used in Pomme Blanche.
" glandulosa,	Paraguay Jesuits' tea,	Ls. vul. pec.; see Ilex.
" pentaphylla,	Mexican contrayerva,	Rt. as Contrayerva.
Psychotria crocea,	West India ipecac,	Rt. eme.
" emetica,	Peruvian Black ipecac,	Rt. used as Ipecacuanha.
" Ipecacuanha,*	" " "	see Psychotria emetica.
Psyllium pulicaria,	Fleawort,	see Plantago Psyllium.
Ptarmica vulgaris,*	German pellitory,	see Achillea Ptarmica.
Ptelea trifoliata,	Wafer ash,	Bk. tonic, sti. alt. a-per. ant.
Pteris aquilina,	Com'n brake, Female fern,	Rt. ant.; in tapeworm.
" atropurpurea,	Rockbrake,	Pl. ast. ant. f-com.; in night sweats
" palustris,*	Female fern,	see Aspidium Filix Fœmina.
Pterocarpus Draco,	W. I. kino, Dragons' blood,	Ex. powerfully astringent.
" erinaceus,	African kino,	Ex. powerfully astringent.
" Marsupium,	East Indian kino,	Ex. " "
" santalinus,	Red sanders wood,	Wd. ton. ast.; for color'g tinctures
Pterocaulon pycnostachyum,	Indian black root,	Rt. see Conyza pycnostachyum.

BOTANICAL.	COMMON.	PROPERTIES, PRODUCTIONS, USES, ETC.
Pterospora andromeda,	Albany beech drops,	Rt. bitter, ant. diu. emm.
" elatior,	High stalk'd beech drops,	varieties of Pterospora andromeda.
" flavicaulis,	Yellow " - "	" " "
" leucorhiza,	White " "	" " "
" · pauciflora,	Few flowered "	" " "
Pulegium vulgare,*	European pennyroyal,	see Mentha pulegium.
Pulicaria dysenterica,*	Small fleabane,	see Inula dysenterica.
Pulmonaria maculata,*	Lungwort,	see Pulmonaria officinalis.
" officinalis,	"	IIb. dem. muc. pec.
" Virginica,	Am'n lungwort, (Cowslip),	see Mertensia Virginica.
Pulsatilla nigricans,*	Meadow anemone,	see Anemone pratensis.
" patens,*	Pulsatilla,	" - " Pulsatilla.
" pratensis,*	Meadow anemone,	" " pratensis.
" vulgaris,*	Pasque flower,	" " Pulsatilla.
Punica Granatum,	Pomegranate,	Fl. and Bk. ast. ant. f-com.
Punicum malum,	" fruit,	Rind of Ft. ant.; in tapeworm.
Pycnanthemum aristatum,	Wild basil,	Pl. dia. sti. a-spa. car. ton.
" incanum,	" " .	" " " "
" linifolium,	Flax leaved basil,	Pl. dia. ton. sed.
" montanum,	Mountain mint,	. " " Ol. sti.
" pilosum,	a species of thyme,	" " "
" Virginicum,	Narrow lv'd Va. thyme,	Pl. dia. sti. a-spa. car.
Pyrethrum carneum,	Persian insect powder,	Fl. in powder; insecticide.
" cinerariæfolium,	Dalmatian insect powder	" " "
" parthenium,	Feverfew,	Pl. ton. car. emm. sti. ver.
" roseum,	Caucasian insect powder	Fl. in powder; insecticide.
" sylvestre,*	German pellitory,	see Achillea Ptarmica.
" tanacetum,	Costmary,	Ls. sto. cor. cep. f-com.; Sd. ver.
Pyrola elliptica,	Shin leaf,	Ls. diu. ton. ast.
" maculata,*	Spotted pyrola,	see Chimaphila maculata.
" rotundifolia,	Round leaved pyrola,	Ls. ton. ast. diu. a-spa.
" umbellata,*	Pipsissewa,	see Chimaphila umbellata.
" uniflora,	One flow'd wintergreen,	Pl. anal. to Pyrola rotundifolia.
Pyrularia oleifera,	Buffalo (oil) nut,	Nu. oily; eaten by cattle.
Pyrus Americana,	Mountain ash,	see Sorbus Americana.
" arbutifolia,	Chokeberry,	Ft. very astringent.
" aria,	Beam tree, Wild pear,	Ft. ast.; in alvine fluxes.
" communis,	Wild pear tree,	Ft. ast. acerb. a-sec.
" coronaria,	Crab apple,	cultivated for its fruit; esc.
" Cydonia,*	Quince tree,	see Cydonia vulgaris.
" domestica,	Sorb apple,	see Sorbus domestica.
" Malus,	Apple tree and fruit,	Ft. esc. nut. lax. diu.; Bk. ton. feb.
" sorbus,	True service tree,	Ft. ast.
Pyxidaria macrocarpa,		Bk. ast.
Quamoclit vulgaris,	Cypress vine,	see Ipomœa quamoclit.
Quassia amara,	Bitter wood, Jamaica,	see Simarouba.
" excelsa,	Surinam Quassia,	see Simarouba excelsa.
" Simarouba,	Simarouba, Mt. damson,	Bk. bit. ton.
" Simaruba,*	" "	see Quassia Simarouba.
Quercula minor,*	Germander,	see Teucrium chamædrys.
Quercus alba,	White oak, Officinal,	Bk. ast. ton. a-sep.; Nu. roasted.
" ægilops,	Holm oak,	yields an inferior gall.
" æsculus,	Italian oak,	has properties anal. to Q. alba.
" aquatica,	Water oak,	Bk. ast. ton.; wash for ulcers.
" Catesbæi,	Scrub oak,	Shrub oak; Bk. ast. bit.
" cerris,	Bitter (Turkey) oak,	yields galls; ast.
" coccifera,	Kermes oak,	yields Kermes dye.
" coccinea,	Scarlet oak,	Bk. used by tanners.
" " var. tinctoria,	Quercitron, Black oak,	" "
" excelsa,	Gall oak,	yields Nut galls.

BOTANICAL.	COMMON.	PROPERTIES, PRODUCTIONS, USES, ETC.
Quercus falcata,	Red (Spanish) oak,	Bk. quercitron; in tanning.
" ilex,	Evergreen oak,	Ft. esc. far.; yields galls.
" illcifolia,	Scrub (Bear) oak,	Shrub, Bk. ast.
" imbricaria,	Laurel, Shingle oak.	
" infectoria,	Dyers' oak,	yields officinal Galls.
" Lusitanica, var.	" "	see Quercus infectoria.
" montana,	Rock chestnut oak,	Bk. ast. ton. feb.; for tanning.
" niger, nigra,	Black oak (Jack),	Bk. quercitron; bit.
" obtusifolia,	Iron oak.	
" obtusiloba,	American Turkey oak,	Bk. bit. ast.
" occidentalis,	Corkwood oak,	Bk. known as cork.
" palustris,	Pin oak,	Bk. used as Quercus alba.
" pedunculata,*	Common British oak,	see Quercus tinctoria.
" Phellos,	Willow leaved oak.	
" prinoides,	Chinquapin oak,	Shrub oak.
" prinus,	White chestnut oak,	Bk. used as Quercus alba.
" robur,	Common European oak,	Bk. ast. ton. feb.
" rubra,	Red oak,	Bk. ast.
" sessiliflora,	Durmast oak.	
" suber,	Corkwood oak,	yields corkwood.
" tinctoria,	Black (Gall) oak,	Bk. officinal; ast. ton. feb. bit.
" . virens,	Live oak,	Wd. for building, etc.
Quillaya saponaria,	Soap tree bark,	Bk..sap. feb.; in coryza.
Quinquinia bicolor,	Bicolorata bark,	see Cinchona Condaminea.
" pinckneya	Florida (Georgia) bark,	see Pinckneya pubens.
Randia dumetorum,	Malabar ipecac,	Rt. eme.; Ft. fish poison; eme.
" latifolia,	Indigo berry,	Ft. a West India blue dye.
Ranunculus abortivus,	Chicken pepper,	Pl. acr. c-irr. ves.
" aconitifolia,	Bachelors' buttons,	Pl. and Fl. orn.
" acris,	Crowfoot buttercup,	Pl. anal. to Ranunculus bulbosus.
" albus,*	Wood anemone,	see Anemone nemorosa.
" aquatica,	Water crowfoot, .	see Ranunculus sceleratus.
" arvensis,	Corn crowfoot, Hungerweed,	Pl. acr. pol.
" bulbosus,	Crowfoot, (bulbous),	Pl. acr. rub. c-irr. nar.
" digitatus,*	Marsh crowfoot,	see Ranunculus sceleratus.
" ficaria,	Lesser celandine,	see Chelidonium minus.
" flammula,	Spearwort,	distilled water an emetic.
" laetus,*	Crowfoot, Buttercup,	see Ranunculus bulbosus.
" palustris,*	Marsh crowfoot,	" " sceleratus.
" Pennsylvanicus,	Bristly.crowfoot,	" " bulbosus.
" pubescens,	a So. African species,	Ju. in cancerous ulcers.
" repens,	Common (Creeping) crowfoot,	see Ranunculus bulbosus.
" sceleratus,	Marsh crowfoot,	Pl. acr.
" vernus,	Lesser celandine,	see Chelidonium minus.
" Virginianus,*	Blue cardinal, (Lobelia),	see Lobelia syphilitica.
Rapa,	Cabbage and Turnip,	see Brassica.
Raphanus hortensis,*	Garden radish,	see Raphanus sativus.
" marinus,*	Horse radish,	see Cochlearia Armoracia.
" sativus,	Garden radish,	Rt. a common esculent; diu. a-sco.
" sylvestris,*	Horse radish,	see Cochlearia Armoracia.
Rapuntium galeatum, var.*	Red lobelia,	see Lobelia cardinalis.
" maximum,*	" "	" " "
Renealmia cardamomum,*	Cardamom,	see Elettaria Cardamomum.
Reseda luteola,	Dyers' weed,. Weld,	Pl. yields a yellow dye; dep.
" odorata,	Mignonette,	Pl. and Fl. orn. fragrant.
Reticularia officinalis,	Lung moss,	see Sticta pulmonaria.
Rhabarbarum,	Rhubarb,	see Rheum.
Rhamnus alaternus,	Evergreen privet,	Ft. alt. lax.; yield sap green.
" catharticus,	Purging buckthorn,	Ft. powerfully cathartic; dep.
" frangula, .	Europ'n (Bl'k) alder buckth'n,	Bk. bit. eme. pur.; Ft. pur.

BOTANICAL.	COMMON.	PROPERTIES, PRODUCTIONS, USES, ETC.
Rhamnus Hispanicus,*	Evergreen privet,	see Rhamnus alaternus.
" infectorius,	French yellow berries,	Ft. yellow dye; Bk. ast. bit.
" Jujuba,	Jujube,	Ft. sty.; Bk. ast.; Sd. feb.
" Lotus,	"	analogous to Rhamnus Ziziphus.
" lycioides,	Black ramthorn,	Ft. in gouty pains.
" Ziziphus,	Jujube tree,	Ft.-pulp muc. pec.; in paste.
Rhaphanis,*	Garden radish,	see Raphanus sativa.
Rhaphanus,*	" "	" " "
Rheum,	Rhubarb,	Rt. cat. ast. ton. vul.
" Australe,*	Garden rhubarb,	see Rheum Emodi.
" compactum,	Chinese rhubarb,	Rt. used as Rheum palmatum.
" Capsicum,	Altai mountain rhubarb,	" " "
" crassinervium,	European rhubarb,	Rt. lax. ast.; stalks as pie plant.
" Emodi,	" (Garden) rhubarb	Rt. " " " "
" hybridnm,	" (Am'n) rhubarb,	Rt. " " " "
" leucorrhizum,	Tartarian rhubarb,	Rt. pur. ast. sto. ton. vul.
" officinale,	Indian (China) rhubarb,	" " "
" Moorecraftianum,	Himalaya Mt. rhubarb,	Rt. as Rheum palmatum.
" palmatum,	Chinese rhubarb,	Rt. pur. ast. sto. ton. vul.
" Rhabarbarum,*	India rhubarb,	see Rheum undulatum.
" Rhaponticum,	French (Crimea) rhubarb,	Rt. bit. ast. aro. muc.
" Sinense,	India rhubarb,	see Rheum palmatum.
" speciforme,	Himalaya Mt. rhubarb,	Rt. see Rheum palmatum.
" Tartaricum,	Tartarian rhubarb,	" " "
" undulatum,	India rhubarb,	" " "
" Webbianum,	Himalaya Mt. rhubarb,	" " "
Rhexia Virginica,	Deer grass,	Pl. and Fl. orn.
Rhinacanthus communis,*	Braid root,	see Justicia nasuta; for ringworm.
Rhinanthus crista-galli,	Penny grass, Yel. rattle,	Pl. alt. det.; to kill lice.
Rhizophora mangle,	Mangrove, (root bearer)	Ft. and Bk. ast.; Ls. det. vul.
Rhododendron,	Rose tree,	•
" chrysanthemum,	Yel'w fl'd rhododendron,	Ls. acr.-nar. sti. ast. a-rhe.
" chrysanthum,*	" " "	see Rhododendron chrysanthemum
" ferrugineum,	Dwarf rose bay,	Ls. bit. nar. sti. dia.
" hirsutum,	Rose of the Alps,	Pl. and Fl. orn.
" maximum,	Mt. laurel, Wild rose bay	Ls. nar. sti. ast. a-rhe.; in gout.
" Ponticum,	Rose bay,	Ls. nar. ast. bit.
Rhodanthe maculata,	Rose flow'd everlasting,	Pl. and Fl. orn. •
" manglesii,	" " "	" " "
Rhodorrhiza scoparia,	Canary rosewood,	see Genista canariensis.
Rhodymenia palmata,	Dulse, Dillisk,	Algæ ant. esc.
Rhus Belgica,*	Meadow fern,	see Myrica Gale.
" copallina,*	Mt. sumach, W. I. copal,	see Rhus glabrum. [tanning.
" coriaria,	Tanners' sumach,	Bk. and Ls. ast.; for dyeing and
" cotinus,	False fringe tree,	Wd. yields young fustic; ast.
" diversiloba,	California pois'n sumach	Pl. an acrid, irritating poison.
" glabra,	Smooth (Sleek) sumach,	see Rhus glabrum.
" glabrum,	" " "	Bk. and Ls. ton. ast. a-sep.; Ft. diu.
" lobata,*	California pois'n sumach	see Rhus diversiloba. [ref. aci.
" metopium,	W. I. hog gum tree,	said to yield Hog gum.
" pumilum,	So. States pois'n sumach	extremely poisonous.
" radicans,	a variety of	Rhus toxicodendron.
" semi-alata,		yields Chinese galls.
" succedanea,	Red lac sumach,	yields Japan wax.
" succedanum,	Chinese sumach,	" " "
" sylvestris,	Meadow fern,	see Myrica Gale.
" Toxicodendron,	Poison oak (sumach),	Pl. poi. rub. ves.; Ls. in paralysis.
" " var. radicans,	Poison vine (ivy),	Pl. poi. ves. diu. dia. lax. sti.
" typhina,	Stags' horn sumach,	analogous to Rhus glabrum.
" venenata,*	Poison sumach,	see Rhus vernix.

BOTANICAL.	COMMON.	PROPERTIES, PRODUCTIONS, USES, ETC.
Rhus vernix,	Poison swamp sumach,	Pl. poi.; anal. to R. toxicodendron.
" vernicifera,	Japanese sumach,	yields Japanese lacquer varnish.
Ribes album,	White currant,	a variety of Ribes rubrum.
" aureum,	Buffalo currant,	Fl. orn.; spicy scented.
" Floridanum,	Wild black currant,	Ft. diu. dia. a-sco.; Bk. lit.
" grossularia,	Garden gopseberry,	Ft. esc. aci. diu. ref. feb.
" nigrum,	Black currant,	Ft. diu. dia. a-sco.; Bk. lit.; Ls.
" officinale,*	Red currant,	see Ribes rubrum. [diu.
" olidum,*	Black currant,	" " nigrum.
" rubrum,	Red currant,	Ft. aci. ref. feb.; Rt. diu.
" triflorum,*	Gooseberry,	see Ribes grossularia.
" Uva crispa,	"	" " "
" vulgare,*	Red-currant,	" " rubrum.
Richardia Africana,	Calla lily, Trumpet lily,	Pl. and Fl. orn.; Rt. acr. ves.
" scabra,*	White ipecac,	see Richardsonia scabra.
Richardsonia Brasiliensis,*	" "	" " " "
" emetica,*	" "	" " " "
" scabra,	" " (undulated)	Rt. eme. sud.; as ipecac.
Riciuoides elæagnifolia.*	Cascarilla,	see Croton eleuteria.
Ricinus Africanus,	Castor oil plant,	Sd. yields oil; cat.
" communis,	" " "	Sd. yields castor oil; cat.
" "	" " "	Ls. gal. emm. f-com.
" " var. viridis,	Lamp oil seed,	a variety of Ricinus communis.
" lividus,	So. Af'n castor oil plant,	Sd. dra. cat. irr.; Ol. cat.
" major,*	Physic nuts,	see Jatropha curcas.
" minor,*	Cassava (Tapioca) plant,	." " maulhot.
" viridis,*	Castor oil plant,	see Ricinus communis.
" vulgaris,*	. " " "	" " "
Riviana octandra,	Hoop withy,	Pl. and Fl. orn.
" tinctoria,	Rouge plant,	" " Ju. scarlet.
Robinia amara,	Bitter locust,	Bk. bit.; in diarrhœa and dyspepsia
" hispida,	Bristly rose acacia,	Bk. of Rt. eme. cat. exp.
" pseudo-acacia,	Locust tree,	" " " ton.
" viscosa,	Clammy locust,	" " " nar.
Roccella fusiformis,	Angola weed,	yields a dyestuff.
" tinctoria,*	Archel, Litmus plant,	see Lichen roccella.
Ronabea emetica,*	Black ipecac,	see Psychotria emetica.
Rosa afzeliana,*	Dog rose,	see Rosa canina.
" alba,	White rose,	Fl. orn. fragrant.
" armata,*	Dog rose,	see Rosa canina.
" astriaca,*	Red rose,	" " Gallica.
" arvensis,	Ayrshire rose,	Fl. orn.
" blanda,	Wild rose,	Fl. ast. ton. opt.
" canina,	Dog rose, Hip tree,	Ft. in conserve of rose; pill mass.
" Carolina,	Swamp rose,	Fl. orn.
" centifolia,	Hundred lv'd rose (red),	Petals ape.; distilled for rose water
" cinnamomea,	Cinnamon rose,	Fl. fragrant, orn. [and otto.
" Damascena,	Damask (Red) rose,	see Rosa centifolia.
" Gallica,	French red rose,	Petals ton. ast. opt.; in conserve.
" Indica,	Bourbon rose,	Fl. orn.
" leucautha,*	White rose,	see Rosa alba.
" moschata,	Musk "	yields otto of rose.
" muscosa,	Moss "	Pl. orn.
" pallida,	Pale "	see Rosa centifolia.
" parviflora,	Wild "	Fl. ast. ton. opt.
" rubra,* *	Red " French,	see Rosa Gallica.
" " vulgaris,*	Dog "	" " canina.
" rubiginosa,	Sweet brier,	Pl. climbing; Fl. fragrant.
" sempervirens,	Evergreen rose,	the Ayrshire rose is a variety.
" setigera,	Prairie rose, climbing,	Pl. and Fl. orn.

BOTANICAL.	COMMON.	PROPERTIES, PRODUCTIONS, USES, ETC.
Rosa sinica,	Cherokee rose,	Pl. for garden hedges; Fl. white.
" sylvestris,*	Dog rose,	see Rosa canina.
" spinosissima,	Scotch rose, (Burnet),	Fl. orn. fragrant.
" sulphurea,	Yellow rose,	"
" usitatissima,*	White rose,	see Rosa alba.
Rosella vulgaris,*	Sundew,	see Drosera rotundifolia.
Rosmarinus hortensis,*	Rosemary plant,	see Rosmarinus officinalis.
" officinalis,	" "	Ls. and Fl. aro. bit. ccp. sti. cmm.
" "	" "	Ol. sti. rub. a-rhc. per.
" stœchadis facie,*	Poley mountain, of Candy,	see Teucrium Cretlcum.
" sylvestris,*	Marsh tea,	see Ledum palustre.
Rottlera Schimperi,	Musena bark,	Bk. with Koosso for tapeworm.
" tinctoria,	Monkey face tree, Kamecla,	Po. of Ft.-Ca. ver. against tape-[worm.
Roupellia grata,	Cream fruit,	Pl. orn.
Royena,	Bladder nut, African,	Wd. resembles ebony.
Rubia cordifolia,	Bengal madder,	Rt. dco. dct. dlu. cmm. ; a dyestuff.
" Iberica,*	Dyers' madder,	see Rubia tinctoria.
" peregrina,*	" " English,	" " "
" relbun,	Chili madder,	" " "
" tinctoria,	Dyers' madder,	Rt. dco. diu. cmm.; a dyestuff.
" sylvestris,*	" "	see Rubia tinctoria.
Rubus affinis,*	Blackberry,	see Rubus fruticosus.
" arcticus,	Shrubby strawberry,	Ft. a-scp. ref. a-sco.
" batus,*	Dewberry plant,	see Rubus Cœsius.
" Cœsius,	" "	Ft. resembles the blackberry.
" Canadensis,	Low blackberry, Dewberry,	Ft. csc. a-sco.; Bk. of Rt. ast.
" Chamœmorus,	Cloud berries, Knot berries,	Ft. a-sco. feb· ref.
" flagellaris,*	Dewberry,	see Rubus Cœsius.
" frondosus,	Common blackberry,	" " fruticosus.
" fruticosus,	" " Bramble,	Ft. edi. diu. ast.; Bk. ast.
" idæus,	Raspberry, Hindberry,	Ft. csc. lax.; Ls. ast. ton. sti. par.
" niger,*	Blackberry,	see Rubus fruticosus. [a-cmc.
" occidentalis,	Thimbleberry,	Ft. lax.; Ls. mild astringent.
" odoratus,	Rose flower'g raspberry,	Pl. and Fl. diu,; in dropsy.
" procumbens,*	Low blackberry,	see Rubus Canadensis.
" rosæfolius,	Brier rose,	Pl. orn; for its flowers.
" sexatilis,	" herb,	St. and Ls. ast. vul.
" strigosus,	Wild Red raspberry,	Ft. csc. lax.; Ls. ast. a-cmc. pur.
" trivialis,*	Low blackberry, Dewberry,	see Rubus Canadensis.
" villosus,	High (Am'n) blackberry,	Bk. of Rt. ast.; in dysentery.
Rudbeckia laciniata,	Thimbleweed,	Pl. diu. ton. bal.; in kidney compl'ts
" purpurea,	Red sunflower,Black sampson,	Rt. a-syp. dep.; see Stillingia
Rudolphia frondosa,*	Dhak tree, Bengal kino,	see Butea frondosa.
Ruellia tuberosa,	Many root,	Rt. used as ipecacuanha.
Rumex acetosa,	Common sorrel,	Ls. aci. ref. diu. a-sco.; Rt. ast.
" acetosella,	" field sorrel,	Ls. " " Ex. in cancer.
" acetosus,*	" " "	see Rumex acetosella.
" acutus,	Sharp leaved Wild dock,	Rt. dep. alt. ast.; for itch.
" Ætinensis,	French sorrel,	see Rumex scutatus.
" Alpinus,	Mt. (Monk's) rhubarb,	Rt. lax. ton. dep.
" aquaticus,	Great water dock,	Rt. ton. ast. a-sco.
" Britannica,	Water dock,	Ls. sub-aci. lax.; Rt. ton. ast. a-sco.
" crispus,	Yellow dock,	Rt. alt. ton. dep. ast. a-sco.
" gigantea,*	Vegetable mercury,	see Ascleplas gigantea.
" glaucus,*	French sorrel,	see Rumex scutatus.
" hastatus,*	" "	" " " "
" hydrolapathum,	Water dock,	" " Britannica.
" intermedius,*	Common field sorrel,	" " acetosella.
" lunaria,	Moonwort.	[frice.
" obtusifolius,	Blunt leaved dock,	Rt. ast. a-sco. dep.; Po. as denti-

BOTANICAL.	COMMON.	PROPERTIES, PRODUCTIONS, USES, ETC.
Rumex patientia,	Garden patience dock,	Rt. dep. lax.; mild cathartic.
" purpurea,	Blunt leaved dock,	see Rumex obtusifolius.
" sanguineus,	Olcott rt., Bloody dock,	Rt. a slow cathartic; dep.
" scutatus,	French sorrel,	Ls. aci. ref. a-sco.
" tuberosa,*	Common sorrel,	see Rumex acetosa.
Ruppia maritima,	Ditch grass,	Pl. vul. dep.
Ruscus aculeatus,	Sweet broom root,	Rt. ape. diu.
" hypoglossum,	Horse tongue,	formerly used in relaxed uvula.
Ruta graveolens,	Garden rue,	Hb. aro. bit. pun. ton. emm. acr.
" hortensis,*	" "	" " emm. ton. sti.
" muraria,*	Wall rue,	see Asplenium ruta muraria.
" sylvestris,	Harmel wild rue,	Hb. hyp.
Sabadilla officinalis,	Cevadilla seed,	see Veratrum Sabadilla.
Sabbatia angularis,	Am'n (red) centaury,	Pl. a-bil. ton. bit. ver. emm. feb.
Sabal Adansonii,	Dwarf palmetto,	Bd. eaten as salad.
" Palmetto,	Cabbage palmetto,	" "
" serrulata,	Saw palmetto,	Ls. in making hats.
Saccharum officinarum,	Sugarcane,	yields most of the sugar of com-
" officinale,*	"	see Saccharum officinarum. [merce.
Sagina procumbens,	Pearlwort.	
Sagittaria saggittifolia,	Arrowhead, Arrow weed	Rt. yields fecula; ref. ast. det.
" variabilis,	" " "	" " Ls. dispel milk.
Sagittarium Alexipharmacum,	Malacca root,	see Canna Indica; for arrow pol.
Sagus lævis (inervis*),	Spineless sago palm,	Pl. yields sago; esc. nut. far.
" Raphia,*	Sago palm,	see Sagus Rumphii.
" Ruffia,*	" "	" " "
" Rumphii,	Prickly sago palm,	Pl. yields sago; esc. nut. far.
" vinifera,	Sago palm,	" " "
Salicornia annua,	Saltwort,	yields Barilla by burning.
" herbacea,	Marsh samphire,	St. pickled eaten as samphire.
Salisburia adiantifolia,	Ginkgo tree (ornament'l)	Sd. edible, ast. ton.; yields oil.
Salix alba,	White willow, (salacin),	Bk. bit. feb. a-per.; Ls. ast.
" Ægyptiaca,	Ban, Calaf,	Bk. a-sep. cor. a-aph.
" Babylonica,	Weeping willow,	Tree orn.
" fragilis,	Crack willow,	Bk. bit. ton.
" helix,	Yellow dwarf willow,	a variety of Salix purpurea.
" latifolia,	Broad leaved willow.	
" humilis,	Low bush willow.	[a-aph.
" nigra,	Black (Pussy) willow,	Bk. bit. ast. a-sep. det. a-per.; Bd.
" pentandra,	European willow,	Bk. ast. ton. feb. aro.; y'ds Salacin.
" purpurea,	Purple willow,	Bk. bitter.
" Russeliana,	Bedford willow,	Bk. ast. ton. feb.
" viminalis,	Basket willow, (Osier),	Bk. ast. ton. muc.
Salsala Kali,	Saltwort,	Pl. emm. sti. diu. hyd.; y'ds Barilla.
" tragus,	"	yields Barilla.
" soda,	Glasswort,	yields soda of Alicant.
Salvadora Persica (Indica),	Scripture mustard tree,	Bk. of Rt. acr. ves.; Sd. pun.
Salvia Africana,	Wild African sage,	Hb. analogous to Salvia officinalis.
" argentea,	White sage,	Pl. orn.
" Claytoni,	Vervain leaved sage,	Pl. sud. ton. a-aph.
" colorata,	Purple topped sage,	see Salvia horminum.
" fulgens,	Mexican sage,	Pl. and Fl. orn
" Hispanica,	Chia seed, wild,	Sd. opt. muc.
" horminum,	Purple topped clary,	Hb. exc.; Sd. opt. aph.
" hortensis,*	Garden sage,	see Salvia officinalis.
" lyrata,	Wild sage, Cancer weed,	Ls.-Ju. for warts and cancers.
" minor,*	Garden sage,	see Salvia officinalis.
" officinalis,	" " officinal,	Ls. ton. sud. ast. exp. aro. a-aph.
" pomifera,	Crete sage, Sage apple,	Galls with honey form a confection.
" pratensis,	Meadow sage,	analogous to Salvia officinalis.

14*

BOTANICAL.	COMMON.	PROPERTIES, PRODUCTIONS, USES, ETC.
Salvia sclarea,	Clary,	IIb. a-spa. bal. cor.
" splendens,	Scarlet flowered sage,	Pl. and Fl. orn.
" urticifolia,	Nettle leaved sage,	IIb. as Salvia officinalis.
" verbenaca,	Clear eye,	Sd. in ophthalmia.
" vitæ,*	Wall rue,	see Asplenium ruta muraria.
" vulgaris,*	Garden (Common) sage, see Salvia officinalis.	
Sambucus arborea,*	Elder (Elderberry),	see Sambucus Canadensis. [Ft. diu.
" Canadensis,	American elder,	Bk. cat. eme.; Fl. diu. dia. exa.;
" Ebulus,	Dwarf elder, Danewort,	Rt. pur.; Ls. in gout; Ft. for wine.
" helion,*	" "	see Sambucus Ebulus.
" herbacea,*	" "	" " "
" humilis,*	" "	" " "
" nigra,	Bl'k berried elder, Europ'n,	Bk. cat.; Ft. diu.; Fl. exp. dep.
" pubens,*	Red berried elder,	see Sambucus pubescens.
" pubescens,	" "	Bk. hyd.-cat. eme.
" racemosa,*	Mountain elder,	see Sambucus pubescens.
" vulgaris,*	Elder,	" " Canadensis.
Samolus valerandi,	Water pimpernel,	Ls. vul. pur.; for chaps.
Sandalum cœruleum,*	Behen (Ben) nuts,	see Moringa pterygosperma.
" rubrum,	Red sanders wood,	see Pterocarpus santalinus.
Sanguinalis corrigiola,*	Knot grass, (Blood staunch), see Polygonum aviculare.	
Sanguinaria acaulis,*	Blood root,	see Sanguinaria Canadensis.
" Canadensis,	" " (Red puccoon)	Rt. eme. sed. feb. sti. ton. diu. emm.
" "	" " "	Rt. res. a-sep. err. e-sch. det. exp.
Sanguisorba Canadensis,	Canada burnet (Saxifrage)	Pl. ast. ton.; in hemorrhage. [lax.
" media,	Great burnet,	Pl. ast. ton.; in hemorrhage.
" officinalis,	Garden (Italian) burnet, Pl.	" "
" rubra,*	Italian pimpernel "	see Sanguisorba officinalis.
Sanicula Canadensis,	American sanicle (black)	Rt. ncr. ano. feb. ast. exa.; in cho-
" Europæa,	European sanicle,	Ls. mild ast. vul. exa. [rea.
" eborascens,*	Butterwort, Rot grass,	see Pinguicula vulgaris.
" Marilandica, .	Bl'k sanicle (snakeroot),	Rt. ncr. ano. a-per. ast.; in chorea.
" montana,*	Butterwort, Rot grass,	see Pinguicula vulgaris.
" officinarum,*	European sanicle,	see Sanicula Europæa.
Sanseviera Guineensis,	African hemp,	used for cords and bowstrings.
Santalum album,	White sandal (Young wood), Wd. fragrant, per. sti. sud.	
" "	Yellow sandal (Old wood), Wd. fragrant, per.; Ol. a-syp.	
" citrinum,*	" "	see Santalum album.
" freycinetianum,	Sandwich Id. Yel. sandal	Wd. fragrant, per. sti.
" myrtifolium,*	Yel. and White sandal,	see Santalum album.
" pallidum,	" "	" " "
" rubrum,*	Red (sanders) saunders,	see Pterocarpus santalinus.
Santolina chamæcyparissus,	Lavender cotton (cypress),	Tw. ver. ins. aro.
Sanvitalia procumbens,	Sanvitalia,	Pl. orn.; see Rudbeckia.
Saphora tinctoria,*	Wild indigo weed,	see Baptisia tinctoria.
Sapindus emarginatus, .		Sd. exp. sap. a-cpi. .
" saponaria,	Indian soap nut (berry),	Ft. a-rhe. det. acr. sap.; fish poi.
Sapota Mulleri,	Bullet tree,	y'lds a substance like gutta percha.
" achras,*	Sapodilla (Marmalade) plum, see Acharas sapota.	
Saponaria dioica,*	White soapwort,	see Lychnis dioica.
" officinalis,	Soapwort, Bouncing bet,	Pl. ton. dia. alt. a-syp. exa. sap.
" vaccaria,	Cow herb,	Pl. gal. to cows; Sd. diu.
Sargassum bacciferum,	Gulf weed,	Seaweed, apc. diu. a-sco.
Sarothamnus Scoparius,—	Broom herb,	see Cytisus Scoparius.
Sarracenia flava, .	Water cup, yellow flow'd	Pl. sti. ton. diu. lax. ast.
" purpurea,	Pitcher plant, Huntsman's cup,	Pl. bit. ast. sti. ton. diu.
" variolaris, var.	Small-pox plant,	Pl. in small-pox.
Sarothra gentianoides,	Nit weed, False Johnswort,	Pl. aro. res. ncr.
Sassafras officinale,	Sassafras tree (bark),	Bk. aro. sti. alt. diu. dia. [sti. a-sep.
" "	" "	Pl. dem. emo. opt.; Ol. rub. a-rhe.

BOTANICAL.	COMMON.	PROPERTIES, PRODUCTIONS, USES, ETC.
Sassafras parthenoxylon,	Oriental sassafras,	Ft. bal.; Ol. a-rhe. sti.
Satureia,*		see Satureja.
Satureja capitata,	Cilliated savory,	Hb. ton. car. emm. a-spa.
" hortensis,	Summer savory,	Hb. aro. sti. car. emm. aph. con.
" montana,	Winter savory,	Hb. anal. to Satureja hortensis.
" origanoides,*	Mountain dittany,	see Cunila mariana.
" viminea,	Pennyroyal tree,	Ls. aro. car. sud. emm.
Saururus cernuus,	Breast weed,	Rt. emo. dis.; for inflamed breasts.
Sauvagesia erecta,	St. Martin's herb (wort)	Hb. vul. ast.
Saxifraga alba,*	White saxifrage,	Pl. diu. lit. ape.
" crassifolia,		Rt. a-sep.
" granulata,	White saxifrage,	see Saxifraga alba.
" rubra,	Dropwort, ·	see Spiræa filipendula.
" sarmentosa,	Strawberry geranium,	Pl. and Fl. orn.
" tridactylites,	Whitlow grass,	Pl. diu. dep.
" umbrosa,	London pride,	Pl. and Fl. orn. diu. ˙
" Virginiensis,	Saxifrage,	" "
" vulgaris,*	Meadow saxifrage,	see Peucedanum Silaus.
Scabiosa arvensis,	Field scabious,	Ls. dep. exa. pec. exp.
" atropurpurea,	Sweet scabish,	Pl. and Fl. orn.; in itch.
" stellata,	Star scabious,	" " "
" succissa,	Devil's bit,	Rt. a-syp.
Scandix bulbocastanum,* ·	Earth (Kipper) nut,	see Bunium bulbocastanum.
" cerefolium,	Chervil,	a culinary herb; ape. diu. aro.
" odorata,*	Sweet cicily,	see Myrrhis odorata.
" pecten veneris,*	Shepherds' needle,	Pl. vul.; shoots eaten; diu.
Schinus arveira,	Arveira,	Bk. aro.; resinous, astringent.
" molle,	Peruvian mastich,	Ju. pur.
Schizanthus pinnatus,	Schizanthus,	Pl. and Fl. orn. vis.
Schœnus mariscoides,	Water bogrush,	Pl. dep. alt.
Schrankia uncinata,	Sensitive brier,	Pl. orn.; Ls. sensitive.
Scilla esculenta,	Wild hyacinth, Quamash	Rt. esc.; diet of Western Indians.
" Fraseri,	" "	Rt. esc.; emo. to inflamed breasts.
" maritima,	Officinal squill,	Bu. irr. emc. exp. diu. cat. nar.
Scirpus maritimus, ` ⟋	Spurt grass,	Pl. diu. ast.
" lacustris,	Club rush, Bulrush,	Sd. ast. emm. diu. hyp.
Scleranthus annuus,	Gravel chickweed,	Pl. ast. emo. diu.
Sclerotium clavus,	a fungus, Ergot of rye,	Fu. poi. emm. par. abo. f-com.
" giganteum,	Indian bread (loaf),	Fu. an aliment; aph.
Scolopendrium lingua,*	Harts' tongue, Spleenw't	see Asplenium Scolopendrium.
" officinarum,	" " fern, "	Pl. ast. pec. vul.; spleen compl'ts.
" ruta muraria,*	Wall rue,	see Asplenium ruta muraria.
" vulgare,	Spleenwort,	" " Scolopendrium.
Scolymus Hispanicus,	Spanish cardoon,	see Cynara cardunculus.
ˌ " sativus,*	Garden artichoke,	" " scolymus.
Scopolina atropoides,*	Night henbane,	see Hyosciamus Scopolia.
Scorodosma fœtidum,	Assafœtida,	see Narthex Assafœtida.
˙ Scorzonera Hispanica,	Winter asparagus,	Rt. esc.
" humilis (lanata*)	Vipers' grass,	Rt. alc. deo.; in hypochondria.
Scrophularia aquatica,	Water figwort,	Ls. correct the bad flavor of Senna.
" fœtida,*	Figwort, Healall,	see Scrophularia nodosa.
" lanceolata,*	" ".	" " "
" Marilandica,	Scrofula plant,	Ls. and Rt. alt. diu. ano. hep. deo.
" nodosa,	Figwort, Healall,	Pl. vul. dep. emm. f-com. exa.
" vernalis,	Yellow figwort,	Pl. diu. nar. alt. ano. vul.
" vulgaris,	Figwort,	see Scrophularia nodosa.
Scutellaria alba,	White flowered scullcap,	var. of Scutellaria laterifiora.
" Caroliniana,*	Large " "	see Scutellaria integrifolia.
" galericulata, ˌ	Europ'n (Am'n) scullcap,	Pl. bitter; in tertian ague.
" hyssopifolia, var.	Hyssop leaved scullcap,	Pl. bit. ton. ast.

BOTANICAL.	COMMON.	PROPERTIES, PRODUCTIONS, USES, ETC.
Scutellaria integrifolia,	Large flowered sculleap,	Pl. an intense bitter, ton.
" lateriflora,	Blue sculleap,	Pl. ner. ton. diu. a-spa.
Secale cereale,	Rye plant, Rye,	Ft. (Sd.) an article of diet; nut. lax.
" cornutum,*	Ergot of rye,	see Sclerotium clavus.
" luxurians,*	" "	" " "
Sceamone Ægyptiaca,*	Smyrna scammouy,	see Periploca sceamone.
Sedum acre,	Small houseleek,	Pl. ner. cmc. cat. ves. diu.
" album,	White stonecrop,	Pl. ref. ast.; in salads.
" roseum,	Rosewort,	Rt. cep. ast.
" rupestre,	Stonecrop,	Pl. acr. cmc. cat.
" telephium,	Orpine livelong,	Pl. ast. vul. irr. ref.
Segeretia Theezan,	Theezan tea,	Ls. used as tea.
Selaginella lepidophylla,	Resurrection plant,	Pl. a hygrometric plant.
Selenium Angelica,*	European Wild angelica,	see Angelica sylvestris.
" Canadense,	Hemlock parsley,	Pl. anti-dysenteric.
" Galbanum,*	So. African Galbanum,	see Bubon Galbanifera.
" Imperatoria,*	Masterwort,	see Imperatoria ostruthium.
" Opopanax,*	Opopanax, Rough parsnip,	see Pastinaca Opopanax.
" Oreoselinum,*	Speedwell,	see Athamanta oreoselinum.
" Ostruthium,*	Masterwort,	see Imperatoria Ostruthium.
" palustre,	Marsh parsley,	Pl. emm. diu. a-spa.
" peucedanum,*	Swamp sow fennel,	see Peucedanum palustre.
" pubescens,*	European Wild angelica,	see Angelica sylvestris.
Semecarpus Anacardium,	Cashew (Marking) nuts,	Ft.-Ju. cau. poi. a-syp.; Nu. edi.
Sempervivum acre,*	Small houseleek,	see Sedum acre. [cep.
" tectorum,	Common houseleek,	Pl. ref. cmo. a-ven. vul. ast. aci.
Senecio aureus,	Life root plant,	Pl. diu. dia. emm. feb. exp.
" balsamitæ,	Groundsel balsam,	Pl. diu. vul. des. ast.
" cineraria,	Dusty miller,	Pl. orn. vul. det.
" elongata,	Long stem'd groundsel,	Pl. a-sco. det. vul. ast.
" gracilis,	Unkum, Female regulator,	Pl. emm. diu. dia. pec. alt. f-com.
" hieracifolius,*	Fireweed,	see Erechthites hieracifolius.
" Jacobea,	Ragwort,	Ls. bit. nau. a-rhe. vul.
" lobatus,	Butter weed,	Ls. vul. ton. ast.
" obovatus,	Squaw weed,	var. Senecio aureus, which see.
" saricenicus,	Saracen's woundwort,	Ls. vul. det. des.
" scaudens,	"German Ivy,".	Pl. orn.; house plant.
" sonchifolia,	Tassel flower,	Pl. and Fl. orn.
" tomentosa,	Ashwort,	Pl. orn.; ashy white.
" vulgaris,	Common groundsel,	Pl. a-sco. diu. vul.
Sequoia gigantea,	Calif'nia (Giant) redw'd,	the giant trees of California.
" sempervirens,	" Common "	
Serpentaria nigra,*	Black cohosh,	see Cimicifuga racemosa.
Serratula tinctoria,	Saw-wort,	Pl. vul.; dyes yellow.
Sesteria cærulea,	Moor grass.	
Sesamum Indicum,	Benne plant,	Ls. cmo. mue. vul.
" orientale,	" Oily grain,	Sd. yields oil; esc. nut. lax.
Seseli montanum,	Bastard spignel,	Rt. pur.
Setaria glauca,	Bristly foxtail grass,	Pl. for.
" Italica,	Italian millet, Bengal grass,	Pl. and Sd. for.
" viridis,	Green foxtail grass,	Pl. for.
Shepherdia argentea,	Buffalo berry,	Pl. orn.; Ft. aci. esc.
Shorea robusta,	Dammar-resin tree,	Re. in plasters; Sd. yields oil.
Sicyos angulata,—	Single seeded cucumber,	Sd. diu. pur.; in dropsy.
Sida abutilon,	Velvet leaf,	Ls. mue. dem.; in diarrhœa.
" canariensis,	Canary tea,	Ls. dem. vul.; as tea.
" spinosa,	False mallows,	Ls. diu. dem. vul.
Sideritis hirsuta,	Ironwort,	Hb. vul. ast. aro.
" thœnzan,	Theezan tea,	Ls. as tea; aro. vul.
Sigillaria multiflora,*	Solomon seal,	see Convallaria.

BOTANICAL.	COMMON.	PROPERTIES, PRODUCTIONS, USES, ETC.
Silene acaulis,	Moss campion,	Pl. and Fl. orn.
" armeria,	Limewort catchfly,	Pl. vis.; Rt. cor.
" inflata,*	Spatling poppy,	see Cucubalus behen.
" maritima,	Witches' thimbles,	Fl. orn.
" Pennsylvanica,	Pink catchfly,	Pl. vis. cor.
" Virginica,	Catchfly, Wild pink,	Pl. ver. ner.; see Spigelia:
Silibium marianum,*	Mary thistle,	see Carduus marianus.
Silphium gummiferum,	Rosin weed,	Rt. eme. feb.; in asthma; Gn. diu.
" laciniatum,	Compass weed (plant),	Pl. eme. feb.; heaves in horses.
" perfoliatum,	Ind'n (ragged) cup plant	Rt. dia. sti. diu. pec.; Gn. sti. a-spa.
" terebinthinaceum,	Prairie burdock,	Gn. fragrant, sti. a-spa. pec.
Simaba Cedron,	Cedron seed,	Sd. ton. a-per. a-ven. a-cpl.
" Quassioides,	Nepal bitterwood,	Wd. used as quassia.
Simarouba amara,*	Mountain damson,	see Simarouba officinalis.
" excelsa,	Quassia, Bitterwood,	Wd. a bitter tonic; feb.
" Guaianensis,*	Mountain damson,	see Quassia Simarouba.
" officinalis,	Simarouba bark,	Bk. bit. ton. feb.
Sinapis alba,	White mustard,	Sd. pun. lax. sti. acr. con.
" arvensis,	Wild mustard, Cherlock,	Sd. as black mustard seed.
" cernua,*	Black mustard,	see Sinapis nigra.
" Chinensis,	Oriental mustard,	Sd. sti. lax. sto. ton.
" eruca,*	Garden rocket,	see Brassica eruca.
" nigra,	Black mustard,	Sd. pun. lax. diu. sti. acr. ves. con.
" tuberosa,*	Turnip,	see Brassica rapa.
Siphisia glabra,*	Pipe vine,	see Aristolochia Sipho.
Siphonia cahuchu,*	India rubber tree,	see Siphonia elastica.
" elastica,	" " "	yields Caoutchouc.
Sirion Myrtifolium,*	White sandal,	see Santalum album.
Sison ammi,		Sd. aro. car.
" amomum,	Field honewort,	Sd. aro. car. pun. diu.
" "	Bastard stone parsley,	Sd. " "
" anisum,*	Anise plant (seed),	see Pimpinella anisum.
" podograria,*	Gout weed,	see Ligusticum podograria.
Sisymbrium Alliaria,*	Hedge garlic,	see Alliaria officinalis.
" Barbarea,*	Winter watercress,	see Erysimum barbarea.
" menthastrum,*	Watermint,	see Mentha aquatica.
" muralis,	Flix weed,	Pl. a-sco. dep.
" nasturtium,	Watercress,	see Nasturtium officinale.
" officinale,	Hedge mustard,	Pl. det. a-sco. lit.
" parviflorum,*	" " "	see Sisymbrium sophia.
" sophia,	Flix weed,	Pl. ast. det. vul.
" palustre,	Marsh watercress,	see Nasturtium palustre.
Sisyrinchium Bermudianum,	Blue eyed lily (grass),	Rt. acr. cat.
Sium angustifolium,	Creeping water parsnip,	see Sium nodiflorum.
" apium,*	Wild celery,	see Apium graveolens.
" Berula,*	Creeping water parsnip,	see Sium nodiflorum.
" bulbocastanum,*	Earth (Kipper) nut,	see Bunium bulbocastanum.
" Carvi,*	Caraway plant,	see Carum carui.
" erectum,*	Creeping water parsnip,	see Sium nodiflorum.
" graveolens,*	Wild celery,	see Apium graveolens.
" latifolium,	Common water parsnip,	Pl. a-sco. diu. pol.
" lineare,	" " "	" "
" ninsi,	Ninsing root,	see Panax schinseng.
" nodiflorum,	Creeping water parsnip,	Pl. sti. diu. lit.
" sisarum,	Skirret,	Pl. aro. sad. pec. *
Smilacina racemosa,	False spikenard,	Rt. alt. diu. dia. dep.
Smilax aspera Chinensis,*	China root,	see Smilax China.
" " Peruviana,*	Sarsaparilla, Jamaica,	" " officinalis.
" China,	China root,	Rt. dep. exa. a-syp. diu.
" camauensis,*	Spanish (Honduras) sarsaparilla,	see Smilax officinalis.

BOTANICAL.	COMMON.	PROPERTIES, PRODUCTIONS, USES, ETC.
Smilax glauca,	American smilax,	analagous to Smilax officinalis.
" glycyphylla,	Botany Bay tea,	Ls. used as tea; Rt. alt.
" herbacea,	Carrion flower,	see Smilax peduncularis.
" hederæfolia,*	American China root,	see Smilax pseudo-China.
" Indica spinosa,	" " "	" " "
" medica,	Mexican sarsaparilla,	Rt. dep. alt. a-syp. emc.
" officinalis,	Jamaica "	Rt. alt. diu. dem. a-syp.
" papyracea,	Brazilian "	Rt. " "
" peduncularis,	Jacob's ladder vine,	Rt. and St. diu. alt. emm.
" pseudo-China,	American China root,	Rt. dep. alt.; in cutaneous compl'ts
" rotundifolia,	Green (horse) brier,	Rt. diu. alt. emm.
" Sarsaparilla,	Sarsaparilla common to the U. S., Rt. alt. dep. dis.
" syphilitica,	Brazilian sarsaparilla,	Rt. see Smilax papyracea.
" Tamnoides,*	American China root,	see Smilax pseudo-China.
Smyrnium olusatrum,	Alexanders,	Rt. con.; Sd. res. diu. emm.
Solamen intestinorum,*	Anise plant,	see Pimpinella anisum.
Solandra,	Trumpet flower,	Fl. and foliage orn.
Solanum bacciferum,	Susumber berries,	Ft. said to be poisonous.
" Carolinensis,	Horse nettle,	Pl. nar. ano.
" Dulcamara,	Bittersweet, Woody nightshade,	Pl. nar. dep. deo. her. ano.
" laciniatum,	Kangaroo apple,	ripe fruit, esculent. [res. emc.
" lycopersicum,	Tomato, Love apple,	Tp. nau. ano.; Ft. nut. lax. a-sco.
" esculentum,*	Potato plant,	see Solanum tuberosum.
" fœtidum,*	Thorn apple plant,	see Datura Stramonium.
" furiosum,*	Belladonna,	see Atropa Belladonna.
" hortense nigrum,*	"	" " "
" lethale,*	"	" " "
" lignosum,	Bittersweet,	see Solanum Dulcamara.
" magnum Va. rubrum,*	Poke, Garget,	see Phytolacca decandra.
" maniacum,*	Belladonna,	see Atropa Belladonna.
" "	Thorn apple plant,	see Datura Stramonium.
" melongena,	Egg plant,	cultivated for its fruit; nut.
" nigrum,	Garden nightshade,	Fl. nar. poi. dia.; in oil ano. dis.
" niveum,	a So. African species,	Ls. and Ju. in ointment; det.
" paniculatum,	Jurubeba,	Pl. and Fl. ton.
" pseudo-capsicum,	Jerusalem cherry,	Pl. ano.; Ft. nar.
" racemosum Am'n,*	Poke, Garget,	see Phytolacca decandra.
" sanctum,	Palestine nightshade,	Ft. eaten as food.
" somniferum,	Sleepy nightshade,	Pl. ano. poi.
" scandens,	Bittersweet, Nightshade	see Solanum Dulcamara.
" tuberosum,	Potato plant,	Tp. nar. a-spa.; Bu. as food.
" urens,*	Capsicum,	see Capsicum annuum.
" vesicarium,*	Winter cherry,	see Physalis alkekengi.
" Virginianum,	Virginian nightshade,	resembles S. nigrum in properties.
" vulgare,	Garden nightshade,	see Solanum nigrum.
Solenostemma Argel,*	Argel, Arghel,	see Cynanchum Argel.
Solidago odora,	Sw't scented goldenrod,	Ls. and Ol. car. sti. dia. diu. ast.
" rigida,	Hard leaved "	Pl. ast. sty.
" Saracenica,*	Saracens' woundwort,	see Solidago vigaurea, and Senecio
" virgaurea,	European goldenrod,	Hb. vul. diu.; in hemorrhage.
" vulgaris,*	Common goldenrod,	see Solidago odora.
Sonchus oleraceus,	Sow thistle,	Ju. bit. diu.
Sophora heptaphylla,	Semen anti-cholerica,	Sd. bit.; in cholera, colic, etc.
" Japonica,	Japan sophora,	Pl. dra.; " "
" leucantha,	Tall white false indigo,	analogous to Baptisia tinctoria.
" tinctoria,*	Wild indigo plant,	see Baptisia tinctoria.
Sorbus Americana,	Mountain ash,	Bk. ast. ton. det.; Ft. esc. a-sco.
" aria,*	White beam tree,	see Cratægus aria.
" aucuparia,	European mountain ash,	Bk. ast. ton. feb.; Ft. esc.
" Cydonia,*	Quince tree,	see Cydonia vulgaris.

BOTANICAL.	COMMON.	PROPERTIES, PRODUCTIONS, USES, ETC.
Sorbus domestica,	Sorb apple,	Ft. ast. a-sco.; see Pyrus.
" malus,*	Apple tree,	see Pyrus malus.
Sorghum nigrum,	Chinese sugarcane, Sorgho,	St. yields sugar.
" nutans,	Wood grass,	Pl. for.
" saccharatum,	Chinese sugarcane,	St. yields sugar.
" "	Broom corn,	Sd. diu. lit.; panicles for brooms.
" vulgare,	Ind'n millet, Small maize	Sd. esc. far.
" " var. cernuum,	Guinea corn,	Sd. nut.; used as food.
Soymoida febrifuga,	Bahama redwood,	Bk. ast. ton. feb. a-per.
Sparganum racemosum,	Burr reed.	
Spartium junceum,	Spanish broom,	Fl. orn.; Sd. diu. ton. eme. cat.
" scoparium,*	Broom herb,	see Cytisus Scoparius.
" tinctorium,*	Dyers' broom,	see Genista tinctoria.
Spartinia stricta,	Cord grass,	Pl.
Spathyema fœtida,*	Skunk cabbage,	see Ictodes fœtidus.
Specularia hybrida,	Corn violet,	Pl. and Fl. orn.
" speculum,	Venus' looking-glass,	" "
Spergula arvensis,	Spurry,	Pl. used as forage.
Spermœdia clavus,	Ergot (Smut) of rye,	see Sclerotium clavus.
" maydis,	" (Smut) of corn,	Fu. abo. poi. par.
Sphacelia segetum,*	"	see Sclerotium clavus.
Sphæria Sinensis,	Vegetable caterpillar,	a fungus; esteemed a tonic.
Sphærococcus crispus,*	Irish moss,	see Chondrus crispus.
" helminthocortus,*	Corsican moss,	see Fucus helminthocorton.
" lichenoides,*	Ceylon moss,	see Gracilaria lichenoides.
Sphagnum commune,	Bog moss,	Pl. used in packing green bulbs.
Sphenogyne speciosa,	Sphenogyne,	Pl. and Fl. orn.
Sphondylium Branca-ursina,*	Cow parsley,	see Heracleum spondylium.
Spigelia anthelmia,	Demerara pink rt. (W. I.),	Rt. ant.
" Lonicera,*	Carolina pink root,	see Spigelia Marilandica.
" Marilandica,	" " "	Rt. ant. nar. acr.
Spilanthes oleracea,	Para cress,	see Spilanthus oleraceus.
Spilanthus acmella,	Balm leaved spilanthus,	Pl. vis. bit. fragrant, diu. emm.
" oleraceus,	Para cress,	Tincture used for toothache.
Spinacea oleracea,	Spinage, Spinach,	Ls. emo. lax. alt. a-sco. sad.
Spiranthes autumnalis,	Ladies' tresses,	Fl. orn.; Rt. aph.
Spiræa deuudata,*	Meadow sweet,	see Spiræa ulmaria.
" filipendula,	Dropwort,	Hb. ast. diu.; Rt. ton. esc.
" hypericifolia,	May wreath,	Fl. ast. ton. orn.
" lobata,	Queen of the prairie,	said to yield Oil anal. to Gaultheria.
" opulifolia,	Nine bark,	Pl. analagous to Spiræa tomentosa.
" palmata,*	American meadow sweet	see Spiræa lobata.
" tomentosa,	Hardhack,	Pl. ast. ton. vul.
" trifoliata,*	Indian physic,	see Gillenia trifoliata.
" ulmaria,	Meadow sweet,	Ls. ast. diu.; Fl. a-spa.
Spondias dulcis,	Otaheite apple,	Ft. edl. aci. ref.
" entra,	Hog plum,	Bk. as. fomentation in dropsy.
Squilla pulcherrima,*	Squill,	see Scilla maritima.
Stachys betonica,*	Wood betony,	see Betonica officinalis.
" palustris,	Clown heal,	Pl. nau. exp. emm. a-hys. vul.
" sylvatica,	Hedge woundwort,	" " "
Stachyta Jamaicensis,	Brazilian tea,	Ls. pur. deo. emm. ant.
Stagmaria verniciflua,	Japan lacquer,	Re. acr. ves. poi.; as varnish.
Stalagmitis cambogioides,	Gamboge tree,	said to yield gamboge.
Staphylea pinnata trifolia,	Bladder nut tree,	Nu. edible.
Statice Armeria,	Thrift,	Fl. diu.
" Caroliniana, var.	Am'n thrift, Marsh rosemary,	Rt. ast.; in canker and spongy
" Limonis,*	" " "	see Statice Limonium. [gums.
" Limonium,	Eurp'n ' "	Rt. ast. ton.; in diarrhœa.
" maritimum,*	" " "	see Statice Limonium.

BOTANICAL.	COMMON.	PROPERTIES, PRODUCTIONS, USES, ETC.
Statice vulgare,*	Marsh rosemary,	see Statice Limonium.
Stellaria media,	Chickweed,	Pl. ref. dem. muc. dis. alt.
" palustris,	Stitchwort, Meadow star	Pl. alt. ast. ton. bit.
Stephanotis floribunda,	Stephanotis,	Fl. fragrant, orn.
Stephensia elongata,	Matico,	see Artanthe elongata.
Stercularia Tragacantha,	African tragacanth,	Gu. resembles Tragacanth.
Sterculia acuminata,*	Kola nut, (Caffein nut),	see Cola acuminata.
" foetida,	Karil root,	Sd. yields a solid oil (Stearine).
Sticta pulmonaria,	Tree lungw't, Lungmoss,	Pl. bit. ton.; in pulmonary compl't.
Stillingia ligustina,	Stillingia,	Rt. see Stillingia sylvatica.
" sebifera,	Vegetable tallow tree,	Sd. yields Japan wax.
" sylvatica,	Queen's root, Yaw root,	Rt. alt. res. diu. sti. ton. a-syp. exp.
Stipa avenacea,	Black oat grass,	Pl. and Sd. for.
" pennata,	Feather grass,	Pl. orn.
Stizolobium pruriens,*	Cowhage,	see Dolichos pruriens.
Stobaea rubricalis,	a So. African plant,	Rt. diu.; in gravel.
Stoechas Arabica,*	Arabian Lavender,	see Lavandula stoechas.
Strambo carpa-pubescens,	Screwpod mimosa,	see Algarobia glandulosa.
Stramonium foetidum,*	Thornapple,	see Datura Stramonium.
" majus-album,*	"	" " "
" vulgatum,*	"	" " "
Stratiotes aloides,	Water soldier,	Pl. vul. ref.
Strychnos colubrina,	Snakewood,	Wd. bit. poi. a-per.
" Ignatia,	St. Ignatius bean,	Sd. bit. poi. feb. ver.
" Ligustrina,*	Snakewood,	see Strychnos colubrina.
" Nux Vomica,	Poison nut, Nux Vomica,	Sd. poi. diu. dia. lax.
" potatorum,	Clearing (Drinkers') nut,	Xn. used to purify drinking water.
" pseudo-quinia,	Copalchi bark,	Bk. used as Cinchona.
" tiente,	Upas tree of Java,	see Antiaris toxicaria.
" toxifera (toxicaria*)	Woorari poison,	Ju. a powerful arrow poison.
Stylophorum diphyllum,	Celandine (Horn) poppy,	Ju. foet. nar. diu. lit. vul.
Styrax Americana,	Spring orange,	Ba. aro. exp. pec. vul.
" Benzoin,	Benzoin tree, (Laurel),	Gum Benzoin aer. irr. sti. pec. err.
" calamita,	Oriental sweet gum tree,	Gu. aro. fragrant, bal. sti. exp.
" officinalis,	Cane storax tree,	Ba. aro. fragrant, sti. exp.
Subularia aquatica,	Awlwort,	Pl. ref. vul.
Succissa pratensis,*	Devil's bit,	see Scabiosa succissa.
Swertia difformis,*	American Columbo,	see Frasera Walteri.
Swietenia chloroxylon,	Satin wood,	said to yield wood oil.
" febrifuga,	Indian redwood tree,	Bk. feb. ast.; in jungle fever.
" · Mahogani,	Mahogany tree,	Bk. feb. ast.; as Cinchona.
" Senegalensis,	African mahogany,	Bk. bitter, feb.; in fevers.
" Soymida,*	Redwood tree,	see Swietenia febrifuga.
Sycios angulatus,*	One seeded cucumber,	Rt. and Sd. bit. pur. diu. see Sicyos
Symphoricarpus occidentalis	Wolfberry,	Rt. ast. ton.
" racemosus,	Snowberry,	Rt. alt. ton.; in ague; a-syp.
" vulgaris,	Indian currant,	Rt. ast. ton.
Symphytum album,*	Comfrey,	see Symphytum officinale.
" hirsutum,	Western States comfrey,	" " "
" minus,*	Healall,	see Prunella vulgaris.
" officinale,	Comfrey,	Rt. dem. ast. muc. pec. vul. f-com.
" Petraea,	Montpellier coris,	Pl. bit. nau. a-syp.; see Sanicula.
Symplocarpus angustispatha	Nar. spath'd sk. cabbage,	see Ictodes foetidus.
" foetidus,	Skunk cabbage,	" " "
Syringia Persica,	Persian lilac,	Pl. and Fl. orn.
" vulgaris,	Common lilac,	Ls. Bk. and Ft. bit. aer. a-per.
Tabernaemontana utilis,	Hya-Hya,	Bk. and Pi. yield a refreshing drink
Tacca fecula,	South Sea arrow root,	yields arrow starch.
" oceanica,	Tahiti arrow root,	" " "
" pinnatifida,	South Sea arrow root,	" " "

BOTANICAL.	COMMON.	PROPERTIES, PRODUCTIONS, USES, ETC.
Tagates erecta,	African marigold,	Fl. opt.; dye yellow.
" patula,	French marigold,	Fl. orn. opt.
Tamarindus Indica,	Tamarind tree,	Ft.-pulp lax. ref. aci.; Ls. ant.; Bk.
Tamarix Anglica,	Tamarisk tree,	Bk. bit. ast. ton. [ton.
" Gallica,	French Tamarisk (Manna), Bk. ape. hep.; Ls. exa.	
" Germanica,*	German Tamarisk,	see Myricaria Germanica.
" orientale,		Bk. ast. ton.; yields Atlee galls.
" pentandra,*	German Tamarisk,	see Myricaria Germanica.
Tamnus communis,	Black bryony,	see Tamus communis.
Tamus communis,	" "	Rt. diu. lit.; for bruise marks.
Tanacetum balsamita,*	Costmary,	see Pyrethrum Tanacetum.
" crispum,	Double tansy,	see Tanacetum vulgare.
" hortense,*	Costmary,	see Pyrethrum Tanacetum.
" multiflorum,	a So. African species,	Pl. ton. a-spa. ant. dia. diu.
" vulgare,	Common tansy,	Ilb. aro. ton. emm. dia. vul.; Sd.
Tanghinia lactaria,	Milk tree.	[ver.
" venifera, or	ʃʃ Ncɴiſʕ ʑ A	see Cerbera Tanghiu.
Tapsus barbatus,*	Black mullein,	see Verbascum nigrum.
Taraxicum Dens-leonis,	Dandelion,	Rt. sto. ton. diu. ape. dep. hep.
" officinale,*	"	see Taraxacum Dens-leonis,
" officinalis,*	"	" " "
Taxodium distichum,	Virginian (Bald) cypress Re. diu. car. vul.	
Taxus baccata,	Yew tree,	Ls. in epilepsy; poi. to cattle.
" Canadensis,	Dwarf yew, Shinwood,	Ls. sed.; said to act as Digitalis.
Tecoma pentaphylla,	Jamaica boxwood,	Wd. W. I. whitewood.
" radicans,*	Trumpet flower, Va. creeper, see Bignonia radicans.	
Tectaria calahuala,*	Calaguala root,	see Polypodium Calaguala.
Tectonia grandis,	Teakwood tree,	Ls. diu.; in thrush.
Telfaria pedata,	Zanzibar species, Joliffia, Ft.-pulp bit.; Sd. yields bland oil.	
Telephium,	Orpine,	see Sedum telephium.
Tephrosia Apollinea,	Silver leaved senna,	Ls. used as senna.
" cineria,	W. I. Fish poison,	Ls. a fish poison; as Digitalis.
" purpurea,		Rt. bit. a-eme. dep.
" toxicaria,	W. I. Fish poison,	Ls. a fish poison.
" Virginiana,	Hoary pea,	Rt. a-syp. ver. stl. ton.
Terminalia bellirica,	Belleric myrobalans,	Ft. ast.; Bk. yields gum. [cense.
" Benzoin,	Incense tree,	yields a fragrant juice used as in-
" chebula,	Chebullic myrobalans, Ink nut, Ft. ast.; in making paint.	
" citrina,	Yellow myrobalans,	Ft. pur.; also pickled.
" vernix,	Chinese varnish tree,	yields Chinese black lacquer.
Testudinaria elephantipes,	Hottentot bread,	Pl. orn.; "Elephants' foot."
Tetragonia expansa,	New Zealand spinach,	Pl. a-sco. muc.; as spinach.
Teucrium Ægyptiacum,*	Polymountain of Montpellier, see Teucrium capitatum.	
" Belium,*	" " see Teucrium capitatum.	
" Canadensis,	Wood sage, Germander, Pl. stl. ton. aro. bit. a-rhe.	
" capitatum,	Polymountain of Montpellier, Pl. alexilerial.	
" chamædrys,	Germander,	Pl. aro. ton. bit. feb. hep.
" chamæpitys,*	Ground pine,	see Ajuga chamæpitys.
" creticum,	Polymountain of Candy, Pl. ape. ton.	
" flavum,	Germander,	see Teucrium chamædrys.
" hyssopifolium,*	Polymountain of Candy, " " creticum.	
" iva,	French Ground pine,	Pl. aro. bit. ton. feb.
" marum,	Syrian Herb mastich,	Pl. emm.; attractive to cats.
" officinalis,*	Germander,	see Teucrium chamædrys.
" palustre,*	Water germander,	" " scordium.
" polium,	Yellow polymountain,	Pl. ton. bit. hep.
" scordium,	Water germander,	Pl. bit. pun. ton.
" scorodonia,	Wood sage, (Garlic),	Pl. ver. ton.; in fomentations.
Thalictron,*	Poor man's rhubarb,	see Thalictrum.
Thalictrum anemonoides,	" " " Rue anemone, see Thalictrum dioicum.	

BOTANICAL.	COMMON.	PROPERTIES, PRODUCTIONS, USES, ETC.
Thalictrum dioicum,	Meadow rue,	Rt. pur. diu.
Thamnochortus,	Shrubby grass.	
Thapsia Garganica,	Resin of Thapsia, Deadly carrot,	Rt. eme. pur.; Re. irr.
" silphium,	No. African species, (Asa dulcis),	Rt. an active purgative.
" trifoliata,	Round heart plant,	Pl. vul. alt. dia. a-syp.
Thapsium actæifolium,*	American lovage,	see Ligusticum actæifolium.
" atropurpureum,	Round heart,	Pl. vul. dia. a-syp. a-ven.
" barbinode,	Meadow parsnip,	Pl. " "
Thea Assamica,	Assam tea,	Ls. used as beverage.
" Bohea,	Black tea,	Ls. ast. sti.; in nervous headache.
" Germanica,		see Veronica.
" Sineusis,	Chinese tea plant and its varieties.	
" viridis,	Green tea,	Ls. ast. sti.; in headache.
Theobroma Cacao.	Chocolate, Cacao,	Nu. used as a drink.
" "	" "	Cacao butter; in suppositories.
Thesium linophyllum,	Bastard toad flax,	Pl. ast.
Thevetia Ahouai,	Ahouai, Snake nuts,	Ke. a violent poison.
" nerifolia,	Exile tree, Yellow oleander,	Bk. acr. poi. a-per.
Thlaspi arvense,	Treacle mustard, Pennycress, resembles mustard; Sinapis.	
" bursa-pastoris,*	Shepherds' purse,	see Capsella bursa-pastoris.
" campestris,*	False (Bastard) cress,	see Lepidium campestre.
Thlapsus bursa-pastoris,*	Shepherds' purse,	see Capsella bursa-pastoris.
Thuja articulata,	Arar tree, Sandarach,	Gu.-re. used in pounce.
" occidentalis,	Arbor vitæ,	Ls. sti. dia. ant. feb. a-spa.
" orientalis,	Chinese arbor vitæ,	resembles Thuja occidentalis.
Thuya,*	Arbor vitæ,	see Thuja.
Thymus acinos,*	Mountain thyme.	see Acinos vulgare.
" Alpinus,*	" "	see Calamintha alpina.
" calamintha,*	" balm,	see Melissa Calamintha.
" citriodorus,	Lemon thyme,	Pl. aro. sti. cor. car. cmm.
" Creticus,	Cretian balm,	Hb. aro. car. sti. cmm.
" mastichina,	Herb mastich,	Hb. see Teucrium marum.
" multiflorus,*	Field calamint,	see Melissa Nepeta.
" Serpyllum,	Mother (Wild) thyme,	analogous to Thymus vulgaris.
" sylvaticus,*	Dogmint, Field thyme,	see Clinopodium vulgare.
" sylvestris,*	Cilliated savory,	see Satureja capitata.
" vulgaris,	Common garden thyme,	Hb. res. ton. car. cmm. a-spa.
Thysselinum palustre,*	Swamp sowfennel,	see Peucedanum palustre.
Tiarella cordifolia,	Coolwort, Mitrewort,	Hb. diu.; in strangury.
Ticorea febrifuga,	Ticorea,	Bk. intensely bitter, ast. a-per.
" Jasminifolia,	" Jasmin flowered,	Ls. in frambœsia.
Tiglium officinale,	Croton oil plant,	see Croton Tiglium.
Tigridia parvonia,	Tiger flower,	Pl. and Fl. orn.; Bu. acr.
Tilia alba,	White linden (wood),	see Tilia Americana.
" Americana,	Basswood,	Bk. emo. muc. vul.; Fl. cep. sti.
" Canadensis,*	"	see Tilia Americana. [sed.
" Europæa,	Linden, Lime tree,	Fl. ano. a-spa. cep.; Bk. ast. diu.
" glabra,*	Basswood,	see Tilia Americana. [cmm.
" grandiflora,*	European linden,	" " Europæa.
" pauciflora,*	" "	" " "
" ulmifolia,*	" "	" " "
Tillæa ascendens,	Pigmy weed,	Pl. dem. ref.
Tillandsia usneoides,	Spanish (Long) moss,	Po. in lard for piles.
Tinospora cordifolia,	Guilancha,	see Cocculus cordifolia.
Toddalia aculeata,	Lopez root,	Bk. of Rt. sti. ton.; in "hill fever."
Toluifera balsamina,*	Balsam Tolu tree,	see Myrospermum Toluiferum.
Tormentilla erecta,	Tormentilla, Septfoil,	see Potentilla Tormentilla.
" officinalis,	" "	" " "
Torreya Californica,	California nutmeg,	Ke. ast.; yields oil.
" myristica,	" stinking yew,	" "

BOTANICAL.	COMMON.	PROPERTIES, PRODUCTIONS, USES, ETC.
Trachelium Americanum,*	Red lobelia (cardinal),	see Lobelia cardinalis.
Trachylobium gœrtnerianum	Brazilian copal,	see Hymenœa courbaril.
" martianum,	Copal, Anime,	" " "
Tradescantia Virginica,	Spiderwort,	Rt. dem.
Tragium anisum,*	Anise,	see Pimpinella anisum.
Tragopogon porrifolium,	Vegetable oyster,	Rt. eaten as food.
" pratensis,	Meadow salsify,	Rt. diu.; eaten as food.
Tragoselinum angelica,*	Gout weed,	see Ligusticum podograria.
" Magnum,*	Greater (False) Pimpinella,	see Pimpinella magna.
" saxifraga,*	" " "	see Pimpinella magna.
Trapa bicornis,	Ling,	Sd. an article of food, (China).
" nutans,	Jesuits' (Water) nuts,	Sd. far.; an article of food.
Tremella nostoc,	Will-o'-the-wisp,	Fu. on dead trees.
Tricuspis purpurea,	Sand grass,	Pl. aci.; on sandy shores.
Trientalis Europœa,	Chick wintergreen.	
Trifolium Alpinum,	Mountain liquorice,	Rt. as liquorice.
" arvense,	Hares' foot,	Ls. pec.; anti-dysenteric.
" aquaticum,*	Bogbean,	see Menyanthes trifoliata.
" aureum,*	Kidney liver leaf (wort),	see Hepatica Americana.
" cervinum,*	Hemp agrimony,	see Eupatorium cannabinum.
" fibrinum,*	Bogbean,	see Menyanthes trifoliata.
" hepaticum,*	Liverwort,	see Hepatica Americana.
" Melilotus,	Blue melilot,	see Melilotus cœrulea.
" " cœruleum,	Common blue melilot,	" " officinalis.
" pratense,	Red clover,	Fl. det. dep.; in ointment.
" repens,	White clover,	Fl. " "
Triglochin maritinum,	Seaside arrowgrass,	Pl. sac.; as forage.
Trigonella Fœnum,*	Fœnugreek seed,	Sd. in veterinary practice.
" " grœcum,	"	Sd. far. muc.; in poultices.
" ornithorhyncus,	Birds' bill,	Pl. and Fl. orn.
Trillium erectum,	Beth root,	Rt. see Trillium pendulum.
" latifolium,	"	" " " [alt. f-com.
" pendulum,	Nodding wake robin,	Rt. ast. ton. a-sep. emm. exp. dia.
" purpureum,	Beth root,	Rt. see Trillium pendulum.
Trinia vulgaris,	Honewort,	Rt. pur. acr.
Triosteum angustifolium,		anal. to Triosteum perfoliatum.
" perfoliatum,	Fever root, Horse gentian,	Rt. eme. dia. ton. lax. feb.
Tripsacum dactyloides,	Sesame (Buffalo) grass,	Pl. for.; also known as Gama.
Triticum Æstivum,	Summer wheat,	see Triticum hybernum.
" caninum,*	Dog grass,	" " repens.
" cereale,*	Summer wheat,	" " Æstivum.
" cevallos,*	" "	" " "
" compactum,*	" "	" " "
" compositum,	Egyptian wheat,	Sd. yields flour; far. esc.
" faginum,*	Common buckwheat,	see Polygonum fagopyrum.
" glaucum,	see Triticum Æstivum, and Triticum repens.	
" hybernum,	Winter wheat,	Sd. ground is the Flour of wheat.
" repens,	Dog grass, Couch grass,	Rt. diu. ape.; in dropsy.
" spelta,	Spelt (German) wheat,	Sd. used as food.
" vulgare,* (sativum*),	Common wheat,	see Triticum hybernum.
Tritoma uvaria,	Red-hot-poker plant,	Pl. and Fl. orn.
Trollius Americanus,	Globe flower,	see Trollius laxus.
" Europœus,	" · ranunculus,	see Ranunculus.
" laxus,	" flower,	Pl. aer.
Tropœolum majus,	Nasturtium,	Pl. a-aco. sad.; Ju. in itch.
" minus.	Indian cress,	Pl. " " "
" peregrinum,	Canary bird flower,	Pl. and Fl. orn. a-sco.
" tuberosum,	Peruvian Ulluco,	Rt. used as food.
Tuber Æstivum,	English truffles,	Fu. used as sauce to soups.
" griseum,	Black truffles,	" " "

BOTANICAL.	COMMON.	PROPERTIES, PRODUCTIONS, USES, ETC.
Tulipa Gesneriana,	Tulip,	Fl. orn.
Tulipifera Liriodendron,*	" tree,	see Liriodendron.
Turritis glabra,	Tower mustard,	used as Sinapis.
Tussaca reticulata,*	Adders' violet,	see Goodyera pubescens.
Tussilago farfara,	Coltsfoot,	Pl. emo. dem. ton. err. exp. pec.
" frigida,*	"	see Tussilago farfara.
" petasites,	Pestilence weed,	Ls. det.; Fl. dia.; Rt. ver.
" vulgaris,*	Coltsfoot,	see Tussilago farfara.
Tylophora asthmatica,*	Isle of France ipecac,	see Asclepias asthmatica.
Typha angustifolia,	Reed mace,	Rt. ast. emo.; Down applied to
" aromatica,*	Sweet flag,	see Acorus calamus. (burns.
" elephantiana,	Elephant grass.	
" latifolia,	Cat tail flag, Reed mace,	Rt. ast. emo. det.
Ulmus alata,	Lynn Wahoo, Winged elm,	Bk. as poultice to inflamed parts
" Americana,	Common white elm,	Bk. ast. ton. alt. diu.; in lepra.
" campestris,	English (European) elm,	Bk. ast. dem. diu.; Ls. vul.
" effusa,	European elm,	see Ulmus campestris.
" fulva,	Slippery elm, Red elm,	Bk. muc. nut. exp. diu. dem. emo.
" glabra,*	European elm,	see Ulmus Campestris. [len.
" montana,	Wych (Scotch) elm,	Tw. as divining rods.
" rubra,*	Slippery elm, Red elm,	see Ulmus fulva.
" sativa,*	European elm,	" " campestris.
" scabra,*	" "	" " "
" suberosa,	" cork elm,	a var. of Ulmus campestris.
Ulva crispa,*	Irish moss,	see Chondrus crispus.
" latissima,	Green laver,	Algæ esc. a-sco.
" umbilicalis,	Purple laver,	" pickled called laver.
Umbilicaria,	Rock tripe lichen,	Lichen ast. sty.
Umbilicus pendulinus,*	Kidneywort,	see Cotyledon umbilicus.
Uncaria acida,*	Catechu,	see Nauclea longiflora.
" Gambir,*	Gambir, Terra Japonica,	" " Gambir.
Unisema deltifolia,	Pickerel weed.	Rt. emo. ast. det.; Sd. far.
Unona Æthiopica,	Ethiopian pepper,	Ft. aro. pun.; used as spice.
" odoratissima,	Ylang Ylang,	Fl. a favorite perfume.
" polycarpa,	Yellow dyetree of Soudan, see Cœloeline polycarpa.	
Upas antiar,	Vegetable poison,	see Antiaris toxicaria.
" tieute,	Upas tree of Java,	see Strychnos tieute.
Uraspermum Claytoni,*	Sweet cicily,	see Osmorrhiza longistylis.
Urginea maritima,*	Squill,	see Scilla maritima.
" Scilla,*	"	" " "
Urtica balearica,	Pill bearing nettle,	see Urtica pilulifera.
" Canadensis,	Canada nettle,	Rt. and Ls. diu. ano. ton. ast. pec.
" capitata,	Common nettle,	Ls. irr. acr.; Rt. diu. ton. ast.
" crenulata,	East Indian nettle,	see Urtica dioica.
" dioica,	Common stinging nettle,	Pl. diu. pec. ast. ton.
" major,	" " "	see Urtica dioica.
" minor,	Dwarf nettle (stinger),	" " urens.
" pilulifera,	Pill bearing nettle,	Sd. in diseases of the chest.
" pumila,	Stingless nettle, Coolweed,	Ls. dem.; in Rhus poisoning.
" stimulans,	East Indian nettle,	Fresh Ls. powerfully acrid.
" urens,	Dwarf stinging nettle,	Pl. in hemorrhage.
" urentissima,	Devil's stinging leaf,	Fresh Ls. powerfully acrid.
Usnea hirta,*	Death head moss,	see Lichen saxatilis.
" plicata,*	" " "	" " "
Ustilago Maydis,	Ergot of corn,	properties as Sclerotium clavus.
Utricularia macrorhiza,	Bladderwort,	Pl. diu.
Uvaria triloba,*	Papaw, (Custard apple),	see Carica papaya.
Uvularia grandiflora,	Large flower'd bellwort,	Pl. ton. dem. ner. alc. her.
" perfoliata,	Bellwort,	" " "
Vaccaria vulgaris,*	Field soapwort,	see Saponaria officinalis.

BOTANICAL.	COMMON.	PROPERTIES, PRODUCTIONS, USES, ETC.
Vaccinium arboreum,	Farkleberry,	Bk. and Ft. ast. ; in apthæa.
" corymbosum,	Giant whortleberry,	Ft. esc. diu. ; Bk. diu. ; Ls. ast.
" dumosum,	Bush "	Ft. " " "
" frondosum,	Blue (Whortle) berry,	Ft. esc. sac. diu. a-sco.
" macrocarpa,*	Cranberry,	see Vaccinium oxycoccos.
" myrtillus,	Whortle (Huckle) berry,	Ft. esc. a-sco. a-sop.
" nigrum,*	" " "	see Vaccinium myrtillus.
" oxycoccos,	Cranberry, American,	Ft. aci. ref. con. a-sco. ; Ls. diu.
" Pennsylvanicum,	Low blue (Whortle) berry,	Ft. esc. a-sco diu.
" punctatum,*	Red bilberry,	see Vaccinium Vitis Idæa.
" resinosum,	Huckleberry,	Ft. esc. sac. diu. ; Bk. ast. diu.
" stamineum,	Dangleberry,	Ft. esc. a-sco. diu. ast. ; Rt. diu.
" Vitis Idæa,	Red bilberry, Cowberry,	Ls. ast. lit. diu. ; Ft. aci. ref.
Valantia aparine,*	Cleavers, Goose grass,	see Galium aparine.
Valeriana Capensis,	a Cape of G. H. species,	resembles V. officinalis in prop'rties
" celtica,	Celtic nard,	Rt. aro. sti. per. sto. car. diu.
" dentata,	European corn salad,	Pl. sad. a-sco. apc.
" dioica,	Swamp valerian,	Rt. ton. a-spa. sto. ner.
" edulis,	Oregon tobacco, •	Rt. cooked, used as food.
" Jatamansi,	Spikenard of the ancients,	see Nardostachys Jatamansi.
" locusta,*	European corn salad,	see Valeriana dentata.
" major,*	Great valerian,	" " phu.
" minor,*	Valerian, wild,	" " officinalis.
" montana,	Mountain valerian,	Rt. aro. sti. a-hys.
" officinalis,	Officinal valerian,	Rt. aro. sti. ton. ano. ner. a-spa.
" pauciflora,	American wild valerian,	analogous to Valeriana officinalis.
" phu,	Spikenard of Crete,	Rt. a-epi. a-rhc.
" saxatilis,*	Celtic nard,	see Valeriana celtica.
" spica,*	Spikenard of the ancients,	see Nardostachys Jatamansi.
" sylvatica,*	American wild valerian,	see Valeriana pauciflora.
" tuberosa,	Mountain valerian,	" " montana.
Valerianella olitoria,*	Lambs' lettuce,	see Fedia olitoria.
Valisneria spiralis,	Eel grass,	Pl. ref. dem.
Vandellia diffusa,	Bitter blain,	Hb. a-bil. emc. a-per. hep. feb.
Vanilla aromatica,	Vanilla plant (bean),	Ps. aro. per. sti. aph. ; as flavor.
" Guianensis,	" "	see Vanilla aromatica.
" palmarum,	" "	" " "
" planifera,	" "	" " "
" pompona,	" "	" " "
" sativa,	" "	" " "
" sylvestris,	" "	" " "
Variolaria faginea,	Maple lungwort,	see Sticta pulmonaria.
Vateria Indica,	Anime, Copal,	Gu.-re. in varnish.
Veratrum album,	White (Europ'n) hellebore,	Rt. poi. emc. cat. err. ins. ; in
" Lobelianum,*	American hellebore,	see Veratrum viride. [itch.
" luteum,*	False unicorn root,	see Helonias dioica.
" officinale,	Cevadilla seed,	Sd. pol. ant. ins.
" Sabadilla,	" "	" " " Veratria.
" viride,	Am'n white hellebore,	Rt. nar. emc. dia. exp. sed. feb.
" "	" " "	Rt. ner. a-spa. alt. res. ins.
Verbascum album,*	White mullein,	see Verbascum Thapsus.
" alatum,*	" "	" " "
" blattaria,	Moth mullein,	" " " Pl. ins.
" nigrum,	Black mullein,	Rt. ast. ; Ls. and Fl. pec. ano. ; Sd.
" thapsiformis,	Mullein,	see Verbascum Thapsus. [nar.
" Thapsus,	Common (White) mullein,	Pl. dem. diu. ano. a-spa. vul.
" "	" " "	Sd. nar. pur. ; in asthma.
Verbena hastata,	American vervain,	Pl. emc. ton. exp. sud. vul.
" officinalis,	European vervain,	Pl. feb. emc. vul. rub.
" spuria, var.	Blue vervain,	analogous to Verbena hastata.

BOTANICAL.	COMMON.	PROPERTIES, PRODUCTIONS, USES, ETC.
Verbena triphylla,*	Sweet verbena,	see Aloysia citriodora.
" urticifolia, var.	Nettle leaved vervain,	Pl. with Quercus alba in Rhus poi.
Verbesina sativa,*	India ramtil (Niger) seed, see Guizotia oleifera.	
" sinuata,	Yellow crown beard,	Rt. aro. spicy, sud. a-syp. dep.
" Virginica,	Crown beard,	Rt. dia. dep.
Vernonia angustifolia,	Iron weed,	Rt. bit. ton. dco. alt. f-com.
" fasciculata,	"	Rt. see V. angustifolia; feb. dep.
" tomentosa,	"	" "
Veronica Americana,*	Brooklime,	see Veronica Beccabunga.
" aquatica,*	"	" " "
" Beccabunga,	"	Hb. a-sco. diu. emm. exa. feb.
" hederæfolia,	Winterweed,	Hb. a-sco. exa. vul.
" officinalis,	Speedwell, Paul's betony	Hb. exp. alt. ton. diu.
" peregrina,	Purslane, Speedwell,	Hb. dep; wash to scrofulous ulcers
" purpurea,*	Wood betony,	see Betonica officinalis.
" quinquefolia,	an East Indian plant,	Pl. in leprosy.
" Virginica,	Black (Culver's) root,	see Leptandra Virginica.
Vetiveria odorata,*	Vettivert,	see Andropogon muricatus.
Viburnum acerifolium,	Dockmackie,	externally to tumors.
" cassinoides,*	Paraguay tea,	see Ilex Paraguaiensis.
" dentatum,	Arrow wood, Mealy tree, Bk. diu. det.; in cancer.	
" edule,		Ft. used as cranberries.
" lantana,	Lithy tree,	Bk. diu. det.; wash in ulcers.
" lantanoides,	Hobble bush,	Bk. a-per. diu.
" lævigatum,*	Paraguay tea,	see Ilex Paraguaiensis.
" lentago,	Nanny bush,	Bk. a-per.
" opulus,	Cramp bk., High Cranberry, Bk. a-spa. a-per. exp. alt. ton.	
" oxycoccos,*	" " " see Viburnum opulus. [alt. opt.	
" prunifolium,	Black haw, Sweet viburnum, Ls. as tea; Bk. ton. ast. diu.	
" "	" " " Bk. of Rt. to prevent abortion.	
" roseum,	Guelder rose, Snowball,	Pl. and Fl. orn.
Vicia ervilla,*	Lentils,	see Ervum lens.
" faba,	Horse bean (Common),	Ls. cmo. a-phl. dis.; Sd. esc. car.
" sativa,	Tare plant, Vetch,	Sd. det. ast.; as food.
" vulgaris,	Common white bean,	see Phaseolus vulgaris.
Victoria Regia (Regina*),	Amazon water lily, (Maize), Fl. fragrant; Sd. eaten roasted.	
Victorialis fœmina,*	Sword lily, Round mandrake, see Gladiolus communis.	
" longa,*	Allerman's root,	see Allium victorialis.
" rotunda,*	Sword lily,	see Gladiolus communis.
Vilfa,	Rush grass,	Pl. some species orn.
Vinca major,	Periwinkle, large,	Pl. orn. ton. ast. dis.
" minor,	" small,	Pl. orn. ton. bit. ast. anti-gal.
" rosea,	" Madagascar, Pl. and Fl. orn.	
Viola arvensis,	Violet,	see Viola cuculata.
" Canadensis	American sweet violet,	Pl. eme. alt.; Fl. per.
" canina,	Dog violet,	Rt. eme. cat. vul.
" cuculata,	Common violet (Blue),	Pl. muc. lax. eme. alt.
" blanda,	Am'n sweet violet (White), Pl. eme. alt.; Fl. per.	
" Ipecacuanha	White ipecac,	see Ionidium Ipecacuanha.
" lutea,*	Wall flower,	see Cheiranthus cheiri.
" martia,*	Sweet violet,	see Viola odorata.
" odorata,	" "	Fl. and Sd. eme. cat. lit.
" ovata,	Rattlesnakes' violet,	Pl. a-ven. opt. dep.
" palustris,	Marsh violet,	see Pinguicula vulgaris.
" pedata,	Birdsfoot violet (Blue),	Pl. muc. pec. emo. lax. eme.
" pumila,*	Dog violet,	see Viola canina.
" rostrata,	Canker (Beaked) violet, Pl. in canker; pec. eme.	
" sylvestris,*	Dog violet,	see Viola canina.
" tricolor,	Pansy, "Heartsease,"	Pl. pec. muc. dis. lax. vul.
Virgilia lutea,*	Kentucky yellow ash (wood), see Cladrastis tinctoria.	

BOTANICAL.	COMMON.	PROPERTIES, PRODUCTIONS, USES, ETC.
Viscum album,*	Mistletoe, (Birdlime),	see Viscum flavescens.
" flavescens,	" "	Ls. nar. a-spa. emc. ton. ner.
" verticillatum,*	" "	see Viscum flavescens.
Vismia Guianensis,*	So. American Gamboge,	see Hypericum bacciferum.
Vitex agnus castus,	Chaste tree,	Bk. sti. ton.; Sd. aro. sti. car.; Ls.
" verticillatus,*	"	see Vitex agnus castus. '[dis.
Vitis alba sylvestris,*	White bryony,	see Bryonia alba.
" cordifolia,	Frost grape vine,	Ft. aci. ref. a-sco.
" corinthiaca,	Corinthian currants,	Ft. dried, known as currants.
" Idæa myrtillus,*	Huckleberry,	see Vaccinium myrtillus.
" " palustris,*	Cranberry,	" " oxycoccus.
" sativa,*	Grapevine.	see Vitis vinifera.
" vinifera,	"	Ls. ast. sty.; Sap lit. opt.
" "	"	Ft. dried forms raisins; muc. dem.
" vulpina,	Foxgrape, (Wild),	Ft. aci. a-sco. nut. ref. [nut.
Voandzeria subterranea,	Bambarra nut,	Ft. called ground nuts; esc.
Volkameria inermis,	Volkameria,	Fl. yield a rich perfume.
Vouacapoua Americana,	Yellow cabbage tree,	see Andira inermis.
Warnera Canadensis,*	Goldenseal,	see Hydrastis Canadensis.
Wellingtonia gigantea,*	California redwood (Giant),	see Sequoia gigantea.
Whitlavia grandiflora,*	Whitlavia,	Pl. orn.
Wiegela rosea,*	Bush honeysuckle,	see Diervilla rosea.
Winterana aromatica,	Winter's bark,	Bk. aro. fragrant, sti. pun. a-sco.
Winterania Canella,*	Canella bark,	see Canella alba.
Wistaria frutescens,	American wistaria,	Fl. orn.
" Sinensis,	Chinese wistaria, (Glycine),	Fl. orn.
Woodsia hypoborea,	Flower cup fern.	
Wrightia anti-dysenterica,	Conessi bark,	Bk. ton. feb.; in dysentery.
" tinctoria,	Pala indigo,	Wd. resembles ivory.
Xanthium strumarium,	Sea burdock, Clotbur,	see Arctium Lappa.
Xanthorhymus ovalifolius,	Painters' Gamboge,	Ju. yellow; Ft. esc.
" pictorius,	" "	" " "
Xanthorrhiza Apiifolia,	Yellow root shrub,	Rt. a pure bitter tonic.
" marbolea,*	" "	see Xanthorrhiza Apiifolia.
" tinctoria,*	" "	" " "
" simplicissima,*	" "	" " "
Xanthorrhœa,	Grass tree,	yields grass tree gum.
" arborea,	Blackboy gum (red),	Gu. ast.; as kino.
" hastilis,	Acaroid, (Acroid),	Yellow gum; resembles Storax.
Xanthoxylon,		see Xanthoxylum.
Xanthoxylum alatum,	Japanese pepper,	Bk. Ls. and Ft. aro.; as spice.
" Americanum,*	Prickly ash,	see Xanthoxylum fraxineum.
" Carolinianum,	Southern prickly ash,	Bk. more pun. than X. fraxineum.
" Clavus Herculis	Yellow prickly ash, W. I.	Bk. sia. sud. diu., pun. a-spa.
" fraxineum,	Northern prickly ash,	Bk. pun. sti. sia. alt. ton.
" fraxinifolium,*	" " "	Ft. pun. sti. sia.; in cholera.
" glandulosum,		resembles Aralia spinosa.
" mantekaricum,	Chinese pepper,	Ft. used as pepper.
" piperitum,	Japanese pepper,	see Xanthoxylum alatum.
" ramiflorum,*	Prickly ash,	" " fraxineum.
" tricarpum,	"	" " "
Xeranthemum annuum,	Purple everlasting,	Pl. and Fl. dried orn.
Xylopia glabra,	Bitter wood,	Wd. bit. ton.; in colic.
Xylosteum ciliatum,	Twinberry,	see Lonicera.
Xyris Caroliniana,	Yellow eyed grass,	Hb. in cutaneous complaints.
Yucca aloifolia,	Spanish dagger (bayonet),	Pl. in conservatories; orn.
" filimentosa,	Silk grass,	Rt. a-syp.
" gloriosa,	Adam's needle,	Rt. yields Cassava.
Zamia integrifolia,	Narrow leaved zamia,	yields Florida arrow root.
" pumila,	Pigmy zamia,	" Bahama " "

BOTANICAL.	COMMON.	PROPERTIES, PRODUCTIONS, USES, ETC.
Zanthorrhiza apiifolia,*	Yellow root shrub,	see Xanthorhiza Apiifolia.
Zanthoxylum,*		see Xanthoxylum.
Zapania nodiflora,	Fog fruit,	Pl. orn.; as verbena.
Zea Mays,	Indian corn, Maize,	Ft. an article of diet.
Zedoaria,		see Kœmpferia.
Zingiber cassumuniar,	Zerumbet, Bengal root,	Rt. aro. bit. ton. sti.
" officinale,	Ginger,	Rt. aro.-sti. pun. car. sia. err. con.
" Germanicum,*	Spotted arum,	see Arum maculatum.
" purpureum,	Zerumbet,	see Zingiber cassumuniar.
" Zerumbet,	Wild Zerumbet,	" " "
Zinnia elegans,	Garden zinnia,	Pl. and Fl. orn.
" violacea,	Blood marigold,	" "
Zizania aquatica,	Indian (Canada) rice,	Sd. furnish food.
Zizia aurea,	Meadow parsnip, Alexanders,	Pl. diu.
Zizyphus Jujuba,	Jujube,	see Rhamnus Jujuba.
" lotus,	"	Ft. esc.
" œnoplia.	"	" Bk. of Rt. vul.
" sativa,		"
" vulgaris,	Jujube,	Ft.-pulp esc. pec. muc.; in paste.

C. E. HOBBS'

BOTANICAL HAND-BOOK.

PHARMACOPŒIAL.

U. S.   Indicates the Pharmacopœial name in the United States Pharmacopœia of 1870.
U. S. '60.   Indicates the Pharmacopœial name in the United States Pharmacopœia of 1860.
B.   Indicates the Pharmacopœial name in the British Pharmacopœia of 1867; reprint of 1874.
G.   Indicates the Pharmacopœial name in the German Pharmacopœia of 1872.
*.   Indicates a synonym in the Pharmacopœia to which it is affixed.
†.   Indicates that the name is common to both the United States and British Pharmacopœias.
‡.   Indicates that the name is common to the United States, British and German Pharmacopœias.

PHARMACOPŒIAL.		COMMON.	BOTANICAL.
Absinthium,	U.S.	Wormwood,	Artemisia Absinthium.
Acacia,	U.S.	Gum Arabic,	Acacia vera and species.
Acacia gummi,	B.	Gum Acacia (Arabic),	One or more species of Acacia.
Aconiti folia,	†.	Aconite leaves,	Aconitum Napellus.
" folium,	U.S.'60.	" leaf,	" "
" radix,	†.	" root,	" "
Adstringens Fothergilli,		Kino,	Pterocarpus marsupium.
Æthiops vegetabilis,		Charcoal of sea weeds.	
Achillea,	U.S.	Yarrow,	Achillea Millefolium.
Agaricus albus,	*G.	White agaric,	Boletus Laricis.
" chirurgorum,		Spunk,	" igniarius.
Agresta,		Green grapes.	Vitis vinifera.
Alcœ Ægyptica,		Musk seed,	Abelmoschus moschatus.
Aletris,	U.S.'60.	Unicorn root,	Aletris farinosa.
Allium,	U.S.	Garlic bulbs,	Allium sativum.
Aloe,	G.	Aloe spicata, and other species of aloe.	
" caballina,		Horse aloes,	Aloe Guineusis.
" Barbadensis,	†.	Barbadoes aloes,	" vulgaris.
" Capensis,	U.S.	Cape aloes,	" spicata and other species.
" hepatica,		Hepatic aloes,	" vulgaris.
" lucida,		aloes,	" "
" Socotrina,	†.	Socotrine aloes,	" Socotrina.
Althæa,	U.S.	Marsh mallow root,	Althæa officinalis.
Ambra liquida,		Sweet gum,	Liquidambar styraciflua.
Amenta lupuli,		Hops,	Humulus Lupulus.
Ammoniacum,	‡.	Gum resin ammoniac,	Dorema ammoniacum.
Amygdala amara,	†.	Bitter almond,	Amygdalus communis var amara
" dulcis,	†.	Sweet almond,	" " " dulcis
Amygdalæ amaræ,	G.	Bitter almonds,	" " " amara
" dulces,	G.	Sweet almonds,	" " " dulcis

15*

PHARMACOPŒIAL.		COMMON.	BOTANICAL.
Amylum,	†.	Starch of wheat,	Triticum vulgare.
" curcumæ Ind.,		E. India arrow root,	Curcuma angustifolia.
" marantæ,	G.	W. " " "	Maranta arundinaceæ.
" Tritici,	G.	Starch of wheat,	Triticum vulgare.
Anethi fructus,	B.	Dill seed,	Anethum graveolens.
Angelica,		Angelica root,	Angelica archangelica.
Angustura,	U.S.	Angustura bark,	Galipea officinalis.
Anime,		Anime resin,	Hymenaea courbaril.
" articulorum,		Hermodactyle.	
Anisum,	†.	Anise (fruit) seed,	Pimpinella anisum.
Anthemis,	†.	Chamomile flowers,	Anthemis nobilis.
Antheræ lilii,		White lily anthers,	Lilium candidum.
Anthodia cinæ,	*G.	Levant wormseed,	Artemisia contra.
Antophylli,		Cloves,	Caryophyllus aromaticus.
Apocynum androsæmifolium	U.S.	Dogsbane root,	Apocynum androsæmifolium.
" cannabinum,	U.S.	Bl'k Ind'n hemp root,	" cannabinum.
Aralia nudicaulis,	U.S.	American spikenard,	Aralia nudicaulis.
" spinosa,	U.S.	Prickly elder bark,	" spinosa.
Arillus myristica,	*G.	Mace,	Myristica fragrans.
Armoraciæ,	B.	Horse radish root,	Cochlearia Armoracia.
Arnica,	U.S.	Arnica flowers,	Arnica montana.
Arnotta,		Annotto,	Bixia orellana.
Arum,	U.S.'60	Indian turnip,	Arum triphyllum.
Asa dulcis,		Gum Benzoin,	Styrax Benzoin.
Asa fœtida,	G.	Gum assafœtida,	Ferula Asa fœtida.
Asclepias,	U.S.'60.	Butterfly weed,	Asclepias tuberosa.
" incarnata,	U.S.	White Indian hemp,	" incarnata.
" Syriaca,	U.S.	Silk weed root,	" Syriaca.
" tuberosa,	U.S.	Pleurisy root,	" tuberosa.
Asarum,	U.S.	Wild ginger,	Asarum Canadense.
Assafœtida,	†.	Assafœtida,	Narthex Assafœtida.
Aurantia immatura,	*G.	Orange berries,	Citrus vulgaris.
Aurantii amari cortex,	U.S.	Bitter orange peel,	" "
" dulcis cortex,	U.S.	Sweet " "	" aurantium.
" flores,	U.S.	Orange flowers,	" "
Azedarach,	U.S.	Azedarach bark,	Melia Azedarach.
Bablah,		Bablah pods,	Acacia Bambolah.
Bacca Phytolaccæ,	U.S.	Pokeberry,	Phytolacca decandra.
Baccæ alkekengi,		Com. winter cherries,	Physalis alkekengi.
" berberidis,		Barberries,	Berberis Canadensis.
" cubebæ,	*G.	Cubebs,	Piper Cubeba.
" ebuli,		Dwarf elder berries,	Sambucus Ebulus.
" hederæ arboreæ,		Ivy berries,	Hedera Helix.
" Juniperi,	*G.	Juniper berries,	Juniperus communis.
" lauri,		Laurel berries,	Laurus nobilis.
" mororum,		Mulberries,	Morus alba and nigra.
" myrtilli,	*G.	Myrtle berries,	Vaccinium myrtillus.
" myrtillorum,		Huckleberries,	" "
" oxycoccos,		Cranberries,	Oxycoccos macrocarpus.
" paradisi,		Fox grape berries,	Paris quadrifolia.
" pimentæ,		Pimenta,	Eugenia Pimenta.
" piperis nigri,		Black pepper,	Piper nigrum.
" phytolaccæ,	U.S.'60.	Pokeberries,	Phytolacca decandra.
" rhamni cathartica,	*G.	Buckthorn,	Rhamnus cathartica.
" rhois glabra,		Sumach berries,	Rhus glabrum.
" ribium rubrum,		Currants, red,	Ribes rubrum.
" spinæ cervinæ,	*G.	Buckthorn berries,	Rhamnus cathartica.
" sumach,		Sumach berries,	Rhus glabrum.
Balsamum Ægypticum,		Balsam of Gilead,	Amyris Gileadensis.
" Benzoin,		Gum Benzoin,	Styrax Benzoin.

PHARMACOPŒIAL.		COMMON.	BOTANICAL.
Balsamum Brasiliense,		Copaiva,	Copaifera bijuga.
" Canadense,		Balsam of fir,	Abies balsamea.
" Carpathicum,		Riga balsam,	Pinus cembra.
" Copaiba,		Copaiva,	Copaifera multijuga.
" Copaivæ,	G.	"	" and other
" de Mecca,		Mecca balsam,	Amyris Gileadensc. [species.
" de Tolu,	*G.	Balsam Tolu,	Myroxylon Toluiferum.
" Dipterocarpi,		Wood oil,	Dipterocarpus turbinatus.
" Gileadense,		Mecca balsam,	Amyris Gileadensc.
" Gurjunæ,		Wood oil,	Dipterocarpus turbinatus.
" Hungaricum,		Hungarian balsam,	Pinus pumilo.
" Indicum,	*G.	Balsam of Peru,	Myroxylon Sonsonatensc.
" " album, .		White balsam of Peru	Myrospermum Peruiferum.
" " nigrum,		Balsam of Peru,	" "
" Indicum sic,		Hard balsam Peru,	Myrospermum Toluiferum.
" Libani,		Riga balsam,	Pinus cembra.
" mariæ,		Gum Tackamahac,	Amyris tomentosa.
" nucistæ,	*G.	Fixed oil of mace,	Myristica fragrans.
" Peruvianum,	B.	Balsam of Peru,	Myroxylon Pereiræ.
" "	U.S.	" "	Myrospermum Peruiferum.
" "	G.	" "	Myroxylon Sonsonatense.
" " album,		" " white,	" Pereiræ.
" " nigrum, *G.		" "	" Sonsonatensc.
" " siccum,		" " dry,	Myrospermum Peruiferum.
" Tolutanum,	U.S., G.	Balsam of Tolu,	" Toluiferum.
" "	B.	" "	Myroxylon Toluifera.
" Styracis,		Liquid styrax,	Liquidambar orientale.
Barras,		Wh. turpentine (Europ'n),	Pinus sylvestris and other
Baume de Copalme,		Liquid styrax,	Liquidambar orientale. [species.
" de San Salvador,		Balsam of Peru,	Myrospermum Peruiferum.
Bdellium,		Gum bdellium,	Amyris commiphora and other
Bedeguar,		Dog rose gall,	Rosa canina. [species.
Bela,	B.	Bengal quince (unripe),	Ægle marmelos.
Belladonnæ folia,	U.S.	Belladonna leaves,	Atropa Belladonna.
" radix,	U.S.	" root,	" "
Benzoe,	G.	Gum benzoin,	Styrax Benzoin.
Benzoinum,	†.	Benzoin, (Balsam),	" "
Berberis,	U.S.	Barberry bk. (of rt.),	Berberis vulgaris.
Berberidis Indicus,		Indian barberry bark,	Berberis aristata and other spe-
Bezetta cœrulea,		Blue patch,	Croton tinctorium. [cies.
" rubra,		Red patch,	" "
Boletus cervinus,		Puff ball,	Lycoperdon cervinum and pro-
" chirurgorum,	*G.	Agaric of the oak,	Boletus fomentarius. [teus.
" salacis,		Willow sponge, .	" suaveolens.
Bombax,		Cotton,	Gossypium herbaceum and other
Bovista,		Puff balls,	Lycoperdon bovistis. [species.
Brayera,	U.S.	Koosso,	Brayera anthelmintica.
Buchu,	U.S.	Buchu leaves,	Barosma crenata and other spe-
Bulbus Colchici,		Colchicum root,	Colchicum autumnale. [cies.
" Scillæ,	G.	Squill,	Scilla maritima.
Butyrum Cacao,		Cacao butter,	Theobroma Cacao.
" nucistæ,	*G.	Fixed oil of nutmeg,	Myristica fragrans,
Caffea,	U.S.	Coffee seed,	Coffea Arabica.
Calamus,	U.S.	Calamus, (Sweet flag)	Calamus aromaticus. [species.
Cahuchu,		India rubber, (Gum elastic),	Siphonia elastica and
Calumba,	U.S.	Columbo root,	Cocculus palmatus and other
Cambodia,		Gamboge,	Garcinia morella. [species.
Cambogia,	B.	" Gum resin,	" "
Camelli,		St. Ignatius bean,	Strychnos Ignatia.
Camphora,	‡.	Camphor, (Concrete oil),	Camphora officinarum.

PHARMACOPŒIAL.		COMMON.	BOTANICAL.
Camphora Borneo,		Camphor, (oil),	Dryobalanops aromatica.
" Japonica,		Japan camphor,	Cinnamomum Camphora.
" Sumatra,		Camphor, (oil),	Dryobalanops aromatica.
Canella,	U.S.	Canella bark,	Canella alba.
" alba,	B.	" white,	" "
" Cubana,		Clove bark,	Laurus caryophyllata.
" dulcis,		Canella bark,	Canella alba.
" Zeylanica,		Ceylon cinnamon,	Cinnamomum Zeylanicum.
Canna,	U.S.	Canna, (Tous les Mois),	Canna, undetermined species.
Cannabis Americana,	U.S.	American hemp,	Cannabis sativa var. Americana
" Indica,	†.	Indian hemp,	" " Indica.
Caoutchouc,		Gum elastic, India rubber,	Siphonia elastica and other
Capita papaveris,	*G.	Poppy heads, (Capsules),	Papaver somniferum. [spec.
Capsici fructus,	B.	Capsicum, Guinea pepper,	Capsicum fastigiatum.
Capsicum,	U.S.	Cayenne (Af'n) pepper,	Capsicum annuum and other
" Africanum,		African pepper,	Capsicum fastigiatum. [species.
Capsulæ papaveris,	*G.	Poppy heads, (Capsules),	Papaver somniferum.
Cardamomum,	†.	Cardamom, (Cardamoms, B.),	Elettaria cardamomum.
" longum,		Large (wild) cardamom,	Elettaria major.
" Malabaricum,	*G.	Small cardamom,	Elettaria Cardamomum.
" medium,		" "	" "
" minus,	*G.	" "	" "
" rotundum,		Round "	Amomum racemosum.
" majus,		Wild "	Elettaria Cardamomum major.
Caricae,	G.	Figs,	Ficus carica.
Carobe,		St. John's (honey) bread,	Ceratonia siliqua.
Carota,	U.S.	Carrot seed,	Daucus Carota.
Carpobalsamum,		Balsam of Gilead,	Amyris Gileadensis.
Carrageen,	G.	Irish moss,	Chondrus crispus.
Carui fructus,	B.	Caraway fruit (seed)	Carum Carui.
Carum,	U.S.	" " "	" "
Carthamus,	U.S.	Safflower,	Carthamus tinctorius.
Caryophyllum,	B.	Cloves,	Caryophyllus aromaticus.
Caryophyllus,	U.S.	Cloves, unexpanded flowers,	Caryophyllus aromaticus.
Cascarilla,	U.S.	Cascarilla bark,	Croton Eleuteria.
" cortex,	B.	" "	" "
Cassia caryophyllata,		Clove bark,	Laurus caryophyllata.
" Cinnamomea,		Cinnamon bark,	" Cassia.
" Fistula,	U.S.	Purging cassia,	Cassia Fistula.
" lignea,		Chinese cinnamon,	Cinnamomum aromaticum.
" Marilandica,	U.S.	American senna,	Cassia Marilandica.
Cassiæ pulpa,	B.	Cassia pulp,	" Fistula.
Castanea,	U.S.	Chestnut leaves,	Castanea vesca.
Castanæa,		"	" "
" equinae,		Horse chestnut,	Æsculus Hippocastanum.
Cataria,	U.S.	Catnep,	Nepeta cataria.
Catechu,	G., U.S.	Catechu, (extract),	Acacia Catechu.
" nigrum,		Black catechu,	" "
" pallida,	B.	Pale catechu,	Uncaria (Nauclea) Gambir.
Cerasa acida,		Sour cherries,	Cerasus.
" dulcis,		Sweet cherries,	"
Cetraria,	†.	Iceland moss,	Cetraria Islandica.
Chenopodium,	U.S.	Wormseed, (Am'n),	Chenopodium anthelminticum.
Chimaphila,	U.S.	Pipsissewa leaves,	Chimaphila umbellata.
Chirata,	B.	Chiretta, (plant),	Ophelia Chirayta.
Chiretta,	U.S.	" "	Agathotes Chirayta.
Chondrus,	U.S.	Irish moss,	Chondrus crispus.
Cimicifuga,	U.S.	Bl'k snakeroot,(Cohosh),	Cimicifuga racemosa.
Cinchona,	U.S.	The bark of all species of Cinchona containing at least	

"   two per cent. of the proper Cinchona alkaloids, which yield crystallizable salts.

PHARMACOPŒIAL.		COMMON.	BOTANICAL.
Cinchona flava,	U.S.	Yellow Cinchona,	Cinchona Calisaya.
" " cortex,	B.	Yellow Cinchona bk.,	" "
" pallida,	U.S.	Pale Cinchona,	" Condaminea and Mi-
" " cortex,	B.	Pale Cinchona bark,	" " [crantha.
" rubra,	U.S.	Red Cinchona, (Red bk.),	" succirubra.
" " cortex,	B.	Red Cinchona bark,	" "
Cinnamomum,	U.S.	Cinnamon bk.(prep'd)	Cinnamomum Zeylanicum.
"	U.S.	" " "	" aromaticum.
" acutum,	*G.	Ceylon cinnamon,	" Zeylanicum.
Cinnamomi cortex,	B.	Cinnamon bark,	" "
Clavelli cinnamomi,		Cassia buds,	Laurus cinnamomum.
Clavus secaliua,		Ergot of rye,	Sclerotium clavus.
Cocculi Indici,		Oriental berries,	Anamirta cocculus.
" Levantici,		" "	" "
" piscatorii,		" "	" "
Cocculus,		Coccnlus Indicus,	" "
Colchici cormus,	B.	Colchicum corm (rt.),	Colchicum autumnale.
" radix,	U.S.	" root,	" "
" semen, -	U.S.	" seed,	" "
" semina,	B.	" seeds,	" "
Colocynthides,		Colocynth apple,	Citrullus Colocynthis.
Colocynthidis pulpa, .	B.	" pulp,	. " "
Colocynthis,	U.S.	Colocynth,	" "
Colophonium,	G.	Resin, (Rosin).	
Conium,	U.S.'60.	Conium, (Hemlock ls)	Couium maculatum.
Conii folia,	†.	" " "	" " "
" fructus,	†.	Conium fruit, (seed U. S.),	Conium maculatum.
Copal,		Gum copal,	see Copal in First part.
Copaiba,	†.	Copaiba, (Copaiva, B.),	Copaifera multijuga and other
Corallina,		Worm moss,	Fucus helminthocortou. [spec.
Coriandri fructus,	B.	Coriander fruit (seed)	Coriandrum sativum.
Coriandrum,	U.S.	Coriander fruit,	" "
Cornus circinata,	U.S.	Round lv'd dogwood,	Cornus circinata.
" Florida,	U.S.	Dogwood, (Boxwood)	" Florida.
" sericea,	U.S.	Swamp dogwood,	" sericea.
Cortex Adansoniæ,		Baobab tree bark,	Adansonia digitata.
" adstringens, Brazil,		Adstringens bk. of Brazil,	Acacia Jurema.
" Alcornoque, .		Alcornoque bk. (Cork)	Byrsouima crassifolla.
" alni glutinosa,		Europ'n alder bark,	Alnus glutinosa.
" " spuria,		Am'n Black alder,	Prinos verticillatus.
" Alstouiæ,		Devil's tree bark,	Alstonia scholaris.
" alyxia aromatica,		Scentwood,	Alyxia aromatica.
" amygdali Persicæ,		Peach tree bark,	Amygdalus Persica.
" Angosturæ,		Angustura bark,	Galipea officinalis.
" " spuria,		Nux Vomica bark,	Strychnos Nux Vomica.
" Aralia spinosa,		Aralia, Prickly elder,	Aralia spinosa.
" Aurantii,	B.	Bitter orange peel,	Citrus Bigaradia.
" " amari,	U.S.	" " "	" vulgaris.
" " dulcis,	U.S.	Sweet orange peel,	" Aurantium.
" Azedarachi,		Pride of India tree bk.	Melia azedarach.
" Bebeeru,		Bebeeru bark,	Nectaudra Rodiæi.
" Berberidis,	U.S.	Barberry bk. (of rt.),	Berberis vulgaris.
" " Indicus,		Ind'n barberry bark,	" lycium and other spec.
" Betula,		Birch bark, (white),	Betula alba.
" Cabaggii,		Cabbage tree bark,	Audira iuermis.
" Canellæ alba,	B.	Canella alba bark,	Canella alba.
" Capparadis,		Caper bush bark,	Capparis spinosa.
" Caryophyllatus,		Clove bark,	Laurus caryophyllata.
" Caryophylloides,		Culilawan (Clove) bk.	Cinnamomum Culilawan.
" Cascarilla,	B., G.	Cascarilla bark,	Croton Eleuteria.

PHARMACOPŒIAL.		COMMON.	BOTANICAL.
Cortex Castanæ Americana,		Chestnut tree bark,	Castana vesca, var. Americana.
"　　" equinæ,		Horse chestnut bark,	Æsculus Hippocastanum.
"　Cassia albæ,		Canella bark,	Canella alba.
"　　" lignea,		Cassia wood and bark,	Laurus Cassia.
"　Cedrelæ febrifugæ,		Febrifuge bk. (Redwood),	Swietenia febrifuga.
"·　Chabarro,		Am'n cork tree bark,	Bowdichia virgilioides.
"　Chacarilla,		Cascarilla bark,	Croton Eleuteria.
"　Chinæ,		Cinchona bark,	Cinchona.
"　　" Calisaya,	G.	Calisaya bk. (Yel. bk.)	"　Calisaya.
"　　" flavus,		Yellow Peruvian bark	"　lancecolata.
"　　" fuscus,	G.	Brown (Pale) Cinchona,	"　micrantha and other spe.
"　　" grisea,	*G.	"　　"　　"　　"	"　　"　　"　　"
"　　" Loxa,		Loxa bark, (Pale bk),	"　Condaminea, var. vera.
"　　" regius,	*G.	Royal yellow bark,	"　Calisaya.
"　　" ruber,	G.	Red Cinchona, (Red bk.),	"　succirubra and other sp.
"　　" spuria,		False Peruvian bark.	
"　Cinchona,		Peruvian bark, ·	see Cinchona, U. S.
"　　" flavæ,	B.	Yel. Cinchona,(Calisaya),	Cinchona Calisaya.
"　　" pallidæ,	B.	Pale　"	Cinchona Condaminea.
"　　" rubræ,	B.	Red　"	"　succirubra.
"　Cinnamomi,	B.	Ceylon Cinnamon,	Cinnamomum Zeylanicum.
"　　" acuti,	?G.	"　　"	"　　"
"　　" Cassiæ,	G.	Cassia bark,	"　Cassia.
"　　" Chinensis,	*G.	Chinese Cinnamon, (Cassia),	Cinnamomum Cassia.
"　　" Zeylanici,	G.	Ceylon　"	Cinnamomum Zeylanicum.
"　Citri,		Lemon peel,	Citrus Limonum, var.
"　Clutcæ,		Cascarilla bark,	Croton Eleuteria.
"　Coccognidii,		Mezereon bark,	Daphne Mezereum.
"　Copalchi,		Copalchi bk. (Mexican),	Croton pseudo-China.
"　Copalki,		"　　"	Strychnos pseudo-quinia.
"　Culilaban,		Culilawan (Clove) bk.	Cinnamomum Culilawan.
"　Culilawani,		Papuan bark,	"　　Xantheneurum.
"　　" Papuanus,		"　　"	"　　"
"　　" verus,		Culilawan bark,	"　Culilawan.
"　Cupressi,		Cypress bark,	Cupressus sempervirens.
"　Curacoa,		Bitter orange peel,	Citrus vulgaris (Bigaradia).
"　Cuspariæ,	B.	Cusparia (Angustura)	Galipea Cusparia.
"　Eleuteriæ,		Cascarilla bark,	Croton Eleuteria.
"　Eleutheræi,		"　　"	"　　"
"　Fagi,		Beech tree bark,	Fagus ferruginea.
"　Frangulæ,	G.	Alder buckthorn,	Rhamnus Frangula.
"　Fraxini,		Ash bark, (Manna),	Fraxinus excelsior.
"　Fructus aurantii,	G.	Bitter orange peel,	Citrus vulgaris, (var. amara).
"　　" citri,	G.	Lemon peel,	"　Limonum, (Medica).
"　　" Granati,	U.S.	Pomegranate rind,	Punica Granatum.
"　　" Juglandis,	G.	Green walnut hulls,	Juglans regia.
"　Geoffroyæ flavus,		Yel. cabbage tree bk.	Andira inermis.
"　·　" fuscus,		Brown　"　　"	"　retusa.
"　　" ·Jamaicensis,		Yellow　"　　"	"　inermis.
"　　" Surinamensis,		Brown　"　　"	"　retusa.
"　Guaiaci ligni,		Lignum vitæ bark,	Guaiacum officinalis.
"　Granati fructus,	U.S.	Pomegranate rind,	Punica Granatum.
"　　" radicis,	U.S.	"　root bark,	"　　"
"　Granatum radicum,		"　　"	"　　"
"　　" pomorum,		"　fruit peel,	"　　"
"　Hippocastani,		Horse chestnut bark,	Æsculus Hippocastanum.
"　Juglandis,		Butternut bark,	Juglans cinerea.
"　Kinæ aromaticæ,		Cascarilla bark,	Croton Eleuteria.
"　Laricis, ·	B.	Larch bark (inner),	Larix Europæa, (Abies Larix).
"　Laureola,		Mezereon bark,	Daphne Mezereum.

PHARMACOPŒIAL.		COMMON.	BOTANICAL.
Cortex Ligni sancti,		Lignum vitæ bark,	Guaiacum officinalis.
" Ligni Cassiæ,		Cassia (Cinnamon) bark, Cinnamomum Cassia.	
" Limonis,	†.	Lemon peel,	Citrus Limonum.
" Magellanicus,		Winter's bark,	Drimys (Winteri) aromatica.
" Magnoliæ,		Sweet bay tree bark,	Magnolia fragrans (glauca).
" Malabathri,		Tamala Cinnamon,	Cinnamomum Tamala.
" Malambo,		Matias bark,	Croton Malambo.
" Mali,		Apple tree bark,	Pyrus malus.
" Malicorium,		Pomegranate peel,	Punica Granatum.
" Margosæ,		Nim bark,	Melia Indica.
" Massoy,		Massoy bark,	Cinnamomum Kiamis.
" Matias,		Matias (Malambo) bk.	Croton Malambo.
" Mezerii,	G., B.	Mezereon bark,	Daphne Mezereum (Laureola).
" Mezereum,		" "	" " "
" Mudar,		Mudar bark,	Asclepias gigantea.
" Musenæ,		Musenna bark,	Rottlera Schimperi.
" Myricæ ceriferæ,		Bayberry bark,	Myrica cerifera.
" narcaphtum,		Cane storax tree,	Styrax officinalis.
" Nectandræ,	B.	Bebeeru (Greenheart) bk. Nectandra Rodiæi.	
" nucum Juglandis,	*G.	Green walnut hulls,	Juglans regia.
" " " viridis,		" " "	" " "
" Olea,		Olive tree bark,	Olea Europœa.
" Paratoda,		Paratoda bark,	Canella axillaris.
" Pareiræ,		Pareira brava bark (rt.), Cissampelos Pareira.	
" Pereiræ,		Braziletto bark,	Cæsalpina Brasiliensis.
" Persicorum,		Peach bark,	Amygdalus Persica.
" Pini,		Pine bark (white),	Pinus strobus and other species.
" " Canadensis,		Hemlock bark,	Abies (Pinus) Canadensis.
" Pomorum Aurantii,	*G.	Bitter orange peel,	Citrus vulgaris.
" " Granatorum,		Pomegranate fruit peel, Punica Granatum.	
" Populi,		Poplar bark (Eurp'n),	Populus tremula.
" Pruni padi,		Bird cherry tree bark, Prunus Padus.	
" " serotinæ,		Wild cherry tree bk.,	" Virginiana.
" " Virginicæ,		" " "	" " (serotina).
" Psydii,		Pomegranate rind,	Punica Granatum.
" Pyrus malus,		Apple tree bark,	Pyrus malus.
" Quassia,		Quassia,	Simaruba excelsa.
" Quercus,	B.	Oak bark,	Quercus pedunculata.
" "	G.	Oak bark,	" " and Q. sessiliflora
" " albus,	U.S.	Wh. oak bark (inner)	" alba.
" " ruber,		Red oak bark,	" rubra.
" " tinctoriæ,		Bl'k oak bark, Quercitron, Quercus tinctoria.	
" Radicis Azedarach,		Bead tree root bark,	Melia Azedarach.
" " Gossypii,		Cotton root bark,	Gossypium herbaceum and other
" " Granati,	G.	Pomegranate rt. bark, Punica Granatum. [species.	
" " "	U.S.	" " "	" " "
" " rubri villosi,		Blackberry root bark,	Rubus villosus.
" Rhamni frangulæ,	*G.	Alder buckthorn,	Rhamnus frangula.
" salicis,		Willow bark,	Salix of several species.
" " nigri,		Black (Pussy) willow, · " nigra.	
" Sambuci,		Elder bark,	Sambucus Canadensis.
" Sassafras,		Sassafras bark of rt.	Sassafras officinale.
" Simarubæ,		Simaruba bark of rt.	Simaruba officinalis.
" Sintoc,		Sintoc bark of Java,	Cinnamomum sintoc.
" Soymoidæ,		Jamaica redwood bk.	Soymoida febrifuga.
" Sumach,		Sumach bark,	Rhus coriaria.
" " radicis,		" root bark,	" glabrum.
" Swieteniæ,		Indian redwood bark, Swietenia febrifuga.	
" Tabernæ montanæ,		Devil's tree bark,	Alstonia scholaris.
" Tamarisci Gallica,		yields False manna,	Tamarix Gallica.

PHARMACOPŒIAL.		COMMON.	BOTANICAL.
Cortex Tamarisci Germanica,		German Tamarisk,	Tamarix Germanica.
" Taxi,	.	Yew tree bark,	Taxus baccata.
" Thuris,		Cane storax tree,	Styrax officinalis.
" Thymeleœ,		Mezereon bark,	Daphne Mezereum.
" Thymiamatis,		Cane storax bark,	Liquidambar orientale.
" Tiliæ,		Basswood bark,	Tilia Americana.
" Tulipiferæ,		Tulip tree bark,	Liriodendron Tulipifera.
" Ulmi,	B.	Elm bk. (British) in'r,	Ulmus Campestris.
" " Americana,	.	Am'n elm bark,	" Americana.
" " fulvæ,		Slippery elm bark,	" fulva.
" " interior,		British elm bark,	" Campestris.
" Winteri,		Winter's bark,	Drimys Winteri.
" Winteranus,		" "	" "
" " spuria,		Canella bark,	Canella alba.
Cormus Colchici,	B.	Colchicum corms,	Colchicum autumnale.
Costus amarus, .		Sweet costus,	Costus Arabicus.
" corticosus,		Canella bark,	Canella alba.
" dulcis,		" "	" "
Cotula,	U.S.	Mayweed herb,	Anthemis Cotula.
Croci stigmata,		Saffron, .	Crocus sativus.
Crocus,	‡.	Saffron (Spanish),	" "
" Americanus,		American saffron,	see Carthamus.
" Hispanicus,		Spanish saffron,	Crocus sativus.
" Lilii,		White lily anthers,	Lilium candidum.
" Orientalis,		Ceylon saffron,	Crocus orientalis.
Cubeba,	†.	Cubeb, (Cubebs, B.),	Cubeba officinalis. (Piper Cu-
Cubebæ,	G.	Cubebs, .	" " [beba.)
Cucumis agrestis,		Wild sq'rt'g cucumber	Momordica Elaterium.
" asinus,		" "	" "
Curcuma,	U.S.	Turmeric root,	Curcuma longa.
Cudbear,		Cudbear,	Lecanora tartarea.
Cusso,	B.	Kousso,	Brayera anthelmintica.
Cuspariæ cortex,	B.	Cusparia bark,	Galipea Cusparia.
Cydonia exsiccata,		Dried quince fruit,	Cydonia vulgaris.
Cydonium,	U.S.	Quince seed,	" "
Cynosbata,		Dog rose, Hips,	Rosa canina.
Cypripedium,	U.S.	Cypripedium root,	Cypripedium pubescens and par-
Dactyli,	.	Dates,	Phœnix dactylifera. [viflorum.
Delphinium,	U.S.	Larkspur seed,	Delphinium Consolida.
Dextrinium,	G.	Dextrin, (Artificial gum). .	
Digitalis,	U.S.	Foxglove leaves,	Digitalis purpurea.
" 'folia,	B.	Digitalis leaf,	" "
Diospyros,	U.S.	Persimmon fruit,	Diospyros Virginiana.
Dolichi pubes,		Cowhage down W. I.	Mucuna pruriens.
Dracontium,	U.S.	Skunk cabbage,	Ictodes (Symplocarpus) fœtidus
Dulcamara,	†.	Bittersweet,	Solanum Dulcamara.
Ecbalii fructus,	B.	Squirt'g cucumber ft.	Ecbalium officinarum.
Elaterium,	B.	Elaterium (sediment of ft.),	Ecbalium officinarum.
"	U.S.	" "	Momordica Elaterium.
Elemi,	G., B.	Elemi, (Gum resin),	Canarium commune. [purea.
Ergota,	†.	Ergot of rye,	Sclerotium of Caviceps pur-
Erigeron,	U.S.	Fleabane ls. and tops,	Erigeron heterophyllum and
" Canadense,	U.S.	Can. " " "	" Canadense. [Philad.
Eupatorium,	U.S.	Thoroughwort, (Boneset),	Eupatorium perfoliatum.
Euphorbia Corollata,	U.S.	Large flow'g spurge,	Euphorbia corollata.
" Ipecacuanha,	U.S.	Ipecacuanha spurge,	" Ipecacuanha.
Euphorbium,	G.	Gum Euphorbium,	" resinifera.
Euonymus,	U.S.	Wahoo bark,	Euonymus atropurpurea.
Extractum Glycyrrhizæ,	†.	Ex. of Liquorice root,	Glycyrrhiza glabra.
Faba Calabarica,	G.	Calabar bean,	Physostigma venenosum.

PHARMACOPŒIAL.		COMMON.	BOTANICAL.
Faba Physostigmatis,	B.	Calabar bean,	Physostigma venenosum.
" Sancti Ignatii,		St. Ignatius' bean,	Strychnos Ignatia.
Fabæ Coffeæ,		Coffee seed,	Coffea Arabica.
" Dividivi,		Dividivi, Libidibi,	Cæsalpina coriaria.
" de Tonca,		Touka bean,	Dipterix odorata.
" febrifugæ,		St. Ignatius' bean,	Strychnos Ignatia.
" Iudica,		" "	" "
" Libidibi,		Dividivi,	Cæsalpina coriaria.
" Pichurim,		Pichurim (Sassafras)	nuts, Nectandra puchury.
" Sanctæ Ignatiæ,	.	Iguatia bean,	Strychnos amara.
Farinæ Avenæ,	U.S.	Oat flour (meal),	Avena sativa.
" Fagopyri,		Buckwheat meal,	Polygonum Fagopyrum.
" Lini,	†.	Linseed (Flax) meal,	Linum usitatissimum.
" Secalis,		Rye flour,	Secale cereale.
" Tritici,	B.	Wheaten flour,	Triticum vulgare and var.
Festucæ Carophyllorum,		Clove stems,	Caryophyllus aromaticus.
Fibrilla Artemisiæ,		Mugwort root,	Artemisia vulgaris.
Fici,	*G.	Figs, (dried fruit),	Ficus Carica.
Ficus,	†.	Fig, "	" "
" passæ,		" "	" "
Filix Mas,	†.	Male fern rhizome,	Aspidium Filix Mas.
Flores Absinthii,		Wormwood flowers,	Artemisia Absinthium.
" Acaciæ,		Sloe tree flowers,	Prunus spinosa.
" Achillæ,		Yarrow flowers,	Achillea Millefolium.
" Agnicasti,		Chaste tree flowers,	Vitex Agnus castus.
" Alceæ,	*G.	Hollyhock flowers,	Althæa rosea.
" Althæa,		Marshmallow flowers,	" officinalis.
" Amygdali,		Almond blossoms,	Amygdalus communis.
" Anthos,		Rosemary flowers,	Rosmarinus officinalis.
" roismarini,		" "	" "
" Anthemidis,	B.	Chamomile flowers,	Anthemis nobilis.
" Aquilegiæ,		Columbine flowers,	Aquilegia vulgaris.
" Arnicae,	G.	Arnica flowers,	Arnica montana.
" Aurantii,	G.	Orange blossoms,	Citrus Aurantium and amara.
" Aurantiorum,		" "	" "
" Balaustines,		Pomegranate flowers,	Punica Granatum.
" Bellidis,		Daisy flowers,	Bellis perennis.
" Bismalvæ,		Marshmallow flowers,	Althæa officinalis.
" Boraginis,		Borage flowers,	Borago officinalis.
" Brayeræ Anthelmintica,	*G.	Koosso, (Kosso),	Brayera anthelmintica.
" Buglosi,		Bugloss flowers,	Anchusa Italica.
" Calcitrapæ,		Knapweed flowers,	Centaurea calcitrapa.
" Calendulæ,		Marigold flowers,	Calendula officinalis.
" Calthæ,		Marsh marigold fl'rs,	Caltha palustris.
" Caprifoliæ,		Woodbine (Honeysuckle) fl'rs, Lonicera periclymenum	
" Carthami,		Safflower,	Carthamus tinctorius.
" Cassiæ,		Cassia buds,	Laurus Cinnamomum.
" Cassie,		Sponge tree flowers,	Acacia Farnesiana.
" Chamæmell,		German Chamomile fl.	Matricaria Chamomilla.
" Chamomillæ,		" " " "	" "
" " foetidæ,		Mayweed flowers,	Anthemis Cotula.
" " Romanae,	G.	Roman Chamomile fl.	" nobilis.
" " spurlæ,		Mayweed flowers,	" Cotula.
" " vulgaris,	G.	German Chamomile fl.	Matricaria Chamomilla. [spec.
" Chrysanthemi,		Persian feverfew fl'rs	Pyrethrum carneum and other
" Cinæ,	G.	Levant wormseed,	Artemisia, undetermined species
" Clematidis,		Virgins' bower fl'rs,	Clematis erecta.
" Colchici,		Colchicum flowers,	Colchicum flowers.
" Cyani,		Blue centaury flowers	Centaurea cyanus.
" Digitalis,		Foxglove flowers,	Digitalis purpurea.

PHARMACOPŒIAL.	COMMON.	BOTANICAL.
Flores Doronica Germanica,	Arnica flowers,	Arnica montana.
" Farfaræ,	Coltsfoot flowers,	Tussilago farfara.
" Flammulæ,	Virgins' bower fl'rs,	Clematis erecta.
" Gnaphalii,	Life everlasting fl'rs,	Gnaphalium polycephalum.
" Granatorum,	Pomegranate flowers,	Punica Granatum.
" Hageniæ,	Koosso flowers,	Brayera anthelmintica.
" Ilispidulæ,	Life everlasting fl'rs,	Gnaphalium polycephalum.
" Hyperici,	Johnswort flowers,	Hypericum perforatum.
" Jasmini,	Jasmine flowers,	Jasminum officinale.
" Koso,	Koosso flowers,	Brayera anthelmintica.
" Kosso,	G. " "	" "
" Koosso,	" "	" · "
" Lamii albi,	Blind nettle flowers,	Lamium album.
" Lavandulæ,	G. Lavender flowers,	Lavandula officinalis (vera).
" Lavendulæ,	*G. " "	" " "
" Libanotidis,	Rosemary flowers,	Rosmarinus "
" Liliorum convallaria,	Lily of the valley fl'rs,	Convallaria majalis.
" Lillia albi,	White lily flowers,	Lilium candidum.
" Lupuli,	Hops,	Humulus Lupulus.
" Malvæ arborea,	G. Hollyhock flowers,	Althæa rosea.
" " alceæ,	*G. " "	" "
" " hortense,	*G. " "	" "
" " rosea,	" "	" "
" " sylvestris,	*G. High mallow flowers,	Malva sylvestris.
" " vulgaris,	G. " "	" "
" Martii,	Sweet violet flowers,	Viola odorata.
" Meliloti,	White melilot flowers,	Melilotus officinalis.
" Millefolii,	G. Yarrow flowers,	Achillea Millefolium.
" " nobilis,	Noble yarrow flowers,	" nobilis.
" Moschatæ,	Mace,	Myristica fragrans.
" Naphæ,	Orange blossoms,	Citrus Aurantium and amara.
" Oxycanthæ,	White thorn flowers,	Cratægus Crus-galli.
" Pæoniæ,	Peony flowers,	Pæoniæ officinalis.
" Papaveris albus,	White poppy flowers,	Papaver somniferum var. album
" " Rhœades,	Red poppy flowers,	" Rhœas.
" " rubri,	" " "	" "
" Persicorum,	Peach tree flowers,	Amygdalus Persica.
" Pilosellæ,	Life everlasting fl'rs,	Gnaphalium dioicum.
" Primulæ,	G. Primrose (Cowslip) fl.	Primula officinalis (vera).
" " veris,	*G. " " "	" " "
" Pruni sylvestri,	Sloe tree flowers,	Prunus spinosa.
" Ptarmica,	Sneezewort flowers,	Achillea Ptarmica.
" Punicæ,	Pomegranate flowers,	Punica Granatum.
" Pyrethri Dal.	Persian feverfew fl'rs,	Pyrethrum of several var.
" " Germanica,	German pellitory fl'rs,	" Ptarmica.
‹. " rosci,	Rose fl'd feverfew fl'rs	" roseum.
" Ranunculi,	Wood anemone fl'rs,	Anemone nemorosa.
" Rhœades,	G. Red poppy flowers,	Papaver Rhœas.
" Roris marini,	Rosemary flowers,	Rosmarinus officinalis.
" Rosae,	G. Pale rose (Hundred l'vd),	Rosa centifolia.
" Rosæ incarnata,	Pale rose petals,	Rosa centifolia.
" " pallida,	" " "	" "
" " rubræ,	Red rose petals,	" Gallica.
" Rosarum incarnatum,	Pale rose petals,	Rosa centifolia.
" " pallidum,	" " "	" "
" " . rubrum,	Red rose petals,	" Gallica.
" Sambuci,	B., G. Elder flowers,	Sambucus nigra.
" " Canadensis,	Am'n elder flowers,	" Canadensis.
" Scopariæ,	Broom tops,	Cytisus Scoparius.
" Spartii,	" "	" "

PHARMACOPŒIAL.		COMMON.	BOTANICAL.
Flores Spicæ,		Spike lavender flowers,	Lavandula spica.
" Stœchados Arableæ,		Arabian lavender flow'rs,	" stœchas.
" " citrinæ,		German golden locks,	Gnaphalium arenarium.
" " Neapoliti,		Eternal flower,	" stœchas.
" Tanaceti,	·	Tansy flowers,	Tanacetum vulgare and crispum
" Thapsi,		Mullein flowers,	Verbascum Thapsus.
" Tilae,	G.	Linden flowers,	Tilia ulmifolia and platyphyllos.
" " cum braet,		" " with ls.	" " " "
" Trifolii albi,		White clover flowers,	Trifolium repens.
" " odorati,		Melilot clover flowers,	Melilotus leucanthe.
" " rubri,		Red clover flowers,	Trifolium pratense.
" Tussilaginus,		Coltsfoot flowers,	Tussilago farfara.
" Urticae,		Nettle flowers,	Urtica dioica.
" " mortuœ,		White nettle flowers,	Lamium album.
" Verbasci,	G.	Mullein flowers,	Verbascum thapsiformis and
" Violarum,		Sweet violet flowers,	Viola odorata. [other species.
" Viola odorata,		" " "	" " "
Fœniculum,	U.S.	Fennel (seed),	Fœniculum dulce.
Fœniculi fructus,	B.	" fruit,	" "
Folia Abelmoschi,		Musk plant leaves,	Abelmoschus esculentus.
" Acetosæ,		Common sorrel leaves,	Rumex acetosa.
" Aceris rubri,		Red maple leaves,	Acer rubrum.
" " striati,		Striped maple leaves,	" striatum.
" Aconiti,	†.	Aconite leaves,	Aconitum Napellus.
" Alni,		Alder leaves,	Alnus rubra.
" " glutinosæ,		" " European,	" glutinosa.
" Althææ,	G.	Marshmallow leaves,	Althæa officinalis.
" Anthos,		Rosemary leaves,	Rosmarinus officinalis.
" Arctostaphyli,	*G.	Bearberry (Uva Ursi) ls.	Arctostaphylos (Arbutus) Uva
" Aurantii,	G.	Orange leaves,	Citrus vulgaris. [Ursi.
" Aurantiorum,		" "	" aurantium.
" Belladonnæ,	B., G.	Belladonna leaves,	Atropa Belladonna.
" Buccu,		Buchu leaves,	Barosma crenata and other spe.
" Buchu,	B.	" "	" betulina.
" "	B.	" "	" crenulata.
" "	B.	" "	" serratifolia.
" Cardui benedicti,	*G.	Cardus leaves,	Centaurea benedicta.
" Castanæ,		Chestnut leaves,	Castana Americana.
" Cicutæ,	_	Hemlock leaves, poison,	Cicuta maculata.
" Coca,		Coca leaves,	Erythroxylon Coca.
" Coluteæ,		Bladder senna,	Colutea arborescens.
" Conii,	B.	Hemlock ls. (spotted),	Conium maculatum.
" Daturæ alba,		Indian Datura leaves,	Datura alba.
" Digitalis,	B., G.	Digitalis (leaf, B.) l'ves,	Digitalis purpurea.
" Diosmæ,		Buchu leaves,	see Folia Buchu.
" Fagi,		Beech leaves,	Fagus sylvestris.
" Farfaræ,	G.	Coltsfoot leaves	Tussilago farfara.
" Fraxini,		Ash leaves,	Fraxinus ornus.
" Guaco,		Guaco leaves,	Mikania Guaco.
" Hederæ,		English Ivy leaves,	Hedera Helix.
" Hippocastani,		Horse chestnut leaves,	Æsculus Hippocastanum.
" Hyosciami,	ʄ.	Henbane leaves,	Hyosciamus niger.
" Ilicis aquifolii,		European holly leaves,	Ilex aquifolium.
" " Americanæ,		American holly leaves,	" opaca.
" Iudi,		Tamala cinnamon leaves,	Cinnamomum nitidum.
" Juglandis,	G.	European walnut leaves,	Juglans regia.
" " regiæ,		English walnut leaves,	" "
" Juniperi,		Juniper leaves,	Juniperus communis.
" " Sabinæ,		Savin leaves,	" Sabina.
" Lauri,		Laurel leaves,	Laurus nobilis.

PHARMACOPŒIAL.		COMMON.	BOTANICAL.
Folia Laurocerasi,	B., G.	Cherry laurel leaves,	Prunus Laurocerasus.
" Libanotidis,		Rosemary leaves,	Rosmarinus officinalis.
" Malabathri,		Tamala cinnamon leaves,	Cinnamomum nitidum.
" Malvæ,	G.	Com. (Low) mallow ls.	Malva vulgaris (rotundifolia).
" Maticæ,	B.	Matico leaves,	Artanthe elongata.
" Mate,		Paraguay tea.	Ilex Paraguaiensis.
" Melissæ,	G.	Balm leaves,	Melissa officinalis.
" Menthae crispæ,	G.	Curled mint leaves,	Mentha crispa.
" " piperitæ,	G.	Peppermint leaves,	" piperita.
" " viridi,		Spearmint leaves,	" viridis.
" Myrciæ,		Jamaica bayberry leaves	Myrcia acris.
" Myricae ceriferæ,		Bayberry (Am'n) leaves,	Myrica cerifera.
" Millefolii,	*G.	Milfoil (Yarrow) leaves,	Achillea Millefolium.
" Nicotianæ,	G.	Tobacco leaves,	Nicotiana Tabacum.
" Papaveris,		Poppy leaves,	Papaver somniferum.
" Patchouli,		Patchouly leaves,	Pogostemon Patchouly.
" Persicorum,		Peach tree leaves,	Amygdalus Persica.
" Pini,		Pine leaves,	Pinus strobus and other species?
" " Canadensis,		Hemlock tree leaves,	Abies Canadensis.
" Populi,		Poplar tree leaves,	Populus tremula.
" Quercus,		Oak leaves,	Quercus robur and alba.
" Rhododendri,		Yel. fl'd rhododendron,	Rhododendron chrysanthemum?
" Rhois Toxicodendri,		Poison oak leaves,	Rhus Toxicodendron.
" Ricinii,		Castor oil leaves,	Ricinus communis.
" Rosmarini,	G.	Rosemary leaves,	Rosmarinus officinalis.
" Rubi fruticosi,		Blackberry leaves,	Rubus fruticosus.
" " strigosi,		Raspberry leaves,	" strigosus.
" Rutæ,	G.	Rue leaves,	Ruta graveolens.
" Sabinæ,		Savin leaves,	Juniperus Sabina.
" Salviæ,	G.	Sage leaves,	Salvia officinalis.
" Savinæ,		Savin leaves,	Juniperus Sabina.
" Sennæ,	G.	Alexandria or Tripoli senna,	Cassia lenitiva (acutifolia).
" " Alexandrinæ,		" "	" " "
" " Indica,		Indian senna,	Cassia lanceolata.
" " Marilandicaæ,		American senna,	" Marilandica.
" Sesami,		Benne leaves,	Sesamum Indicum and orientale
" Stramonii,	‡.	Stramonium leaves,	Datura Stramonium.
" Sumach,		Sumach leaves,	Rhus coriaria.
" Tabaci,	B.	Tobacco leaves,	Nicotiana Tabacum.
" Tamala pathri,		Tamala cinnamon leaves	Cinnamomum nitidum.
" Taxi,		Yew tree leaves,	Taxus baccata.
" Theæ,		Chinese tea leaves,	Thea Chinensis.
" Toxicodendri,	G.	Poison oak leaves,	Rhus Toxicodendron.
" Trifolii fibrini,	G.	Buckbean leaves,	Menyanthes trifoliata.
" Tussilago,		Coltsfoot leaves,	Tussilago farfara.
" Tylophoræ,		Indian Ipecac,	Asclepias asthmatica.
" Uvæ Ursi,	G.	Bearberry leaves,	Arctostaphylos (Arbutus) Uva
" Verbasci,		Mullein leaves,	Verbascum Thapsus. [Ursi.
" Vitis viniferæ,		Grapevine leaves,	Vitis vinifera.
Folium aconitum,	U.S.'60.	Aconite leaf,	Aconitum Napellus.
" Belladonnæ,	U.S.'60.	Belladonna leaf,	Atropa Belladonna.
" Hyosciami,	U.S.'60.	Henbane leaf,	Hyosciamus niger.
" Sesami,	U.S.'60.	Benne leaf,	Sesamum Indicum and orientale
" Stamonii,	U.S.'60.	Stramonium leaf,	Datura Stramonium.
Folliculi sennæ,		Husks of Senna,	Cassia of several species.
Frasera,	U.S.	American columbo,	Fraseri Walteri.
Fructus Ajowan,		Ajava seed,	Ammi copticum.
" Anethi,	B.	Dill fruit (seed),	Anethum graveolens.
" Anisi,		Anise fruit (seed),	Pimpinella Anisum.
" " stellati,	G.	Star anise,	Illicium anisatum.

PHARMACOPŒIAL.		COMMON.	BOTANICAL.
Fructus Anisi vulgaris,	G.	Anise fruit (seed),	Pimpinella Anisum.
" Auranti immaturi,	G.	Orange berries,	Citrus vulgaris. (Aurautium
" Aurantii,	B.	Bitter orange, (fruit),	" Bigaradia. [amara.)
" Belæ,	B.	Bengal quince,	Ægle marmelos.
" Cacao,		Cacao beans,	Theobroma Cacao.
" Cannabis,	G.	Hemp seed,	Cannabis sativa.
" Capsici,	G.	Cayenne pepper,	Capsicum annuum (longum).
" Cardamomi,		Cardamom seed,	Elettaria Cardamomum.
" " minores,	G.	Small (Malabar) Cardamom, Elettaria Cardamomum.
" Caricæ,	*G.	Figs,	Ficus carica.
" Carui,	B.	Caraway fruit,	Carum Carui.
" Carvi,	G.	" " (seed),	" "
" Cassia Fistulæ,		Purging Cassia,	Cassia Fistula.
" Ceratoniæ,	G.	St. John's bread,	Ceratonia siliqua.
" Citri,		Lemons,	Citrus Limonum.
" Cocculi,		Oriental berries,	Anamirta Cocculus.
" Colocynthidis,	G.	Colocynth,	Citrullus (Cucumis) Colocynthis
" Conii,	†.	Hemlock fruit,	Conium maculatum.
" Coriandri,	B., G.	Coriander fruit,	Coriandrum sativum.
" Cubebæ,	*G.	Cubebs,	Cubeba officinalis. (Piper Cube-
" Cydoniæ,		Quince, (fruit),	Cydonia vulgaris. [ba.)
" Cumlui,		Cummin,	Cuminum Cyminum.
" Cynobasti,		Dog rose gall,	Rosa canina.
" Diospyri,		Persimmon fruit,	Diospyros embryopteris.
" Ecballi,	B.	Squirt'g cucumber fruit, Ecbalium officinarum.
" Elaterii,		" " "	" "
" Fœniculi,	B., G.	Fennel seed (fruit),	Fœniculum officinale (dulce).
" Hibisei esculenti,		Okra, Gombo,	Abelmoschus esculentus.
" Hordei,		Barley,	Hordeum distichon.
" Juniperi,	G.	Juniper berries,	Juniperus communis
" Lauri,	G.	Bay (Laurel) berries,	Laurus nobilis.
" Limonis,		Lemons,	Citrus Limonum.
" Malus,		Apples, (fruit),	Pyrus malus.
" Mori,		Mulberries,	Morus nigra.
" Mororum,		"	" "
" Moschatæ,		Nutmegs,	Myristica fragrans.
" Myrtilli,	G.	Whortleberries,	Vaccinium Myrtillus.
" Papaveris,	G.	Poppy heads,	Papaver somniferum.
" Petroselini,	G.	Parsley seeds,	Petroselinum sativum.
" Piperis longi,		Long pepper,	Piper longum.
" " nigri,		Black pepper,	" uigrum.
" Phellandrii,	G.	Fine lv'd water-hemlock seed, Œnanthe Phellandrium.
" Pruni,		Prunes,	Prunus domestica.
" Pyrus,		Pears,	Pyrus vulgaris.
" Rhamni,		Persian (Buckthorn) bs. Rhamnus infectorius.
" " catharticæ,G.		Buckthorn berries,	" catharticus.
" Rosæ caninæ,	B.	Dog rose fruit,	Rosa canina.
" Sabadillæ,	G.	Cevadilla seeds,	Sabadilla officinalis.
" Tamarindorum,	*G.	Tamarinds,	Tamarindus Indica.
" Theobromæ,		Chocolate nuts,	Theobroma Cacao.
" Vanillæ,	G.	Vanilla beans,	Vanilla planifolia.
Fucus Amylaceus,		Ceylon moss,	Gracillaria lichenoides.
" Crispus,	*G.	Irish moss,	Chondrus crispus.
" Ceylonensis,		Ceylon moss,	Gracillaria lichenoides
" Hibernicus,		Irish moss,	Chondrus crispus.
" Versiculosus,		Sea (Bladder) wrack,	Fucus vesiculosus.
Fungus Albus saligneus,		Willow sponge,	Dædalea suaveolens.
" Bedeguar,		Dog rose galls,	Rosa canina.
" Civinus,		Willow sponge,	Dædalea suaveolens.
" Chirurgorum,		Puff ball,	Lycoperdon bovista.

16

PHARMACOPŒIAL.		COMMON.	BOTANICAL.
Fungus Cynosbati,		Dog rose galls,	Rosa canina.
" Faginosus,			Morchella esculentus.
" Ignarius,		Oak agaric, Spunk,	Boletus ignarius.
" " præparatus,	G.	Agaric of the oak,	" "
" Laricis,	G.	Larch (White) agaric,	" Laricis.
" Melitensis,		Maltese sponge,	Cynomorum coccineum.
" Quercinus,		Agaric of the oak,	Boletus ignarius.
" Rosarum,		Dog rose galls,	Rosa canina.
" Salacis,		Willow sponge,	Dædalia suaveolens
" Sambucinus,		Jews' ear, (Elder fungus)	Peziza auricula.
Fusti,		Clove stalks (stems),	Eugenia caryophylla.
Galbanum,	U.S.	Galbanum, Gum resin of	an undetermined plant.
"	B.	" " "	unascertained plant.
"	G.	"	Ferula erubescens.
Galipot,		Turpentine gum,	Pinus pinaster and other species
Galla,	†.	Nutgall, (Galls, B.),	Quercus infectoria.
Gallæ,	G.	" Galls,	" "
" Aleppo,		"	" "
" Chinensis,		Chinese galls,	Distylium racemosum.
" Halepenses,	*G.	Nutgall, Galls,	Quercus infectoria.
" Levanticæ,	*G.	" "	" "
" Turcicæ,	*G.	" "	" "
Gamba,		Gamboge, Gum-resin,	Garcinia morella.
Gambier,		Pale Catechu,	Nauclea Gambir.
Gambir,		" "	" "
Gambogia,	U.S.	Gamboge, Gum-resin,	Garcinia morella.
Gaultheria,	U.S.	Checkerberry leaves,	Gaultheria procumbens.
Gelsemium,	U.S.	Yel. Jessamine(Jasmine)	Gelsemium sempervirens.
Gemmæ abietis,		Pine buds,	Pinus sylvestris.
" Balsamæ,		Balm Gilead tree buds,	Populus candicans.
" Capparidis conditæ,		Caper bush buds,	Capparis spinosa.
" Pini,	*G.	Young pine shoots,	Pinus silvestris.
" Populi,	G.	Poplar buds (black),	Populus nigra and other species.
Gentiana,	U.S.	Gentian root,	Gentiana lutea.
" Catesbæi,	U.S.	Blue gentian root,	" Catesbæi.
Gentianæ radix,	B.	Gentian root,	" lutea.
Geranium,	U.S.	Geranium, Cranesbill,	Geranium maculatum.
Geum,	U.S.	Water avens,	Geum rivale.
Gillenia,	U.S.	Gillenia,	Gillenia trifoliata and G. stipu-
Glandes Quercus,		Acorns,	Quercus. [lacea.
Glandulæ Lupuli,	G.	Lupulin,	Humulus Lupulus.
" Rottleræ,	*G.	Kameela,	Rottlera tinctoria.
Glycyrrhiza,	U.S.	Liquorice root,	Glycyrrhiza glabra.
Glycyrrhizæ radix,	B.	" "	" " [species.
Gossypii radicis cortex,	U.S.	Cotton root bark,	Gossypium herbaceum and other
Gossypium,	†.	" wool,	Gossypium of various species.
Grana Avenionensis,		French yellow berries,	Rhamnus infectorius.
" gnidi,		Mezereon berries,	Daphne gnidium.
" lycii Gallici,		French yellow berries,	Rhamnus infectorius.
" Maleguettæ,		Paradise seed,	Amomum Melegueta.
" Moschata,		Musk plant seed,	Abelmoschus moschatus.
" Paradisi,		Paradise seed,	Amomum Melegueta.
" Tiglia,		Croton oil bean,	Croton Tiglium.
" Tiglii,		" "	" "
" Tilli,		" "	" "
Granati fructus cortex,	U.S.	Pomegranate rind,	Punica Granatum.
" radicis cortex,	†.	" bark of root,	" "
Guaiaci lignum,	†.	Guaiacum wood,	Guaiacum officinale.
" resina,	†.	" resin, (Guaiac, U.S.)	" "
Guarana,		Guarana paste,	Paullinia sorbilis.

PHARMACOPŒIAL.		COMMON.	BOTANICAL.
Gummi Acaciæ,	B.	Gum Acacia (Arabic),	see Acacia gummi.
" Achariari,		Carauna gum,	Icica Caranna.
" Ammoniacum,		Gum Ammoniacum,	Dorema Ammoniacum.
" Acaroides,		Acaroid resin,	Xanthorrhœa hastilis.
" Astragali,		Tragacanth,	Astragalus verus and other spe.
" Anime,		Anime,	Hymenia courbaril.
" Arabicum,	G.	Gum Arabic,	Acacia Nilotica, Seyal and torti-
" Asafœtida,		Assafœtida,	Narthex Assafœtida. [lis.
" Assafœtida,		"	" "
" Australe,		Australian gum,	Acacia decurrens.
" Barbaricum,		Barbary gum,	" gummifera.
" Bassora,		Gum Bassora,	" leucophlea.
" Bdellii,		Bdellium, Indian,	Amyris commiphora.
" " Africani,		" African,	Heudelotia Africana.
" Benzoe,		Benzoin,	Styrax benzoin.
" Benzoini,		"	" "
" Camphora,		Camphor,	Camphora officinarum.
" Capensis,		Cape gum Arabic,	Acacia Karroo.
" Carannæ,		Caranna gum,	Bursera gummifera.
" Cerasum,		Cherry tree gum,	Cerasus and Prunus.
" Copallinum,		Copal,	see Copal in First part.
" Damarra,		Damar gum,	Agathis (Pinus) Damarra.
" Elasticum,		India rubber,	Siphonia elastica and other spe.
" Elemi,	*G.	Elemi,	undetermined plant of Yucatan.
" Euphorbii,		Gum Euphorbium,	Euphorbia officinarum and other
" Galbanii,		Galbanum,	Ferula Galbaniflua. [species.
" Gambiense,		Gum Kino,	see Kino in First part.
" Gambir,		Gambier, Catechu,	Nauclea Gambir.
" Gedda,		Gum Gedda (Arabic),	Acacia gummifera.
" Geddah,		" " "	" "
" Guaiaci,		Gum-resin Guaiac,	Guaiacum officinale.
" Gutta,		Gamboge,	Garcinia morella.
" " Ceylonense,		"	" "
" Hederæ,		Ivy gum,	Hedera Helix.
" de Jemu,		Gamboge,	Garcinia morella.
" Juniperi,		Sandarach resin,	Thuja articulata.
" Kino,	*G.	Kino,	Pterocarpus Marsupium.
" Kutera,		Bassora gum,	Acacia leucophlea.
" Laccæ in massis,		Lac, Lump lac,	see Lac in First part.
" " in ramulis,		Stick (Twig) lac,	" "
" " in granis,		Seed lac,	" "
" " in tabulis,		Shellac, (Lump lac),	" "
" Ladanum,		Labdanum, Gum cistus,	Citrus Ladaniferus.
" Lamac,		Acacia, Gum Arabic,	Acacia vera and other species.
" Laricis,		Larch gum, (Manna),	Pinus Larix.
" Leucum,		Acacia, Gum Arabic,	Acacia vera and other species.
" Mimosæ,	*G.	" " "	" " "
" Myrrha,		Myrrh,	Balsamodendron Myrrha.
" Nutt,		Acaroid gum,	Xanthorrhœa hastilis.
" Oleæ,		Olive gum,	Olea Europæa.
" Orenburgense,		Manna of Briancon,	Larix Europæa.
" Panacis,		Opopanax,	Pastinaca Opopanax.
" Rubrum adstringens,		Kino, Red astring't gum,	Pterocarpus Marsupium.
" " " Gambiense,		"	" erinaccus.
" " adstringens,		"	see Kino in First part.
" Sandarac,		Sandarach resin,	Thuja articulata.
" Sagapenum,		Gum Sagapenum,	Ferula Persica.(?)
" Sanctum,		Guaiac resin,	Guaiacum officinale.
" Sarcocolla,		Sarcocolla, product of	Penæa Sarcocolla.
" Scorpionis,		Gum Arabic,	Acacia of several species.

PHARMACOPŒIAL.		COMMON.	BOTANICAL.
Gummi Seneca,		Senegal gum,	Acacia Senegal.
" Senega,		" "	" "
" Senegalense,		" "	" "
" Thebaicum,		Gum Arabic,	" gummifera.
" Tragacantha,	*G.	Tragacanth,	Astragalus of several species.
" Uralense,		Orenburgh gum,	Pinus Larix.
Gummi-resina Asa fœtida,	*G.	Assafœtida,	Ferula (Scorodosma) fœtida.
" Ammoniacum,	*G.	Ammoniac,	Dorema Ammoniacum.
" Galbanum,	*G.	Galbanum,	Ferula erubescens.
" Gutti,	*G.	Gamboge,	Garcinia morella.
" Myrrha,	*G.	Myrrh,	Balsamodendron Myrrha.
" Olibanum,	*G.	Olibanum,	Boswellia papyrifera.
Gutta Gambir,		Gambier,	Nauclea Gambir.
" Gemu,		Gamboge,	Garcinia morella.
" percha,	†.	Guttapercha,	Isonandra Gutta.
" Tuban,	*G.	" purified,	" "
Gutti,	G.	Gamboge,	Garcinia morella.
Hæmatoxylon,	U.S.	Logwood, (heartwood),	Hæmatoxylon Campechianum.
Hæmatoxyli lignum,	B.	" "	Hæmatoxylum "
Hemidesmi radix,	B.	Hemidesmus root,	Hemidesmus Indicus.
Hedeoma,	U.S.	American pennyroyal,	Hedeoma pulegioides.
Helianthemum,	U.S.	Frostwort,	Helianthemum Canadense.
Helleborus,	U.S.	Black hellebore root,	Helleborus niger.
Helmintochortos,		Worm moss,	Fucus helmintochorton.
Hepatica,	U.S.	Liverwort,	Hepatica Americana.
Herba Abrotini,		Southernwood herb,	Artemisia abrotanum.
" " fœmini,		Lavender cotton herb,	Santolina chamæcyparissus.
" " hortensis,		Southernwood herb,	Artemisia abrotanum.
" " marl,		" "	" "
" Absinthi,	G.	Wormwood herb,	" Absinthium.
" Acanthi,		Cotton (Musk) thistle,	Onopordon acanthium.
" Acetosellæ,		Wood sorrel,	Oxalis acetosella.
" Achilleæ,		Yarrow (Milfoil) herb,	Achillea Millefolium.
" Acinos,		Mountain thyme herb,	Acinos vulgare.
" Aconiti,		Aconite leaves,	Aconitum Napellus.
" Adianthi albi,		Wall rue herb,	Asplenium ruta-muraria.
" " Americani,		Haircap moss,	Polytrichium Juniperum.
" " nigri,		Black spleenwort,	Asplenium Adiantum nigrum.
" " rubri,		Common maidenhair,	" trichomanes.
" Ægyptiaca,		Blue melilot herb,	Melilotus cœrulea.
" Agerati,		Maudlin tansy,	Achillea Ageratum.
" Agrimoniæ,		Agrimony herb,	Agrimonia Eupatoria.
" Alchemilli,		Ladies' mantle,	Alchemilla vulgaris.
" Alexandrina,		Alexanders,	Smyrnium olusatrum.
" Allii Ursini,		Ransoms,	Allium ursinum.
" Althææ,	*G.	Marshmallow leaves,	Althæa officinalis.
" Alliariæ,		Hedge garlic,	Alliaria officinalis.
" Amarici,		Sweet marjoram herb,	Origanum Marjorana.
" Anagallis,		Pimpernel, scarlet,	Anagallis arvensis.
" Andrographidis,		Creyat, (Chiretta, E. I.),	Andrographis paniculata.
" Anethi,		Dill plant,	Anethum graveolens.
" Anthos,		Rosemary leaves,	Rosmarinus officinalis.
" " sylvestris·		Marsh tea,	Ledum palustre.
" Antirrhini,		Snapdragon leaves,	Antirrhinum majus.
" Apocyni andro.,		Dogsbane leaves,	Apocynum androsæmifolium.
" " cannabini,		Bl'k Indian hemp leaves,	" cannabinum.
" Arbor vitœa,		Arbor vitæ leaves,	Thuja occidentalis.
" Arctii,		Burdock leaves,	Arctium Lappa.
" Armoraciæ,		Horse radish leaves,	Cochlearia Armoracia.
" Arnica,		Arnica leaves,	Arnica montana.

PHARMACOPOEIAL.	COMMON.	BOTANICAL.
Herba Artemisiæ,	Wormwood herb,	Artemisia Absinthium.
" " albæ,	Mugwort herb,	" vulgaris.
" " moxa,	Moxa herb (Chinese),	" Chinensis.
" " vulgaris,	Mugwort herb,	" vulgaris.
" Asari,	Asarabacca leaves,	Asarum Europæum.
" Asperulæ,	Woodroof,	Asperula odorata.
" Athanasia,	Tansy herb,	Tanacetum vulgare.
" Atriplicis fœtida.	Stinking goosefoot,	Chenopodium vulvaria.
" " Mexicana,	Mexican (Jerusalem) tea,	" ambrosioides.
" Ariculæ muris,	Mouse ear hawkweed,	Hieracium pilosella.
" Ballota, .	Wolfsballote,	Ballota lanata.
" Balsaminæ,	Balsam weed,	Impatiens pallida.
" Balsamitæ,	Costmary herb,	Pyrethrum Tanacetum.
" Barbæ capræ,	Meadowsweet herb,	Spiræa ulmaria.
" Barbaræ,	Barberry leaves,	Berberis vulgaris.
" Bardanæ,	Burdock leaves,	Arctium Lappa, minor.
" Basilici,	Sweet basil herb,	Ocimum basilicum.
" Beccabungæ,	Brooklime herb,	Veronica Beccabunga.
" Belladonnæ, *G.	Belladonna leaves,	Atropa Belladonna.
" Bellidis,	Garden daisy. herb,	Bellis perennis.
" Benedicta,	Herb Bennet,	Geum urbanum.
" Betonica,	Betony herb,	Betonica officinalis.
" Bidensis tripartita,	Swamp beggars' tick,	Bidens tripartita.
" Bismalvæ,	Marshmallow leaves,	Althæa officinalis.
" Boraginis,	Borage herb,	Borago officinalis.
" Botrys Mexicana, *G.	Mexican tea,	Chenopodium ambrosioides.
" " vulgaris,	"	" "
" Branca,	Cow parsley,	Heracleum sphondylium.
" " ursina Germanica,	"	" "
" Britannica,	Water dock leaves,	Rumex Britannica.
" Buglossi,	Bugloss herb, small,	Anchusa officinalis.
" Buglæ,	Bugle weed, European,	Lycopus Europæus.
" Bursæ-pastoris,	Shepherds' purse,	Capsella bursa-pastoris.
" Calamintha,	Calamint, Mount'n balm,	Melissa Calamintha.
" Calendulæ,	Marigold leaves,	Calendula officinalis.
" Calthæ sativæ,	" "	" "
" Cannabis,	Indian hemp-plant,	Cannabis sativa var. Indica.
" " Americanæ,	Am'n "	" " var. Americana
" " Indicæ, G.	Indian "	" " var. Indica.
" Canni,	Levant wormseed, ·	Artemisia Santonica (Cina.).
" Camphoratæ,	Stinking ground vine,	Camphorosma Monspelica.
" Capilli Americani,	American maidenhair,	Adiantum pedatum.
" " veneris,	European maidenhair,	" Capillus veneris.
" Cardiaca,	Motherwort herb,	Leonurus cardiaca.
" Cardui benedicti, G.	Blessed thistle herb,	Cnicus benedictus.
" " sancti,	" "	" "
" " veneris,	Fullers' teasel leaves,	Dipsacus fullonum.
" Cedronellæ,	Moldavian balm,	Dracocephalum Moldavica.
" Centaurii, G.	European centaury,	Erythræa centaurium.
" " Americanæ,	American "	Sabbatia angularis.
" " benedicti,	Blessed thistle,	Cnicus benedictus.
" " minoris, *G.	European centaury,	Erythræa centaurium.
" Cerifolii,	Chervil herb,	Scandix cerefolium.
" " Hispanici, *G.	European sweet cicily,	Myrrhis odorata.
" Ceterachi, ·	Ceterach herb,	Ceterach officinarum.
" Chamæcyparissus, ·	Lavender cotton,	Santolina chamæcyparissus.
" Chamædrys,	Germander herb,	Teucrium Chamædrys.
" Chamæpityos,	Ground pine,	Ajuga chamæpitys.
" Chelidonii, G.	Celandine,	Chelidonium majus.
" Cheloni glabræ,	Balmony herb,	Chelone glabra.

PHARMACOPŒIAL.		COMMON.	BOTANICAL.
Herba Chenopodii,		Oak of Jerusalem herb,	Chenopodium anthelminticum.
" " ambrosioides,	G.	Mexican tea,	" ambrosioides.
" Chimaphilæ,		Pipsissewa herb,	Chimaphila umbellata and other
" Chiraytæ,		Chiretta herb,	Agathotes Chirayta. [species.
" Chiratæ,		" "	" "
" Cichorei,		Chicory herb,	Cichorium intybus.
" Cicutæ,	*G.	Conium leaves,	Conium maculatum.
" " aquatica,		Poison water hemlock,	Cicuta aquatica.
" " majoris,		Conium leaves,	Conium maculatum.
" " maculatæ,		Spotted Am. hemlock ls. ·	" "
" " minoris,		Small hemlock leaves,	Æthusa cynapium.
" " virosæ,		Water hemlock leaves,	Cicuta aquatica.
" Citronellæ,		Balm lemon herb,	Melissa officinalis.
" Clematidis,		Virgins' bower herb,	Clematis erecta.
" Clinopodii,		Mountain thyme,	Calamintha alpina.
" " vulgaris,		Dogmint, Field thyme,	Clinopodium vulgare.
" Cochleariæ,	G.	Scurvy grass,	Cochlearia officinalis.
" Collinsoniæ,		Stoneroot plant leaves,	Collinsonia Canadensis.
" Conii,	G.	Conium leaves,	Conium maculatum.
" " maculati,	*G.	" "	" "
" Consolidæ,		Goldenrod leaves,	Solidago virgaurea.
" Conyzæ,		Marsh fleabane,	Conyza squarrosa.
" " mediæ,		Small fleabane,	" media.
" Costa,		Cats' ear herb,	Hypocharis maculata.
" Costi,		Costmary herb,	Pyrethrum Tanacetum
" cum floribus Linariæ	*G.	Common toadflax,	Linaria vulgaris.
" Cunilæ,		Wild marjoram,	Origanum vulgare.
" " Americanæ,		Mountain dittany,	Cunila mariana.
" Cupressi,		Lavender cotton,	Santolina chamæcyparissus.
" Cuscutæ,		Dodder, Hellweed,	Cuscuta Europæa.
" Cynapii,		Small hemlock leaves,	Æthusa cynapium.
" Cynoglossi,		Hounds' tongue herb,	Cynoglossum officinale.
" Datiscæ,		False hemp leaves,	Datisca cannabina.
" Daturæ,		Thorn apple leaves,	Datura Stramonium.
" " Stramonii,		" "	" "
" Dauci,		Carrot leaves,	Daucus Carota.
" Delphinii,		Larkspur herb,	Delphinium consolida.
" Dens-leonis,		Dandelion herb,	Taraxacum Dens-leonis.
" Digitalis,		Foxglove leaves,	Digitalis purpurea.
" " purpurea,	*G.	" "	" "
" Dorea,		European goldenrod,	Solidago virgaurea.
" Doronici,		Arnica leaves,	Arnica montana.
" Dracocephali,		Sweet balm herb,	Dracocephalum canariense.
" Dracunculi,		Taragon,	Artemisia dracunculus.
" Dulcamaræ,		Bittersweet,	Solanum Dulcamara.
" Equiseti,		Scouring rush,	Equisetum hyemale.
" Erigeronitis,		Fleabane leaves,	Erigeron heterophyllum and
" " Canadensis,		Canada fleabane,	" Canadense. [Philad.
" Erysimi,		Hedge mustard,	Sisymbrium officinale.
" Eupatorii,		Boneset,	Eupatorium perfoliatum.
" " cannabini,		Hemp agrimony,	" cannabinum.
" " perfoliata,		Boneset, Thoroughwort,	" perfoliatum.
" Euphrasiæ,		Eyebright,	Euphrasia officinalis.
" Farfaræ,	*G.	Coltsfoot leaves,	Tussilago farfara.
" Felis,		Catnep, (Catmint)	Nepeta cataria.
" Flamulæ,		Virgins' bower,	Clematis erecta.
" Fœniculi,		Fennel herb,	Fœniculum dulce.
" Fragariæ,		Strawberry leaves,	Fragaria vesca.
" Fumitoræ,		Fumitory herb,	Fumaria officinalis
" Fumariæ,		" "	" "

PHARMACOPŒIAL.		COMMON.	BOTANICAL.
Herba Galegœ,		Goats' rue herb,	Galega officinalis.
" .Galeopsidis,	G.	Hemp nettle,	Galeopsis ochroleuca.
" " grandiflora,		Hollow tooth herb,	. " grandiflora.
" Galli,		Cleavers, Goosegrass,	Galium aparine.
" Gaultheria,		Checkerberry leaves,	Gaultheria procumbens.
" Genista,		Dyers' broom,	Genista tinctoria.
" Genipi veri,			Achillea atrata.
" Geranii,		Cranesbill leaves,	Geranium maculatum.
" " Robertiani,		Herb Robert leaves,	" Robertianum.
" Glastium,		Dyers' woad, .	Isatis tinctoria.
" Glechomœ,		Ground Ivy herb,	Nepeta Glechoma.
" Gnaphalii,		Life everlasting,	Gnaphalium dioicum.
" Gratiolœ,	G.	Hedge hyssop,	Gratiola officinalis.
" Hederœ,		European Ivy leaves,	Hedera Helix.
" Helianthi,		Common sunflower,	Helianthus vulgare.
" Hepaticœ nobilis,		Noble liverwort,	Hepatica Americana.
" " stellata,		Woodroof,	Asperula odorata.
" " trilobœ,		Liverwort,	Hepatica Americana.
" Hispidulœ,		Mouse bloodwort,	Hieracium pilosella.
" Hydrocotyle,		Indian pennywort,	Hydrocotyle Asiatica.
" Hyosciami,	*G.	Black henbane leaves,	Hyosciamus niger.
" " alba,		White henbane,	" alba.
" Hyperici,		St. Johnswort herb,	Hypericum perforatum.
" Hyssopi,		Hyssop herb,	Hyssopus officinalis.
" Iberis,		Bitter candytuft,	Iberis amara.
" Ibisci,		Marshmallow leaves,	Althœa officinalis.
" Ignis,		Scarlet cup lichen,	Lichen cocciferus.
" Impatiens,		Wild celandine,	Impatiens pallida.
" Impia,		Cotton rose, Cudweed,	Filago Germanica.
" Isatis,		Dyers' woad,	Isatis tinctoria.
" Jaceœ,	*G.	Pansy, "Heartscase,"	Viola tricolor.
" Lactucœ,	G.	Acrid lettuce leaves,	Lactuca virosa.
" " scariolœ,		Wild lettuce leaves,	" scariola.
" " virosæ,	*G.	Acrid lettuce leaves,	" virosa.
" Lamii,	.	Blind nettle,	Lamium album.
" " sylvatici fœt.,		Hedge woundwort,	Stachys sylvatica.
" Lappa,		Burdock leaves,	Arctium Lappa.
" . " majoris,		" large,	" "
" " minoris,		" small,	" "
" Ledi palustri,		Marsh tea,	Ledum palustre.
" " latifolii, .		Labrador tea,	" latifolia.
" Levistici,		Lovage leaves,	Ligusticum levisticum.
" Linariœ,	G.	Common toadflax,	Linaria vulgaris.
" " cum floribus,	*G.	" "	" "
" Leonuri,		Motherwort herb,	Leonurus cardiaca.
" Lini cathartici,		Purging flax,	Linum catharticum.
" Lobellœ,	G.	Lobelia, Indian tobacco,	Lobelia inflata.
" " inflatæ,	*G.	" "	" "
" Lunariœ,		Moonwort,	Botrychium lunaria.
" Luteolœ,		Dyers' weed, Woad,	Reseda luteola.
" Lycopi.		Bugle weed,	Lycopus Europæus.
" Lycopodii,		Club moss,	Lycopodium clavatum.
" Lysimachlœ,		Loosestrife, yellow,	Lysimachia vulgaris.
" Malvœ,	*G.	Common Low mallow,	Malva rotundifolia.
" " vulgaris,		" "	" " (vulgaris).
" " visci,		Marsh mallow,	Althæa officinalis.
" Majoranœ,	G.	Sweet marjoram,	Origanum Marjorana.
" Marl veri,		Syrian herb mastich,	Teucrium marum.
" Marrubii,		Com. white horehound,	Marrubium vulgare.
" " albi,		" "	" "

PHARMACOPŒIAL.		COMMON.	BOTANICAL.
Herba Marrubii aquatici,		European bugle weed,	Lycopus Europæus.
" " nigri,		Black horehound,	Ballota nigra (fœtida).
" Matico,		* Matico leaves,	Artanthe elongata.
" Matricariæ,		Feverfew herb,	Pyrethrum Parthenium.
" Meliloti,	G.	Meliot herb,	Melilotus officinalis.
" Melissæ,	*G.	Lemon balm,	Melissa officinalis.
" Menthæ acutæ,		Spearmint herb,	Mentha viridis.
" " aquaticæ,		Watermint herb,	" aquatica.
" " arvensis,		Water calamint,	" arvensis.
" " catariæ,		Catnep, Catmint,	Nepeta cataria.
" " crispæ,	*G.	Curled balm mint,	Mentha crispa.
" " equinæ,		Europ'n horsemint herb,	" gratissima (sylvestris).
" " piperitæ,	*G.	Peppermint herb,	" piperita.
" " Romanæ,		Spearmint herb,	" viridis.
" " sylvestris,		Europ'n horsemint herb,	" gratissima (sylvestris).
" " viridis,		Spearmint herb,	" viridis.
" " vulgaris,		" "	" " " "
" Mercurialis,		Mercury herb,	Mercurialis annua.
" Mesembryanthemi crystallini,		Europ'n ice plant,	Mesembryanthemum crystalli-
" Millefolii,	G.	Milfoil, Yarrow,	Achillea Millefolium. [num.
" " nobilis,		Noble yarrow,	" nobilis.
" Moldavicæ,		Moldavian balm,	Dracocephalum Moldavica.
" Monardæ,		Mountain rose balm,	Monarda coccinea.
" Myrti,		Meadow fern leaves,	Myrica Gale.
" Nardi sylvestris,		Asarabacca plant,	Asarum Europæum.
" Nasturtii,		Water cress,	Nasturtium officinale.
" Nepetæ,		Catnep, Catmint,	Nepeta cataria.
" Nicotianæ,	*G.	Tobacco leaves,	Nicotiana Tabacum.
" Nummulariæ,		Moneywort herb,	Lysimachia Nummularia.
" Œnanthæ,		Water heml'k dropwort,	Œnanthe crocata.
" " aquaticæ,		Water hemlock leaves,	" fistulosa.
" Ononidis,		Rest harrow herb,	Ononis spinosa.
" Origani,		Wild marjoram herb,	Origanum vulgare.
" " Cretici,		Dittany of Crete,	" Creticum.
" Osmundæ,		Buckhorn fern leaves,	Osmunda regalis.
" Paradisi,		Herb Paris, Foxgrape,	Paris quadrifolia.
" Parietariæ,		Wall pellitory herb,	Parietaria officinalis.
" Patchouli,		Patchouly herb,	Pogostemon Patchouly.
" Pedicularis,		Wood (Head) betony,	Pedicularis Canadensis.
" Pentaphylli,		Cinquefoil herb,	Potentilla reptans.
" Perfoliati,		Thorough wax,	Bupleurum perfoliatum.
" Persicariæ,		Heartsease,	Polygonum Persicaria.
" Petroselini,		Parsley herb,	Petroselinum sativum.
" Pilosellæ,		Mouse ear hawkweed,	Hieracium pilosella.
" Pimpernellæ hortense,		Salad burnet, Bibernel,	Poterium sanguisorba.
" Plantaginis lanceolata,		Ribwort plantain,	Plantago lanceolata.
" " latifolia,		Common plantain,	" major.
" Polygalæ,	G.	Europ'n bitter milkwort,	Polygala amara.
" " amaræ,	*G.	" " "	" "
" " vulgaris,		Bitter polygala,	" vulgaris.
" Portulacæ,		Garden purslane,	Portulaca oleracea.
" Prunellæ,		Healall herb,	Prunella vulgaris.
" Ptarmicæ,		Sneezewort,	Achillea Ptarmica.
" Pulegii,		European pennyroyal,	Mentha pulegium.
" Pulmonaria,		Lungwort herb,	Pulmonaria officinalis.
" " arborea,		Lungmoss, (Lichen),	Sticta pulmonaria.
" " maculata,		Spotted lungwort,	Pulmonaria officinalis. [la.
" Pulsatillæ,	G.	Pulsatilla herb,	Anemone pratensis and Pulsatil-
" " nigricantis,	*G.	" "	" " " "
" Pyrethri Germanica,		Sneezewort herb,	Achillea Ptarmica.

PHARMACOPŒIAL.		COMMON.	BOTANICAL.
Herba Pyrola maculata,		Spotted wintergreen,	Chimaphila (Pyrola) maculata.
" Ranuncull,		Wood anemone, (Crowfoot),	Anemone nemorosa.
" Rhois Toxicodendri	*G.	Poison oak (ivy),	Rhus Toxicodendron.
" Ribli,		Black currant leaves,	Ribes nigrum.
" Rorellæ,		Sundew herb,	Drosera rotundifolia.
" Roris marini,	*G.	Rosemary leaves,	Rosmarinus officinalis.
" " solis,		Sundew herb,	Drosera rotundifolia.
" Rosmarini, .	*G.	Rosemary leaves,	Rosmarinus officinalis.
" Rosmarini sylvestri,		Marsh tea,	Ledum palustris.
" Rubii fruticosa,		Blackberry leaves,	Rubus fruticosus.
" " idæi,		Raspberry leaves,	" idæus.
" " strigosi,		" "	" strigosus.
" Rutæ,	*G.	Rue leaves,	Ruta graveolens.
" " murariæ,	.	Wall rue leaves,	Asplenium Ruta-muraria.
" Sabinæ,	*G.	Savine tops,	Juniperus Sabina.
" " Virginianæ,		Red (Pencil) cedar,	" Virginiana.
" Salviæ,	*G.	Garden sage leaves,	Salvia officinalis.
" " pratense,		Meadow sage leaves,	" pratense.
" " sclareæ,		Clary leaves,	" sclarea.
" Saniculi,	.	Sanicle leaves,	Sanicula Europæa.
" Santolinæ,		Lavender cotton,	Santolina chamæcyparissus.
" Saponariæ,		Soapwort, Bouncing bet,	Saponaria officinalis.
" Saturejæ,		Summer savory,	Satureja hortensis.
" Savinæ,		Savin tops,	Juniperus Sabina.
" Schœnanthi,		Ginger grass,	Andropogon schœnanthus.
" Sclareæ,		Clary leaves,	Salvia sclarea.
" Scolopendriæ,		Spleenwort fern,	Asplenium scolopendrium.
" Scoparii,		Broom tops,	Cytisus Scoparius.
" Scrophulariæ,		Scrofula plant, Figwort,	Scrophularia nodosa.
" " aquaticæ,		Water flgwort,	" aquatica.
" " Marilandicæ,		Scrofula plant,	" Marilandica.
" Scutellariæ,		Scullcap,	Scutellaria lateriflora.
" " galericulatæ,		European scullcap,	" galericulata.
" Sempervivi,		Houseleek,	Sempervivum tectorum.
" Senecio,		Life root plant,	Senecio aureus.
" Serpylli,	G.	Wild (Mother of) thyme,	Thymus serpyllum.
" Sideritis,		Ironwort,	Sideritis hirsuta.
" Solani furiosi,		Deadly nightshade,	Atropa Belladonna.
" " lethalis,		" "	" "
" " nigri,		Bl'k nightshade (garden)	Solanum nigrum.
" " maniaci,		Deadly nightshade,	Atropa Belladonna.
" " Dulcamari,		Bittersweet,	Solanum Dulcamara.
" Sonchi,		Sow thistle,	Sonchus oleraceus.
" Sparti,		Broom tops,	Cytisus Scoparius.
" Spilanthis,	G.	Spear-leaved spilanthus,	Spilanthes oleracea.
" " oleraceæ,		" "	" "
" Stramonii,	*G.	Thornapple leaves,	Datura Stramonium.
" Spiræœ,		Hardhack herb,	Spiræa tomentosa.
" Symphiti,		Comfrey leaves,	Symphytum officinale.
" Tabaci,	*G.	Tobacco leaves, .	Nicotiana Tabacum.
" Tanaceti,		Tansy herb,	Tanacetum vulgare.
" Taraxici,		Dandelion herb,	Taraxacum Dens-leonis.
" Theæ,		Chinese tea,	Thea Chinensis.
" Thujæ,		Arbor vitæ leaves,	Thuja occidentalis.
" Thymi,	G.	Thyme, garden,	Thymus vulgaris.
" " vulgaris,		" "	" "
" Trifolii albi,		White clover flowers,	Trifolium repens.
" " acetosi,		Wood sorrel,	Oxalis acetosella.
" " odorati,		Melilot clover,	Melilotus officinalis.
" " fibrini,	*G.	Buckbean leaf,	Menyanthes trifoliata.

PHARMACOPŒIAL.		COMMON.	BOTANICAL.
Herba Trifolii palustri,		Buckbean herb,	Menyanthes trifoliata.
" " rubri,		Red clover flowers,	Trifolium pratense.
" Tussilaginis,	*G.	Coltsfoot leaves,	Tussilago farfara.
" Urticœ,		Common nettle herb,	Urtica dioica.
" " majoris,		" " "	" " "
" " minoris,		Dwarf nettle,	" minor (urens).
" " mortuæ,		Dead nettle, Coolweed,	" pumila.
" Uvæ Ursi,	*G.	Uva Ursi, Bearberry ls.,	Arctostaphylos (Arbutus) Uva
" Verbasci,		Mullein herb,	Vesbascum Thapsus. [Ursi.
" Verbenæ,		Vervain herb,	Verbena officinalis and hastata.
" Veronicæ,		Speedwell herb,	Veronica officinalis.
" Vincæ,		Periwinkle herb,	Vinca major and Vinca minor.
" Violœ,		Violet plant,	Viola odorata.
" " tricoloris,	G.	Pansy, "Heartsease,"	" tricolor.
" Violarum,		Violet plant,	" odorata.
" Virgæ aureæ,		European goldenrod,	Solidago virgaurea.
" Xanthii,		Sea burdock leaves,	Xanthium strumarium.
Heuchera, .	U.S.	Alum root,	Heuchera Americana.
Hordeum,	U.S.	Barley (decorticated sd.)	Hordeum distichon.
" decorticatum,	B.	Pearl barley (husked sd.)	. " "
" perlatum,		" "	" "
" preparatum,		Prepared barley,	" "
Hordei semeni,		Barley,	" "
Humuli Lupuli,		Hops, (strobiles),	Humulus Lupulus.
Humulus,	U.S.	" "	" "
Hydrastis,	U.S.	Hydrastis, (Goldenseal),	Hydrastis Canadensis.
Hyosciami folia,	†.	Hyosciamus (Henbane) ls.	Hyosciamus niger.
" folium,	U.S.'60.	" leaf,	" "
" semen,	U.S.	" (Henbane) sd.	" "
Ignatia,	U.S.	Bean of St. Ignatius,	Strychnos Ignatia.
Indigo,		Indigo,	Indigofera anil.
Inula,	U.S.	Elecampane root,	Inula Helenium.
Ipecacuanha,	†.	Ipecac root,	Cephaelis Ipecacuanha.
Iris Florentina,	U.S.	Florentine Orris root,	Iris Florentina.
" versicolor	U.S.	Blue flagroot,	" versicolor.
Jalapa,	†.	Jalap,	Exogonium purga.
Juglans,	U.S.	Butternut (inner bark),	Juglans cinerea.
Jujubæ,		Jujube, fruit pulp,	Rhamnus Ziziphus.
Juniperus,	U.S.	Juniper fruit (berries),	Juniperus communis.
" Virginianus,	U.S.	Red cedar tops,	" Virginiana.
Kamala,	B., G.	Kamala, Kameela,	Rottlera tinctoria.
Kino,	†.	Kino (inspissated juice),	Pterocarpus Marsupium.
" Australe,		Red (Brown) gum,	Eucalyptus resinifera.
" Indicum, E.		East Indian kino,	Pterocarpus Marsupium.
" " W.		Jamaica kino,	Coccoloba uvifera.
Kousso,		Koosso,	Brayera anthelmintica.
Krameria,	U.S.	Rhatany root,	Krameria triandra.
Krameriæ radix,	B.	" "	" "
Labdanum,		Ladanum, (Gum resin),	Cistus ladaniferus and other spe.
Lacca alba,		White lac resin,	Croton lacciferum.
" in baculis,		Stick lac,	see Lac, Page 62.
" in globulis,		Lac dye, in balls,	" "
" in granis,		" in grains, (Seed lac)	" "
" in massis,		" in lump,	" "
" in placentis,		" "	" "
" in ramulis,		Stick lac,	" "
" in tabulis,		Shellac,	" "
" musica,		Lacmus,	Lichen tartareus.
Lacmus,		" "	" "
Lactuca,	B.	Lettuce herb,	Lactuca virosa.

PHARMACOPŒIAL.		COMMON.	BOTANICAL.
Lactucarium,	U.S.	Lactucarium,	Lactuca sativa.
"	G.	" (Lettuce opium)	" virosa.
" Germanicum,	*G.	" "	" "
Ladanum,		Labdanum,	Cistus ladanifera and other spec.
Laminaria,	G.	Laminaria Cloustoni and (partly) Laminaria digitata.	
Lappa,	U.S.	Burdock root,	Lappa minor.
Laurocerasi folia,	B.	Cherry laurel leaves,	Prunus Laurocerasus.
Lavandula,	U.S.	Lavender flowers,	Lavandula vera.
Leptandra,	U.S.	Leptandra (Black) root,	Leptandra Virginica.
Lichen caragheen,		Irish moss,	Chondrus crispus.
" cocciferus,		Scarlet cup moss,	Lichen cocciferus.
" Islandicus,	G.	Iceland moss,	Cetraria Islandica.
" Tartareus,		Cudbear moss,	Lichen (Lecanora) Tartareus.
Lignum Agallochi,		Aloes wood,	Aloexylon Agallochum.
" Aloes,		"	" "
" Anacahuitœ,		Anacahuite wood,	Cordia Boisseri.
" Aspalati,		Aloes wood,	Aloexylon Agallochum.
" Benedictum,		Guaiac wood,	Guaiacum officinale.
" Brasile,		Sappan wood,	Cæsalpina Sappan.
" Brasiliense,		Braziletto wood,	" Brasiliensis.
" " rubrum,		Pernambuco wood,	" echinata.
" Campechianum,	G.	Logwood,	Hæmatoxylon Campechianum.
" Citrinum,		Fustic, (old),	Morus tinctoria.
" Colubrinum,		Snakewood,	Strychnos colubrina.
" Fernambuci,		Pernambuco wood,	Cæsalpina echinata.
" Guaiaci,	†.	Guaiacum wood,	Guaiacum officinale.
" Guajaci,	G.	" "	" "
" Hæmatoxyli,	B.	Logwood,	Hæmatoxylum Campechianum.
" Hederæ arboreæ,		Ivy wood,	Hedera Helix.
" Indicum,		Guaiacum wood,	Guaiacum officinale.
" "		Logwood,	Hæmatoxylon Campechianum.
" Nephriticum,		Bonduc tree wood,	Moringa aptera.
" Pavanœ,		Croton wood,	Croton Tiglium.
" Pterocarpi,	B.	Red sandalwood,	Pterocarpus santalinus.
" Quassiœ,	B.	Quassia wood,	Picræna (Quassia) excelsa.
" "	G.	" "	Quassia amara.
" " Surinamensis,	*G.	" "	" "
" Rhodii,		Rhodium (Rose) wood,	Convolvulus Scoparius.
" Rhodium,		" "	" "
" Sanctum,	*G.	Guaiacum wood,	Guaiacum officinale.
" Santa Marthœ,		Pernambuco wood,	Cæsalpina echinata.
" Santali albi,		White sandalwood,	Santalum album.
" " citrinum,		Yellow "	" "
" Santalinum rubrum,		Red saunders wood,	" rubrum.
" Sappan,		Sappan wood,	Cæsalpina Sappan.
" Sassafras,	G.	Sassafras root (wood),	Sassafras officinale.
" Taxi,		Yew tree wood,	Taxus baccata.
" Vitœ,	*G.	Guaiacum wood,	Guaiacum officinale.
Limones,		Lemons,	Citrus Limonum.
Limonis cortex,	†.	Lemon peel,	" "
" succus,	†.	" juice,	" "
Lini farina,	U.S.	Flaxseed meal,	Linum usitatissimum.
" "	B.	Linseed meal,	" "
" semina,	B.	Linseed,	" "
Linum,	U.S.	Flaxseed,	" "
Liquidambar,		Sweet gum, Styrax,	Liquidambar styraciflua.
Liriodendron,	U.S.	Tulip tree bark,	Liriodendron Tulipifera.
Lobelia,	†.	Lobelia herb,	Lobelia inflata.
Lupulin,	*G.	Lupulin (of hop),	Humulus Lupulus.
Lupulina,	U.S.	"	" "

PHARMACOPŒIAL.		COMMON.	BOTANICAL.
Lupulus,	B.	Hop,	Humulus Lupulus.
Lycopodium,	U.S., G.	Lycopodium sporules,	Lycopodium clavatum and other
Lycopus,	U.S.	Bugle weed,	Lycopus Virginicus. [species.
Macis,	U.S., G.	Mace,	Myristica fragrans.
Magnolia,	U.S.	Magnolia bark,	Magnolia glauca, acuminata and
Mala aurantii,		Orange, bitter,	Citrus vulgaris. [tripetala.
" citræ,		Lemon,	" Limonum.
Maltum,		Malt,	Hordeum vulgare.
Malum Persicum,		Peach fruit,	Amygdalus Persica.
Mauna,	G.	Manna,	Fraxinus ornus.
"	†.	" " and rotundifolia.	
" Alhagi,		Alhagi manna,	Alhagi camelorum.
" Australe,		Australian manna,	Eucalyptus viminalis.
" Laricinæ,		Briancon manna,	Larix Europæa,
" Quercinæ,		Oak manna,	Quercus Persica.
" Tamarissina,		Tamarisk manna,	Tamarix Gallica, var. maunifera
Maranta,	U.S., *G.	Arrow root,	Maranta arundinacea.
Marrubium,	U.S.	Horehound, ls. and tops,	Marrubium vulgare.
Mastiche,	†, *G.	Mastiche,	Pistacia Lentiscus.
Mastix,	G.	"	" "
Matico,	U.S.	Matico leaves,	Artanthe elongata.
Matricaria,	U.S.	German chamomile,	Matricaria Chamomilla.
Medulla sambuci,		Elder pith, of stem,	Sambucus Canadensis aud nigra.
" Sassafras,	U.S.	Sassafras pith, of stem,	Sassafras officinalis.
Meconium,	*G.	Opium,	Papaver somniferum.
Melissa,	U.S.	Balm,	Melissa officinalis.
Mentha piperita,	U.S.	Peppermint,	Mentha piperita.
" viridis,	U.S.	Spearmint,	" viridis. [um.
Mezereum,	U.S.	Mezereon bark,	Daphne Mezereum and D. Gnidi-
Mezerei cortex,	B.	" "	" " " "
Monarda,	U.S.	Horsemint herb,	Monarda punctata.
Monesia,		Monesia extract, ·	Chrysophyllum glycyphlœum.
Mori albi,		White mulberries,	Morus alba.
" nigri,		Black mulberries,	" nigra.
" succus,	B.	Mulberry juice,	" "
Moxæ,		Pith of elder or sunflower,	Sambucus and Helianthus.
Mucuna,	U.S.	Cowhage,	Mucuna pruriens.
Museus,		Lichen,	see Lichen.
Myristica,	U.S.	Nutmeg,	Myristica fragrans.
"	B.	"	" officinalis.
" adeps,	*B.	Oil of nutmegs, expressed	" "
Myrrha,	‡.	Myrrh, (Gum resin),	Balsamodendron Myrrha.
Myxæ,		Sebesten plum,	Cordia myxa.
Nectandra,	U.S.	Bebeeru bark,	Nectandra Rodici.
Nectandræ cortex, ·	B.	" "	" Rodiæi.
Nuces Americanæ,		Physic nuts,	Jatropha curcas.
" Aquaticæ,		Water nuts,	Trapa nutans.
" Areeæ,		Areca nuts,	Areca Catechu.
" Behen,		Beheu (Ben) nuts,	Moringa pterygosperma.
" Barbadensis,		Purging nuts,	Jatropha curcas.
" Betel (Betle),		Betel (Areca) nuts,	Areca Catechu.
" Cathartica,		Purging nuts,	Jatropha curcas.
" Cupressi,		Cypress nuts,	Cupressus sempervirens.
" Juglandis,		English waluuts,	Juglans regia.
" " cinereæ,		Butternuts,	" cinerea.
" Moschatæ,		Nutmegs,	Myristica fragrans.
" Persicorum,		Peach "stones,"	Amygdalus Persica.
" Pineæ,		Stone pine nuts,	Pinus pinca.
" Quercus,		Acorus, ·	Quercus of several species.
" Vomicæ,		Nux Vomica seed,	Strychnos Nux Vomica.

PHARMACOPŒIAL.		COMMON.	BOTANICAL.
Nuces Pistachiæ,		Pistacia nut,	Pistacia vera.
Nuclei Myristicæ,		Nutmegs,	Myristica fragrans.
Nux Moschata,	*G.	"	"
" Vomica,	*G., †.	Nux Vomica (seed),	Strychnos Nux Vomica.
Oculi populi,	*G.	Black poplar buds,	Populus nigra.
Oculis Christi,		Clary seed,	Salvia verbenaca.
Oleum Abietis,		Oil of Hemlock,	Abies (Pinus) Canadensis.
" Absinthii,		" Wormwood,	Artemisia Absinthium.
" Abrotani,		" Southernwood,	" abrotanum.
" Acori,		" Sweet flag root,	Acorus Calamus.
" Aloeticum,		" Aloes,	Aloe vulgaris and spicata.
" Amygdalæ,	B.	Almond oil, expressed from bitter and sweet almonds.	
" " amaræ,	U.S.	Oil of Bitter almonds,	Amygdalus communis, var.
" " expressum	U.S.	Expressed oil of almond	" " [amara.
" " dulcis,	U.S.'60.	" " "	" "
" Amygdalarum,	G.	" " "	" "
" Anamirtæ cocculi,		Oil of Cocculus Indicus,	Anamirta Cocculus.
" Anethi,	B.	" Dill,	Anethum graveolens.
" Andropogonis,		Indian grass oil,	Andropogon of several species.
" " Nardi,		Oil of Citronelle,	" Nardus.
" " Citrati,		" Lemon grass,	" Citratus.
" " Schœnanthi,		" Ginger grass,	" Schœnanthus.
" Anisi,	‡.	" Anise,	Pimpinella Anisum.
" " stellati,		" Star anise,	Illicium anisatum.
" Anthemidis,	B.	" Chamomile,	Anthemis nobilis.
" Anthos,	*G.	" Rosemary,	Rosmarinus officinalis.
" Arachis,		Peanut oil,	Arachis hypogæa.
" Armoraciæ,		Oil of Horseradish,	Cochlearia Armoracia.
" Arnicæ flores,		" Arnica flowers,	Arnica montana.
" " radicum,		" . " root,	" " [dense.
" Asari,		" Snakeroot,	Asarum Europæum and Cana-
" Aurantii,		" Orange (Neroli),	Citrus vulgaris and aurantium.
" " corticis,	G.	" " peel,	" " " "
" " florum,	G.	" " flowers,	" " " "
" " folii,		" " leaves,	see oil of Petit grain. [tera.
" Balanium,		" Ben nuts,	Moringa pterygosperma and ap-
" Balatinum,		" "	" " "
" Barosmæ,		" Buchu leaves,	Barosma of several species.
" Belladonnæ,		" Belladonna fruit,	Atropa Belladonna.
" Benzoini,		" Benzoin gum,	Styrax Benzoin.
" Bergamii,	U.S.	" Bergamot,	Citrus Limetta.
" Bergamottæ,	G.	" "	" Bergamia.
" Betulæ,		" Birch,	Betula alba and other species.
" Buxi,		" Box,	Buxus sempervirens.
" Cacao,	G.	Cacao butter,	Theobroma Cacao.
" Cadinum,	*G.	Oil of Cade,	Juniperus oxycedrus.
" Cajeputi,	G.	" Cajeput,	Melaleuca leucadendon (minor).
" Cajuputi,	†.	" "	" Cajuputi (minor, B.).
" Calami,	G.	" Sweet flag,	Acorus Calamus.
" Camphoræ,	U.S.	" Camphor, volatile,	Camphora officinarum.
" Cannabis,		Hemp seed oil,	Cannabis sativa.
" Capsici,		Oil of Cayenne pepper,	Capsicum annuum.
" Cardamomi,		" Cardamom seed,	Elettaria Cardamomum.
" Cari,	U.S.	" Caraway seed,	Carum carui.
" Carlinæ,		" Carline thistle rt.,	Carlina acaulis.
" Carui,	B.	" Caraway seed,	Carum carui.
" Carvi,	G.	" " "	" "
" Caryophylli,	†.	" Cloves,	Caryophyllus aromaticus.
" Caryophyllorum,	G.	" "	" "
" Cascarillæ,		" Cascarilla bark,	Croton eluteria.

PHARMACOPŒIAL.		COMMON.	BOTANICAL.
Oleum Cassiæ,	*G.	Oil of Cassia bark,	Cinnamomum (Laurus) cassia.
" Castoris,	*G.	Castor oil, (Ricini),	Ricinus communis.
" Cebadillæ,		Oil of Cevadilla seed,	Veratrum Sabadilla.
" Cedri,		" Cedrat (Citron),	Citrus medica.
" Chærophylli,		" Chervil herb,	Anthriscus cerefolium.
" Chamomillæ,		" German chamomile,	Matricaria Chamomilla.
" Chamœmeli œth.,		" Chamomile,	Anthemis nobilis.
" Chenopodii,	U.S.	" Am'n wormseed,	Chenopodium anthelminticum.
" Cinnamomi,	†.	" Ceylon cinnamon,	Cinnamomum Zeylanicum.
" "	*G.	" Cassia bark,	" (Laurus) Cassia.
" " cassiæ,	G.	" "	" " "
" " veri,		" Ceylon cinnamon,	" Zeylanicum.
" " Zelanici,	G.	" " "	" "
" Citri,	G.	" Lemon,	Citrus Limonum.
" " florum,		" Citron flower	" medica.
" Cocois,	G.	Cocoa-nut oil,	Cocos nucifera.
" " butyraceæ,		" "	" "
" " nuciferæ,		" "	" "
" Conii,		Oil of Poison hemlock,	Conium maculatum.
" Copaibæ,	†.	" Copaiba(Copaiva)	Copaifera multijuga and other
" Coriandri,	B.	" Coriander,	Coriandrum sativum. [species.
" Corticis Aurantii,	*G.	" Orange peel,	Citrus aurantium.
" Coryli,		Hazelnut oil,	Corylus avellana.
" Croci,		Oil of Saffron,	Crocus sativus.
" Crotonis,	B., G.	Croton oil,	Croton Tiglium, T. officinale.
" Cubebæ,	†.	Oil of Cubeb (Cubebs,B.)	Cubeba officinalis (Piper).
" Cumini,		Oil of Cummin seed,	Cuminum Cyminum.
" Curacoa,		" Bit. orange peel,	Citrus vulgaris.
" Cymini,		" Cummin seed,	Cuminum Cyminum.
" de Cedro,	*G.	" Lemon, .	Citrus Limonum.
" de Colza,		Rape seed oil,	Brassica campestris (napus).
" de Kerva,		Castor oil,	Ricinus communis.
" Digitalis,		Oil of Foxglove,	Digitalis purpurea.
" Diosmæ,		" Buchu leaves,	Diosma (Barosma) of several
" Ergotæ,		" Ergot,	Sclerotium clavus. [species.
" Erigerontis Canadensis,		" Canada fleabane,	Erigeron Canadense.
" Erechthites,		" Fireweed,	Erechthites hieracifolius.
" Fabarum pichurim,		" Pichurim beans,	Nectandra puchury.
" Fagi,		Beech nut oil,	Fagus sylvatica and ferruginea.
" Filicis,		Oil of Male fern,	Aspidium Filix Mas.
" " maris,		" "	" "
" Florum naphæ,	*G.	" Orange flowers,	Citrus aurantium.
" Fœniculi,	U.S., G.	" Fennel,	Fœniculum officinale.
" " dulcis,		" "	" dulce.
" " vulgaris,		" "	" vulgare.
" Fructum Juniperum,		" Juniper berries,	Juniperus communis.
" Galbani,		" Galbanum,	Ferula Galbanifera.
" Garciniæ,		Cocum oil (butter),	Garcinia purpurea.
" Gaultheriæ,	U.S.	Oil of Checkerberry,	Gaultheria procumbens.
" Geranii,		" Rose geranium,	Pelargonium odoratissimum.
" Graminis Indici,		Indian grass oil,	Andropogon of several species.
" Guaiaci,		Oil of Gum guaiac,	Guaiacum officinale.
" Hedeomæ,	U.S.	" Am'n pennyroyal,	Hedeoma pulegioides.
" Helianthi,		Sunflower seed oil,	Helianthus annuus.
" Hyosciami,		Oil of Henbane,	Hyosciamus niger.
" Hyperici,		" St. Johnswort,	Hypericum perforatum.
" Hyssopi,		" Hyssop,	Hyssopus officinalis.
" Infernale,		Jatropha oil,	Curcas purgans (Jatropha).
" Jasmini,		Oil of Jasmine flowers,	Jasminum officinale.
" Jatrophæ curcadis,		Jatropha oil,	Curcas purgans (Jatropha).

PHARMACOPŒIAL.			COMMON.	BOTANICAL.
Oleum	Juglandis,		Walnut oil,	Juglans regia and other species.
"	Juniperi,	‡.	Oil of Juniper berries,	Juniperus communis.
"	" empyreumaticum, G.		" Cade,	" oxycedrus.
"	" Virginianum,		" Cedar,	" Virginiana.
"	Lapidum prunarum,		" Plum stone pitts,	Prunus domestica.
"	Lathyris,		" Garden spurge,	Euphorbia lathyris.
"	Lauri exp.,	G.	" Bays (Laurel),	Laurus nobilis.
"	" Sassafras,		" Sassafras,	Sassafras officinale.
"	" volatil,		" Laurel berries,	Laurus nobilis.
"	Laurinum,	*G.	" " exp.	" " "
"	Lauro cerasi,		" Cherry laurel,	Cerasus Lauro-cerasus.
"	Lavandulæ,	‡.	" Lavender flowers,	Lavandula vera.
"	" flores,		" " "	" "
"	" spicatæ,		" Spike lavender,	" spica.
"	" vera,		" Lavender flowers,	" vera.
"	Liliarum,		" Wh. lily flowers,	Lilium candidum.
"	Limonis,	†.	" Lemon,	Citrus Limonum.
"	Limonium,		" "	" "
"	Limettæ,		" Lime fruit,	" Limetta.
"	Lini,	‡.	Linseed oil,	Linum usitatissimum.
"	Lobeliæ,		Oil of Lobelia,	Lobelia inflata.
"	Lupuli,		" Hop,	Humulus Lupulus.
"	Macidis,	G.	" Mace,	Myristica fragrans.
"	Macis essentiale,		" "	" "
"	Madi,		Madia seed oil,	Madia sativa.
"	Marjoranæ,	G.	Oil of Sweet marjoram,	Origanum Marjorana.
"	Matico,		" Matico,	Artanthe elongata.
"	Melaleucæ,		" Cajeput,	Melaleuca Cajuputi.
"	Melissæ,		" Balm,	Melissa officinalis.
"	Menthæ,		" Mint,	Mentha viridis.
"	" crispæ,	G.	" Curled mint,	" crispa.
"	" piperitæ,	‡.	" Peppermint,	" piperita.
"	" pulegli,		" Europ'n pennyroyal,	" pulegium.
"	" viridis,	†.	" Spearmint,	" viridis.
"	Millefolil,		" Yarrow,	Achillea Millefolium.
"	Monardæ,	U.S.	" Horsemint,	Monarda punctata.
"	Moschi,		" Musk root,	Euryangium Sumbul.
"	Myristicæ,	†.	Volatile oil of Nutmeg,	Myristica fragrans.
"	"	G.	Express'd " "	" "
"	" essentiale, B., G.		Volatile " "	" "
"	" expressum,	B.	Express'd " "	" "
"	Myrti,		Oil of Myrtle blossoms,	Myrtus communis.
"	Narcissi,		" Jonquil,	Narcissus jonquilla.
"	Neroli,	*G.	" Orange flowers,	Citrus aurantium and amara.
"	Nucis moschata,		" Nutmeg,	Myristica fragrans.
"	" pini,		" Pine nuts,	Pinus cembra.
"	Nucistæ expressum *G.		Butter of nutmegs,	Myristica fragrans.
"	Nucum juglandis,		Walnut oil,	Juglans regia and other species.
"	Olivæ,	†.	Olive oil,	Olea Europœa.
"	Olivarum,	G.	"	"
"	" Provincial, G.		" Provence,	"
"	" viridis,	G.	" Common,	"
"	Origani,	U.S.	Oil of Origanum,	Origanum vulgare.
"	" Cretici,		" Dittany,	" dictamnus.
"	Palmæ,		Palm oil,	Elais Guiniensis.
"	" Christi,	*G.	Castor oil,	Ricinus communis.
"	Papaveris,	G.	Poppy seed oil,	Papaver somniferum, var.
"	Petroselini,		Oil of Parsley root,	Petroselinum sativum.
"	Patchouli,		" Patchouly,	Pogostemon Patchouly.
"	Pepo,		Pumpkin seed oil,	Cucurbita Pepo.

PHARMACOPŒIAL.		COMMON.	BOTANICAL.
Oleum Pichurim,		Oil of Sassafras nuts.	Nectandra puchury.
" Picis liquidæ,		" Tar,	Pinus of several species.
" Pimentæ,	†.	" Pimento,	Eugenia Pimenta.
" Pimpinellæ,		" Burnet saxifrage,	Sanguisorba officinalis.
" Pini Canadense,		" Hemlock.	Abies (Pinus) Canadensis.
" " rubrum,		" Tar,	Pinus of several species.
" " sylvestri,		Scotch fir oil,	Pinus sylvestris.
" Piperis,		Oil of Black pepper,	Piper nigrum.
" Populus,		" Poplar,	Populus tremuloides.
" Provinciale,		Olive oil,	Olea Europæus.
" Pulegii,		Oil of Europ'n pennyroyal,	Mentha pulegium.
" Radicis carlinæ,		" Carline thistle,	Carlina acaulis.
" Rapæ,		Rape seed oil,	Brassica campestris.
" Raphini,		Oil of Wild mustard,	Sinapis arvensis.
" Ravensaræ,		" Clove nutmeg,	Agathophyllum aromaticum.
" Resedæ,		" Mignonette,	Reseda odorata.
" Rhodii lignii,		" Rhodium,	Genista Canariensis.
" Rhodiolæ,		" Rosewort,	Rhodiola rosea.
" Ricini,	‡.	Castor oil,	Ricinus communis.
" Rorismari,		Oil of Rosemary,	Rosmarinus officinalis.
" Rosæ,	G.	" Rose (Otto),	Rosa moschata, Damascena and
" "	U.S.	" "	" centifolia. [other species.
" Rosarum,		" " (by infusion)	" "
" Rosmarini,	‡.	" Rosemary,	Rosmarinus officinalis.
" Rutæ,	†.	" Rue,	Ruta graveolens.
" Sabinæ,	‡.	" Savin,	Juniperus Sabina.
" Salviæ,		" Sage,	Salvia officinalis.
" Sambuci,		" Elder,	Sambucus nigra.
" Sanctæ Mariæ,		" Calaba balsam,	Calophyllum Calaba.
" Santalini albi,		" Sandal wood,	Santalum myrtifolium and other
" " ligni,		" " "	" " [species.
" Sassafras,	U.S.	" Sassafras,	Sassafras officinale.
" Satureja,		" Summer savory,	Satureja hortensis.
" Serpylli,		" Lemon thyme,	Thymus serpyllum.
" Sesami,	U.S.	" Benne,	Sesamum Indicum and orientale
" Sinapis,	B.G.	" Mustard seed,	Sinapis alba and nigra.
" " expr.,		Mustard seed oil,	" "
" Spicæ,		Oil of Spike lavender,	Lavandula spica.
" Spiræ ulmariæ,		" Meadow sweet,	Spiræa ulmaria.
" Stœchadis,		" Spike lavender,	Lavandula spica.
" Stillugiæ,		" Stillingia root,	Stillingia sylvatica.
" Strobuli pini,		" Hemlock tops,	Abies (Pinus) Canadensis.
" Succini,	U.S.	" Amber.	
" Tabacæ,		" Tobacco,	Nicotiana Tabacum.
" Tabaci,	U.S.	" "	" " .
" Tanaceti,		" Tansy,	Tanacetum vulgare and crispum
" Templinum,		" Hungarian balsam	Pinus pumilo.
" Terebinthinæ,	‡.	" Turpentine,	" palustris and other species
" "		" "	" Tœda and pinaster.
" Theæ,		" Tea,	Cammelia sesanqua and oleifera
" Theobromæ,	†.	" (Butter of) Cacao,	Theobroma Cacao.
" Thymi,	U.S., G.	" Thyme,	Thymus vulgaris.
" Tiglii,	U.S.	Croton oil, ·	Croton Tiglium.
" Tritici,		Oil of Wheat,	Triticum vulgare.
" Valerianæ,	U.S., G.	" Valerian,	Valeriana officinalis.
" Verbenæ,		" Verbena,	Aloysia citriodora.
" Vitis,	∕	" Vine (Grape),	Vitis vinifera.
" Xanthoxylii,		" Prickly ash,	Xanthoxylum fraxineum.
" Zingiberis,		" Ginger,	Zingiber officinale.
Olibanum,	G.	Frankincense, Thus,	Boswellia papyrifera.

PHARMACOPŒIAL.		COMMON.	BOTANICAL.
Olibanum Americanum,		Juniper gum,	Juniperus oxycedrus.
" Arabicum,		Arabian olibanum,	" lycia.
" ostindi,		Indian olibanum,	Boswellia serrata.
" sylvatici,		Soft turpentine,	Pinus sylvestris.
Opium,	‡.	Opium,	Papaver somniferum.
Opobalsamum,		Balsam of Gilead (Mecca),	Amyris Gileadensis.
" siccum,		Dry balsam of Peru,	Myrospermum Peruiferum.
Opopanax,		Opopanax resin,	Pastinaca Opopanax.
Origanum,	U.S.	Wild marjoram,	Origanum vulgare.
Orleana,		Annotto,	Bixa Orleana.
Orsellla,		Litmus plant,	Lichen roccella,
Panax,	U.S.	Ginseng root,	Panax quinquefolium.
Papaver,	U.S.	Poppy capsules (unripe)	Papaver somniferum.
Papaveris capsulæ	B.	" " white "	" "
Pareira,	U.S.	Pareira Brava,	Cissampelos Pareira.
" radix,	B.	" root,	" "
Passulæ majoris,		Raisins,	Vitis vinifera.
" minoris,		Corinthian currants,	" Corinthiaca.
Pasta guarana,	G.	Guarana,	Paullinia sorbilis and cupana.
Pedunculi cerasorum,		Cherry stems,	Prunus.
Pepo,	U.S.	Pumpkin seed,	Cucurbita Pepo.
Petala rosæ centifolia,		Pale rose petals,	Rosa centifolia.
" " Gallica,		Red rose petals,	" Gallica.
Petroselinum,	U.S.	Parsley root,	Petroselinum sativum.
Physostigma,	U.S.	Calabar bean,	Physostigma venenosum.
Physostigmatis faba,	B.	"	" "
Phytolaccæ bacca,	U.S.	Poke (Garget) berry,	Phytolacca decandra.
" radix,	U.S.	" " root,	" "
Pigmentum Indicum,		Indigo,	Indigofera anil.
Pimenta,	†.	Pimento, Allspice,	Eugenia Pimenta.
Piper,	U.S.	Black pepper,	Piper nigrum.
" album,		White pepper,	" "
" aromaticum,		Black pepper,	" "
" caudatum,		Cubebs, (Tailed pepper)	" Cubeba.
" Hispanicum,	*G.	Cayenne pepper,	Capsicum annuum and longum.
" Jamaicense,		Allspice,	Eugenia Pimenta.
" longum,		Long pepper,	Piper longum.
" nigrum,	B.	Black pepper,	" nigrum.
Pix abietina,		Burgundy pitch,	Abies excelsa.
" alba,	*G.	" "	" of several species.
" arida,		" "	" " "
" Burgundica,	†.	" "	" excelsa.
" Canadensis,	U.S.	Hemlock pitch,	" Canadensis.
" liquida,	U.S·	Tar,	Pinus palustris and other species
" "	*G., B.	"	" sylvestris " "
" navalis,	G.	Black pitch,	} Produced by Abies of several
" nigra,	*G.	"	} species, also from Fagus.
" solida,	*G.	"	
Placenta seminis lini,	G.	Flaxseed oilcake,	Linum usitatissimum.
Podophylli radix,	B.	Podophyllum root,	Podophyllum peltatum.
Podophyllum,	U.S.	Mayapple root,	" "
Polygala rubella,	U.S.	Bit. polygala rt. and hb.,	Polygala rubella.
Poma acidula,		Sour apple,	Pyrus malus.
" aurantii immatura,		Orange bs. (unripe ft.),	Citrus aurantium.
" citri,	●	Lemon (fruit),	" Limonum.
" colycynthidis,	*G.	Colocynth,	Citrullus Colocynthis.
Prinos,	U.S.	Black alder bark,	Prinos verticillatus.
Pruna,		Prunes,	Prunus domestica.
" agresta,		Wild plum, Sloe,	" spinosa.
Prunum,	†.	Prunes, (Dried plums),	" domestica.

17

PHARMACOPŒIAL.		COMMON.	BOTANICAL.
Prunus Virginiana,	U.S.	Wild cherry bark,	Cerasus serotina.
Pterocarpi lignum,	B.	Red sandalwood,	Pterocarpus santolinus.
Pulpa Tamarindorum,	G.	Tamarind pulp,	Tamarindus Indica.
Pyrethri radix,	B.	Pellitory root,	Anacyclus Pyrethrum.
Pyrethum,	U.S.	" "	
Quassia,	U.S.	Quassia wood,	Simaruba excelsa.
Quassiæ lignum,	B.	" "	Picræna (Quassia) excelsa.
Quercus alba,	U.S.	White oak bark,	Quercus alba.
" cortex,	B.	Oak bark,	" pedunculata.
" tinctoria,	U.S.	Black oak bark,	" tinctoria.
Radix Abri,		Indian liquorice root,	Abrus precatorius.
" Acanthi,		Cotton thistle root,	Onopordon acanthium.
" Aconiti,	†.	Aconite root,	Aconitum Napellus.
" " heterophylli,		Yellow wolfsbane root,	" lycotonum.
" " hyemalis,		Winter hellebore root,	Helleborus hyemalis.
" " Indicæ,		Indian aconite root,	Aconitum ferox.
" " racemosæ,		Baneberry root,	Actæa spicata.
" Acori,		Sweet flag root,	Acorus Calamus.
" Actæ racemosa,		Black cohosh root,	Cimicifuga racemosa.
" Adonidis,		False hellebore root,	Adonis vernalis.
" Alcannæ,		Alkanet root,	Anchusa tinctoria.
" Alkannæ,	G.	" "	Alkanna tinctoria.
" Allii,		Onion bulbs,	Allium cepa.
" " sativa,		Garlic bulbs,	" sativum.
" Althææ,	G.	Marshmallow root,	Althæa officinalis.
" Anethi ursini,		Bearswort root,	Meum athamanticum.
" Angelicæ,	G.	Garden angelica root,	Archangelica officinalis.
" " sylvestris,		Wild angelica root,	Angelica sylvestris.
" Anti-cholerica,		Sophora root,	Sophora Japonica.
" Apii montani,		Speedwell root,	Athamanta Oreoselinum.
" Apocyni andro.,		Bitter (Dogsbane) root,	Apocynum androsæmifolium.
" " cannabini,		Bl'k Indian hemp root,	" cannabinum.
" Araliæ nudicaulis,		Am'n sarsaparilla root,	Aralia nudicaulis.
" " racemosæ,		Am'n spikenard root,	" racemosa.
" " spinosæ,		Prickly elder root,	" spinosa.
" Archangelicæ,	*G.	Garden angelica root,	Archangelica officinalis.
" Ari maculati,		Spotted arum root,	Arum maculatum.
" " triphylli,		Dragon root,	" triphyllum.
" Aristolochiæ,		Virginia snakeroot,	Aristolochia Serpentaria.
" " clematidis,		Common birthwort root,	" clematitis.
" " longæ,		Long " "	" longa.
" " rotundæ,		Round " "	" rotunda.
" " solidæ,		Thick " "	" "
" " vulgaris,		Common birthwort,	" clematitis.
" Armoraciæ,	B.	Horseradish root,	Cochlearia Armoracia.
" Arnicæ,	G.	Arnica root,	Arnica montana.
" Aronis,		Spotted arum root,	Arum maculatum.
" Artemisiæ,	G.	Mugwort root,	Artemisia vulgaris.
" Arundinis,		Great reed, Spanish cane,	Arundo donax.
" " vulgaris,		Common reed,	" phragmites.
" Asari,	G.	Asarabacca,	Asarum Europæum.
" " Canadensis,		Canada snakeroot,	" Canadense.
" " Europæus,		European snakeroot,	" Europæum.
" Asclepiadis,		Bastard ipecac root,	Asclepias Curassavica.
" Asclepias incarnata	U.S.	Flesh colored asclepias,	" incarnata.
" " Syriaca,	U.S.	Milkweed root,	" Syriaca.
" " tuberosa,	U.S.	Pleurisy root,	" tuberosa.
" Asparagi,		Asparagus root,	Asparagus officinalis.
" Asphodeli,		Asphodel root,	Asphodelus ramosus.
" Astrantiæ,		Imperial masterwort rt.,	Astrantia major.

PHARMACOPŒIAL.		COMMON.	BOTANICAL.
Radix Bardanæ,	G.	Burdock root,	Lappa officinalis and other spec.
" Belladonnæ,	B., G.	Belladonna root,	Atropa Belladonna.
" Berberidis,		Barberry root,	Berberis Canadensis and vulgare
" Bismalvæ,		Marshmallow root,	Althæa officinalis.
" Bistortæ,		Bistort root,	Polygonum Bistorta.
" Brancæ ursinæ,		Cow parsley root,	Heracleum sphondylium.
" Brasiliense,		Ipecac root,	Cephaelis Ipecacuanha.
" Bryoniæ,		Bryony root, white,	Bryonia alba.
" Buglossi,		Garden alkanet root,	Anchusa officinalis.
" Cahincæ,		Snowberry (Cahinca) rt.	Chiococca racemosa.
" Calagualæ,		Callahuala root,	Polypodium calaguala.
" Calami,	*G.	Sweet flag root	Acorus Calamus.
" Caltha alpinæ,		Arnica root,	Arnica montana.
" Calumbæ,	B.	Columba root,	Jateorrhiza Calumba.
" Cardui,		Cotton (Musk) thistle rt.	Onopordon acanthium.
" Caricis,	*G.	Sea sedge (Sarsaparilla) rt.,	Carex arenaria.
" Carlinæ,	G.	Carline thistle root,	Carlina acaulis.
" Caryophyllatæ,		European avens root,	Geum urbanum.
" " aquaticæ,		Chocolate rt. Water avens,	" rivale.
" " urbana,		European avens root,	" urbanum.
" Cassamuuar,		Zerumbet (Bengal) root,	Zingiber cassumuniar.
" Cepæ,		Onion bulbs,	Allium cepa.
" Cervariæ,		Black gentian root,	Libanotis vulgaris.
" Chinæ,	*G.	China root,	Smilax China.
" " Americanæ,		American China root,	" pseudo-China.
" " Orientalis,		China root,	" China.
" " spuriæ,		American China root,	" pseudo-China.
" Chirayta,		Chiretta root and top,	Agathotes Chirayta.
" Cichorei,		Chicory root,	Cichorium intybus and endiva.
" Cicutæ,		Water hemlock root,	Cicuta virosa (aquatica).
" " virosæ,		" " "	" " "
" Cimicifuga,		Black cohosh root,	Cimicifuga racemosa.
" " serpentariæ,		" " "	" "
" Colchici,		Colchicum root,	Colchicum autumnale.
" Collinsoniæ,		Stone root,	Collinsonia Canadensis.
" Colombo,	G.	Columbo root,	Jateorrhiza Calumba.
" Columbo,	*G.	" "	" "
" " Americanum,		Am'n columbo root,	Frasera Walteri.
" " spuria,		Columbo wood,	Coscinium fenestratum.
" Consolida majoris,		Comfrey root,	Symphytum officinale.
" Contrayerva,		Lisbon contrayerva root,	Dorstenia contrayerva.
" Corallorhiza,		Crawley root,	Corallorhiza odontorhiza.
" Curcumæ,	*G.	Turmeric root,	Curcuma longa (viridiflora.)
" " longæ,		" "	" " "
" " rotundæ,		" "	" rotunda (viridiflora.)
" Cyperi,		English galingale root,	Cyperus officinalis.
" " rotundæ,		Round rooted cyperus,	" rotundus.
" Datura,		Thorn apple root,	Datura Stramonium.
" Dauci,		Wild carrot root,	Daucus Carota.
" " sativi,		Garden carrot root,	" " var. sativa.
" Dentallaria,		Leadwort, (Tooth root),	Plumbago Europæa.
" Delphinium,		Larkspur root,	Delphinium consolida.
" Dens-leonis,		Dandelion root,	Taraxacum Dens-leonis.
" Dictamni,		White fraxinella root,	Dictamnus albus (alba).
" Donacis,		Spanish cane root,	Arundo donax.
" Doronica,		Roman leopardsbane rt.,	Doronicum pardalianches.
" Dysentericæ,		Ipecac root,	Cephaelis Ipecacuanha.
" Elebori,		Black hellebore root,	Helleborus niger.
" Enulæ,	*G.	Elecampane root,	Inula Helenium.
" Eryngii,		Wild eryngo root.	Eryngium campestre.

PHARMACOPŒIAL.	COMMON.	BOTANICAL.
Radix Eupatorii,	Hemp agrimony root,	Eupatorium Cannabinum.
" Farfaræ,	Coltsfoot root,	Tussilago farfara.
" Filicis maris, *G.	Male fern root, '	Polystichum Filix Mas.
" Filipendulæ,	Dropwort root,	Spiræa filipendula.
" Fœniculi ursini,	Spignel, Bearswort root,	Meum athamanticum.
" Fragaria,	Strawberry rt.,(comm'n)	Fragaria vesca.
" Fraxinellæ,	White fraxinella,	Dictamnus albus (alba).
" Galangæ, *G.	Galangal root,	Alpina Galanga.
" Gentianæ, B., G.	Europ'n gentian root,	Gentiana lutea.
" " albæ,	White " "	Laserpitium latifolium.
" " nigræ,	Black " "	Libanotis vulgaris.
" " rubræ,	Europ'n " "	Gentiana lutea. .
" Ginseng,	Ginseng root, Asiatic,	Panax schinseng,
" " Americana,	Am'n Ginseng root,	" quinquefolium.
" Glaucii,	Horn poppy root,	Chelidonium glaucum.
" Glycyrrhiza, B.	Liquorice root,	Glycyrrhiza glabra.
" " echinata, *G.	Peeled Russian liquorice rt.	" echinata, var.
" " Hispanica,*G.	Spanish liquorice root,	" glabra.
" Gossypii,	Cotton root,	Gossypium herbaceum.
" Graminis, *G.	Couch grass root,	Agopyrum repens.
" " rubri,	German sarsaparilla,	Carex arenaria.
" Graniti, B.	Pomegranate root (bk.),	Punica Granatum.
" Guaco,	Guaco root,	Mikania Guaco.
" Helenii, G.	Elecampane root,	Inula Helenium.
" Helianthi tuberosi,	Artichoke root,	Helianthus tuberosus.
" Hellebori albi, *G.	White hellebore root,	Veratrum album.
" " fœtidi,	Stink'g " "	Helleborus fœtidus.
" " hyemalis,	Winter " "	" hyemalis.
" " nigri,	Black " "	" nigra.
" " viridis, G.	Europ'n green hellebore,	" viridis.
" Helonias,	False unicorn root,	Helonias dioica.
" Hemidesmi, B.	E. I. sarsaparilla,	Hemidesmus Indicus.
" Heraclei,	Cow parsnip root,	Heracleum lanatum.
" Hermodactyli,	Chequer flower root,	Colchicum variegatum.
" Hirundinariæ,	Wh. swallow-wort root,	Asclepias vincetoxicum.
" Hydrastis Canadensis,	Goldenseal root,	Hydrastis Canadensis.
" Ibisci,	Marshmallow root,	Althæa officinalis.
" Ictodes fœtidi,	Skunk cabbage root,	Ictodes fœtidus.
" Imperatoriæ, G.	Masterwort root,	Imperatoria ostruthium.
" " nigra,	" " Imperial, Astrantia major.	
" Indica Lopeziana,	Lopez root,	Toddalia aculeata.
" Inulæ,	Elecampane root,	Inula Helenium.
" Ipecacuanha, G.	Ipecac root,	Cephaelis Ipecacuanha.
" " albæ,	White Ipecac root,	Richardsonia scabra.
" " amylaceæ,	" " "	" "
" " annualatæ,	Ipecac root,	Cephaelis Ipecacuanh.
" " nigræ,	Peruvian Black ipecac,	Psychotria emetica.
" " striatæ,	" " "	" "
" " undulatæ,	White ipecac root,	Richardsonia scabra.
" 'Iridis Florentina, *G.	Florentine orris root,	Iris Florentina.
" " nostatis,	German orris root,	" Germanica..
" " versicolor,	Blue flag root,	" versicolor.
" Iwarancusæ,	Vetivert root,	Andropogon muricatus.
" Jaborandi,	Jaborandi root,	Piper reticulatum.
" Jalapæ, *G.	Jalap root (tuber),	Convolvulus (Jalapa) purga.
" " albæ,	Man root, (Wild potato),	" panduratus.
" " lævis,	Male (Light) jalap root,	Ipomœa orizabensis.
" Krameriæ,	Rhatany root,	Krameria triandra.
" Lapathi acuti,	Broad leaved dock root,	Rumex obtusifolius.
" Lappæ minor,	Burdock root,	Arctium Lappa, (Lappa minor).

PHARMACOPŒIAL.		COMMON.	BOTANICAL.
Radix Laserpitii Germanici,		German Lovage root,	Ligusticum Levisticum.
" Levistici,	G.	" " "	" "
" Lilli albi,		Common white lily root,	Lilium candidum.
" Liliorum convallium,		Lily of the valley root,	Convallaria majalis.
" Liquiritiæ,		Liquorice root,	Glycyrrhiza glabra.
" " glabræ,	G.	Spanish liquorice root,	" "
" " mundata,	G.	Peeled Russian liquorice	" echinata.
" " Russica,	*G.	Russian liquorice root,	" "
" Lopeziana,		Lopez root,	Toddalia aculeata.
" Malva visca,		Marshmallow root,	Althœa officinalis.
" Mandragoræ,		Europ'n mandrake root,	Atropa Mandragora.
" Maniaca,		Belladonna root,	" Belladonna.
" Mechoacannæ,		Wild potato (Man) root,	Convolvulus panduratus.
" " nigra,		Jalap root,	Ipomœa Jalapa.
" Melampodii,		Black hellebore root,	Helleborus niger.
" Metalistæ,		Metalista root,	Mirabilis longiflora.
" Meu,		Bearswort root,	Meum Athamanticum.
" Mezerei,		Mezereon root,	Daphne Mezereum.
" Morus diaboli,		Devil's bit root,	Scabiosa succissa.
" Mustela,			Ophioglossum serpentium.
" Nardi,		Spikenard of the ancients	Valeriana (Nardostachys) Jata-
" " Indici,		Indian spikenard root,	Nardostachys Jatamansi.(mansi
" Ninsi,		Ninsin root,	Sium sisarum, var. ninsi.
" Nymphœa,		White pond lily root,	Nymphœa odorata and alba.
" Œnanthes,		Water heml'k dropw't rt.	Œnanthe crocata.
" Ononidis,	G.	Rest harrow root,	Ononis spinosa.
" Oreoselini,		Speedwell (fuellin) root,	Athamanta oreoselinum.
" Osmundæ,		Buckhorn fern root,	Osmunda regalis.
" Ostruthii,		Masterwort root,	Imperatoria ostruthium.
" Pæoniæ,		Peony root,	Pæonia officinalis.
" Pareira,	B.	Pareira (Brava) root,	Cissampelos Pareira.
" "			see Chondodendron.
" " Bravæ,		Pareira Brava root,	Cissampelos Pareira.
" Paridis,		Herb Paris root,	Paris quadrifolia.
" Pastinacæ,		Parsnip root,	Pastinaca sativa.
" Patientiæ,		Patience dock root,	Rumex patientia.
" Personatæ,		Arnica root,	Arnica montana.
" Petroselini,		Parsley root,	Petroselinum sativum.
" Peucedani,		Sulphurwort root,	Peucedanum officinale.
" Phytolaccæ,	U.S.	Poke (Garget) root,	Phytolacca decandra.
" Pimpinella,	G.	Burnet saxifrage root,	Pimpinella Saxifraga.
" Pimpinellæ alba,		" " "	" magna.
" " Italicæ,		Garden (Italian) burnet rt.,	Sanguisorba officinalis.
" " " minoris,		Small bibernel root,	Poterium Sanguisorba.
" " majoris,		Great pimperuel root,	Pimpinella magna.
" hortensis,		Garden burnet root,	" "
" nigra,		Europ'n burnet saxifrage,	" Saxifraga, var. nigra.
" Pistolochiæ,		French birthwort,	Aristolochia pistolochia.
" Plantaginis,		Water plantain root,	Alisma Plantago.
" Podophylli,	B.	Am'n mandrake root,	Podophyllum peltatum.
" Polemonii,		Abscess root,	Polemonium reptans.
" Polygalæ amaræ,		Bitter polygala root,	Polygala amara.
" " Virginicæ,		Senega snakeroot,	" Senega.
" " vulgaris,		Milkwort root,	" polymorpha (incarnata)
" Polypodii,		Polypody (Rockbrake) rt.	Polypodium vulgare.
" Pseud-acori,		False sweet flag root,	Iris pseudo-acorus.
" Ptarmicæ,		German pellitory root,	Achillea Ptarmica.
" " montanæ,		Arnica root,	Arnica montana.
" Pyrethri,	B.	Pellitory root, Levant,	Anacyclus Pyrethrum.
" "	G.	" " German,	" officinarum.

17*

PHARMACOPŒIAL.		COMMON.	BOTANICAL.
Radix Pyrethri communis,		Pellitory root,	Anacyclus officinarum.
" " Germanica,	*G.	" "	" "
" " Italici,		Italian pellitory,	Pyrethrum Italicum.
" " Romanæ,		Pellitory of Spain,	Anacyclus Pyrethrum.
" Ranunculi,		Crowfoot root,	Ranunculus bulbosus.
" Raphani,		Horseradish root,	Cochlearia Armoracia.
" Rapontica,		Crimea (Krimea) rhubarb,	Rheum Rhaponticum.
" Ratanhæ,	G.	Rhatany root,	Krameria triandra.
" Restæ bovis,		Rest harrow root,	Ononis spinosa.
" Rhabarb,		Rhubarb root,	Rheum undulatum.
" Rhapontici,		Crimea rhubarb,	" Rhaponticum.
" Rhei,	G.	Rhubarb root,	undetermined sp.of Chinese rhu-
" " albi,		White rhubarb root,	Rheum leucorrhizum. [barb.
" " Asiatica,		Asiatic (India) rhubarb,	" undulatum.
" " Austriacæ,		Austrian rhubarb,	" Rhaponticum.
" " Buchari,		Indian rhubarb,	" palmatum.
" " Chinensis,		Chinese rhubarb,	" officinale (palmatum).
" " Indici,		Indian rhubarb,	" palmatum.
" Rhinacanthi,		Braid root,	Justicia nasuta.
" Rhodia,		Rosewort root,	Sedum roseum.
" Rosæ regiæ,		Peony root,	Pæonia officinalis
" Rubi fruticosi,		Blackberry root,	Rubus fruticosus.
" " villosi,		" "	" villosus.
" Rubiæ tinctoria,		Madder plant root,	Rubia tinctoria.
" Salep,	*G.	Salep,	Orchis mario and other species.
" Salicis nigra,		Black willow root,	Salix nigra.
" Salsaparillæ,	*G.	Sarsaparilla root,	Smilax medica and other species
" Sanguinariæ,		Blood root,	Sanguinaria Canadensis.
" Saniculæ,		Europ'n sanicle root,	Sanicula Europæa.
" " Americanæ,		Am'n " "	" Canadensis.
" " nigra,		Black " "	" Marilandica.
" Saponariæ,	G.	Soapwort root,	Saponaria officinalis.
" " albæ,		White soapwort root,	Lychnis dioica.
" " Hispanicæ,		Spanish " "	Gypsophila struthium.
" " rubræ,		Soapwort root,	Saponaria officinalis.
" Sarsæ,	B.	Jamaica sarsaparilla,	Smilax officinalis.
" Sarsaparillæ,	G.	Sarsaparilla root,	" medica and other species
" " Germanica,		Sea sedge root,	Carex arenaria.
" " Indici,		Indian sarsaparilla,	Hemidesmus Indicus.
" Sarzæ,		Sarsaparilla root,	Smilax sarsaparilla.
" Sassafras,	B.	Sassafras root,	Sassafras officinale.
" Sassaparillæ,	*G.	Sarsaparilla root,	Smilax medica and other species
" Satyrii,		Salep root,	Orchis mario and other species.
" Saxifragæ,		Dropwort root,	Spiræa filipendula.
" Scammoniæ,	B., G.	Scammony root,	Convolvulus Scammonia.
" Scillæ,		Squill,	Scilla maritima.
" Scorzoneræ,		Winter asparagus root,	Scorzonera Hispanica.
" Scrophulariæ,		Scrofula root,	Scrophularia Marilandica.
" " aquaticæ,		Water figwort root,	" aquatica.
" " vulgaris,		Figwort root,	" nodosa.
" Senegæ,	B., G.	Senega root,	Polygala Senega.
" Senekæ,		" "	" "
" Serpentariæ,	B. G,	Serpentary root,	Aristolochia Serpentaria.
" " Virginianæ,		Virginia snakeroot,	" "
" Sisaris,		Skirret,	Sium sisarum.
" Solani,		Belladonna root,	Atropa Belladonna.
" " quadrifolia,		Herb Paris (Foxgrape) rt.,	Paris quadrifolia.
" Spigelia,		W. I. pink root,	Spigelia anthelmia.
" " Marilandica,		Carolina pink root,	" • Marilandica.
" Squillæ,		Squill,	Scilla maritima.

PHARMACOPŒIAL.		COMMON.	BOTANICAL.
Radix Stillingia,		Queen's (Delight) root,	Stillingia sylvatica.
" Succisæ,		Devil's bit root,	Scabiosa succissa.
" Sumbul,	B.	Sumbul (Musk) root,	Euryangium Sumbul.
" Symphytii,		Comfrey root,	Symphytum officinale.
" Taraxaci	G.	Dandelion root,	Taraxacum officinale.
" "	B.	" "	Dens-leonis.
" " cum herba,	G.	Rt. and hb. of Dandelion	" officinale.
" Tinosporæ,		Gulancha root,	Cocculus(Tinospora)cordifolius
" Tormentillæ,	*G.	Tormentilla root,	Potentilla Tormentilla.
" Toddaliæ,		Lopez root,	Toddalia aculeata.
" Trolli,		Globe ranunculus,	Trollius Europæus.
" Turpethi,		Turpeth root,	Convolvulus Turpethum.
" Tussilaginis,		Coltsfoot root,	Tussilago farfara.
" Tylophoræ,		Indian (Ceylon) Ipecac,	Tylophora (Asclepias) asthmat-
" Urticæ,		Nettle root,	Urtica dioica. [ica.
" Valerianæ,	B., G.	Valerian root,	Valeriana officinalis.
" " Americanæ,		American valerian root,	" pauciflora.
" " minoris,	*G.	Valerian root,	" officinalis.
" " montanæ,	*G.	" "	" "
" Veratri albi,		White hellebore root,	Veratrum alba.
" " viridis,		American hellebore root,	" viride.
" Vettivar,		Vettivert, Khus khus rt.	Andropogon muricatus.
" Victorialis,		Round mandrake, Sword lily,	Gladiolus communis.
" Vincetoxici,		Wh. swallow-wort root,	Asclepias vincetoxicum.
" Viola,		Violet root,	Viola odorata.
" Xanthii,		Small burdock root,	Arctium Lappa (Lappa minor).
" Zarsaparilla,		Sarsaparilla root,	Smilax officinalis.
" Zarzæ,		" "	" "
" Zedoariæ,	*G.	Zedoary root,	Curcuma Zedoaria.
" Zerumbet,		Wild Zerumbet root,	Zingiber Zerumbet.
" Zingiberis,		Ginger root,	Amomum Zingiber (Z. officinale)
" " Africanis,		African ginger root,	" " "
" " condita,		Preserved "	" " "
" " Jamaicensis,		Jamaica "	" " "
Ranunculus,	U.S.	Crowfoot herb and root,	Ranunculus bulbosus.
Resina,	†.	Resin, (Rosin),	Pinus and Abies, various species
" Acaroides,		Gum acroid,	Xanthorrhœa hastilis and other
" alba,		White resin,	Pinus of several species. [spec.
" Benzoe,	*G.	Benzoin,	Styrax Benzoin.
" Caraunæ,		Carauna,	see Amyris and Burscra.
" citrini,		Common rosin,	Pinus of several species.
" Colophonium,	*G.	Resin,	" " "
" communis,		Common resin,	" " "
" Copal,		Gum copal,	see Copal, Page 26.
" Dammaræ,		Damarra gum,	Pinus (Agathis) Damarra.
" Draconis,	G.	Dragon's blood,	Dæmonorops Draco.
" Elastica,		Gum elastic,	Siphonia elastica.
" Elemi,	*G.	Elemi,	undetermined plant of Yucatan.
" empyreumatica solida,	*G.	Black pitch,	see Pix navalis.
" " liquida,	*G.	Tar,	" liquida.
" flava,		Yellow rosin,	Pinus and Abies of several spe.
" Gambiensis,		Kino, African,	Pterocarpus erinaceus.
" Guaiaci,	B., G.	Guaiac resin,	Guaiacum officinale.
" Kino,	*G.	East Indian kino,	Pterocarpus Marsupium.
" Ladanum,		Gum cistus,	Cistus creticus.
" Lentiscium,		Mastich,	Pistacia Lentiscus.
" Lutea,		Gum acroid,	Xanthorrhœa hastilis and other
" Mastiche,	*G.	Mastich,	Pistacia Lentiscus. [species.
" Mastix,		"	" "
" Nutt,		Gum nut, (Blackboy),	Xanthorrhœa arborea.

PHARMACOPŒIAL.		COMMON.	BOTANICAL.
Resina Oleœ,		Olive gum,	Olea Europœa.
" Pini,	G.	Burgundy pitch,	Abies of several species.
" " Burgundica,	*G.	" "	. " " "
" Sandaraca,	*G.	Sandarach,	Callitris quadrivalis. [mined spe.
Rhei Radix,	B.	Rhubarb root,	Rheum, one or more undeter-
Rheum,	U.S.	" "	" palmatum and other spe.
Rhizoma Asari,	*G.	Asarabacca root,	Asarum Europæum.
" Calami,	G.	Calamus, Sweet flag rt.,	Acorus Calamus.
" " aromatici,		" " "	" "
" Caricis,	G.	Sea sedge root,	Carex arenaria.
" Chinœ,	G.	China root,	Smilax China.
" Coptidis,		Chinese goldthread root,	Coptis teeta.
" Curcumæ,	G.	Turmeric root,	Curcuma longa (viridiflora).
" Filicis,	G.	Male fern root,	Polystichum Filix Mas.
" Galangœ,	G.	Galangal root,	Alpina officinarum.
" Graminis,	G.	Couch-grass root,	Agropyrum repens.
" Imperatoriœ,	G.	Masterwort root,	Imperatoria Ostruthium.
" Iridis,	G.	Florentine orris root,	Iris Florentina.
" Podophylli,		Mayapple root,	Podophyllum peltatum.
" Tormentillœ,	G.	Tormentilla root,	Potentilla Tormentilla.
" Veratri,	G.	White hellebore root,	Veratrum album.
" " viridis,		American hellebore root	" viride.
" Zedoariæ,	G.	Zedoary root,	Curcuma Zedoaria.
" Zingiberis,	G.	Ginger root,	Zingiber officinale.
Rhœas,		Red poppy petals,	Papaver Rhœas.
Rhœados petala,	B.	" "	" "
Rhus glabrum,	U.S.	Sumach berries,	Rhus glabrum.
Rob Rhamni,		Buckthorn berry juice,	Rhamnus cathartica.
" Sambuci,		Elderberry juice,	Sambucus nigra and Canadensis
Rosa canina fructus,	B.	Dog-rose fruit, (Hips),	Rosa canina.
" centifolia,	U.S.	Pale rose petals,	" centifolia.
" centifoliœ petala,	B.	Cabbage rose petals,	" "
" Gallica,	U.S.	Red rose petals,	" Gallica.
" " petala,	B.	" "	" "
Rosmarinus,	U.S.	Rosemary leaves,	Rosmarinus officinalis.
Rottlera,	U.S.	Kameela,	Rottlera tinctoria.
Rubia,	U.S.	Madder root,	Rubia tinctorium.
Rubus,	U.S.	Blackberry root (bark),	Rubus Canadensis and villosus.
Rumex,	U.S.	Yellow dock root,	Rumex crispus.
Ruta,	U.S.	Rue leaves,	Ruta graveolens.
Sabadilla,	†.	Cevadilla seed,	Veratrum sabadilla (Asagrœa).
Sabbatia,	U.S.	American centaury herb,	Sabbatia angularis.
Sabina,	U.S.	Savine tops,	Juniperus Sabina.
Sabinœ cacumina,	B.	Savin tops,	" "
Saccharum,	G., U.S.	Refined sugar,	Saccharum officinarum.
" Acerinum,		Maple sugar,	Acer saccharinum.
" album,	B.	White sugar,	Saccharum officinarum.
Sagapenum,		Sagapenum,	Ferula Persica(?).
Sago,	U.S.	Sago (prepared pith),	Sagus Rumphii and other spec.
" perlata,		Pearl sago,	" " " "
Salix,	U.S.	Willow bark, white,	Salix alba.
Salvia,	U.S.	Sage leaves,	Salvia officinalis.
Sambucus,	U.S.	Elder flowers,	Sambucus Canadensis.
Sambuci flores,	B.	" " Europ'n,	" nigra.
Sandaraca,	G.	Sandarach,	Callatris quadrivalis.
" Germanica,		" (Juniper gum)	Thuja articulata.
Sandaracha,		"	" "
Sanguinaria,	U.S.	Blood root,	Sanguinaria Canadensis.
Sanguis Carthagena,		So. Am'n Dragon's blood	Pterocarpus Draco.
" Draconis,	*G.	Dragon's blood resin,	Dæmonorops Draco.

PHARMACOPŒIAL.		COMMON.	BOTANICAL.
Santalum,	U.S.	Red saunders wood,	Pterocarpus santalinus.
Santonica,	U.S.	Levant wormseed,	Artemisia Cina.
"	B.	Santonica (flower heads)	" undetermined species
Sarcocolla,		Sarcocolla, product of,	Pœna Sarcocolla.
Sarsœ radix,	B.	Jamaica sarsaparilla,	Smilax officinalis.
Sarsaparilla,	U.S.	Sarsaparilla root,	" " and other species
Sassafras,	U.S.	Sassafras, bark of root,	Sassafras officinale.
" medulla,	U.S.	" pith (of stems),	" "
" radix,	B.	" root (with bk.),	" "
" radicis cortex,	U.S.'60.	" " bark,	" "
Scammoniœ radix,	B.	Scammony root,	Convolvulus Scammonia.
Scammonium,	†.	" resin,	" "
" Gallicum,		False French scammony,	Cynanchium Monspeliacum.
" Smyrnœum,		Smyrna scammony,	Periploca secamone.
Scilla,	†.	Squill, (bulb),	Scilla maritima.
Scoparius,	U.S.	Broom tops,	Sarothamnus Scoparius.
Scoparii cacumina,	B.	" "	" "
Scorbuli Lupuli,		Lupuline (of hops),	Humulus Lupulus.
Scutellaria,	U.S.	Scullcap herb,	Scutellaria lateriflora.
Sebestenœ,		Sebesten plum,	Cordia myxa.
Secale clavatum,		Ergot of rye,	Claviceps purpurea.
" cornutum,	G.	" "	" "
" luxurians,			
Semen Abelmoschi,		Musk seed,	Abelmoschus moschatus.
" Adjowœn,		Bishop's weed seed,	Ammi copticum.
" Agni casti,		Chaste tree fruit,	Vitex Agnus castus.
" Ajavœ,		Bishop's weed seed,	Ammi copticum.
" Ammi,		" " "	" "
" Amomi,		Pimento, Allspice,	Eugenia Pimenta.
" Amygdali amarum,	*G.	Bitter almonds,	Amygdalus communis var amara
" " dulce,	*G.	Sweet almonds,	" " var dulcis
" Anacardii,		Cashew nuts,	Semecarpus anacardium.
" Andœ,		Anda seed,	Anda Brasiliensis and Gomesei.
" Anethi,		Dill seed,	Anethum graveolens.
" Angelicœ,		Angelica seed,	Archangelica officinalis.
" Anguriœ,		Watermelon seed,	Cucumis (Cucurbita) Citrullus.
" Anisi stellati,	*G.	Star anise seed,	Illicium anisatum.
" " vulgaris,	*G.	Anise seed,	Pimpinella anisum.
" Anti-cholerica,		Sophora seed,	Sophora heptaphylla.
" Apii graveolens,		Celery seed,	Apium graveolens.
" Arecœ,		Areca (Betel) nuts,	Areca Catechu.
" Athanasiœ,		Tansy seed,	Tanacetum vulgare.
" Avellanœ,		Filbert (Hazel) nut,	Corylus avellana.
" " purgatrix,		French physic nuts,	Curcas multifidus.
" Badiani,		Star anise seed,	Illicium anisatum.
" Bardanœ,		Burdock seed,	Arctium Lappa.
" Berberidis,		Barberry seed,	Berberis vulgaris and Canaden-
" Bonducellœ,		Bonduc nuts,	Guilandina bonduc. [sis.
" Cacao,		Chocolate nuts,	Theobroma Cacao.
" Canariensis,		Canary seed,	Phalaris Canariensis.
" Cannabis,	*G.	Hemp seed,	Cannabis sativa.
" Cardamomi.		Cardamom seed,	Elettaria Cardamomum.
" " majoris,		" " large,	" major.
" " minoris,	*G.	" (Malabar) small	" Cardamomum.
" " rotundis,		" seed, round,	Amomum racemosum.
" Cardui,		Thistle seed,	Carduus marianus.
" " mariœ,		Mary thistle seed,	" "
" Carthami,		Saffron seed,	Carthamus tinctorius.
" Carui,		Caraway seed,	Carum Carvi.
" Carvi,	*G.	" "	" "

17**

PHARMACOPŒIAL.		COMMON.	BOTANICAL.
Semen Carvi montani,		Cummin seed,	Cuminum Cyminum.
" Cataputiæ,		Castor oil seed,	Ricinus communis.
" Cebadillæ,		Cevadilla seed,	Veratrum Sabadilla.
" Chenopodii,		Wormseed, American,	Chenopodium anthelminticum.
" Chervil,		Chervil seed,	Scandix cerefolium.
" Cicutæ,		Cicuta (Hemlock) seed,	Cicuta maculata. [misia.
" Cinæ,	*G.	Levant wormseed,	undetermined species of Arte-
" " Americanæ,		American wormseed,	Chenopodium anthelminticum.
" " Barbaricum,		Barbary wormseed,	Artemisia Sieberi and other spe-
" " Levanticum,		Levant wormseed,	" Contra. [cies.
" Citrulli,		Watermelon seed,	Cucumis (Cucurbita) citrullus.
" Coccognidii,		Mezereon seed,	Daphne Mezereum.
" Cocculi Indici,		Oriental berries,	Anamirta Cocculus.
" Cochleariæ,		Scurvy grass seed, ·	Cochlearia officinalis.
" Colchici,	G.	Colchicum seed,	Colchicum autumnale.
" Conii maculati,		Conium seed,	Conium maculatum.
" Contra,		Levant wormseed,	Artemisia Contra (maritima).
" Coriandri,	*G.	Coriander seed,	Coriandrum sativum.
" Crotonis,		Croton oil seed,	Croton Tiglium.
" Cucumis,		Cucumber seed,	Cucumis sativus.
" Cucurbitæ,		Pumpkin seed,	Cucurbita Pepo.
" " Lagenaria,		Gourd seed,	" Lagenaria.
" Cumini,		Cummin seed,	Cuminum Cyminum.
" " nigri,		Black caraway seed,	Nigella sativa.
" Cydoniæ,	G.	Quince seed, ·	Cydonia vulgaris.
" Cymini,		Cummin seed,	Cuminum Cyminum.
" Cynæ Levanticæ,		Levant wormseed,	Artemisia Contra (maritima).
" Cynobasti,		Dog rose seed, (Hips),	Rosa canina.
" Daturæ,	*G.	Stramonium seed,	Datura Stramonium.
" " albæ,		Indian Datura seed,	" alba.
" " stramonii,		Thorn-apple seed,	" Stramonium.
" Dauci sativæ,		Garden carrot seed,	Daucus Carota, var. sativa.
" " sylvestris,		Wild carrot seed,	" "
" Delphinii,		Larkspur seed,	Delphinium consolida.
" Erucæ,		White mustard seed,	Sinapis alba.
" et folia Daturæ albæ,		Seed and flowers of	Datura alba.
" Euphorbiæ lathyris,		Caper spurge seed,	Euphorbia lathyris.
" Fœniculi,	*G.	Fennel seed,	Fœniculum.
" " aquatici,		Water fennel seed,	Œnanthe Phellandrium.
" " Cretici,		Wild (large) fennel,	Fœniculum officinalis.
" " dulcis,		Sweet fennel seed,	" dulce.
" " Romani,		Wild (large) fennel,	" officinalis.
" " vulgaris,		Common fennel seed,	" vulgare.
" Fœni-Græci,	G.	Fenugreek seeds,	Trigonella Fœnugræcum.
" Fœnugræci,		" "	" "
" Guilandinæ,		Bonduc seed,	Guilandina bonduc.
" Gynocardiæ,		Chaulmugra seed,	Gynocardia odorata.
" Helianthi,		Sunflower seed,	Helianthus annuus.
" Hippocastani,		Horse chestnut,	Æsculus Hippocastanum.
" Hordei,		Pearl barley,	Hordeum distichon.
" Hyosciami,	G.	Henbane seed,	Hyosciamus niger.
" Ignatii,		St. Ignatius' bean,	Strychnos Ignatia.
" Ispaghulæ,		Spogel seed,	Plantago decumbens.
" Kaladanæ,		Kaladana seed,	Convolvulus Nil.
" Laserpitii,		Lovage seed,	Ligusticum Levisticum.
" Levistici,		" "	" "
" Lini,	G.	Flaxseed, Linseed,	Linum usitatissimum,
" Lithospermi,		Gromwell sd.(Stone sd.)	Lithospermum arvense.
" Lobeliæ,		Lobelia seed,	Lobelia inflata.
" Lycopodii,	*G.	Lycopodium,	Lycopodium clavatum.

PHARMACOPŒIAL.		COMMON.	BOTANICAL.
Semen Melanthii,		Black caraway seed,	Nigella sativa.
" Milii,		Millet seed,	Panicum miliaceum.
" " solis,		Gromwell seed,	Lithospermum arvense.
" Myristicæ,	G.	Nutmeg,	Myristica fragrans.
" Nigellæ,		Black caraway seed,	Nigella sativa.
" Nucis vomicæ,		Nux Vomica seed,	Strychnos Nux Vomica.
" Nicotianæ,		Tobacco seed,	Nicotiana Tabacum.
" Oryzæ,		Rice,	Oryza sativa.
" Pœoniæ,		Peony seed,	Pæonia officinalis.
" Papaveris,	. G.	Maw (Blue poppy) seed,	Papaver somniferum.
" " album,		White poppy seed,	" " var. alba.
" " cœruleum,		Blue poppy (Maw) seed,	" "
" " nigrum,		Black poppy seed,	" " var. nigra.
" Pedicularis,		Stavesacre seed,	Delphinium Staphisagria.
" Pepo,		Pumpkin seed,	Cucurbita Pepo.
" Petroselini,	*G.	Parsley seed,	Petroselinum sativum.
" Phalaris,		Canary seed,	Phalaris canariensis.
" Phellandrii aquat.,	*G.	Fine lv'd water heml'k sd.	Œnanthe Phellandrium.
" Physostigmatis,	*G.	Calabar bean,	Physostigma venenosum.
" Psyllii,		Fleawort (Plantain) sd.,	Plantago psyllium.
" Rapa,		Rape seed,	Brassica campestris.
" Ricini,		Castor oil beans,	Ricinus communis.
" " majoris,		Barbadoes nuts,	Curcas purgans.
" Rosa Benedictæ,		Peony seed, `	Pæonia officinalis.
" Sabadillæ,	*G.	Cevadilla seed,	Sabadilla officinalis.
" Sanctum,	*G.	Levant wormseed,	Artemisia undetermined species
" Santonici,	*G.`	" "	" " "
" Sinæ, (Cinæ),		" "	" Cina and other species.
" Sinapis,	B.	Bl'k and white mustard sd.	Sinapis niger and Sinapis alba
" "	G.	Black mustard seed,	Brassica nigra.
" " albæ,		White "	Sinapis alba.
" " nigræ,		Black "	" nigra.
" Staphidis agriæ,		Stavesacre seed,	Delphinium Staphisagria.
" Staphisagriæ,		" "	· " "
" Stramonii,	G.	Thornapple seed,	Datura Stramonium.
" Strychni,	G.	Nux Vomica,	Strychnos Nux Vomica.
" Tabaci,		Tobacco seed,	Nicotiana Tabacum.
" Tanaceti,		Tansy seed,	Tanacetum vulgare and crispum
" Theobromæ,		Chocolate nuts,	Theobroma Cacao.
" Tiglii,		Croton oil beans,	Croton Tiglium.
" Urticæ,		Nettle seed,	Urtica urens.
" vel anthrodi cinæ,	*G.	Levant wormseed,	Artemisia undetermined species
" Zedoariæ,		" "	" Contra and other species
" Zinæ,		" "	" " " "
Semina Stramoni,	B.	Stramonium seed,	Datura Stramonium.
Senega,	U.S.	Seneka root,	Polygala Senega.
" radix,	B.	Senega root,	" " [elongata.
Senna,	U.S.	Senna, leaflets of	Cassia acutifolia, obovata and
" Alexandrina,	B.	Alexandrian senna,	" lanceolata and obovata.
" Indica,	B.	Tinnivelly senna,	" elongata.
Serpentaria,	U.S.	Virginia snakeroot,	Aristolochia Serpentaria and
Serpentariæ radix,	B.	Serpentary root,	" " other spe.
Sesamum,	U.S.	Benne leaves,	Sesamum Indicum and orientale
Sesami folium,	U.S.'60.	" "	" " "
Setæ siliqua hursutæ,		Cowhage, (Cowitch),	Mucuna pruriens.
" mucuna,		" "	" (Dolichos) pruriens.
Siliqua Bablah,		Bablah (Mimosa) pods,	Mimosa cineraria.
" Dividivi,		Dividivi, Libidibi,	Cæsalpina coriaria.
" Dulcis,	*G.	St. John's bread,	Ceratonia siliqua.`
" hursuta,		Cowhage, Cowitch,	Mucuna (Dolichos) pruriens.

PHARMACOPŒIAL.		COMMON.	BOTANICAL.
Siliqua Libidibi,		Dividivi, Libidibi,	Cæsalpina coriaria.
" purgativa,		Purging cassia,	Cassia Fistula.
" vanillæ,	*G.	Vanilla beans,	Vanilla planifolia (aromatica).
Sinapis,	B.	Mustard seed,	Sinapis niger and alba.
" alba,	U.S.	White mustard seed,	" alba.
" nigra,	U.S.	Black mustard seed,	" nigra.
Simaruba,	U.S.	Simaruba bark,	Simaruba officinalis.
Solidago,	U.S.	Goldenrod herb,	Solidago odora.
Spica Celtica,		Spikenard of the ancients,	Valeriana Celtica.
" Indica,		" " "	" . (nardus) Jatamansi.
Spigelia,	U.S.	Carolina pink root,	Spigelia Marilandica.
Spiræa,	U.S.	Hardhack root,	Spiræa tomentosa.
Sporulæ lycopodii,		Lycopodium,	Lycopodium clavatum.
Statice,	U.S.	Marsh rosemary root,	Statice Limonium, var. Carolin-
Stillingia,	U.S.	Stillingia (Queen's) root,	Stillingia sylvatica. [iana.
Stipites amaræ dulcis,		Bittersweet stems,	Solanum Dulcamara.
" cerasorum,		Cherry stems,	Prunus Cerasus.
" caryophylli,		Clove stems,	Caryophyllus aromaticus.
" Dulcamaræ,	G.	Bittersweet twigs,	Solanum Dulcamara.
Storax,		Storax, (balsam),	Liquidambar orientale.
" calamita,		"	" "
" liquidi,		Liquid storax,	. " "
Stramonii folia,	†.	Stramonium leaves,	Datura Stramonium.
" folium,	U.S.'60.	" "	" "
" semen,	U.S.	" seed,	" "
" semina,	B.	" "	" "
Strobuli Lupuli,		Lupuline (of hops),	Humulus Lupulus.
Styrax,	U.S.	Storax balsam,	Liquidambar orientale.
" liquidus,	G.	Liquid storax,	" "
Suber,		Cork,	Quercus suber.
Succus Absinthii,		Wormwood juice,	Artemisia Absinthium.
" Acacia vera,		from the green pods,	Acacia Arabica and vera.
" Aconitii,		Aconite juice,	Aconitum Napellus.
" Belladonnæ,	B.	Belladonna juice,	Atropa Belladonna.
" Catechu,		Gum Catechu,	Acacia Catechu.
" Citri,		Lemon juice,	Citrus Limonum.
" Conii,	B.	Juice of Hemlock,	Conium maculatum.
" Dauci,		Carrot juice,	Daucus Carota.
" Digitalis,		Foxglove juice,	Digitalis purpurea.
" Glycyrrhizæ crudus,	*G.	Extract of Liquorice,	Glycyrrhiza glabra.
" Hyosciami,	B.	Juice of Hyosciamus,	Hyosciamus niger.
" Juniperi,		" Juniper,	Juniperus communis
" Lactucæ,		" Lettuce,	Lactuca sativa.
" Limonis,	†.	Lemon juice,	Citrus Limonum.
" Liquiritiæ crudus,	G.	Extract of Liquorice,	Glycyrrhiza glabra.
" Mori,	B.	Mulberry juice,	Morus nigra.
" Rhamni,	B.	Buckthorn berry juice,	Rhamnus catharticus.
" Sambuci,	G.	Elderberry juice,	Sambucus nigra.
" Taraxici,	B.	Dandelion juice,	Taraxacum Deus-leonis.
" Thebaicus,		Opium,	Papaver somniferum.
" Scoparii,	B.	Juice of Broom,	Sarothamnus Scoparius.
" Viridis,		Sap green, (Buckthorn),	Rhamnus catharticus.
Sumbul radix,	B.	Sumbul root, (Musk rt.)	Euryangium Sumbul.
Summitates Absinthi,	*G.	Wormwood tops,	Artemisia Absinthium.
" Juniperi,		Juniper tops,	Juniperus communis.
" Meliloti,	*G.	Melilot,	Melilotus officinalis.
" Millefoli,	*G.	Yarrow flowers,	Achillea Millefolium.
" Patchouli,		Patchouly,	Pogostemon Patchouli.
" Sabinæ,	G.	Savine tops,	Juniperi Sabina.
" Scopariæ,		Broom tops,	Sarothamnus Scoparius.

PHARMACOPŒIAL.		COMMON.	BOTANICAL.
Tabacum,	U.S.	Tobacco,	Nicotiana Tabacum.
Tacamahac,		Tacamahac resin	Fagara octandra.
Tamarindi,	*G.	Tamarinds,	Tamarindus Indica.
Tamarindus,	†.	"	" "
Tanacetum,	U.S.	Tansy herb,	Tanacetum vulgare.
Tapioca,	U.S.	Tapioca (fecula of)	Janipha Manihot.
Taraxaci radix,	B.	Dandelion root,	Taraxacum Dens-leonis.
Taraxacum,	U.S.	" "	" "
Terebinthina,	U.S.	Turpentine, (Oleo resin)	Pinus palustris and other species
"	G.	" European,	" pinaster " "
" Canadensis,	†.	Canada balsam (of fir),	Abies balsamea.
" alba,		White turpentine,	Pinus of several species.
" Argentoratensis,		Strasburgh turpentine,	Abies taxifolia and picea.
" Chia,		Chian turpentine,	Pistacia terebinthus.
" Communis,	G.	Common turpentine gum	Pinus of several species.
" Cyprian,		Chian (Scio) turpentine,	Pistacia terebinthus.
" Gallica,		French turpentine,	Pinus pinaster.
" Germanica,		Com. Europ'n turpentine	" sylvestris and other species
" Hungarica,		Hungarian pitch,	" pumilo.
" Laricis,	*G.	Venice turpentine,	Abies (Pinus) Larix.
" Laricina,	G.	" "	Larix decidua.
" Strassburgensis,		Strasburgh turpentine,	Abies taxifolia and picea.
" Veneta,	*G.	Venice turpentine, ·	Larix decidua.
" vulgaris,		Com. Am'n turpentine,	Pinus palustris and other species
Thea Bohea,		Black tea,	Thea Bohea.
" Chinensis,		Chinese tea leaves,	" Chinensis.
" Hispanica,		Mexican tea,	Chenopodium ambrosioides.
" nigra,		Black tea leaves,	Thea Bohea.
" Romanæ,		Mexican tea,	Chenopodium ambrosioides.
" Paraguay,		Paraguay tea, (Mate),	Ilex Paraguaiensis.
" viridis,		Green tea,	Thea viridis.
Theriaca,	B.	Treacle, Molasses.	
Thridace,		Lettuce opium,	Lactuca virosa.
Thus,	*G.	Olibanum, Frankincense,	Boswellia papyrifera.
" Americanum,	B.	Am'n turpentine, (hard),	Pinus tæda and Pinus palustris.
" vulgaris,		Common gum turpentine,	" " "
Tormentilla,	U.S.	Tormentil,	Potentilla Tormentilla.
Toxicodendron,	U.S.	Poison oak,	Rhus Toxicodendron.
Tragacantha,	†.	Tragacanth gum,	Astragalus verus and other spe.
"	G.	" "	" creticus " "
Tridace,		Lettuce opium,	Lactuca virosa.
Triosteum,	U.S.	Fever root,	Triosteum perfoliatum.
Tubera Aconiti,	G.	Aconite root,	Aconitum Napellus.
" China,		China root,	Smilax China.
" Colchici,		Colchicum root,	Colchicum autumnale.
" Jalapa,	G.	Jalap root,	Convolvulus (Ipomœa) purga.
" Salep,	G.	Salep,	Orchis mario and other species.
Turiones Abietis,		Spruce tops,	Abies nigra.
" Juniperi,		Juniper tops,	Juniperus communis.
" Pini,	G.	Young pine shoots,	Pinus sylvestris and Palustris.
Ulmi cortex,	B.	Broad leaved Elm bark,	Ulmus campestris.
Ulmus,	U.S.	Slippery Elm bark,	" fulva.
" fulva,	U.S.'60.	" "	" "
Uva passa,	U.S.	Raisins, (Dried grapes),	Vitis vinifera.
Uva Ursi,	U.S.	Uva Ursi leaves,	Arctostaphylos Uva Ursi.
Uvæ Ursi folia,	B.	Bearberry leaves,	" "
Uvæ,	B.	Raisins, (Dried grapes),	Vitis vinifera.
Valeriana,	U.S.	Valerian root,	Valeriana officinalis.
Valerianæ radix,	B.	" "	" "
Vanilla,	U.S.	Vanilla, (unripe fruit),	Vanilla aromatica.

PHARMACOPŒIAL.	COMMON.	BOTANICAL.
Verati viridis radix,	B. Green hellebore root,	Veratrum viride.
Veratrum album,	U.S. White hellebore root,	" album.
" viride,	U.S. American hellebore root,	" viride.
Viola,	U.S. Violet root,	Viola pedata.
Xanthorrhiza,	U.S. Yellow root,	Xanthorrhiza apiifolia. [oliniana
Xanthoxylum,	U.S. Prickly ash bark,	Xanthoxylum fraxineum and Car-
Zingiber,	†. Ginger root,	Zingiber officinale (Amomum Z.)

THE FOLLOWING ARE NAMES OF THE CRUDE VEGETABLE SUBSTANCES IN THE GERMAN PHARMACOPŒIA.

GERMAN	PHARMACOPŒIAL.	BOTANICAL.
Alantwurzel,	Radix Helenii,	Inula Helenium.
Alexandrinische sennesblatter*	Folia Sennæ,	Cassia lenitiva (acutifolia).
Alkannawurzel,	Radix Alkannæ,	Alkanna tinctoria.
Aloë,	Aloë,	Aloe spicata and other species of
Altheeblätter,	Folia Althææ,	Althæa officinalis. [aloe.
Altheewurzel,	Radix "	" "
Ammoniakgummi,	Ammoniacum,	Dorema Ammoniacum.
Anis,	Fructus Anisi vulgaris,	Pimpinella Anisum.
Anis gemeiner,	" " "	" "
Anisöl,	Oleum Anisi,	" "
Arabisches Gummi,	Gummi Arabicum,	Acacia Nilotica (Seyal).
Arnikablüthen,*	Flores Arnicæ,	Arnica montana.
Arnikawurzel,	Radix "	" "
Bärentraubenblätter	Folia Uvæ Ursi,	Arctostaphylos Uva Ursi.
Bärlappsamen,	Lycopodium,	Lycopodium clavatum.
Baldrian,	Radix Valerianæ,	Valeriana officinalis.
Baldrianöl,	Oleum "	" "
Beifusswurzel,	Radix Artemisiæ,	Artemisia vulgaris.
Belladonnawurzel	" Belladonnæ,	Atropa Belladonna.
Benzoë,	Benzoë,	Styrax Benzoin.
Bergamottöl,	Oleum Bergamottæ,	Citrus Bergamia.
Bernstein,	Amber,	
Bertramwurzel	Radix Pyrethri,	Anacyclus officinarum.
Bilsenkraut,	Folia Hyosciami,	Hyosciamus niger.
Bilsensamen,	Semen "	" "
Bitterklee,	Folia Trifolii fibrini,	Menyanthes trifoliata.
Bittersüssstengel,	Stipites Dulcamaræ,	Solanum Dulcamara.
Blauholz,	Lignum Campechianum,	Hæmatoxylon Campechianum.
Bockshornsamen,	Semen Fœni Græci,	Trigonella Fœnum Græcum.
Brechnuss,*	" Strychni,	Strychnos Nux Vomica.
Brechwurzel,	Radix Ipecacuanhæ,	Cephaelis Ipecacuanha.
Cajeputöl,	Oleum Cajeputi,	Melaleuca Leucadendron and
" gereinigtes,	" " rectificatum,	" " [minor.
Centifolicurose,	Flores Rosæ,	Rosa centifolia. [species.
Chinarinde braune,	Cortex Chinæ fuscus,	Cinchona micrantha and other
" rothe,	" " ruber,	" succirubra and other spe.
Chinawurzel,	Rhizoma Chinæ,	Smilax China.
Citronenöl,	Oleum Citri,	Citrus Limonum.
Citronenschale,	Cortex Fructus Citri,	" " (medica).
Cubeben,	Cubebæ,	Cubeba officinalis, (Piper Cubeba
Drachenblut,	Resina Draconis,	Dæmonorops Draco.
Dreiblatt,	Folia Trifolii fibrini,	Menyanthes trifoliata.

GERMAN.	PHARMACOPŒIAL.	BOTANICAL.
Eberwurzel,	Radix Carlinæ,	Carlina acaulis.
Eibischkraut,	Folia Althææ,	Althæa officinalis.
Eibischwurzel,*	Radix "	" "
Eichenrinde,	Cortex Quercus,	Quercus pedunculata and sessili-
Eisenhutknollen,	Tubera Aconiti,	Aconitum Napellus. [flora.
Elemi,	Elemi.	
Engelwurzel,	Radix Angelicæ,	Archaugelica officinalis.
Enzianwurzel,	" Gentianæ,	Gentiana lutea.
Euphorbium,	Euphorbium,	Euphorbia resinifera.
Faulbaumrinde,	Cortex Frangulæ,	Rhamnus Frangula.
Feigen,	Caricæ, (Fici),	Ficus Carica.
Feldkümmelkraut,*	Herba Serpylli,	Thymus Serpyllum.
Fenchelholz,*	Lignum Sassafras,	Sassafras officinale.
Fenchelöl,	Oleum Fœniculi,	Fœniculum officinale.
Fenchelsamen,	Fructus "	" "
Fichtenharz,	Resina Pini (Burgundica)Abies of several species.
Fichtensprossen,	Turiones Pini,	Pinus sylvestris.
Fieberkleeblätter,	Folia Trifolii fibrini,	Menyanthes trifoliata.
Fingerhutkraut,	" Digitalis,	Digitalis purpurea.
Fliederblumen,	Flores Sambuci,	Sambucus nigra.
Freisamkraut,	Herba Viola tricoloris,	Viola tricolor.
Galgant,	Rhizoma Galangæ,	Alpina officinarum.
Galläpfel,	Gallæ.	Quercus infectoria.
Gartenthymian,	Herba Thymi,	Thymus vulgaris.
Geigenharz,	Colophonium.	
Gewürznelken,	Caryophylli,	Caryophyllus aromaticus.
Giftlattich,	Herba Lactucæ,	Lactuca virosa.
Giftlattichsaft,	Lactucarium,	" "
Giftsumachblätter,	Folia Toxicodendri,	Rhus Toxicodendron.
Gottesgnadenkraut,	Herba Gratiolæ,	Gratiola officinalis.
Granatwurzelrinde,	Cortex Radicis Granati,	Punica Granatum.
Guajakharz,	Resina Guajaci,	Guaiacum officinale.
Guajakholz,	Lignum "	" "
Guarana,	Pasta Guarana,	Paullina sorbilis.
Gummi, Arabisches,	Gummi Arabicum,	Acacia Nilotica (Scyal).
Guttapercha,	Gutta Percha depurata,	Isonandra Gutta.
Gutti,	Gutti, (Gamboge),	Garcinia Morella.
Hanfsamen,	Fructus Cannabis,	Cannabis sativa.
Haselwurzel,	Radix Asari,	Asarum Europæum.
Hauhechelwurzel,	" Ononis,	Ononis spinosa.
Heidelbeeren,	Fructus Myrtilli,	Vaccinium Myrtillus.
Hollunderblüthen,*	Flores Sambuci,	Sambucus nigra.
Hohlzahn,	Herba Galeopsidis,	Galeopsis ochroleuca.
Hopfenmehl,	Glandula Lupuli,	Humulus Lupulus.
Huflattigblätter,	Folia Farfaræ,	Tussilago farfara.
?.discher Hanf,	Herba Cannabis Indicæ,	Cannabis sativa.
Ingwer,	Rhizoma Zingiberis,	Zingiber officinale.
Irländisches Moos,	Carrageen,	Chondrus crispus.
Isländische Flechte,*	Lichen Islandicus,	Cetraria Islandica.
Isländisches Moos,	" "	" "
Jalapenknollen,	Tubera Jalapæ,	Convolvulus Purga.
Jalapenwurzel,*	" "	" "
Johannisbrot,	Fructus Ceratoniæ,	Ceratonia siliqua.
Kadeöl,	Oleum Juniperi empyreumaticum, Juniperus oxycedrus.
Kakaobutter,	" Cacao,	Theobroma Cacao.
Kalabarbohne,	Faba Calabarica,	Physostigma venenosum.
Kalisayarinde,	Cortex Chinæ Calisayæ,	Cinchona Calisaya.
Kalmusöl,	Oleum Calami,	Acorus Calamus.
Kalmuswurzel,	Rhizoma Calami,	" "
Kamala,	Kamala,	Rottlera tinctoria.

GERMAN.	PHARMACOPŒIAL.	BOTANICAL.
Kamille gemeine,	Flores Chamomillæ vulgaris, Matricaria Chamomilla.	
" römische,	" " Romanæ, Anthemis nobilis.	
"	" " vulgaris, Matricaria Chamomilla.	
Kampfer,	Camphora,	Camphora officinarum.
Kardamom kleiner,	Fructus Cardamomi minores, Elettaria Cardamomum.	
Kardobenediktenkraut	Herba Cardui benedicti,	Cnicus benedictus.
Kaskarillrinde,	Cortex Cascarillæ,	Croton Eluteria (Cascarilla).
Katechu,	Catechu,	Acacia Catechu.
Kellerhalsrinde,*	Cortex Mezerei,	Daphne Mezereum.*
Kino,	Kino, .	Pterocarpus Marsupium.
Kirschlorbeerblätter,	Folia Laurocerasi,	Prunus Laurocerasus.
Klatschrosen,	Flores Rhœados,	Papaver Rhœas.
Klettenwurzel,	Radix Bardanæ,	Lappa officinalis and other spe-
Kokosöl,	Oleum Cocois,	Cocos nucifera. [cies.
Kolombowurzel,	Radix Columbo,	Jateorrhiza Calumba.
Koloquinten,	Fructus Colocynthidis,	Citrullus Colocynthis.
Konigschina,*	Cortex Chinæ Calisayæ,	Cinchona Calisaya.
Kopaivabalsam,	Balsamum Copaivæ,	Copaifera multijuga and other
Koriandersamen,	Fructus Coriandri,	Coriandrum sativum. [species.
Kosso,	Flores Kosso,	Hagenia Abyssinica.
Kossoblüthen,*	" "	" "
Krähenaugen,	Semen Strychni,	Strychnos Nux Vomica.
Krauseminzblätter,	Folia Menthæ crispæ,	Mentha crispa.
Krauseminzöl,	Oleum " "	
Kreuzblumenkraut,	Herba Polygalæ,	Polygala amara.
Kreuzdornbeeren,	Fructus Rhamni catharticæ, Rhamnus Cathartica.	
Krotonöl,	Oleum Crotonis,	Tiglium officinale.
Küchenschelle,	Herba Pulsatillæ,	Anemone pratensis and A. Pul-
Kümmel,	Fructus Carvi,	Carum Carvi. [satilla.
Kümmelöl,	Oleum Carvi,	" "
Kurkuma,	Rhizoma Curcumæ,	Curcuma longa (viridiflora).
Lärchenterpenthin,	Terebinthina laricina,	Larix decidua.
Lärchenschwamm,	Fungus Laricis,	Polyporus officinalis.
Lakriz,	Succus Liquiritiæ crudus Glycyrrhiza glabra.	
Lavendelblüthen,	Flores Lavandulæ,	Lavandula officinalis (vera).
Lavendelöl,	Oleum "	" " "
Leinkraut,	Herba Linariæ,	Linaria vulgaris.
Leinkuchen,	Placentæ Seminis Lini,	Linum usitatissimum.
Leinöl,	Oleum Lini,	" "
Leinsamen,	Semen "	" "
Liebstöckelwurzel,	Radix Levistici,	Ligusticum officinale.
Lindenblüthen,	Flores Tiliæ,	Tilia ulmifolia and platyphyllos.
Lobelienkraut,	Herba Lobeliæ,	Lobelia inflata.
Löffelkraut,	" Cochlearia,	Cochlearia officinalis.
Löwenzahnwurzel,	Radix Taraxaci,	Taraxacum officinale.
Lorbeeren,	Fructus Lauri,	Laurus nobilis.
Lorbeeröl,	Oleum Lauri,	" "
Macis,	Macis,	Myristica fragrans.
Macisöl,	Oleum Macidis,	
Malvenblätter,	Folia Malvæ,	Malva vulgaris (rotundifolia).
Malvenblüthen gemeine,	Flores Malvæ vulgaris,	" sylvestris.
Mandeln bittere,	Amygdalæ amaræ,	Amygdalus communis var.amara
" süsse,	" dulces,	" " var. dulcis.
Mandelöl,	Oleum Amygdalarum,	" " "
Manna,	Manna,	Fraxinus ornus.
Mastix,	Mastix,	Pistacia Lentiscus.
Marantastärke,	Amylum Marantæ,	Maranta arundinacea.
Meerzwiebel,	Bulbus Scillæ,	Scilla maritima.
Meiran,	Herba Majoranæ,	Origanum Majorana.
Meiranöl,	Oleum "	" "

GERMAN.	PHARMACOPŒIAL.	BOTANICAL.
Meisterwurzel,	Rhizoma Imperatoriæ,	Imperatoria Ostruthium.
Melissenblätter,	Folia Melissæ,	Melissa officinalis.
Mexikanisches Traubenkraut,	Herba Chenopodil ambrosioides,	Chenopodium ambrosio-
Mohnkopfe,	Fructus Papaveris,	Papaver somniferum. [ides.
Mohnöl,	Oleum Papaveris,	" "
Mohnsaft,*	Opium,	" "
Mohnsamen,	Semen Papaveris,	" "
Muskatblüthe,*	Macis,	Myristica fragrans.
Muskatblüthenöl,	sec Maeisöl,	" "
Muskatbutter,	sec Muskatnussöl,	" "
Muskatnuss,	Semen Myristicæ,	" "
Muskatnussöl,	Oleum "	" "
Mutterharz,	Galbanum,	Ferula erubescens.
Mutterkorn,	Seeale cornutum,	Claviceps purpurea. [num.
Myrrhe,	Myrrha,	Balsamodendron Ehrenbergia-
Nelkenöl,	Oleum Caryophyllorum,	Caryophyllus aromaticus.
Nieswurzel grüne,	Radix Hellebori viridis,	Helleborus viridis.
" weisse,	Rhizoma Veratri,	Veratrum album.
Olivenöl,	Oleum Olivarum,	Olea Europæa.
Opium,	Opium,	Papaver somniferum.
Pappelknospen,	Gemmæ Populi,	Populus nigra and other species
Parak. esse,	Herba Spilanthis,	Spilanthes oleracea.
Perubalsam,	Balsamum Peruvianum,	Myroxylon Sonsonatense.
Petersiliensamen,	Fruetus Petroselini,	Petroselinum sativum.
Pfeffer spanischer,	" Capsiei,	Capsicum annuum and longum.
Pfefferminze,	Folia Menthæ piperitæ,	Mentha piperita.
Pfefferminzöl,	Oleum " "	" "
Pimpinellwurzel,	Radix Pimpinella,	Pimpinella Saxifraga and magna
Pomeranzen, unreife,	Fructus Aurantii immaturi,	Citrus vulgaris (aurantium).
Pomeranzenblätter,	Folia Aurantii,	Citrus vulgaris.
Pomeranzenblüthen,	Flores Aurantii,	" aurantium (amara).
Pomeranzenblüthenöl,	Oleum " florum,	" "
Pomeranzenschale,	Cortex Fruetus Aurantii,	" vulgaris (amara).
Pomeranzenschalenöl,	Oleum Aurantii eorticis,	" " "
Quassia,	Lignum Quassiæ,	Quassia amara.
Quassiaholz,*	" "	
Queckenwurzel,	Rhizoma Graminis,	Agropyrum repens.
Quecke rothe,	" Caricis,	Carex arenaria.
Quendel,	Herba Serpylli,	Thymus Serpyllum.
" römischer,	" Thymi,	" vulgaris.
Quittenkörner,*	Semen Cydonia,	Cydonia vulgaris.
Quittensamen,	" "	
Ratanhawurzel,	Radix Ratanhæ,	Krameria triandri.
Rautenblätter,	Folia Rutæ,	Ruta graveolens.
Rhabarber,	Radix Rhei,	Rheum, undetermined species.
Ricinusöl,	Oleum Ricini,	Ricinus communis.
Römische Kamille,	Flores Chamomilæ Romanæ,	Anthemis nobilis.
Rose,	" Rosæ,	Rosa centifolia. [other species.
Rosenöl,	Oleum Rosæ,	" moschata, Damascena and
Rosmarinblätter,	Folia Rosmarini,	Rosmarinus officinalis.
Rosmarinöl,	Oleum Rosmarini,	" "
Sabadillsamen,	Fructus Sabadillæ,	Sabadilla officinalis.
Sadebaumöl,	Oleum Sabinæ,	Juniperus Sabina.
Sadebaumspitzen,	Summites Sabinæ,	Sabina officinalis (Juniperus S.)
Safran,	Croeus,	Croeus sativus.
Salbeiblätter,	Folia Salviæ,	Salvia officinalis.
Salep,	Tubera Salep,	Orchis mario and other species.
Sandarak,	Sandaraca,	Callitris quadrivalis.
Sandriedgraswurzel,*	Rhizoma Caricis,	Carex arenaria.
Sassaparille,	Radix Sarsaparillæ,	Smilax medica and other species

GERMAN.	PHARMACOPŒIAL.	BOTANICAL.
Scammoniaharz,	Resina Scammoniæ,	Convolvulus Scammonia.
Scammoniawurzel,	Radix "	" "
Schafgarbenblüthen,	Flores Millefolii,	Achillea Millefolium.
Schafgarbenkraut,	Herba "	" "
Schierlingskraut,	Herba Conii,	Conium maculatum.
Schiffspech,	Pix navalis.	
Schlangelwurzel, Virginischt,	Radix Serpentariæ,	Aristolochia Serpentaria.
Schlüsselblumen,	Flores Primulæ,	Primula officinalis.
Schöllkraut,	Herba Chelidonii,	Chelidonium majus.
Seidelbastrinde,	Cortex Mezeiri,	Daphne Mezereum.
Seifenwurzel,	Radix Saponariæ,	Saponaria officinalis.
Senegawurzel,	" Senegæ,	Polygala Senega.
Senfsamen schwarzer,	Semen Sinapis,	Brassica nigra.
Sennesblätter,	Folia Sennæ,	Cassia lenitiva (acutifolia).
Spanisches Süssholz,	Radix Liquiritiæ glabræ,	Glycyrrhiza glabra.
Stechapfelblätter,	Folia Stramonii,	Datura Stramonium.
Stechapfelsamen,	Semen Stramonii,	" ".
Steinklee,	Herba Meliloti,	Melilotus officinalis.
Sternanis,	Fructus Anisi stellati,	Illicium anisatum.
Stiefmutterchenthee,*	Herba Viola tricoloris,	Viola tricoloris.
Stinkasant,	Asa fœtida,	Ferula Asa fœtida.
Stockrosen,	Flores Malva arboreæ,	Althæa rosea.
Storax flüssiger,	Styrax liquidus,	Liquidambar orientale.
Strychnossamen,	Semen Strychni,	Strychnos Nux Vomica.
Süssholz spanisches,	Radix Liquiritiæ glabra,	Glycyrrhiza glabra.
Süssholzwurzel,	" " mundata,	" echinata.
Tabaksblätter,	Folia Nicotianæ,	Nicotiana Tabacum.
Tamarindenmus, rohes,	Pulpa Tamarindorum cruda,	Tamarindus Indica.
Tausendguldenkraut,	Herba Centaurii,	Erythræa Centaurium.
Terpenthin,	Terebinthina,	Pinus pinaster and other species
Terpenthinöl,	Oleum Terebinthinæ,	" " " "
" gereinigtes,	" " rectificatum	" " " "
Theer,	Pix liquida, (Tar).	
Thymianöl,	Oleum Thymi,	Thymus vulgaris.
Tollkirschenblätter,	Folia Belladonna,	Atropa Belladonna.
Tolubalsam,	Balsamum Tolutanum,	Myroxylon Toluiferum.
Tormentilwurzel,	Rhizoma Tormentillæ,	Potentilla Tormentilla. [species
Traganth,	Tragacantha,	Astragalus Creticus and other
Traubenkraut Mexikanisches,	Herba Chenopodii ambrosioides,	Chenopodium ambrosio-
Tripolitanische Sennesblätter,*	Folia Sennæ,	Cassia lenitiva (acutifolia) [Ides
Vanille,	Fructus Vanillæ,	Vanilla planifolia (and other spe-
Veilchenwurzel,	Rhizoma Iridis,	Iris Florentina. [cies).
Virginische Schlangenwurzel,	Radix Serpentariæ,	Aristolochia Serpentaria.
Wachholderbeeren,	Fructus Juniperi,	Juniperus communis.
Wachholderbeeröl,	Oleum Juniperi (fructus),	" "
Wallnussblätter,	Folia Juglandis,	Juglans regia.
Wallnusschale grüne,	Cortex Fructus Juglandis	" "
Wasserfenchel,	Fructus Phellandrii,	Œnanthe Phellandrium.
Weihrauch,	Olibanum,	Boswellia papyrifera.
Weizenstärke,	Amylum Tritici,	Triticum vulgare.
Wermuth,	Herba Absinthii,	Artemisia Absinthium.
Wohlverleihblüthen,	Flores Arnicæ,	Arnica montana.
Wohlverleihwurzel,	Radix "	" " [other species
Wollblumen,	Flores Verbasci,	Verbascum thapsiforme and
Wurmfarnwurzel,	Rhizoma Filicis,	Polystichum Filix Mas.
Wurmsamen,	Flores Cinæ,	Artemisia undetermined species
Zeitlosensamen,	Semen Colchici,	Colchicum autumnale.
Zeylonisches Zimmtöl,	Oleum Cinnamomi Zeylanici,	Cinnamomum Zeylanicum.
Zeylonzimmt,	Cortex "	" " " "
Zimmtkassie,	" "	" Cassiæ, Cinnamomum Cassia.

GERMAN.	PHARMACOPŒIAL.	BOTANICAL.
Zimmtkassienöl,*	Oleum Cinnamomi Cassiœ	Cinnamomum Cassia.
Zimmtöl,	" " "	" "
" Zeylonisches,	" " Zeylanici,	" Zeylanicum.
Zittwersamen,*	Flores Cinæ,	Artemisia undetermined species
Zittwerwurzel,	Rhizoma Zedoariæ,	Curcuma Zedoaria.

OMITTED IN THE ALPHABETICAL ARRANGEMENT.

PHARMACOPŒIAL.	COMMON.	BOTANICAL.
Areca,	B. Areca (Betel) nut,	Areca Catechu.
Caryophylli,	G. Cloves,	Caryophyllus aromaticus.
Coptis,	U.S. Goldthread,	Coptis trifolia.

Jaborandi, a powerful sialogogue is referred to,   Pilocarpus. pinnatus.

Damiana, reputed aphrodisiac, the botanical origin of this not yet determined.